MOTION AND TIME STUDY
Design and Measurement of Work

BY RALPH M. BARNES

MOTION AND TIME STUDY—DESIGN AND MEASUREMENT OF WORK, SIXTH EDITION

WORK SAMPLING, SECOND EDITION

WORK METHODS MANUAL

MOTION AND TIME STUDY APPLICATIONS, FOURTH EDITION

MOTION AND TIME STUDY PROBLEMS AND PROJECTS, SECOND EDITION

MOTION AND TIME STUDY
Design and Measurement of Work

Sixth Edition

RALPH M. BARNES, M.E., Ph.D.

Professor of Engineering and Production Management
University of California, Los Angeles, California

JOHN WILEY & SONS, INC.

NEW YORK   LONDON   SYDNEY

Preface

The major objective of this book is to suggest ways of advancing the goals of an organization by improving human effectiveness. In line with that objective, this revision has been made in order to present a new approach to work design and management systems that will better advance the objectives of the organization, increase the usefulness of the industrial engineer, and bring greater satisfaction and greater rewards to management and nonmanagement people—especially to the latter group.

New knowledge coming out of social and behavioral science research has been assimilated, interpreted, and applied by managers and industrial engineers with great success. It is recognized that while pay is a powerful stimulator, people are motivated by other things as well. They want meaningful work, an opportunity to develop their talents, to learn new skills, to advance into better jobs at higher wages, and to assume greater responsibilities.

The time-tested techniques and problem-solving procedures described in this volume can be used by the industrial engineer, the supervisor and the individual worker, and by teams or groups. It is assumed that management and nonmanagement people working together can learn to consider the goals of the organization and the goals of the individual arriving at mutually satisfactory objectives. Further it is recognized that most people have some creative ability and that they will use this ability as an individual and as a member of a group in order to meet the goals that they have had a part in establishing.

Research findings seem to indicate that the major factors that determine job satisfaction are achievement, recognition, work itself, responsibility, and advancement. If management can provide an environment in which these factors can operate, the people will respond in a positive way. These are the *motivators*. On the other hand, the *maintenance factors* include company policy, supervision, salary, interpersonal relations, and working conditions. These are not motivators in themselves. However, if these factors are reasonably well satisfied the motivators can become effective. In a situation where the satisfiers are absent or are restricted, the employee concentrates on the maintenance factors, which can become dissatisfiers.

Under the new plan, the industrial engineer spends less of his time performing professional duties in the usual sense. More of his time is devoted to assisting management and nonmanagement people. He fully understands the motivation-maintenance theory. He has participated in a number of successful applications, and he knows the benefits that come to the organization and to the people in the form of lower costs, better quality, higher production per man-hour, greater profits to the company; and greater satisfaction, higher wages, and greater self-realization to the people.

The requirements for this more effective human work system will be new to many people and will require considerable time to understand and accept. Implementation can take many different forms. There is no set system or procedure that must be followed. It is the basic theory, the philosophy, that is important—not the mechanism.

In the first chapter of this new section (Chapter 38), there is presented a brief discussion of work organization from early times up to the present. This is followed by a presentation of the research findings leading to the motivation-maintenance theory. The next chapter deals with job enlargement with specific applications. Another chapter describes in detail the procedure used by one large company in changing from conventional wage-incentive plans to a new plan that incorporates stable pay with a form of organization that brings the motivators into operation. The final chapter presents the unique case of a large and successful company that designed, built, staffed, and is now operating a factory that complies with the motivation-management theory. It is hoped that the rather detailed description of these successful applications, with a discussion of the underlying theory, will provide an incentive for other people to make similar applications. The rewards can be very great.

Over the thirty years that this book has been in the process of development, I have had constant assistance and advice from managers, engineers, and educators. To these I would express my great indebtedness. My special thanks for assistance received in connection with the present revision go to George H. Gustat, Robert J. Rohr, Jr., and James A. Richardson of Eastman Kodak Company, Kodak Park Works; to my daughter Elizabeth Barnes Parks; and to the 72 companies that supplied information pertaining to their current industrial engineering practices.

RALPH M. BARNES

Los Angeles, California, 1968

Preface to the Fifth Edition

The great expansion in business and industry in recent years has brought with it wider acceptance and greater use of motion and time study. Many practices in this field that once were accepted in only the most progressive companies have now become commonplace. Moreover, motion and time study is now applied to many different business activities and to areas far removed from industry. Rapid changes are taking place in the field itself. Today the scope of motion and time study is much broader—we are now concerned with the design of work systems and methods. Our objective is to find the ideal method or the method nearest to the ideal that can be practically used, whereas in the past the emphasis was on improving existing methods, rather than defining the problem or formulating the objective and then finding the preferred solution.

The caliber of the people engaged in this field has also changed. Today a large percentage are college graduates or are men from the shop or office who have been carefully selected and thoroughly trained for this work. This specialized staff has better equipment and facilities to work with, is making greater use of mathematics, statistics, and electronic data-processing equipment, and is looking to the future when still more sophisticated techniques and equipment will be generally available to aid in solving complex problems.

The same careful study and consideration that have been given to direct factory labor in the past are now being extended to indirect labor and to the evaluation and control of machines, processes, and materials. This includes the consideration of such factors as yield, quality, waste, and scrap. In striving for high over-all business effectiveness the industrial engineer and all other members of management and labor are coming to better appreciate the value of cooperation and working together for the advancement of common objectives and goals.

The main purpose of this revision of *Motion and Time Study* has been to present new material in the area of methods design and work measurement and to include more and diverse illustrations. In the revision I have adhered to my original purpose of presenting the basic principles that underlie the successful application of motion and time study, supplementing each with illustrations and practical examples. Five new chapters have been added, dealing with the design process and the general problem-solving procedure as applied to work methods design; human engineering, that is, human factors in design; job enlargement and labor effectiveness; measuring work by physiological methods; and work methods design—the broad view. There is new material on the scope and functions of motion and time study and on the evaluation of alternative methods. The section on developing better methods has been enlarged by new material on the use of the elimination approach in solving work design problems. New examples on motion study, mechanization, and automation have been included. I have used the results of my industrial engineering surveys as a guide in presenting and evaluating the most widely used methods and techniques in this country. Likewise, my research in the general field of physiological measurements has furnished much material for the rather complete chapter on work physiology.

I am most grateful to those who have contributed to this volume on motion and time study. I have drawn on the work of many people and have tried to give specific acknowledgments to those whose work is reported. I am expecially indebted to the Maytag Company and to Procter and Gamble. My special thanks for assistance received are extended to the more than one hundred companies that have supplied information pertaining to their industrial engineering practices, and the users of this book who, over the years, have made many helpful suggestions.

Ralph M. Barnes

Los Angeles, California, 1963

Contents

1   Definition and Scope of Motion and Time Study   3

2   History of Motion and Time Study   10

3   The General Problem-Solving Process   21

4   Extent to Which Motion and Time Study May Be Profitably Used   32

5   Work Methods Design—The Broad View   40

6   Work Methods Design—Developing a Better Method   50

7   Process Analysis   63

8   Activity Charts; Man and Machine Charts   97

9   Operation Analysis   110

10 Micromotion Study   128

11 Fundamental Hand Motions   135

12 Motion Study and Micromotion Study Equipment   149

13 Making the Motion Pictures   162

14 Film Analysis   169

15 The Use of Fundamental Hand Motions   190

16 Human Engineering   209

17 Principles of Motion Economy as Related to the Use of the Human
    Body   221

18 Principles of Motion Economy as Related to the Work Place   256

19 Principles of Motion Economy as Related to the Design of Tools
    and Equipment   289

20 Motion Study, Mechanization, and Automation   307

x   Contents

21   Standardization—Written Standard Practice   321

22   The Relation of Motion and Time Study to Wage Incentives   330

23   Time Study: Time Study Equipment; Making the Time Study   342

24   Time Study: Determining the Rating Factor   374

25   Time Study: Determining Allowances and Time Standard   395

26   Mechanized Time Study and Electronic Data Processing   418

27   Determining Time Standards from Elemental Time Data and
     Formulas   428

28   Use of Elemental Time Data and Formulas: Gear Hobbing,
     Soldering Cans   441

29   Determining Time Standards for Die and Tool Work   456

30   Systems of Predetermined Motion-Time Data: Motion-Time Data
     for Assembly Work   471

31   Systems of Predetermined Motion-Time Data: The Work-Factor
     System, Methods-Time Measurement, Basic Motion Timestudy   487

32   Work Sampling   511

33   Measuring Work by Physiological Methods   549

34   Fatigue   563

35   Motion and Time Study Training Programs   580

36   Training the Operator—Effect of Practice   607

37   Evaluating and Controlling Factors Other Than Labor—
     Multi-factor Wage Incentive Plans   639

38   Motivation and Work   660

39   Job Enlargement—Deliberate Change   674

40   The Individual Rate—Performance Premium Payment Plan   688

41   The Lakeview Plan   702

Appendix A　Time Study Manual　715

Appendix B　Wage Incentive Manual　722

Problems　733

Illustrations from Other Books by the Author　768

Bibliography　769

Index　789

1. Methods Design—Finding the Preferred Method

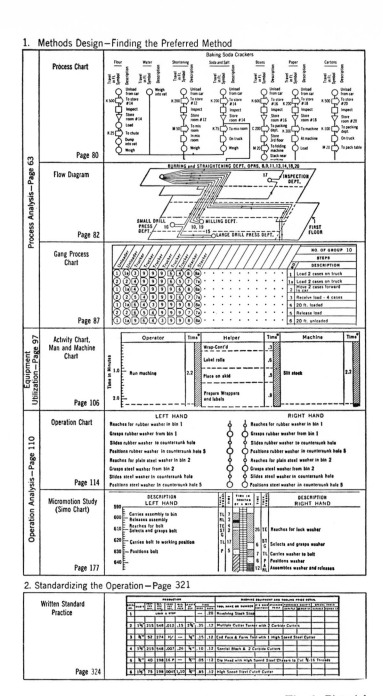

Fig. 1. Pictorial

2. Standardizing the Operation—Page 321

3. Work Measurement—Determining the Time Standard

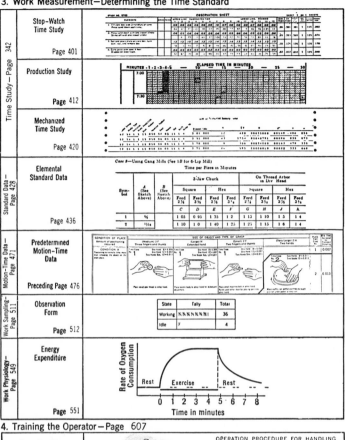

Time Study—Page 342	Stop–Watch Time Study Page 401	OBSERVATION SHEET
	Production Study Page 412	MINUTES ·1·2·3·4·5 — ELAPSED TIME IN MINUTES
	Mechanized Time Study Page 420	
Standard Data—Page 428	Elemental Standard Data Page 436	Case 2—Using Gang Mills (See 1B for 6-Lip Mill) Time per Piece in Minutes
Motion-Time Data—Page 471	Predetermined Motion–Time Data Preceding Page 476	CONDITION OF PLACE ... SIZE OF OBJECT AND TYPE OF GRASP
Work Sampling—Page 511	Observation Form Page 512	State / Tally / Total — Working 36 — Idle 4
Work Physiology—Page 549	Energy Expenditure Page 551	Rate of Oxygen Consumption — Rest / Exercise / Rest — Time in minutes

4. Training the Operator—Page 607

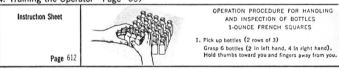

Instruction Sheet Page 612	OPERATION PROCEDURE FOR HANDLING AND INSPECTION OF BOTTLES 1-OUNCE FRENCH SQUARES 1. Pick up bottles (2 rows of 3) Grasp 6 bottles (2 in left hand, 4 in right hand). Hold thumbs toward you and fingers away from you.

5. Controlling Factors Other Than Labor—Page 639

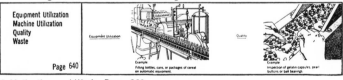

Equipment Utilization Machine Utilization Quality Waste Page 640	Equipment Utilization — Example: Filling bottles, cans, or packages of cereal on automatic equipment. — Quality — Example: Inspection of gelatin capsules, pearl buttons or ball bearings

6 Motivation and Work—Page 660

Motivation Job Enlargement Cases Page 666	Motivation Factors: Achievement Recognition Work Itself	Maintenance Factors: Company Policy Supervision Salary

outline of contents.

CHAPTER 1

Definition and Scope of Motion and Time Study

The terms *time study* and *motion study* have been given many interpretations since their origin. Time study, originated by Taylor, was used mainly for determining time standards, and motion study, developed by the Gilbreths, was employed largely for improving methods. Although Taylor and Gilbreth did their pioneering work around the same time, it seems that in the early days greater use was made of time study and wage incentive than of motion study. It was not until the 1930's that a general movement got under way to study work with the objective of finding better and simpler methods of getting the job done. Then there followed a period during which motion study and time study were used together, the two supplementing each other, and the combined term *motion and time study* came to the forefront. Rapid changes are now taking place in this field. Today the scope of motion and time study is much broader, and the philosophy and practices have changed somewhat from those formerly used. We now are concerned with the design of work systems and methods. Our objective is to find the ideal method or the method nearest to the ideal that can be practically used, whereas in the past the emphasis was too often on improving existing methods, rather than carefully defining the problem or formulating the objective and then finding the preferred solution.

Some have suggested that the term *methods engineering, work design,* or *work study* should be used in place of *motion and time study,* and it may be that eventually these terms will come into wide use in this country. There is, however, at the present time a definite trend toward making *methods design* synonymous with motion study and *work measurement* synonymous with time study. Therefore *motion and time study* and *methods design and work measurement* will be used interchangeably in this book and will have the following broad meaning.

3

Definition of Motion and Time Study. Motion and time study is the systematic study of work systems with the purposes of (1) developing the preferred system and method—usually the one with the lowest cost; (2) standardizing this system and method; (3) determining the time required by a qualified and properly trained person working at a normal pace to do a specific task or operation; and (4) assisting in training the worker in the preferred method.

Motion and time study is composed of four parts, as this definition shows. However, the *two main parts* and those that will be given greatest emphasis in this book are:

Motion study or **methods design**—for finding the preferred method of doing work. That is, the ideal method or the one nearest to it.

Time study or **work measurement**—for determining the standard time to perform a specific task.

1. Developing the Preferred Method—Methods Design. In the broadest sense, every business and industrial organization is concerned with the creation of goods and services in some form or other utilizing men, machines, and materials. In a manufacturing plant, for example, the production process might include the procurement of the raw materials, the machining and fabrication of the parts, and the delivery of the finished product. In designing such a manufacturing process, consideration would be given to the entire system and to each individual operation which would go to make up the system or process. The design of such a process employs the general problem-solving approach. People in the physical sciences and applied sciences refer to the problem-solving procedure as the systematic approach, scientific method, or engineering approach. In concise terms the problem-solving procedure can be stated as follows:

1. *Problem definition.* Prepare general statement of goal or objective —formulate the problem.

2. *Analysis of problem.* Obtain facts—determine specifications and restrictions—describe present method if activity is already in effect.

3. *Search for possible solutions.* Try the elimination approach—use check lists—apply the principles of motion economy—use creative imagination. Use systematic logic.

4. *Evaluation of alternatives.* Determine the preferred solution— method that gives lowest cost or requires least capital—method that permits product to be put into production most quickly—method that gives highest quality or lowest waste.

5. *Recommendation for action.* Prepare written report—make oral presentation—have all supporting data available—anticipate questions and possible objections.

Methods design therefore begins with the consideration of the purpose or goal—to manufacture a specific product, to operate a cash-and-carry cleaning and pressing establishment, or to produce milk on a dairy farm. The objective is to design a system, a sequence of operations and procedures that go to make up the preferred solution. Certain tools and techniques have evolved over the years to assist in developing preferred work methods, and these will be presented in detail on the following pages.

2. Standardizing the Operation—Written Standard Practice. After the best method for doing the work has been determined, this must be standardized. Ordinarily, the work is broken down into specific jobs or operations which are described in detail. The particular set of motions, the size, shape, and quality of material, the particular tools, jigs, fixtures, gauges, and the machine or piece of equipment must be definitely specified. All these factors, as well as the conditions surrounding the worker, must be maintained after they have been standardized. A written standard practice giving a detailed record of the operation and specifications for performing the work is the most common way of preserving the standard.

3. Determining the Time Standard—Work Measurement. Motion and time study may be used to determine the standard number of minutes that a qualified, properly trained, and experienced person should take to perform a specific task or operation when working at a normal pace. This time standard may be used for planning and scheduling work, for cost estimating, or for labor cost control, or it may serve as the basis for a wage incentive plan. In the early days the time standard was sometimes converted into money value and was called a piece rate. Piece rates were usually expressed as so many dollars per hundred pieces, and these piece rates were used as a means of paying workers.

Although elemental data, motion-time data, and work sampling are widely used for establishing time standards, perhaps the most common method of measuring work is stop-watch time study.[1] The opera-

---

[1] A study of 72 companies shows that 99% use time study, 85% use elemental standard data, 49% use predetermined motion time data, and 49% use work sampling for determining time standards. Ralph M. Barnes, *Industrial Engineering Survey,* University of California, 1967.

tion to be studied is divided into small elements, each of which is timed with a stop watch. A selected or representative time value is found for each of these elements, and the times are added together to get the total selected time for performing the operation. The speed exhibited by the operator during the time study is rated or evaluated by the time study observer, and the selected time is adjusted by this rating factor so that a qualified operator, working at a normal pace, can easily do the work in the specified time. This adjusted time is called the normal time. To this normal time are added allowances for personal time, fatigue, and delay, the result being the standard time for the task.

4. Training the Operator. A carefully developed method of doing work is of little value unless it can be put into effect. It is necessary to train the operator to perform the work in the prescribed manner. Where but one or a few persons are employed on a given operation and where the work is relatively simple, it is customary to train the operator at his work place. The foreman, the motion and time study analyst, a special instructor, or a skilled operator may act as the teacher. In most cases it is the foreman who is responsible for training the operator, and the foreman often depends upon the methods and standards department for assistance in this task. The written standard practice or the element breakdown sheet is a valuable aid to the foreman in job training. When large numbers of employees must be trained for a single operation, the training is sometimes carried on in a separate training department. Charts, demonstration units, and motion pictures are frequently used to advantage in such a training program.

Scope. In order to gain perspective and to show relationships, it seems desirable to list in tabular form the tools and techniques of motion and time study (see Fig. 1), and also to show the entire field (see Fig. 2). In the past greatest emphasis has been given to improving existing methods, and it has been customary to begin by making a detailed study of the method in effect. If a better method is developed, it is put into operation, the worker is trained to perform the new method, a written standard practice is prepared, and a time standard is established for the job. However, if a new product is to be manufactured or a new service is to be performed, then a fresh start can be made. There is no "old method" to improve. Here one has a free hand to design the ideal system and method. There is evidence to show that this same approach should be used even though an existing activity is being investigated with the purpose of finding a

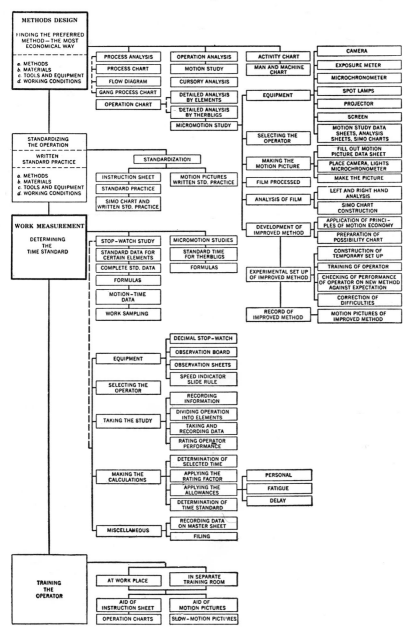

Fig. 2. Scope of motion and time study—tools and techniques.

better work method. Of course, consideration would be given to the present method, but the approach would not be one of *improving* this current method, but rather designing an ideal method. Work methods design then has as its purpose finding the ideal method, or the one nearest to the ideal that can actually be used. We call this the preferred method. Likewise, a systematic study of materials can bring about increased utilization, better quality, and lower costs. Thus, we have a systematic way of developing manpower effectiveness, providing high machine and equipment utilization, and making economical use of materials.

Until fairly recently motion and time study applications were mainly limited to direct factory labor. However, as more people learned about the objectives, methods, and techniques of motion and time study, new uses were found for it. People began to see that its principles are universal and may be equally effective wherever men and machines are employed. Attention is now being focused on the importance of increasing productivity per man-hour and reducing costs, for two main reasons: (1) the rapid increase in hourly wage rates tends to raise labor costs; (2) the rapid increase in capital invested and the increase in operating costs of machines, tools, and equipment tend to raise the "machine-hour rate" or the overhead cost. Moreover, the need for greater output of goods and services provides an additional incentive for increasing the productivity of men and machines. It is only natural that methods and techniques that have proved useful in increasing the effectiveness of direct factory labor should be applied in other areas.

Indirect Factory Labor. With the increasing use of mechanization and automation, the relative importance of direct labor will decrease and greater attention will be given to indirect labor. Routine factory operations will be performed by machines, and these machines will become more complex and will require more highly trained people to operate, service, and maintain them. The advent of the time-lapse camera, electronic data-processing equipment, work sampling, waiting-line theory, and other techniques and procedures for recording, analyzing, and measuring nonrepetitive activities has made it profitable to study nonrepetitive group activities of many kinds. Such studies have brought about increased labor effectiveness and increased machine utilization. Also, in many cases machine speeds have been increased, quality and yield improved, and scrap and waste reduced in substantial amounts.

Office Work. Along with the rise in relative importance of indirect factory labor, there has come a tremendous increase in office work in recent years. In some organizations paper work alone has more than doubled in the past fifteen years. Some companies have expanded their motion and time study activities to include the office; others have established a separate department to study office methods and procedures. The use of paper work simplification, office work measurement, systems and procedures analysis, and office mechanization, and the installation of automatic data-processing equipment are some of the approaches being utilized to increase productivity and reduce costs in the office.

Banks, mail-order houses, hospitals, department stores, and supermarkets are obtaining worthwhile results from the application of principles of motion and time study to their activities. Great strides are being made in simplifying work on farms, as well as in various branches of the government and the military, and builders and contractors are finding these principles highly profitable.

CHAPTER 2

History of Motion and Time Study

In order to understand how motion and time study has come to have the broad meaning presented in Chapter 1, it is necessary to go back and investigate the origin and examine the use that has been made of time study and motion study during the last 75 years.

TAYLOR'S USE OF TIME STUDY

It is generally agreed that time study had its beginning in the machine shop [1] of the Midvale Steel Company in 1881 and that Frederick W. Taylor was its originator. Taylor's employment with the Midvale Steel Company led him to the conclusion that the system under which the factory operated left much to be desired. Therefore, shortly after he became general foreman of the plant, he decided to make a determined effort to change the management system "so that the interests of the workmen and the management should become the same, instead of antagonistic." He further stated that "the greatest obstacle to harmonious cooperation between the workman and the management lay in the ignorance of management as to what really constitutes a proper day's work for a workman." [2] Taylor obtained permission from the president of the Midvale Steel Company "to spend

---

[1] Subcommittee on Administration of the ASME, "The Present State of the Art of Industrial Management," *Transactions of the ASME,* Vol. 34, pp. 1197–1198, 1912.

[2] F. W. Taylor, *The Principles of Scientific Management,* Harper and Bros., New York, 1929, p. 52.

10

some money in a careful, scientific study of the time required to do various kinds of work."

A study of the literature in this general area revealed that two classes of experiments had been made; one by physiologists who were studying the endurance of man, and the other by engineers who were attempting to measure physical work in terms of horsepower. These experiments had been made largely upon men who were lifting loads by means of turning the crank of a winch from which weights were suspended, and upon other men who were engaged in walking, running, and lifting weights in various ways.

Taylor started his study by employing two good steady workers who were physically strong. These men were paid double wages and co-operated fully in the investigation. Taylor states that "in these experiments we were not trying to find the maximum work that a man could do on a short spurt or for a few days, but that our endeavor was to learn what really constituted a full day's work for a first class man; the best day's work that a man could properly do year in and year out, and still thrive."

In this study, Taylor hoped to determine what fraction of a horse-power a man could exert, that is, how many foot-pounds of work a man could do in a day. However, this study and other more carefully designed experiments carried out over a period of several years convinced Taylor that there was no direct relationship between the horse-power that a man exerts and the tiring effect of the work on the man.

However, Taylor did discover that, for very heavy work, the controlling factor in determining how much work a man could do in one day was the percentage of the day that the workman was under load, the percentage of the day he was resting, and the length and frequency of rest periods. As important as this was, Taylor's development and use of stop-watch time study was a far greater contribution. Taylor states, "Time study is the one element in scientific management beyond all others making possible the transfer of skill from management to men. . . ."

Important as is Taylor's contribution in originating time study, this is only one of his many achievements. To him also goes the credit for inventing high-speed steel, discovering and evaluating the variables affecting the cutting of metals, originating the functional type of organization, and developing a system or philosophy commonly referred to as scientific management. These achievements were not accidental, but the result of a systematic study of the factors affecting the problem in each instance. Taylor's real contribution to industry was his scientific method, his substitution of fact-finding for rule-of-thumb procedure. His questioning attitude and his constant search for the facts gave him the high place which he reached and still holds as a proponent of science in management. He was a pioneer in applying the systematic approach to that phase of industry which intimately affects the worker. He understood that he was dealing with a human problem as well as with materials and machines, and he approached the human side of his investigations with an understanding of its psychological aspects.[3]

So great has been Taylor's contribution to the whole problem of effective utilization of human effort in industry that we can profit from a review of some of his work in this field. Taylor came from a well-to-do Philadelphia family, was trained at Phillips Exeter Academy to enter Harvard, and after but a year and a half at Phillips Exeter passed the Harvard entrance examinations with honors but at the cost of seriously impaired eyesight. Forced to give up the idea of further study, at the age of eighteen he obtained a job in a machine shop where he served the apprenticeships of machinist and pattern-maker. In 1878, when he was twenty-two, he went to work at the Midvale Steel Works. As business conditions were bad at that time, he took a job as an ordinary laborer. He was rapidly promoted to time clerk, journeyman, lathe operator, gang boss, and foreman of the machine shop, and at the age of thirty-one was made chief engineer of the works. During his early years at Midvale, Taylor studied at night and in 1883 obtained a degree in mechanical engineering from Stevens Institute.

Taylor's Principles of Management. It was as gang boss and foreman that Taylor first came face to face with such problems as "Which is the best way to do this job?" and "What should constitute a day's work?" Taylor, being very conscientious himself, expected the men under him to do a fair day's work. He set for himself the task of finding the proper method of doing a given piece of work, teaching the worker how to do it in this way, maintaining all condi-

[3] Some maintain that Taylor merely tried to squeeze more work from the employees and that his methods were not scientific. For objections to Taylor's methods, see:

(a) R. F. Hoxie, *Scientific Management and Labor,* D. Appleton & Co., New York, 1915.

(b) Symposium—"Stop-Watch Time Study, an Indictment and a Defense," *Bulletin of the Taylor Society,* Vol. 6, No. 3, pp. 99–135, June, 1921.

(c) E. Farmer, "Time and Motion Study," Industrial Fatigue Research Board, *Report* 14, H. M. Stationery Office, London, 1921.

tions surrounding the work so that the worker could do the task properly, setting a definite time standard for accomplishing the work, and then paying the worker a premium in the form of extra wages for doing the task as specified. Many years later Taylor explained his objectives in the following way:

First. The development of a science for each element of a man's work, thereby replacing the old rule-of-thumb methods.

Second. The selection of the best worker for each particular task and then training, teaching, and developing the workman; in place of the former practice of allowing the worker to select his own task and train himself as best he could.

Third. The development of a spirit of hearty cooperation between the management and the men in the carrying on of the activities in accordance with the principles of the developed science.

Fourth. The division of the work into almost equal shares between the management and the workers, each department taking over the work for which it is the better fitted; instead of the former condition, in which almost all of the work and the greater part of the responsibility were thrown on the men.[4]

Taylor stated many times that scientific management required "a complete mental revolution on the part of the workman—and on the part of those on management's side."[5] "Both sides must recognize as essential the substitution of exact scientific investigation and knowledge for the old individual judgment or opinion."[6]

Although Taylor realized that there is more to the management of an industrial enterprise than conducting investigations on methods of doing work, he stated in no uncertain terms that one of the first duties of management was "to develop a science for each element of a man's work," and he used and advocated the scientific approach in the solution of every problem that arose in this connection.

Mr. Eric Farmer of Great Britain, in a most critical analysis of Taylor's work, states, "Taylor's greatest and lasting contribution to the science of industry is the method he adopted. He approached problems which had been thought either not to exist or to be easily solved by common sense, in the spirit of scientific enquiry."[7]

[4] F. W. Taylor, *The Principles of Scientific Management,* Harper & Bros., New York, 1929, p. 36.

[5] F. B. Copley, *Frederick W. Taylor,* Vol. I, Harper & Bros., New York, 1923, p. 10.

[6] *Ibid.,* p. 12.

[7] E. Farmer, *op. cit.*

During his many years in industry Taylor carried on extended investigations in order to determine the best way to do work and to obtain specific data for standardizing the task. In order to illustrate his approach, one of his well-known studies will be briefly described here.

Taylor's Investigation of Shoveling.   In 1898, when Taylor went to the Bethlehem Steel Works, he undertook to improve methods in various parts of the plant. One task that came to his attention was shoveling. Four hundred to 600 men were employed in the yard, and much of their work was shoveling. More iron ore was shoveled than any other material, and rice coal came next in tonnage. Taylor found that each good shoveler in that yard owned his own shovel; he preferred to do this rather than to have the company furnish it. A foreman supervised 50 to 60 men, and they shoveled a variety of material in the course of a day. The yard was approximately 2 miles long and a quarter of a mile wide, so that the gang moved about over a large area.

With little investigation Taylor found that shovelers were lifting loads of 3½ pounds when handling rice coal and up to 38 pounds to the shovel when moving ore. He immediately set about to determine what shovel load permitted a first-class shoveler to move the most material in a day. Taylor took two good shovelers, set them to work in different parts of the yard, and detailed two time study men with stop watches to study the work of these men. At first large shovels were used so that heavy loads were taken. Then the end of the shovel was cut off to permit a smaller shovel load, and again the tonnage handled was noted. This procedure was continued—from very heavy shovel loads to very light ones. The results of this study showed that with a load of 21½ pounds on the shovel, a man could handle a maximum tonnage of material in a day. Thus, a small spade shovel that would just hold 21½ pounds was provided for the worker when he handled ore, and a large scoop was provided for light material such as ashes.

A toolroom was established, and special shovels were purchased and issued to the workers as needed. In addition Taylor inaugurated a planning department to determine in advance the work to be done in the yard. This department issued orders to the foremen and the workers each morning, stating the nature of the work to be done, the tools needed, and the location of the work in the yard. Instead of the men working together in large gangs, the material handled by each man was measured or weighed at the end of the day, and each man was paid a bonus (60% above day wages) when he did the specified amount of work. If a man failed to earn the bonus, an instructor was sent out to show him how to do his job in the proper way and so earn the bonus.

After 3½ years at the Bethlehem plant Taylor was doing the same amount of work in the yards with 140 men as was formerly done by 400 to 600. He reduced the cost of handling material from 7 to 8 cents to 3 to 4 cents per ton. After paying for all added expenses, such as planning the work, measuring the output of the workers, determining and paying bonuses each day, and maintaining the toolroom, Taylor still showed a saving during the last 6-month period at the rate of $78,000 per year.[8]

One cannot read Taylor's experiments on the art of cutting metals,[9] his study of rest pauses in handling pig iron,[10] or his investigations in shoveling without at once realizing that he was a scientist of high order. With Taylor, as with the factory manager today, time study was a tool to be used in increasing the over-all efficiency of the plant, making possible higher wages for labor and lower prices of the finished products to the consumer.

MOTION STUDY AS IT WAS DEVELOPED BY THE GILBRETHS

Motion study cannot be discussed without constant reference to the work of Frank B. Gilbreth and his wife, Lillian M. Gilbreth. Industry owes a great debt to them for their pioneering work in this field. The fundamental character of their work is indicated by the fact that the principles and techniques which they developed many years ago are being adopted by industry today at an increasingly rapid rate.

The story of the work of the Gilbreths is a long and fascinating one. Mrs. Gilbreth's training as a psychologist and Mr. Gilbreth's engineering background fitted them in a unique way to undertake work involving an understanding of the human factor as well as a knowledge of materials, tools, and equipment. Their activities cover a wide range, including noteworthy inventions and improvements in building and construction work,[11] study of fatigue,[12] monotony,[13]

---

[8] F. B. Copley, *op. cit.*, Vol. II, p. 56.

[9] F. W. Taylor, "On the Art of Cutting Metals," *Transactions of the ASME*, Vol. 28, Paper 1119, pp. 31–350, 1907.

[10] Copley, *op. cit.*, p. 37.

[11] F. B. Gilbreth, *Motion Study*, D. Van Nostrand Co., Princeton, N. J., 1911.

[12] F. B. and L. M. Gilbreth, *Fatigue Study*, Macmillan Co., New York, 1919.

[13] L. M. Gilbreth, "Monotony in Repetitive Operations," *Iron Age*, Vol. 118, No. 19, p. 1344, November 4, 1926.

transfer of skill, and work for the handicapped,[14] and the development of such techniques as the process chart, micromotion study, and the chronocyclegraph.

In this book particular attention is given to their work dealing with the process chart, motion study, and micromotion study.

The Beginning of Motion Study. In 1885, Gilbreth, as a young man of seventeen, entered the employ of a building contractor. In those days brick construction constituted an important part of most structures, so Gilbreth began by learning the bricklayer's trade. Promotions came rapidly, and by the beginning of the century Gilbreth was in the contracting business for himself. From the very beginning of his connection with the building trades Gilbreth noted that each craftsman used his own peculiar methods in doing his work, and that no two men did their job in exactly the same way. Furthermore, he observed that the worker did not always use the same set of motions. The bricklayer, for example, used one set of motions when he worked rapidly, another set when he worked slowly, and still a third set when he taught someone else how to lay brick.[15] These observations led Gilbreth to begin investigations to find the "one best way" of performing a given task. His efforts were so fruitful and his enthusiasm for this sort of thing became so great that in later years he gave up his contracting business in order to devote his entire time to motion study investigations and applications.[16]

It was apparent from the beginning that Gilbreth had a knack for analyzing the motions used by his workmen. He readily saw how to make improvements in methods, substituting shorter and less fatiguing motions for longer and more tiring ones. He made photographs of bricklayers at work, and from a study of these photographs he continued to bring about increased output among his workers. For example, Gilbreth invented a scaffold which could quickly and easily be raised a short distance at a time, thus permitting it to be kept near the most convenient working level at all times. This scaffold was also equipped with a bench or shelf for holding the bricks and mortar at a convenient height for the workmen. This saved the brick-

[14] F. B. and L. M. Gilbreth, *Motion Study for the Handicapped,* George Routledge Sons, London, 1920.

[15] L. M. Gilbreth, "The Quest of the One Best Way," p. 16, a sketch of the life of F. B. Gilbreth published by Mrs. Gilbreth, 1925.

[16] John G. Aldrich. See discussion of Gilbreth's work at the New England Butt Company. "The Present State of the Art of Industrial Management," *Transactions of the ASME,* Vol. 34, Paper 1378, pp. 1182–1187, 1912.

layer the tiring and unnecessary task of bending over to pick up a brick from the floor of the scaffold each time he laid one on the wall.

Formerly, bricks were dumped in a heap on the scaffold and the bricklayer selected the bricks as he used them. He turned or flipped each brick over in his hand in order to find the best side to place on the face of the wall. Gilbreth improved this procedure. As the bricks were unloaded from the freight car, Gilbreth had low-priced laborers sort them and place them on wooden frames or "packets" 3 feet long. Each packet held 90 pounds of brick. The bricks were inspected by these men as they unloaded them. They were then placed on the packet side by side, so that the best face and end were uniformly turned in a given direction. The packets were next placed on the scaffolds in such a way that the bricklayer could pick up the bricks quickly without having to disentangle them from a heap. Gilbreth had the mortar box and the packets of bricks arranged on the scaffold in such relative positions that the bricklayer could pick up a brick with one hand and a trowel full of mortar with the other at the same time. Formerly, when the bricklayer reached down to the floor to pick up a brick with one hand, the other hand remained idle.

In addition, Gilbreth arranged for the mortar to be kept of the proper consistency so that the brick could be shoved into place on the wall with the hand. This eliminated the motion of tapping the brick into place with the trowel. These changes, along with others which Gilbreth developed, greatly increased the amount of work which a bricklayer could do in a day. For example, in exterior brickwork, using the "pick and dip" method, the number of motions required to lay a brick were reduced from 18 in the old method to 4½ in the new method.[17]

On a particular building near Boston, on a 12-inch brick wall with drawn joints on both sides and of two kinds of brick, which is a rather difficult wall to lay, bricklayers were trained in the new method. By the time the building was a quarter to a half of the way up, the average production was 350 bricks per man per hour. The record for this type of work previous to the adoption of the new system had been but 120 bricks per man per hour.[18]

Definition of Micromotion Study. Although Gilbreth was aided greatly in his motion study investigations by photographs which he

---

[17] F. B. Gilbreth, *Motion Study*, D. Van Nostrand Co., Princeton, N. J., 1911, p. 88.

[18] "Taylor's Famous Testimony before the Special House Committee," *Bulletin of the Taylor Society*, Vol. 11, No. 3 and No. 4, p. 120, June–August, 1926.

made of his workers in motion, it was not until he adapted the motion picture camera to his work that he made his greatest contribution to industrial management. In fact, the technique of micromotion study as he and Mrs. Gilbreth developed it was made possible only through the use of motion pictures.

The term *micromotion study* was originated by the Gilbreths, and the technique was first made public [19] at a meeting of the American Society of Mechanical Engineers in 1912. A brief explanation of micromotion study might be given in this way: micromotion study is the study of the fundamental elements or subdivisions of an operation by means of a motion picture camera and a timing device which accurately indicates the time intervals on the motion picture film. This in turn makes possible the analysis of the elementary motions recorded on the film, and the assignment of time values to each.

The Gilbreths made little use of stop-watch study. In fact, concentrating on finding the very best way for doing work, they wished to determine the shortest possible time in which the work could be performed. They used timing devices of great precision and selected the best operators obtainable as subjects for their studies.

The Cyclegraph and the Chronocyclegraph. Gilbreth also developed two techniques, cyclegraphic and chronocyclegraphic analysis, for the study of the motion path of an operator.

It is possible to record the path of motion of an operator by attaching a small electric light bulb to the finger, hand, or other part of the body and photographing, with a still camera, the path of light as it moves through space. Such a record is called a cyclegraph [20] (Figs. 79 through 82).

If an interrupter is placed in the electric circuit with the bulb, and if the light is flashed on quickly and off slowly, the path of the bulb will appear as a dotted line with pear-shaped dots indicating the direction of the motion. The spots of light will be spaced according to the speed of the movement, being widely separated when the operator moves fast and close together when the movement is slow. From this graph it is possible to measure accurately time, speed, acceleration, and retardation, and to show direction and the path of motion in three dimensions. Such a record is called a chronocyclegraph. From the chronocyclegraph it is possible to construct accurate wire models of

[19] F. B. Gilbreth. See his discussion in "The Present State of the Art of Industrial Management," *Transactions of the ASME,* Vol. 34, pp. 1224–1226, 1912.

[20] F. B. and L. M. Gilbreth, *Applied Motion Study,* Sturgis & Walton Co., New York, 1917, p. 73.

the motion paths. Gilbreth used these to aid in improving methods, to demonstrate correct motions, and to assist in teaching new operators.

The Narrower Interpretation of Time Study Is Rapidly Passing. If the development of time study and of motion study is followed carefully and in some detail, it is not difficult to understand how these two terms came to be interpreted by some as having widely different objectives. One group saw time study only as a means of setting rates, using the stop watch as the timing device.[21] Another group saw motion study only as an expensive and elaborate technique, requiring a motion picture camera and laboratory procedure for determining a good method of doing work. At the same time, still others more readily took the best from the work of both Taylor and Gilbreth, and with a proper sense of proportion, used the methods and the devices that seemed to be most applicable to the solution of the particular problem at hand.

Today the controversy over the value of using either the one or the other of the two techniques has largely passed, and industry has found that time study and motion study are inseparable, as their combined use in many factories and offices demonstrates. As explained in Chapter 1, emphasis is now placed on work methods design and on work measurement, using the general problem-solving procedure to find the preferred system, process, or method.

National Organizations. The American Society of Mechanical Engineers (ASME) has played an important part in the development of scientific management, industrial engineering, motion and time study, and related fields during the past half century. It must be remembered that Taylor's "Shop Management" was published under the auspices of the ASME in 1903, and his classic work on "The Art of Cutting Metals" occupied over 200 pages in the 1907 *Transactions*. From that time to this the ASME has been responsible for many outstanding publications in this general field, and the Management Division is one of the most active divisions of the Society.

In 1911 the Amos Tuck School of Dartmouth College sponsored a Conference of Scientific Management, and the following year the Efficiency Society, Inc., was organized in New York City.[22] The Taylor Society came into existence in 1915, and this organization and the Society of Industrial Engineers, with headquarters in Chi-

[21] L. M. Gilbreth, *The Psychology of Management,* Sturgis & Walton Co., New York, 1914, p. 106.

[22] H. B. Drury, *Scientific Management,* Columbia University, New York, 1922, p. 39.

cago, joined forces in 1936 to form the Society for Advancement of Management.

In 1922 the American Management Association was formed by people especially interested in industrial training programs. Over the years the objectives of the AMA have changed, and at the present time this organization is concerned mainly with broad management problems.

The American Institute of Industrial Engineers (AIIE), organized in 1948, has grown rapidly and now serves as the professional engineering society in this field. Although many groups have attempted to define industrial engineering, the AIIE Long Range Planning Committee has arrived at the following tentative definition: "Industrial engineering is concerned with the design, improvement, and installation of integrated systems of men, materials, and equipment; drawing upon specialized knowledge and skill in the mathematical, physical, and social sciences, together with the principles and methods of engineering analysis and design, to specify, predict, and evaluate the results to be obtained from such systems."

CHAPTER 3

The General Problem-Solving Process

The design of the method of performing an operation when a new product is being put into production, or the improvement of a method already in effect, is a very important part of motion and time study. Since methods design is a form of creative problem solving, it seems appropriate to present in some detail the general problem-solving process.[1] In fact, the five steps described here are useful in the logical and systematic approach to solving almost any problem.

1. Problem definition.
2. Analysis of problem.
3. Search for possible solutions.
4. Evaluation of alternatives.
5. Recommendation for action.

1. Problem Definition. Although we state that the definition or formulation of the problem is the first step in the problem-solving procedure, this is often preceded by the need to recognize that a problem exists. Sometimes such statements are made as, "Costs are too high," "Output must be increased," or "There is a bottleneck in order filling in the warehouse." In many cases, it is not easy to determine just what the real problem is. However, the problem must be separated out and clearly stated (Fig. 3). At the same time, one must ascertain

[1] Harold R. Buhl, *Creative Engineering Design,* The Iowa State University Press, Ames, Iowa, 1960. Alex F. Osborn, *Applied Imagination,* Charles Scribner's Sons, New York, 1957. Eugene K. Von Fange, *Professional Creativity,* Prentice-Hall, Englewood Cliffs, N. J., 1959. C. S. Whiting, *Creative Thinking,* Reinhold Publishing Co., New York, 1958. For a description of the problem-solving procedure used by the General Motors Corporation see R. D. McLandress, "Methods Engineering and Operations Research," *Proceedings Twelfth Industrial Engineering Institute,* University of California, Los Angeles-Berkeley, pp. 41–48, February, 1960.

21

whether the problem merits consideration, and, if so, whether this is the proper time to solve the problem. If it is decided to proceed with the formulation of the problem, then information should be obtained as to the magnitude or the importance of the problem and the time available for its solution.

METHODS DESIGN WORKSHEET

Problem Definition—General statement of goal or objective—Formulation of the problem

a. Criteria—Means of judging successful solution of problem

b. Output requirements—(1) Maximum daily output, (2) seasonal variations, (3) annual volume, (4) expected life of product—shape of volume growth and decline curve

c. Completion date—Time available (1) to design, (2) to install and try out facilities, and (3) to bring output up to full production

---

Fig. 3. Methods design worksheet—problem definition.

At the outset it is best to define the problem broadly, and the constraints or restrictions should be as few as possible at this stage. This gives greater freedom for the use of imagination and creativity in finding a solution. Moreover, in those cases where the task or operation is now being performed, undue attention should not be given to the "present method," and the problem should be defined independently of the way it is currently being done. The following case illustrates this matter of problem definition.

Seabrook Farms in southern New Jersey operates some 20,000 acres, approximately 7000 of which are planted to peas each year. Originally the company planted peas during the period from early March to the first of April, and then tried to cope with the harvesting problem as best it could. Sometimes during the harvest season so many acres of peas were ripe at the same time that the pickers, shellers, and quick-freeze crews had to work around the clock. Also, because of the delay in harvesting, some peas were overripe and poor in quality.

Dr. C. W. Thornthwaite, climatologist for Seabrook Farms, after considerable study and experimentation with the rate of growth for peas of different varieties during the different periods of the spring and summer, was able to predict when the crops would be ready for harvest.[2] For example, if a certain division of the farms was equipped to take care of 25 acres per day, Dr. Thornthwaite was able to sched-ule the planting so that just that number of acres would be ready during each of the six days of the week, with no crop maturing on Sunday. This not only made it unnecessary to operate on a crash basis at any time during the summer, but also resulted in a crop of more uniform quality with less waste due to overripe peas.

The problem might have been defined as that of finding a more effective method of harvesting peas during the night—using more and better floodlights and possibly selecting and training crews espe-cially for harvesting peas at night. However, the basic problem was to find ways to have the peas ripen at a rate that would result in the desired loading for the men and equipment in the field and in the shelling and quick-freeze plants. In this case, the solution was not an improvement or a refinement of the present one. It was an original solution resulting from the logical problem-solving process.

Sometimes it is desirable to divide the problem into subproblems, or to determine whether the problem.being considered is part of a larger problem. One may want to go back up the line and examine activities preceding the operation being considered, or possibly the activities that follow. Although a broad formulation of the problem in the early stages of the problem-solving process is desirable, it is usually more difficult to solve a complex problem than a simple one.

2. Analysis of the Problem. The formulation of the problem may have resulted in a broad statement or definition. Now it becomes necessary to obtain data—to sort out the facts and determine how they apply to the problem (Fig. 4). Of course, it is likely that the designer will already have considerable knowledge in the area and will seek out additional information. Evaluation of the facts should not

[2] C. W. Thornthwaite, "Operations Research in Agriculture," *Journal of the Operations Research Society of America,* Vol. 1, No. 2, pp. 33–38, February, 1953.

be made during the analysis stage. Critical judgment should be deferred until later in the problem-solving process.

At the outset, it is desirable to establish the criteria for the evaluation of alternative solutions to the problem. The preferred solution to a manufacturing problem might be the one that has the lowest

---

METHODS DESIGN WORKSHEET

Analysis of Problem (No evaluation is to be made at this step)

    a. Specifications or constraints, including any limits on original capital expenditures

    b. Description of present method if operation is now in effect. This might include (1) process charts, (2) flow diagrams, (3) trip frequency diagrams, (4) man and machine charts, (5) operation charts, and (6) simo charts

    c. Determination of activities that man probably can do best and those that the machine can do best and man-machine relationships

    d. Re-examination of problems—Determination of subproblems

    e. Re-examination of criteria

Fig. 4. Methods design worksheet—analysis of problem.

labor cost, the lowest total cost, or the smallest capital investment, or the one that requires the least floor space or results in the greatest utilization of materials, or the one that permits the organization to get into full production in the shortest period of time.

The specifications or restrictions affecting the problem should also be known. In some cases, restrictions are flexible; in others, as one proceeds with the solution of the problem, he may be compelled to impose specific restraints. Consideration of restrictions is present at every stage in the problem-solving process. However, restrictions should be examined with great care, for they sometimes are fictitious or imaginary. Only real restrictions merit consideration. The packing of citrus fruit in cartons illustrates this point very well. Until recently most of the citrus fruit shipped to market was packed in wood crates. People thought that citrus fruit had to be individually wrapped in tissue paper and packed in even layers in a well-ventilated wood crate, and that the fruit must be held tightly in place by a lid which was forced down under pressure and nailed and strapped at each end. The statements in the preceding sentence are all incorrect. Today nearly all oranges, lemons, and grapefruit are shipped in cardboard cartons. The fruit is not individually wrapped. It is not placed in the

Fig. 5. The improved method of packing lemons in cartons instead of wood crates saves the fruit growers over $5 million per year.

carton in layers. The cartons are not ventilated, and the fruit is not packed under pressure. A telescope-type carton holding only half as much as an orange crate is easier to handle and costs less (Fig. 5). This better method of packing is saving the California lemon growers and packers alone over $5,000,000 per year, and it is estimated that an equal amount in benefits accrues to those who transport the fruit and warehouse it, and to the grocer who sells it.[3]

Also the designer should have information as to the importance of the undertaking, the volume of the product to be produced, the number of people to be employed on an activity, and the probable life of the project.

It is important to have a time schedule. The designer should know what time is available for solving the problem, and if it is a production problem he should know the time available to put the process

[3] Roy J. Smith, "Recent Developments in the Packing of Citrus Fruit," *Proceedings Sixth Industrial Engineering Institute,* University of California, Los Angeles-Berkeley, pp. 92–94.

into operation, "debug" it, and get the specified output of the product of acceptable quality.

In the analysis of a problem, it may be desirable to divide it into small components, examining each of these separately. For example, if the problem is to drill a hole in a small metal plate for the manufacture of a television set, the operation might be broken down into three parts: (1) place part in fixture, (2) drill hole in piece, and (3) remove piece and dispose. The volume might consist of 500,000 parts per year with 60 days being available to develop the method and put it into operation. The first step could be performed manually, using a hand-operated fixture for holding the plate, or the plate could be manually placed in a cavity (with self-actuating clamp) in a dial feed, or the plates could be fed automatically from a magazine into the dial feed. Likewise the hole could be drilled manually, or a power-feed drill could be used. Also, the finished part could be removed from the fixture manually, or it could be released automatically from the rotating table. If this operation had been considered at the time the television set was originally designed, consideration might have been given to eliminating this plate, punching the hole in the plate instead of drilling it, using a washer (which could be purchased) instead of the plate, or combining the plate with some other part. If a bolt were inserted through the hole in the plate to assemble it to other parts of the set, then consideration might have been given to spot-welding these parts together or perhaps using a die casting or a plastic molded piece instead of the metal plate.

3. Search for Possible Solutions. The basic objective of course is to find the preferred solution that will meet the criteria and the specifications that have been established. This suggests that several alternative solutions be found, and then the preferred solution can be selected from these.

Early in the problem-solving procedure one should ask the question, "What is the *basic cause* that has created this problem?" If the basic cause can be eliminated, then the problem no longer exists. For example, a company was considering the replacement of a roof over a row of open tanks containing liquid caustic soda. The present roof was badly corroded and replacement was suggested. When the question was asked, "What is the basic reason for having a roof over the tanks?" it was discovered that the only function for the roof was to keep the rain out—to prevent the dilution of the chemicals. However, an analysis showed that evaporation also affected the concentration of the chemicals; moreover it was found that changes in the concentration were unimportant in the manufacturing process. Therefore, the old roof was removed and was not replaced. Of course the ideal solution to a problem is to remove the basic cause and thus eliminate the problem.[4]

If the problem cannot be completely solved by the elimination approach, perhaps part of the problem can be eliminated. If no way can be found to eliminate the problem, then one should explore the various avenues that may hold possible solutions. At the outset it is wise to take a broad and idealistic view in considering possible solutions to the problem.

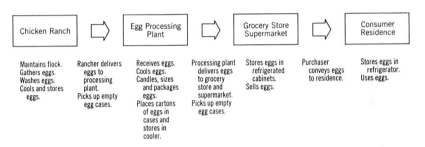

Fig. 6. Egg production, processing, and distribution—from ranch to user.

Let us assume that the problem is to design and build a processing plant for grading and packaging eggs for distribution to grocery stores and supermarkets (Fig. 6). Eggs would be brought to the plant by trucks daily from farms and ranches located 5 to 25 miles from the plant. The usual process consists of (1) candling, determining quality; (2) sizing, determining weight; (3) packaging eggs in cartons; (4) placing cartons into cases and then into a cooler; and (5) delivering eggs to grocery stores and supermarkets.

In thinking of possible solutions, the eggs might be candled, sized, and packaged by hand; candled by hand but sized and packaged by machine; or the entire process of candling, sizing, packaging, and moving to the cooler might be automatic.

By breeding and selection, one might raise chickens that would lay eggs of a given size. Through careful specification of feed and feed supplement and breed of chicken, the color of the yolk, the consistency of the white, and the color of the shell might be uniform, thus eliminating the candling operation. If the eggs were gathered and mar-

_____
[4] From "The Elimination Approach" by Procter and Gamble. See more complete discussion in Chapter 6.

keted each day, freshness would be assured. The eggs might be placed directly into the carton at the time they were gathered on the farm, thus eliminating the need for the candling, sizing, and packaging operations at the plant; or each egg might be removed from the shell and placed in an airtight cube-shaped plastic container automatically, further simplifying the packaging process.

If a small- or medium-sized egg processing plant were to be designed and constructed in the near future, it seems likely that the last two or three alternatives would be quickly dismissed. However, consideration certainly would be given to the manual method versus the automatic method of candling, of sizing, and of packaging the eggs.[5]

Here we are searching for ideas. Imagination, inventive ability, and creative talents are brought to bear on the problem. Some types of problems lend themselves to group effort. There are those who strongly advocate the use of systematic logic, while others believe that the problem-solving conference method often referred to as brainstorming can produce worthwhile ideas.[6] When this latter technique is used, it is essential that the individuals in the group suggest ideas rapidly, that no judgment evaluation whatsoever be made during the brainstorming session, and that the participants be urged to give free rein to ideas even though they may appear highly impractical.

Bernard S. Benson, a strong advocate of using systematic logic in solving problems, finds the following example useful in making his point: [7]

Once upon a time there were two men on an island, who had heard that during the war some soldiers in an army truck had driven to the end of one of the many roads and had buried a fabulous treasure. Both of them, being somewhat materialistically inclined, decided that it would be rather nice to get their hands on it. They decided this individually, however, and this created quite a competitive situation. The first man took a shovel and ran all over the island, digging and prodding at every likely spot. He explored under rocks, he dug at the foot of trees, he stopped at grassy patches and tried his luck. He did a lot of digging, but not much finding. The second man sat himself down and thought. He decided that he must first know of every road on the island; so he proceeded to make himself a map. He did not have the treasure, but at least he had a map of all the possible approaches. Next

---

[5] For a description of a modern egg-processing plant see Chapter 20.

[6] Alex F. Osborn, *Applied Imagination,* Charles Scribner's Sons, New York, 1957.

[7] Bernard S. Benson, "In Search of a Solution—Cerebral Popcorn or Systematic Logic?" *Proceedings Tenth Industrial Engineering Institute,* University of California, Los Angeles-Berkeley, p. 14, February, 1958.

he inspected all of these possibilities against the relevant criteria and sealed off all of the roads which were too narrow for the truck's wheel base; then he sealed off all those where the overhanging rocks were too low to have allowed the truck to pass. Now he looked at every remaining possibility and sealed off all of the roads which ended in large granite areas, where the men could not have dug, and he was left with two roads open. Having first created all of the possibilities, he had eliminated all of those which did not meet the basic requirements. He dug at the end of the first road and found no treasure, but at the end of the second one, he had no trouble finding it; meanwhile, his friend . . . was still jumping around the island, trying here and trying there, frantically looking for the solution to his problem.

4. Evaluation of Alternatives. We have now arrived at several solutions or partial solutions to the problem under consideration. In fact, we may have accumulated a large number of ideas bearing on the problem. Some of these can be eliminated rather quickly and the remaining solutions can be considered more carefully. An examination can be made to determine to what extent each solution meets the criteria and conforms to the original specifications.

In methods design, certainly there is no one "correct answer," but there are usually several possible solutions. Often judgment factors exist that must be considered over and above the quantitative evaluation in arriving at a preferred solution. Although each of several tentative solutions may satisfy the criteria, other solutions might be preferred if some restriction or specification could be changed. It frequently is desirable to select three solutions: (1) the ideal solution, (2) the one that is preferred for immediate use, and (3) possibly another that might be used at some future time—or under different conditions, such as a situation where the annual output might be increased substantially, or the quality of the raw material is more uniform, or more fully trained workers are available.

The evaluation of the preferred solution requires careful consideration of future difficulties that might be encountered, such as time and cost to maintain and repair the equipment, the adjustment to widely varying sizes or product mix, and the effects of wear and tear of equipment on quality of product and down time of equipment. One cannot overlook the human aspects in selecting the preferred solution. It may be that the success of the method selected depends upon getting the division superintendent or the department supervisor to approve it wholeheartedly, or it may be that the inspection department supervisor or the maintenance supervisor has the veto power over the proposed solution. Thus the recommended solution may be the one that is most likely to be accepted and put into effect rather than the ideal solution.

In certain types of problems an evaluation would center around the total capital that would be invested in each of the several proposed methods. In such an analysis we would want to know the initial cost, annual operating cost, expected life of equipment, and its scrap value. Another way to make a comparison is to compute the rate of return on the investment in per cent per year, or the capital recovery period. That is, we determine the number of years that are required for the equipment to pay for itself. In certain types of problems, we may be concerned mainly with finding the method having the lowest direct labor cost. In such cases a comparative analysis could be made, using predetermined motion-time data to determine the total cycle time of each of the several methods. If there is a question as to whether a particular method actually can be performed, it may be necessary to construct a mock-up of the jig, fixture, or work place in the shop or laboratory and try out the method. Some companies have laboratories and workshops especially for such projects.

5. Recommendation for Action. In many cases, the person who solves the problem is not the one who will either use the recommended solution or give final approval for its adoption. Therefore, after the preferred solution has been found, it must be communicated to other persons. The most common form of communication of course is the written or oral report. The written report or the oral presentation then becomes the final step in the problem-solving procedure. Circumstances determine whether the report is mainly a statement of recommendations with the supporting data, or whether an elaborate oral presentation before a group is called for. In some cases, a formal and carefully prepared presentation is needed, including the use of charts, diagrams, photographs, three-dimensional models, or working models. In any event, the presentation should be made in a logical and straightforward manner. It should be easy to follow and to understand. The source of all facts should be indicated, and any assumptions should be clearly stated. A concise written summary should be a part of every report.

Of course in the industrial situation the complete cycle might include a follow-up to ensure that the proposed solution has actually been put into effect. Then an audit or a check from time to time might be made to determine what difficulties were being encountered and to evaluate the over-all results of the installation. It is desirable to know whether the actual operating method is producing the results claimed for it in the proposal. To continue further, a re-evaluation or restudy of the method might be made with the purpose of finding further possibilities for improvement, and so the problem-solving cycle would be repeated. In most business and industrial operations there is no final solution to a problem. A given solution may be put into effect and used until a better one can be found.

CHAPTER 4

Extent to Which Motion and Time Study May Be Profitably Used

At the very outset the cost of a motion and time study application should be considered along with the expected return. If an operation is being considered for improvement, the extent to which the several steps in the problem-solving process will be carried out will depend upon the potential benefits. The problem definition, analysis, and search for possible solutions will be handled in a cursory manner if the operation is a temporary one or if the volume is small or the potential savings insignificant. On the other hand, an exhaustive study might be justified on a job involving many workers, costly materials, and expensive equipment.

If time standards are to be established on an operation and used as the basis for wage incentives, then no short cuts can be permitted on the work measurement phase of motion and time study. The use of work measurement techniques is in a different category from methods design—management must be prepared to guarantee time standards against change, and also a fully documented standard practice is required.

Motion and Time Study Techniques. There are many combinations of the various techniques that may be used, and each of them will be described fully in succeeding chapters. It seems convenient to list in tabular form (see Table 1) five combinations that are very frequently used in motion and time study applications. They range from the most complete, Type A, on the left, to the simplest, Types D and E, on the right.

Four principal factors determine the combination of motion and time study techniques to be used. They are:

32

Table 1. Combinations of Motion and Time Study Techniques

Type	A	B	C	D	E
Methods Design Finding the preferred method—the most economical way considering a. Methods b. Materials c. Tools and equipment d. Working conditions	Process analysis Full micromotion study of operation Application of motion economy principles	Process analysis Motion study Detailed analysis by therbligs Application of motion economy principles	Process analysis Motion study Detailed analysis of elements Application of motion economy principles	…… Motion study Cursory analysis Application of motion economy principles	…… Motion study Cursory analysis Application of motion economy principles
Standardizing the: a. Methods b. Materials c. Tools and equipment d. Working conditions	Standardization of the operation	Standardization of the operation	Standardization of the operation	Standardization of the operation	Standardization of the operation
Written standard practice	Written standard practice Instruction sheet Motion picture record of improved method	Written standard practice Instruction sheet	Written standard practice or Instruction sheet	Written standard practice or Instruction sheet	Written standard practice or Instruction sheet (standardized for each class of work)

Work Measurement
Determining the time standard
	1. Time study
	2. Micromotion study
	3. Standard time data
	  a. Certain therbligs
	  b. Certain elements
	4. Complete standard time data
	5. Motion-time data
	6. Formulas
	7. Work sampling
		1. Time study
		2.
		3. Standard time data
		  a. Certain therbligs
		  b. Certain elements
		4. Complete standard time data
		5. Motion-time data
		6. Formulas
		7. Work sampling
			1. Time study
			2. ……
			3. ……
			4. ……
			5. ……
			6. ……
			7. ……
				1. Time study
				2. ……
				3. ……
				4. ……
				5. ……
				6. ……
				7. ……
					1. ……
					2. ……
					3. ……
					4. Complete standard time data
					5. Motion-time data
					6. Formulas
					7. Work sampling

Training the operator	In separate training department or At work place	In separate training department or At work place	…… At work place	At work place	At work place
Applying the wage incentive	Motion pictures Instruction sheets	Instruction sheets	Instruction sheets	Instruction sheets	Instruction sheets (standardized for each class of work)

This is not a part of motion and time study but often accompanies it.

1. The extensiveness of the job, that is, the average number of man-hours per day or per year used on the work.
2. The anticipated life of the job.
3. Labor considerations of the operation, such as:
   (a) The hourly wage rate.
   (b) The ratio of handling time to machine time.
   (c) Special qualifications of the employee required, unusual working conditions, labor union requirements, etc.
4. The capital investment in the buildings, machines, tools, and equipment required for the job.

An Example of the Most Refined Use of Motion and Time Study. The Type A study would include an analysis of the process and the construction of a chart of the entire manufacturing process of which the operation under consideration is a part. It would require a full micromotion study and the application of the principles of motion economy, which would include a consideration of the most economical use of materials, tools, and equipment, and provisions for satisfactory working conditions. After the preferred method of doing the work had been found, it would be standardized and a standard practice record would be prepared. The Type A study might also involve the making of motion pictures of the old and of the improved method. A time standard would then be set by means of time study, or from data taken from the micromotion study, or from motion-time data or standard data already available. The Type A study would also provide for the training of the operator, either in a separate training department or at the work place, with the aid of motion pictures and instruction sheets. This might be followed by the application of a wage incentive to the job.

An example will be given to show where a Type A study would be used. The job is a semiautomatic lathe operation. The data for this operation, tabulated under the four headings listed above, would appear as follows:

1. More than 100 girls are employed on this operation. They work an 8-hour day and a 40-hour week, which amounts to approximately 200,000 man-hours per year.
2. The job is a permanent one. This operation has been performed for many years, and it is expected that it will be continued indefinitely.
3. Female labor is used.
   a. The basic hourly wage is the going rate in the community.

A 100% premium plan of wage payment is used. Standards are set by time study, and the hourly wage is guaranteed.

b. Each cycle requires 0.25 minute, of which approximately 60% is handling time and 40% machine time.

c. Because special skill is required to perform this operation, each new operator is given 6 weeks of training in a separate training department. Working conditions are normal.

4. The special semiautomatic lathe, fully equipped, costs approximately $3000 when new.

It is apparent that this operation has great potential savings. The fact that 100 girls are employed on this single operation, producing more than 50 million units annually, would at once indicate a Type A study. For every hundredth of a minute saved per piece on this operation, there would be a saving to the company in direct labor cost of over $14,000 per year.

An Example of the Simplest Use of Motion and Time Study. On the other extreme are the Type D and the Type E motion and time studies. These are alike except that the Type E is used where an entire class of work has been previously standardized and where only sufficient analysis need be made to determine into what subdivision a given operation falls. The Type D study would be made on operations of short duration and with little prospect for improvements. This study would involve only a cursory analysis and a very general application of the principles of motion economy, a written standard practice, a time standard set by a stop-watch time study, and an instruction sheet prepared to aid in training the operator.

A Type D study would be used on the following job. The operation is drilling and counterboring a small bracket on a sensitive drill press. The job requires the time of one man for 10 days per month. The operation is expected to last for 6 months, when the model will be changed. In this case a cursory analysis would include a check of the drill speeds, the arrangement of the tote boxes, location of the jig and air hose, and other similar factors. Only a few hours would be required for the analysis and for the execution of the recommended changes. For every hundredth of a minute saved per piece on this operation, there would be a saving to the company in direct labor cost of less than $40 per year. A time study might be made and a time standard established.

The time required for making this Type D study would be short and the cost would be small, whereas months might be required for study of the semiautomatic lathe operation and considerable expense would be involved.

Type A and B studies are used either for individual jobs or for classes of similar work; Type C and D studies are used primarily for individual jobs. In some plants there are many short operations of similar nature which in themselves would warrant only a Type D study, but when considered together as a class would justify a Type A or B study.

The Type E study is used for individual jobs within classes or families, for jobs of a similar nature, and for work already standardized. This type would largely involve the selection of necessary information from standard data on file. Chapter 29 gives an example of such a class of work, that is, hobbing teeth on straight spur gears. The methods, tools, equipment, and working conditions have been standardized. By means of standard time data, motion-time data, and formulas, it is possible to determine time standards synthetically for such work. Instruction sheets are prepared by filling in the necessary machine time (see italics in Fig. 324) on standard forms.[1]

Operating Cost and Capital Cost. Mechanization and automation tend to reduce labor costs but often call for an increase in capital invested in machines and equipment. Therefore in the evaluation of alternatives consideration must be given to both these costs. The following example shows how this may be done.

A Specific Case—Distribution Center. When a new warehouse and distribution center for Eastman Kodak Company was being considered, a "steering committee consisting of personnel on a high management level was appointed to study the problem. They in turn established subcommittees composed of production and staff personnel who examined our existing facilities and considered the alternative possibilities. These committees (1) determined that a real problem did exist and that the need for *new* facilities was valid; (2) they recommended a unit load handling system as being most promising for such facilities; (3) they established the over-all space requirements for the new center, its location, and the general shape thereof; and (4) they evaluated the economies involved." [2]

[1] For additional material on the extent to which motion and time study should be used, see H. B. Maynard, "Methods Engineering Installation: Mapping Out the Program," *Modern Machine Shop*, Vol. 9, pp. 62–70.

[2] R. C. Bryant, S. A. Wahl, and R. D. Willits, "Tractor Train or Dragline Conveyor?" *Modern Materials Handling*, Vol. 6, No. 9, pp. 54–57.

The following five handling methods were considered for the new distribution center: (1) fork truck, (2) conveyor, (3) tractor train, (4) dragline, and (5) combination of tractor train and dragline. Investigations showed that the final selection of handling equipment should be made from these three alternatives: (A) tractor train and dragline, (B) dragline only, and (C) tractor train only.

In making the final evaluation, consideration was given to three factors:

1. Total capital invested in handling equipment.
2. Annual cost to operate the handling equipment.
3. Annual depreciation on equipment.

As shown in Table 2, the tractor train and dragline combination had the highest capital investment of $92,640, as against $69,380 for the tractor train system. However, the tractor train and dragline combination gave the lowest annual operating cost. (See Table 3.) This was $63,300, including depreciation, as against $71,100 for the tractor train. After weighing the two factors of capital investment and operating cost, and considering some intangible factors, the decision was made to install the tractor train and dragline combination.

Table 2. Capital Investment in Equipment

Equipment	A Tractor Train and Dragline	B Drag- line Only	C Tractor Train Only
Tugs	$ 9,200	—	$16,500
Trucks	53,700	$52,500	52,500
Dragline conveyor	28,900	28,900	—
Electrical installation	750	600	380
Total	$92,640	$82,000	$69,380
Less sale of present equipment	—	−4,650	—
Net capital investment	$92,640	$77,350	$69,380

Table 3. Comparative Annual Operating Costs, Including Depreciation on Equipment

Cost Items	A Tractor Train and Dragline	B Dragline Only	C Tractor Train Only
Labor cost—operating personnel	$43,350	$54,800	$57,600
Depreciation on equipment	9,440	8,180	7,790
Space differentials	7,000	0	3,000
Maintenance	3,010	2,800	2,500
Power consumption	500	420	210
Total	$63,300	$66,200	$71,100
Annual cost differentials	0	2,900	7,800

Cost-Reduction Report. It is essential that an estimate be made of the expected savings resulting from improvements in methods before they are made, and also that a report be made after the project is finished and in effect.

The cost-reduction report shown in Fig. 7 is used for presenting proposed changes to management, and also for reporting the savings from new methods after they are installed.

Unit operation times for the old and new methods are based on time studies or on over-all production rates, whichever gives the more representative results for the particular project. Labor costs are based on the average base rate for the particular job, plus the average bonus for the department and a percentage to cover compensation insurance, federal pension, old-age insurance, and other costs that are directly related to labor cost.

Calculated savings do not include fixed overhead costs such as supervision and machine burden, since the annual expenditure for these items would not necessarily be lessened by reducing the labor requirements of a particular job. If a proposed change would increase machine capacity, and if the additional capacity might forestall having to buy more equipment, this fact would be brought out in a note attached to the cost-reduction report.

COST–REDUCTION REPORT

DESCRIPTION OF ITEM INVOLVED
DEPT. Finished Stock & Shipping
OPERATION Marking with name and address of cosignee.
OBJECT OF ANALYSIS To determine possible savings through stamping instead of stenciling.

FILE 11-B
DEPT. NO. 64   DATE
PRODUCT Cartons to be shipped

COMPARISON

PRESENT METHOD	PROPOSED METHOD
MACHINE	MACHINE
TOOLS Fountain stencil brush and pre-cut stencil.	TOOLS Rubber stamp and stamp pad.
DESCRIPTION Stencils are prepared in advance and kept on file for all major cosignees, and name and address is stenciled on each carton.	DESCRIPTION Rubber stamps would be made up for all major cosignees, and name and address is stamped on each carton.

COST OF OPERATIONS INVOLVED	$ PER	COST OF OPERATIONS INVOLVED	$ PER
	carton		carton
LABOR		LABOR	
0.16 minute per carton		0.05 minute per carton	
@ $3.00 per man-hour	0.0080	@ $3.00 per man-hour	0.0025
MATERIALS		MATERIALS	
MISC.		MISC.	
TOTAL OF ABOVE ITEMS	0.0080	TOTAL OF ABOVE ITEMS	0.0025

ESTIMATE OF SAVINGS
SAVING WITH PROPOSED CHANGE ($ 0.0080 – $ 0.0025) EQUALS $ 0.0055 PER carton
PROBABLE YEARLY REQUIREMENTS, 1,250,000 cartons ESTIMATED BY Sales Dept.
ESTIMATED SAVINGS PER YEAR (based on 1,250,000 per year)................................. $ 6875.00

ESTIMATED COST OF CHANGE	PROBABLE SAVINGS PER YEAR........... $ 6500.00
DESIGN $ EST. BY	LESS TOTAL COST OF CHANGE........... $ 500.00
EQUIPMENT $500. ” ”	NET SAVINGS FIRST YEAR................... $ 6000.00
INSTALLATION $ ” ”	NEW METHOD WOULD PAY FOR ITSELF IN 1
$ ” ”	MONTHS.
$ ” ”	NOTE: 100 rubber stamps required at $5.00 each.
TOTAL COST	SUGGESTED BY John Ryan
OF CHANGE $500.00	REPORT PREPARED BY T. A. Wilson

CC TO ATTACHED ARE	DATE DATE
1 Sheets drawings	FIRST CONSIDERED EXPEN. APPR.
Sheets prints	INVESTGN. STARTED INSTALLED
2 Sheets details	REPT. SUBMITTED FINAL REPT.

Fig. 7. Cost-reduction report. Size of form 8½ × 11 inches.

CHAPTER 5

Work Methods Design—The Broad View

In the early days, the production process consisted of a craftsman who possessed the skill, and who by means of simple tools converted materials into a usable product. Gradually, we learned how to transfer certain of the workers' skills to machines, and with the demand for large quantities of identical or similar products the factory system was developed. Division of labor took place, with the worker quickly learning to perform short, repetitive tasks with great speed. The use of jigs, fixtures, and machines further aided in increasing the productivity of the factory worker. Thus, today the production process consists of creating a product through the utilization of men, machines, and materials.

When a new product is to be put on the market, it must be designed, materials for its manufacture must be specified, and the production methods, the tools, and machines must be designated. At an early stage in the design of the product, the materials to be used and the manufacturing process will be considered along with the quality standards and the cost to produce the product. There are almost an infinite number of ways to manufacture a particular product, harvest a crop, or mine coal. The methods designer has at his disposal the systematic problem-solving procedure to aid him in determining the preferred processes and methods to be used.

The over-all process of putting a new product into production can be divided into three parts or phases:

1. Planning.
2. Pre-production.
3. Production.

The General Motors Corporation has shown this graphically [1] in Fig. 8. In this illustration, emphasis is placed on work methods design, or "control of operator method," as it is called by General Motors.

Planning. This is the first step in any production or manufacturing process. As Fig. 8 shows, there are six basic planning functions. (1) The *design of the product* results in drawings showing the size, shape, weight, material, and ultimate use. (2) The *design of the process* consists of determining the production system—the operations required and their sequence; dimensions and tolerances, machines, tools, gauges, and equipment required. (3) The *design of work method* consists of the establishment of man-job relationships by determining how the person is to perform the operation, the work place, flow, and economic evaluation. (4) The *design of tools and equipment* consists of determining the jigs, fixtures, dies, gauges, tools, and machines which will be needed to perform the operations. (5) The *design of the plant layout* consists of determining the total space required in terms of overall location of equipment, stock supply, service centers, work space, material-handling equipment, and the man-machine relationship. (6) The *determination of the standard time* for the operation consists of measuring the work content of the job.

Planning is a decision-making process in that a goal or objective has been determined and a choice has been made from alternatives. The result is a specific product or part and specifications for its actual manufacture. For example, an electrical equipment manufacturer planned to produce and market a line of small electrical appliances. A team composed of a design engineer, an industrial engineer, and manufacturing staff personnel began the project by studying the design and production methods used to manufacture similar appliances already on the market. A product of original design was finally developed with a minimum of components, requiring the fewest operations and making the best possible use of raw materials. The direct labor hours required to produce this product were determined by motion-time data. This enabled the team to compare alternative designs and to select the one that gave the lowest cost. After the best design had been determined and tested, detailed production and assembly methods and inspection procedures were developed, the equipment and

[1] R. D. McLandress, "Methods Engineering and Operations Research," *Proceedings Twelfth Annual Industrial Engineering Institute,* University of California, Los Angeles-Berkeley, pp. 41–48, February, 1960. Reproduced with permission of General Motors Corporation.

CONTROL OF OPERATOR METHOD

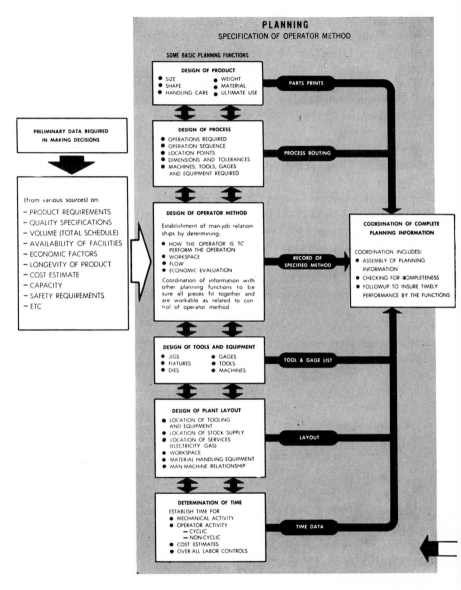

Fig. 8. Factors to be considered in

Developed by the General Motors Work Standards and Methods Engineering Committees and the Work Standards and Methods Engineering Section • Manufacturing Staff.

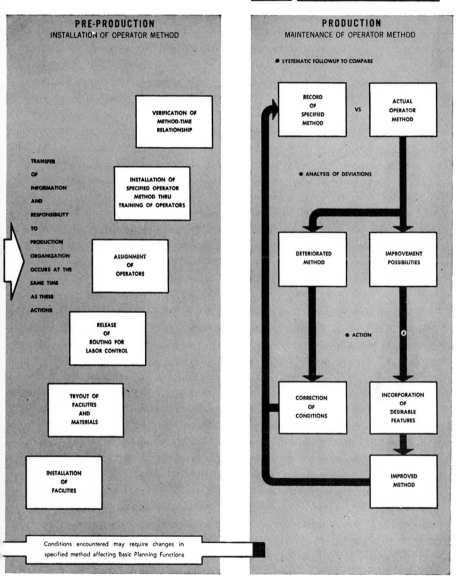

the control of operator method.

facilities needed to manufacture the product were selected, and a factory layout was prepared. This included a three-dimensional layout showing arrangement of machines, inspection, and storage areas, as well as handling-equipment and service areas. It was possible to calculate the direct labor cost and to estimate the indirect labor cost and other overhead costs. Material costs were also determined as a part of the planning activity.

Pre-production. This is the transition phase. The planning information is transferred to the production organization. Tools, machines, and equipment are purchased, installed, and tried out. The routing for labor control is released. Operators are selected and trained for specific tasks. The planned operator method is carefully checked against the method being used, and the actual time taken is checked against the original estimate. This is a period during which the individual operations that go to make up the over-all manufacturing activity are tried out.

Production. This refers to the continuing operation of the manufacturing activity established in the planning and pre-production phases. It involves the use of men, machines, and materials for the most effective manufacture of the part or product. Also, there is the ever-present necessity of (1) preventing the methods from deteriorating or deviating adversely from the planned methods, and (2) constantly examining the current methods for improvement and, when a better method is found, putting it into effect, in which case this then becomes the preferred method.

A SPECIFIC CASE—THE DESIGN OF A PLANT
TO MANUFACTURE FIBERBOARD SHIPPING CARTONS

Assume that a large paper company wishes to construct a plant to manufacture fiberboard shipping cartons. The main objective might be to obtain an adequate return on the capital invested, or to have the plant serve as an outlet for kraft paper which the company makes and markets. Other objectives might be to keep the unit labor cost and the unit material cost as low as possible, and to get the best equipment utilization, that is, the lowest operating cost of the equipment.

At the outset, a market study would be made to determine the nature and extent of the present demand for fiberboard shipping cartons and to forecast future demands. This would be a factor in determining the capacity of the plant that would be constructed, and the provisions that would be made for future expansion. Along with this market survey would be studies to determine a geographic location, and also studies to select a local building site for the factory. A team consisting of an industrial engineer, a process engineer or mechanical engineer, and a market analyst would obtain the information that would be needed to make the decisions referred to above.

Procedure for Design of Manufacturing Process and Production Methods. The industrial engineer, working with a mechanical engineer and a production supervisor from one of the company's paper manufacturing plants, would determine the exact manufacturing process to be used, the general flow pattern, and the methods of handling the raw materials, the materials in process, and the finished product. The industrial engineer would assume the responsibility for designing the detailed operator method and the layout of the work place for each operation. He would also fit this all together into a complete plant layout.

The industrial engineer might submit a detailed analysis of each of two or more methods of performing a specific operation when there appeared to be no clear-cut superiority of one method over another. For example, the finished cartons coming off the folding-gluing machine could be counted and tied into bundles manually, or this operation could be done by an automatic machine. The industrial engineer might submit a separate proposal for each method along with his recommendation as to the preferred method. As a result of the methods design a statement would be prepared showing the number of people required to operate each piece of equipment on each shift. This would include the power lift-truck operators as well as the men in the receiving and shipping departments and the manpower needed to operate the paper bailing machine. In a similar manner the number of supervisors on each shift would be determined.

The design of a manufacturing facility such as this one is very complex. For example, the operation of the combiner, which is over 200 feet long and represents an investment of some $750,000, is not a simple procedure—it is likely that specialists in various areas would be used in designing the plant. Moreover, it is to be expected that during the pre-production stage and during the "debugging" phase some changes and modifications would be made.

Detailed Statement of Problem. The problem is to build a new plant to manufacture corrugated fiberboard cartons and solid fiberboard cartons. All cartons are to be made to customers' orders and specifications. Carton sizes are to range from $4'' \times 4'' \times 4''$ to $45'' \times 45'' \times 40''$. The process will consist of (1) converting kraft paper into corrugated fiberboard or solid fiberboard and cutting to proper size for making cartons; (2) printing and slotting carton blanks; (3) folding and stitching seam, gluing seam, or taping seam, and counting and tying into bundles; and (4) receiving raw materials and shipping finished bundles of cartons (Fig. 9).

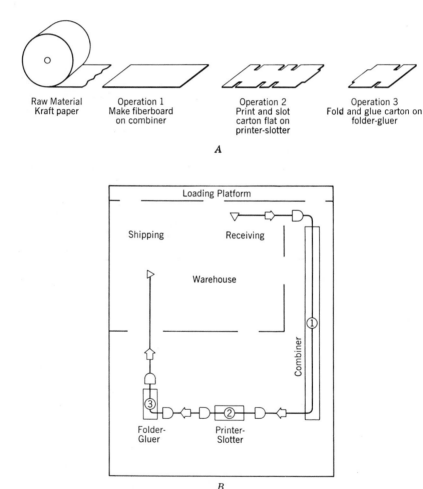

Fig. 9. Manufacture of fiberboard shipping cartons. *A*, Operations required to make carton; *B*, flow diagram for the manufacture of cartons.

Specifications—Raw Materials

1. Kraft paper ranging in thickness from lightweight (18 pounds per 1000 square feet) to heavyweight (90 pounds per 1000 square feet). Roll size: Width 43 inches to 87 inches, diameter 54 inches, weight up to 5000 pounds.

2. Cornstarch adhesive.

Specifications—Equipment Requirements and Output Requirements

A. *Combiner* with a capacity up to 125,000 square feet per hour or 20,000 lineal feet per hour. (See Fig. 345.) Time required to change

Fig. 10. Printer-slotter—machine for printing and slotting fiberboard blanks.

any of the following is 1 minute each: change blank length or width, or change grade of paper.

Combiner should be capable of producing:

1. Two-, three-, or four-ply solid fiberboard.
2. Double-face corrugated fiberboard, and double-wall and triple-wall corrugated fiberboard.

Combiner should run three 8-hour shifts per day, 5 days per week, 50 weeks per year. Capacity of combiner is 15,500,000 square feet per week. Maximum expected down time is 3%. Fiberboard waste expected at combiner is 2%. Total waste expected for all operations in plant, including combiner, is 10%.

B. *Printer-slotter* (see Fig. 10) is capable of handling blanks as shown in Table 4. Printing to be one or two colors. More than two colors requires rerun through printer. Printing ink and dies supplied

Table 4. Specifications for Printer-Slotter

Size of Printer-Slotter	Size of Carton Blank in Inches		Speed of Printer in Pieces per Hour
	Minimum	Maximum	
A	6 × 10	24 × 66	5000
B	20 × 25	35 × 78	5000
C	24 × 30	50 × 100	5000
D	30 × 40	80 × 180	1500

from outside. Setup time for any size printer-slotter is 30 minutes.

C. *Folder-gluer* (see Fig. 11) capable of handling blanks:

Size A, 10″ × 30″ minimum to 35″ × 78″ maximum.
Size B, 12″ × 35″ minimum to 50″ × 103″ maximum.
Speed of folder-gluer is 10,000 cartons per hour.

Cartons may be made by (*a*) stitching seam, (*b*) gluing seam, or (*c*) taping seam. Carton blanks must be stitched or taped if under 10″ × 30″ or over 50″ × 103″ in size. All other sizes may be stitched, glued, or taped. Setup time for either size folder-gluer is 20 minutes.

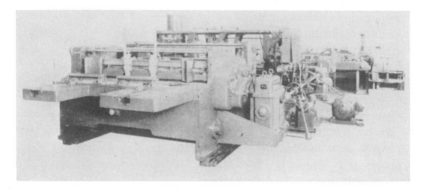

Fig. 11. Folder-gluer, capable of handling folded cartons at the rate of 10,000 feet per hour.

Estimated Cost of Equipment

Name of Equipment	Cost per Unit Installed
Combiner	$750,000
Printer-slotter	70,000
Folder-gluer	60,000
Steam boiler	9,000
Adhesive-mixing equipment	6,000
Fork lift trucks	8,000
Scrap fiberboard bailer	65,000
Building to house equipment	750,000

Time Schedule

Activity	Time in Months
Design plant	3
Build building to house equipment	2
Install equipment	2
"Debug" equipment and get plant running to capacity	5
Total time	12

Work Methods Design. The relatively simple process of manufacturing a fiberboard shipping carton has been used to show how the man, the machine, the materials, and the methods must all be considered together in the design of the factory. The design must meet certain specifications as to volume, scrap, size of cartons, and cost of finished product. Since 12 months is the maximum time allowed to design, build, and put the plant into full operation, it is obvious that this does not permit extended research to determine new or original methods of making shipping containers, such as molding or blowing or forming out of plastic, Fiberglas, or rubber instead of making them out of fiberboard. Also, equipment and processes generally available would probably have to be used. Some novel features might be developed in the equipment and in the layout of the plant and certainly in the design of the operator methods. Although it is true that one has greater freedom in developing work methods when a new process or system is being designed than when improvements are being made in a going activity, numerous specifications and restrictions are always present, including those of time and development costs.

CHAPTER 6

Work Methods Design—Developing a Better Method

At the time a new product or service is being designed or developed, consideration is nearly always given to the system or process that will be required to manufacture the product or provide the service. It is at this stage that one has greatest opportunity to use the design process and to come up with the best production systems and methods. However, experience shows that there is no "perfect method." In fact, there are always opportunities for improvement. Also, conditions may change. Factors such as volume and quality of product, kind and price of raw material, and availability of machines and equipment may be different from those that were present when production first started. Therefore, one is always confronted with the opportunity to improve processes and methods. This may also include the redesign of the product itself, and its components (see Figs. 12 and 13), as well as the standardization and better utilization of raw material (see Fig. 14).

Because this improvement aspect of methods design is so important in every phase of human endeavor, considerable emphasis is given to it in this book. The same problem-solving approach should be used in designing a method for an activity already in operation as in designing a new one. This means the determination of the goal or objective—the formulation of the problem. Then follows the analysis of the problem, obtaining facts, determining specifications and restrictions, and obtaining information about the volume—potential savings per year and savings over the life of the product. However, one usually does not have the same freedom; more constraints are imposed simply because the activity is a going one. There is the added "cost to make a change" that must be considered.

In searching for a better method, the analyst should not be unduly influenced by the current one. He should take a fresh look at all ways

A

B

C

Fig. 12. This product, redesigned for better appearance, has four components instead of six. (From Harold Van Doren, *Industrial Design*, 2nd ed., McGraw-Hill Book Co., New York. Photographs courtesy of Cushing & Nevell.)

A   B

Fig. 13. Carburetor control lever used on Caterpillar tractor.

A, Original Design
Original material, malleable iron casting.
Machining operations required:
 4 Drilling operations
 2 Reaming operations
 2 Tapping operations
 1 Saw-cutting operation

B, Present Design
Present material, steel stamping.

Machining operations required:
 1 Blanking and piercing operation
 1 Burring operation
 1 Forming operation

Note: The original method required a Woodruff keyway cut in both ends, as the shaft was keyed to the lever. The revised design requires a keyway at one end of the shaft only, as the other end is welded into the stamped lever.

Savings: The present cost is 66% of the original cost.

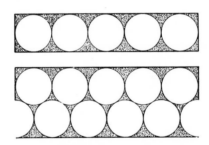

Fig. 14. Savings in raw material. The Detroit Transmission Division of General Motors saved **567** tons of steel in one year by punching double rows of clutch-plate blanks from steel stock, instead of single rows. In addition to saving the equivalent of **10** carloads of steel, there was a saving of **$25,000** in freight and handling costs per year. (From Philip E. Cartwright, "Measured Day Work and Its Relationship to a Continuous Cost Reduction Program," *Proceedings Nineteenth Time and Motion Study Clinic*, IMS, Chicago.)

of achieving the objective rather than merely trying to make an improvement on the present method.

Search for Possible Solutions—Develop the Preferred Method. The following approaches should be considered in developing possible solutions from which the preferred work method will be selected:

> A. Eliminate all unnecessary work.
> B. Combine operations or elements.
> C. Change the sequence of operations.
> D. Simplify the necessary operations.

It is entirely possible that more than one method will be designed. As discussed in Chapter 3, it is usually desirable to design (1) an ideal system or method, (2) a practical method that can be put into immediate use, and (3) a method that might be used if certain restraints or restrictions could be removed.

A. Eliminate All Unnecessary Work. Far too much work is done today that is not necessary. In many instances the job or the process should not be a subject for simplification or improvement, but rather it should be eliminated entirely.

The Procter and Gamble Company has found this matter of work elimination and cost elimination so profitable that it has established a formal procedure which it calls "the elimination approach." [1] Although the company is constantly improving methods and simplifying work, it believes that the ideal solution is to eliminate the cost. Its approach to cost elimination is as follows:

1. Select the cost for questioning. It is suggested that a major cost should be selected first in order to get the greatest money returns. If the major cost is eliminated, this will often lead to the elimination of many smaller operations as well. Labor costs, material costs, clerical costs, and overhead costs of all kinds are possible subjects for elimination. Efficient operations can be eliminated just as easily as those not as well done. The questioning procedure is easy to use. No paper work or calculations are necessary, and in fact complete knowledge of the job or activity is usually not required.

2. Identify the basic cause. A search should be made to determine the basic cause which makes the cost necessary. A basic cause is the reason, purpose, or intent on which the elimination of the cost de-

---

[1] Arthur Spinanger, "The Elimination Approach—A Management Tool for Cost Elimination," paper presented at meeting of American Institute of Industrial Engineers, Cincinnati, Ohio, May 18, 1960. This material reproduced with the permission of the Procter and Gamble Company.

pends. The basic cause is that factor which controls the elimination of the cost. The key question is, "This cost could be eliminated if it were not for what basic cause?" At *this stage* we do *not* ask such a question as "Why is this operation necessary?" or "How could this operation be done better?" These questions are avoided because they tend to justify and defend the job's continued existence. Instead the objective is to find the basic cause. Operations for which there is no basic cause, or for which a basic cause no longer exists, can be eliminated at once. When this is not the case and a basic cause has been identified, it is necessary to proceed to step 3.

3. Question the basic cause for elimination. If the basic cause has been identified, then it can be questioned in two ways.

    a. Disregard the basic cause—consider what would happen if the operation were not done. If the same results or better results can be obtained without the operation, then consideration should be given to eliminating it at once. However, disregarding the basic cause can be dangerous. In this connection it is necessary to consider two points: (1) determine the area of influence of the basic cause—what else might happen if this basic cause were eliminated? and (2) determine the associated "price tag" of the basic cause—is there a proper return on the money spent to obtain the desired results? If the basic cause cannot be disregarded, the second opportunity for elimination is

    b. Apply "why?" questioning. If the job under consideration seems to be necessary, can the job immediately preceding it be eliminated, thus perhaps making all succeeding jobs unnecessary? If complete elimination is not possible, try for partial elimination. Perhaps there are alternate possibilities—try to adopt the lowest cost alternative. Identify the basic cause of each supporting factor and question for elimination or change.

It is often desirable to undertake cost elimination on a department-wide or plant-wide basis. Thus several qualified members of supervision working as a group can help identify basic causes of specific costs selected for study.

Tubeless Tires. The inner tube in the automobile tire has been eliminated. The basic cause for using an inner tube was to contain the air and inflate the tire casing. An alternative possibility was to design the automobile wheel and casing so that the casing would contain the air and the inner tube would not be necessary. This was done, and the inner tube was eliminated.

Packing Lettuce in Cartons. Formerly lettuce was packed and shipped in large wood crates holding a total of approximately 125 pounds. In packing, ice was interspersed between the layers of lettuce.

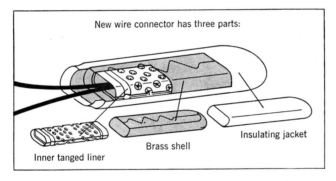

Fig. 15. Splicing insulated wires without removing the insulation or twisting the wires.

A better method of packing has been developed, using a carton which holds approximately 50 pounds. The lettuce is selected, cut, and packed directly into the carton in the field. Shortly thereafter, the packaged lettuce is quickly cooled to 36 to 38° in a vacuum cooling plant, and it is not necessary to place ice in the carton. More than 75% of all lettuce grown in California is now packed in cartons instead of boxes. The result is a saving of approximately $3 per box, and some 60,000 carloads of lettuce are shipped out of California each year. The use of ice in packing lettuce was eliminated, and the substitution of a fiberboard carton for a wood crate further reduced the packing cost.

Code Dating Cartons. Originally four code dates were stamped on each carton of soap coming off the packing line. The basic cause for these dates was the desire of sales people in customers' stores to determine when the carton of soap was manufactured. This seemed necessary, but since one date was satisfactory the other three dates were eliminated. In this case the cost was partially eliminated.

Splicing Insulated Wires.[2] Along the cable routes of the Bell System, wires are spliced at a rate of 250,000,000 a year. Conventionally, connections are made by "skinning" the insulation, twisting the bare wires together, and slipping on an insulating sleeve. Now, with a new connector, splices can be made faster, yet are even more reliable (Fig. 15). The craftsman slips the two wire ends, with insulation intact, into the connector and then flattens the connector with a pneumatic tool. Springy phosphor-bronze tangs inside the connector bite through the insulation to contact the copper wire. Here skinning the insulation and twisting the bare wires together have been eliminated.

Benefits of Work Elimination. If a job can be eliminated, there is no need to spend money on installing an improved method. No interruption or delay is caused while the improved method is being developed, tested, and installed. It is not necessary to train new operators on the new method. The problem of resistance to change is minimized when a job or activity that is found to be unnecessary is eliminated. The best way to simplify an operation is to devise some way to get the same or better results at no cost at all.

B. Combine Operations or Elements. Although it is customary to break down a process into many simple operations, in some instances the division of labor has been carried too far. It is possible to subdivide a process into too many operations, causing excessive handling of materials, tools, and equipment. Also such problems as the following may be created: difficulty in balancing the many operations, accumulation of work between operations when improper planning exists, and delays when inexperienced operators are employed or when operators are off the job.[3] It is sometimes possible to make the work easier by simply combining two or more operations, or by making some changes in method permitting operations to be combined.

Figure 16 shows how two short conveyors installed at the end of a molding machine in a furniture factory replaced the off-bearer and

---

[2] Reproduced with permission of Bell Telephone Laboratories.

[3] For a more detailed discussion of division of labor and job enlargement see Chapter 21.

made it possible for one man to do the work of two. The operator shown in the picture feeds the stock into the machine and places the finished molding strips in the truck as they come back to him on the conveyor. The truck shown is divided into four sections, only three of which are used to bring up raw stock; the fourth receives the finished strips as they come from the machine. This idea reduced the number of trucks needed and also saved floor space.[4]

Fig. 16. The two short belt conveyors eliminate the need for an off-bearer on this molding machine.

C. Change the Sequence of Operations. When a new product goes into production it frequently is made in small quantities on an "experimental" basis. Production often increases gradually, and in time output becomes large, but the original sequence of operations may be kept the same as when production was small. For this and other reasons it is desirable to question the order in which the various operations are performed.

For example, in one plant small assemblies were made on semiautomatic machines in Department A (Fig. 17). They were stored in

[4] Martin S. Meyers, "Evaluation of the Industrial Engineering Program in Small Plant Management," *Proceedings Sixth Industrial Engineering Institute,* University of California, Los Angeles-Berkeley, p. 37.

Department B, inspected in Department C, and packed for shipment in Department D. The manufacturing methods were such that normally only 10% of the finished assemblies were inspected. When an excessive number of defects were found, however, all work was given a 100% inspection until the cause of the trouble was located and corrected.

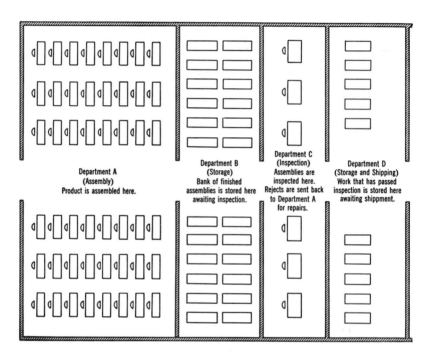

Fig. 17. Layout of building for assembly and inspection of small parts produced on semiautomatic machines—old arrangement of departments. Notice that inspection was done in Department C.

Since there was always a bank of several days' work in Department B awaiting inspection, when trouble was encountered it was necessary to give this entire bank a 100% inspection; moreover, defective assemblies had to be repaired or scrapped. To correct this difficulty the inspectors were placed immediately adjacent to the assembly department, and the bank of finished assemblies awaiting inspection was eliminated, as shown in Fig. 18. Since each unit was inspected as it came from the assembly line, rejects were found within a few minutes after the units were completed, and the cause of the difficulty could be corrected before other "scrap" parts had been made. This simple rearrangement, which was easy and inexpensive to make, saved the company tens of thousands of dollars in inspection costs and greatly reduced the number of scrapped parts.

The process chart and flow diagram described in Chapter 7 serve a useful purpose in pointing out the desirability of changing the sequence of operations to eliminate backtracking, to reduce trans-

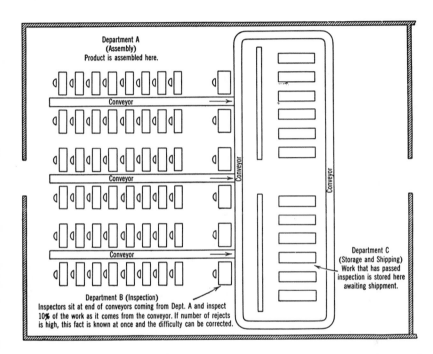

Fig. 18. Layout of building for assembly and inspection of small parts—improved arrangement of departments. Inspection now takes place immediately adjacent to the assembly department.

portation and handling, and to effect a smooth flow of work through the plant.

D. Simplify the Necessary Operations. After the process has been studied and all improvements that seem worthwhile have been made, the next step is to analyze each operation in the process and to try to simplify or improve it. In other words, the over-all picture is studied first and major changes are made; then the smaller details of the work are studied.

One of the best ways to approach the problem of methods improvement is to question everything about the job—the *way* the job is being done now, the *materials* that are being used, the *tools* and *equipment*, the *working conditions*, and the *design* of the product itself. Assume that nothing about the job is perfect. Begin by asking the questions: What? Who? Where? When? How? Why?

1. *What* is done? What is the purpose of the operation? *Why* should it be done? What would happen if it were not done? Is every part of the activity or detail necessary?

2. *Who* does the work? *Why* does this person do it? Who could do it better? Can changes be made to permit a person with less skill and training to do the work?

3. *Where* is the work done? *Why* is it done there? Could it be done somewhere else more economically?

4. *When* is the work done? *Why* should it be done then? Would it be better to do it at some other time?

5. *How* is the work done? *Why* is it done this way? This suggests a careful analysis and the application of principles of motion economy.

Question each element or hand motion. Just as in an analysis of the process we tried to eliminate, combine, and rearrange the sequence of operations, so in the single operation we try to eliminate motions, combine them, or rearrange the sequence of necessary motions in order to make the job easier.

Methods Laboratories. To an increasing extent business and industry are providing methods design laboratories with staff and facilities for the systematic study and improvement of production methods. The early methods laboratories were found mainly in industries having light assembly operations and short-cycle repetitive jobs. Here production methods were developed; temporary jigs and fixtures were designed, built, and tested; and in some cases a mock-up was made of the work place. Sometimes motion picture films were made for teaching the job to new operators, and in some instances the operators were brought into the laboratory for training.

Figure 19 shows part of the motion study laboratory at the Fort Wayne Works of the General Electric Company as it appeared in 1929. It is of special interest because it was one of the first such laboratories in this country. In contrast to this, a large tractor manufacturing company established a methods laboratory and undertook as the initial project the determination of the best methods of performing welding operations as they pertained to the manufacture of the company's products. Twelve carefully selected men from the Industrial Engineering Department were assigned to this task. Approximately one year was needed to develop and standardize welding methods and procedures and to establish standard data for predetermining time standards for such work. After this first investigation there has followed a continuous series of studies to increase productivity and decrease the cost of operating machine tools and production equipment of all kinds used by this company.

Figure 20 shows the methods laboratory of the Work Standards and

Fig. 19. The original motion study laboratory at the Fort Wayne works of the General Electric Co. (Courtesy of General Electric Co.)

Methods Engineering Department at the Packard Electric Division of the General Motors Corporation.[5] This laboratory now has some 9000 square feet of floor space, which includes twelve project areas, a film projection room, and an office area. Project areas may be combined to handle large projects. A permanent staff of salaried personnel and hourly paid people are continually seeking to make improvements in present methods, as well as to plan and develop better techniques for new and existing products. In some cases methods are developed, jigs and fixtures are built, and operators are brought into the laboratory and trained before the new method is installed in the plant.

[5] R. L. McLandress, "Organization and Coordination of Industrial Engineering Functions in a Large Corporation," *Proceedings Eighteenth Time and Motion Study Clinic,* IMS, Chicago, pp. 96–102.

Tools for Methods Improvement. Before better and easier methods of doing a task can be developed, it is necessary to get all the facts pertaining to the job. This involves getting sufficient information to answer the what, who, where, when, how, and why questions, and to answer satisfactorily the four other questions already suggested. Most people find it useful to list the information in tabular or graphic

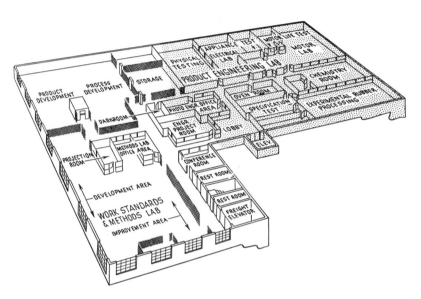

Fig. 20. Methods laboratory at the Packard Electric Division of the General Motors Corporation.

form. Since several different methods of visualizing a process or an operation have been widely used, each of them will be fully described in the next five chapters. Of course, not all of these different methods would be used on any one job. For example, it may be found that a process chart or flow diagram is all that is needed. If a single operation is the subject for study, then the operation chart may be used. The activity chart and the man and machine chart are also useful, and occasionally it may be worthwhile to make a micromotion analysis of the job, particularly if the cycle is short and a large number of people are employed on it.

It should be clearly understood, however, that the process chart, flow diagram, activity chart, man and machine chart, operation chart, and simo chart are merely tools to be used as needed.

CHAPTER 7

Process Analysis

The entire system or process of doing work should be studied before undertaking a thorough investigation of a specific operation in the process. Such an over-all study will ordinarily include an analysis of each step in the manufacturing process.

Process Charts. The process chart is a device for recording a process in a compact manner, as a means of better understanding it and improving it. The chart represents graphically the separate steps or events that occur during the performance of a task or during a series of actions. The chart usually begins with the raw material entering the factory and follows it through every step, such as transportation to storage, inspection, machining operations, and assembly, until it becomes either a finished unit itself or a part of a subassembly. The process chart might, of course, record the process through only one or a few departments.

A careful study of such a chart, giving a graphic picture of every step in the process through the factory, is almost certain to suggest improvements. It is frequently found that certain operations can be eliminated entirely or that a part of an operation can be eliminated, that one operation can be combined with another, that better routes for the parts can be found, more economical machines used, delays between operations eliminated, and other improvements made, all of which go to produce a better product at a lower cost. The process chart assists in showing the effects that changes in one part of the process will have on other parts or elements. Moreover, the chart may aid in discovering particular operations in the process which should be subjected to more careful analysis.

The process chart, like other methods of graphic representation, should be modified to meet the particular situation. For example, it may show in sequence the activities of a person, or the steps that the material goes through. The chart should be either the *man type* or the *material type*, and the two types should *not* be combined.

The process chart may profitably be made by almost anyone in an organization. The foreman, the supervisor, and the process and layout engineers should be as familiar with the process chart as the industrial engineer and should be able to use it.

Many years ago the Gilbreths devised a set of 40 symbols which they used in making process charts.[1] In recent years the abbreviated set of four symbols shown in Fig. 21 has been widely used, and they are all that are needed for many kinds of work. These symbols serve as a special sort of shorthand to aid in listing quickly the steps or activities in a process.

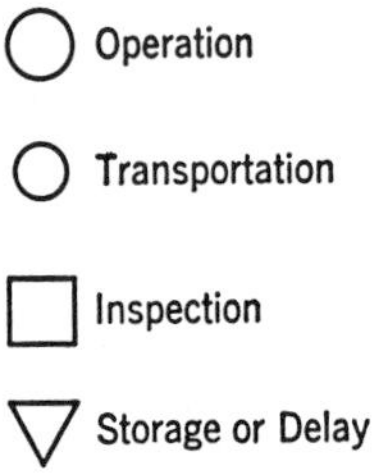

◯ Operation

◯ Transportation

☐ Inspection

▽ Storage or Delay

Fig. 21. The Gilbreth process chart symbols.

In 1947 the American Society of Mechanical Engineers established as standard the five symbols [2] shown in Fig. 22. This set of symbols is a modification of the abbreviated set of Gilbreth symbols in that the arrow replaces the small circle and a new symbol has been added to denote a delay. Although industry has been slow to adopt the ASME symbols, they seem to be gaining in acceptance and will be used in this volume.

Perhaps it is not too important which symbols are used in making process charts and flow diagrams. In fact, an organization may find that it requires a special set of symbols for its particular needs.[3] Experience shows, however, that where foremen and supervisors are expected to take an active part in developing better methods it is desirable to use as few process chart symbols as possible and charts that are simple to construct and easy to understand.

The process chart symbols used in the illustrations in this volume are those shown in Fig. 22, and are described as follows:

◯ *Operation.* An operation occurs when an object is intentionally changed in one or more of its characteristics. An operation represents

---

[1] F. B. and L. M. Gilbreth, "Process Charts," *Transactions of the ASME,* Vol. 43, Paper 1818, pp. 1029–1050, 1921.

[2] *Operation and Flow Process Charts, ASME Standard* 101, published by the American Society of Mechanical Engineers, New York, 1947.

[3] Ben S. Graham, "Paperwork Simplification," *Modern Management,* Vol. 8, No. 2, pp. 22–25.

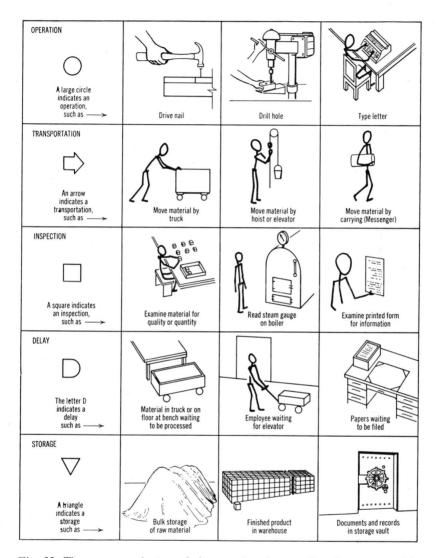

Fig. 22. These process chart symbols save time in recording the steps used in doing work.

a major step in the process and usually occurs at a machine or work station.

▷ *Transportation.* A transportation occurs when an object is moved from one place to another, except when the movement is an integral part of an operation or an inspection.

☐ *Inspection.* An inspection occurs when an object is examined for identification or is compared with a standard as to quantity or quality.

◯ *Delay.* A delay occurs when the immediate performance of the next planned action does not take place.

▽ *Storage.* A storage occurs when an object is kept under control such that its withdrawal requires authorization.

Combined Symbols. Two symbols may be combined when activities are performed at the same work place or when they are performed concurrently as one activity. For example, the large circle within the square ⬚ represents a combined operation and inspection.

Steps Used in Watering Garden. In order to illustrate how these symbols may be used, the process chart shown in Fig. 23 gives the steps followed by Mr. Smith in getting ready to water his garden. Mr. Smith, sitting on his porch, decides to water the garden. He leaves the porch, walks to the garage at the other end of the house, opens the garage door, and walks to the tool locker. There he lifts the reel of garden hose from the locker, carries it to the rear garage door, opens the door, and carries the hose to the faucet at the rear of the garage. He attaches the hose to the faucet, turns on the faucet, and begins to water the garden. An examination of the process chart on the left-hand side of Fig. 23 will show that the nine symbols, five numbers, and nine phrases are all that are needed to describe the entire process fully.

Flow Diagram of Watering Garden. Sometimes a better picture of the process can be obtained by putting flow lines on a plan drawing of the building or area in which the activity takes place. A sketch of the plan view of the house, lawn, and garden is shown in Fig. 24. Lines are drawn on this sketch to show the path of travel, and the process chart symbols are inserted in the lines to indicate what is taking place. Brief notations are included to amplify the symbols. This is called a flow diagram. Sometimes both a process chart and a flow diagram are needed to show clearly the steps in a manufacturing process, office procedure, or other activity.

Recoating Buffing Wheels with Emery. In large factories where heavy polishing and buffing operations are required, it is customary to

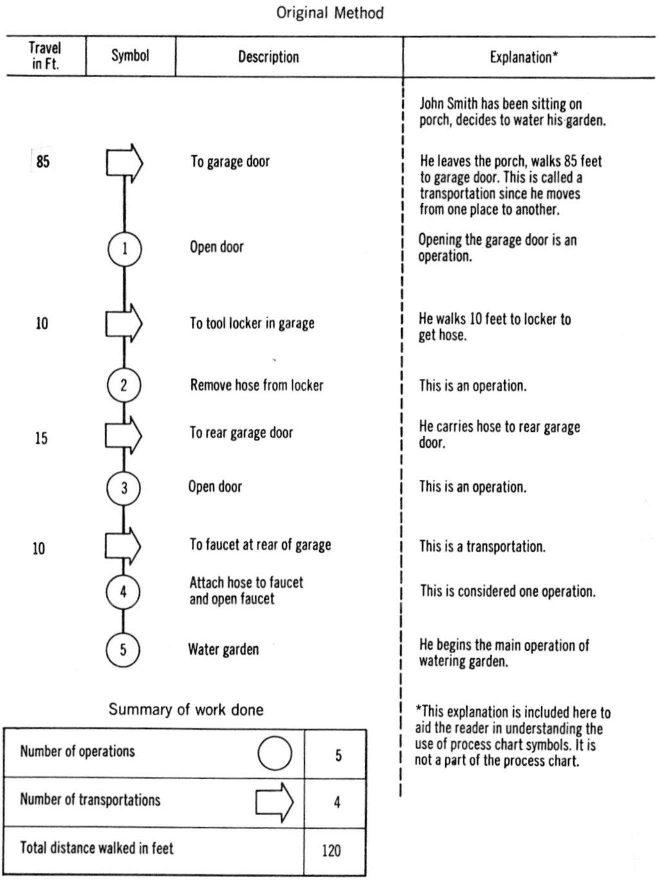

Original Method

Travel in Ft.	Symbol	Description	Explanation*
			John Smith has been sitting on porch, decides to water his garden.
85	⇨	To garage door	He leaves the porch, walks 85 feet to garage door. This is called a transportation since he moves from one place to another.
	1	Open door	Opening the garage door is an operation.
10	⇨	To tool locker in garage	He walks 10 feet to locker to get hose.
	2	Remove hose from locker	This is an operation.
15	⇨	To rear garage door	He carries hose to rear garage door.
	3	Open door	This is an operation.
10	⇨	To faucet at rear of garage	This is a transportation.
	4	Attach hose to faucet and open faucet	This is considered one operation.
	5	Water garden	He begins the main operation of watering garden.

Summary of work done

Number of operations	◯	5
Number of transportations	⇨	4
Total distance walked in feet		120

*This explanation is included here to aid the reader in understanding the use of process chart symbols. It is not a part of the process chart.

Fig. 23. Process chart of watering garden.

recoat buffing wheels (Fig. 25) with emery in the plant, thus keeping a supply of fresh wheels always available. The wheels are made of layers of fabric sewed together, and their average weight is 40 pounds. They vary in diameter from 18 to 24 inches, and in width of face from 3 to 5 inches. The circumference or face of the wheel is coated with glue and emery dust. The first coat of glue is allowed to set approximately one-half hour before the second coat is applied. The temperature in the room where the wheels are cured is maintained between 80 and 90°, and the humidity is also controlled.

Original Method. This was to coat the circumference of the worn wheel with glue (Fig. 26) and then roll the wheel by hand through

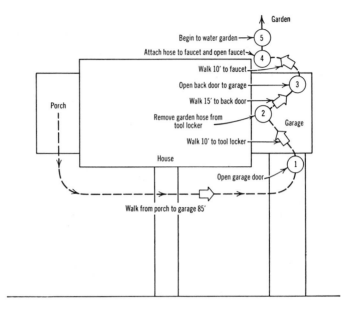

Fig. 24. Flow diagram of watering garden.

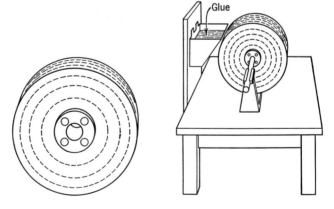

Fig. 25. Buffing wheel.

Fig. 26. Operator applies glue to circumference of worn wheel by means of a brush.

a shallow trough filled with emery dust, thus coating the wheel (Fig. 27). After the glue had dried, a second coat of glue and emery dust was applied in a similar manner. The wheels were then

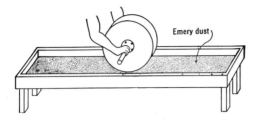

Fig. 27. Old method of recoating wheel. Operator rolls glue-covered wheel back and forth in trough containing emery dust.

hauled to a drying oven, and hung on racks in the oven until the glue was thoroughly dry. Figure 28 shows the flow diagram and Fig. 29 the process chart.

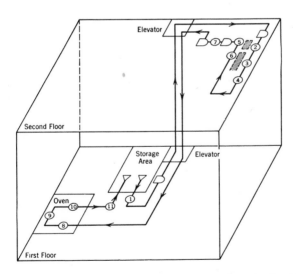

Fig. 28. Flow diagram of old method of recoating buffing wheels with emery.

The following questions might be asked about this job: *Why* coat the wheels by hand? *Why* handle the wheels so often? Could the wheels be coated on the first floor instead of on the second? These questions were answered in the following way.

Travel, feet	Symbol	Description
	▽	Worn wheels on floor (to be recoated)
	①	Load wheels onto truck
40	⇨	To elevator
	D	Wait for elevator
20	⇨	To second floor by elevator
35	⇨	To coating bench
	D	At coating bench
	②	Coat with glue
	③	Coat with emery (1st coat)
	④	On floor to dry
	⑤	Coat with glue
	⑥	Coat with emery (2nd coat)
	D	On floor at coating table
	⑦	Load onto truck
15	⇨	To elevator
	D	Wait for elevator
20	⇨	To first floor by elevator
75	⇨	To drying oven
	⑧	Unload coated wheels onto racks in oven
	⑨	Dry in oven
	⑩	Load wheels onto truck
35	⇨	To storage area
	⑪	Unload wheels onto floor
	▽	Storage

Summary

Number of operations _ _ _ _ _ _ _ _ _ ○		11
Number of delays _ _ _ _ _ _ _ _ _ _D		4
Number of storages _ _ _ _ _ _ _ _ _ ▽		2
Number of inspections _ _ _ _ _ _ _ _☐		1
Number of transportations _ _ _ _ _ _⇨		7
Total travel, in feet _ _ _ _ _ _ _ _ _ _ _		240

Fig. 29. Process chart of old method of recoating buffing wheels with emery.

Improved Method. A special coating machine (Fig. 30) was built, making it possible to apply the glue and emery to the wheel in one operation with much less time and effort than by the old method. Since this machine was located on the first floor between the storage

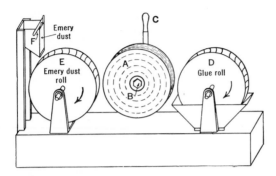

Fig. 30. Schematic drawing showing coating machine. The wheel *A* to be coated is mounted on shaft *B*, which is attached to lever *C*. Swinging the lever to right makes contact with glue roll *D*, which coats circumference of wheel. Lever *C* is then swung to left, making contact with roll *E*, which coats wheel with emery. Rolls *D* and *E* are both power-driven. The lever at *F* controls the amount of emery dust fed onto roll *E*.

area and the drying oven (Fig. 31), it was unnecessary to move the wheels to the second floor. Special truck racks (Fig. 33) were used instead of regular platform trucks, eliminating much unnecessary handling of wheels. The coated wheels remained on the truck racks

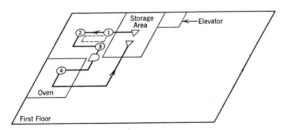

Fig. 31. Flow diagram of improved method of recoating buffing wheels with emery.

while in the drying oven. Figure 32 shows the process chart for the improved method, and a summary of the savings.

Results. The new coating machine, the special truck rack for handling wheels, and the better location of the coating machine reduced the number of *operations* needed to coat the wheels from 11 to 4, the number of *delays* from 4 to 1, and the *length of travel* from 240 to 70 feet. By the old method a crew of four men applied two coats of emery, and their average production was 20 wheels per hour. At the present time a two-man crew applies two coats of emery, producing 45 wheels per hour. Also, the new method of recoating wheels and a new type of synthetic glue seemed to improve the

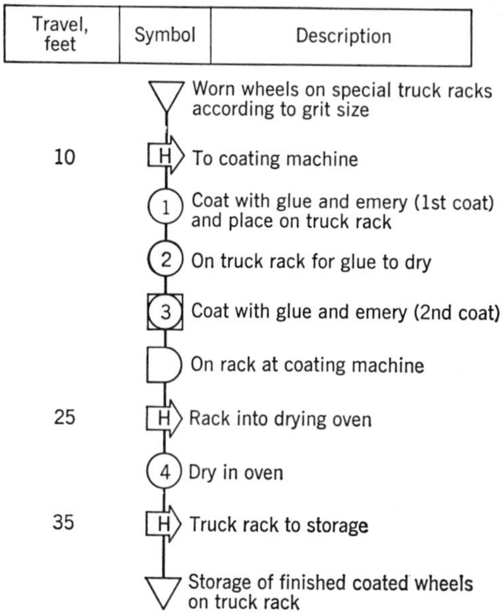

Travel, feet	Symbol	Description
	▽	Worn wheels on special truck racks according to grit size
10	H⟩	To coating machine
	1	Coat with glue and emery (1st coat) and place on truck rack
	2	On truck rack for glue to dry
	3	Coat with glue and emery (2nd coat)
	D	On rack at coating machine
25	H⟩	Rack into drying oven
	4	Dry in oven
35	H⟩	Truck rack to storage
	▽	Storage of finished coated wheels on truck rack

Summary

	Old Method		Improved Method		Difference	
Number of operations _____○	11		4		7	
Number of delays_____D	4		1		3	
Number of storages_____▽	2		2		0	
Number of inspections_____□	1		1		0	
Transportations	No.	Dist.	No.	Dist.	No.	Dist.
By truck_____ H⟩	5	200	3	70	2	130
By elevator_____ E⟩	2	40	0	0	2	40
Total_____	7	240	3	70	4	170

Fig. 32. Process chart of improved method of recoating buffing wheels with emery.

quality of the finished wheels; in fact, the men using the wheels to grind and polish plowshares have increased their production approximately 25%. The wheels seem to cut faster and make the work easier for the operators.[4]

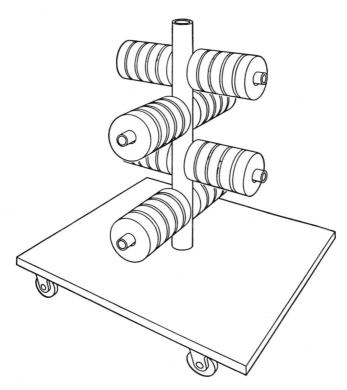

Fig. 33. Special truck rack for holding buffing wheels. Racks are used for wheel storage between operations and also hold wheels while they are in drying oven.

Flow Diagram of Feeding Silage on Small Dairy Farm. Farmers in increasing numbers are finding it profitable to use work methods design. Real savings are being made on small one-man farms as well as on larger ones. For example, on a 22-cow dairy farm in Vermont a systematic study was made of all the farm chores, and changes were designed to make the work easier and to save time. These changes were of four general types:

[4] This project courtesy of James D. Shevlin.

1. Rearrangement of the stables.
2. Improvement of work routines.
3. Provision of adequate and suitable equipment.
4. Convenient location of tools and supplies.

As a result, the time spent on chores was reduced from 5 hours and 44 minutes to 3 hours and 39 minutes daily, a saving of 2 hours and 5 minutes; the travel was reduced from 3¼ to 1¼ miles per day, a saving of 2 miles. Two hours a day is equivalent to more than 90

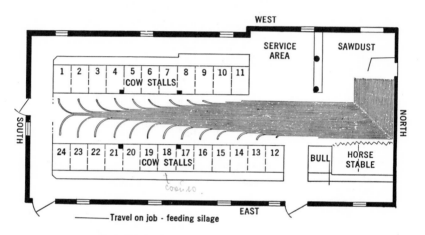

Fig. 34. Flow diagram of feeding silage to cows on small dairy farm—old method. Distance traveled, 2070 feet.

eight-hour working days in a year; 2 miles daily is equivalent to 730 miles yearly.

The lines in Fig. 34 show the amount of walking required to feed silage to the cows when the silage was carried from the silo in a bushel basket. Figure 35 shows the travel required when a two-wheel cart was used for hauling the silage. The total time to throw down the silage and feed 22 cows was reduced from 26.4 minutes to 14.8 minutes, and the travel was reduced from 2070 feet to 199 feet.[5]

Process Chart for an Office Procedure. In the office the process chart might show the flow of a time card, a material requisition, a purchase order, or any other form, through the various steps. The chart might begin with the first entry on the form and show all the

[5] R. M. Carter, "Labor Saving through Job Analysis," University of Vermont and State Agricultural College, *Bulletin* 503, p. 36.

steps until the form is permanently filed or destroyed (Figs. 36–39).

Assembly Process Charts. A special type of process chart, sometimes called an assembly process chart, is useful for showing such situations as the following: when several parts are processed separately and are then assembled and processed together; when a product is disassembled and the component parts are further processed, such as an animal in the packing house; and when it is necessary to show a division in the flow of work, such as separate action on different copies of an office form.

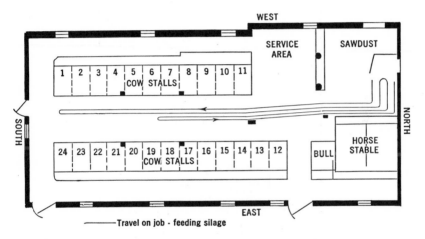

Fig. 35. Flow diagram of feeding silage to cows on small dairy farm—improved method. Distance traveled, 199 feet.

Figure 40 shows a longer and more complicated process, that of baking soda crackers. Figure 41 shows the process chart of making, painting, filling, and closing a rectangular tin can for the export shipment of instruments. Part of this process is described on pages 450 to 455. The raw material goes into stores and then through the various can-making operations. The two parts of the can are sprayed, the product is inserted into the can, the can is soldered shut, and the spraying is completed.

A study of the process chart shows several long moves that should be eliminated. Also, from general observation of the spraying operations it is apparent that some improvement might be possible. The can cover is sprayed on the outside, with the exception of a strip around the edge where it will be soldered to the bottom. In like manner the bottom of the can is sprayed on the outside, with the exception of a strip around the edge for soldering it to the cover. After these spraying operations, the two parts are assembled and moved 2500 feet to a storeroom, and then 570 feet to the packing department to be filled. The filled cans are moved 3000 feet to be soldered shut, and then moved to still another building where the unpainted portion of the outside of the filled can is painted.

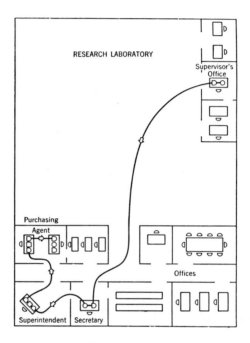

Fig. 36. Flow diagram of an office procedure—present method. Requisition is written by supervisor, typewritten by secretary, approved by superintendent, and approved by purchasing agent; then a purchase order is typewritten by stenographer.

As a result of a careful study of this entire process, the three spraying operations were eliminated entirely and one dipping operation was substituted for them. Cleaning the cans before dipping them in lacquer was also found to be unnecessary—a procedure required in the spraying operations.

The process chart of the improved method is shown in the lower right-hand corner of Fig. 41. A summary gives the savings resulting from the improved method.

An over-all investigation should be the first one made, because entire operations or series of operations may be eliminated in this way.

Present Method ☒			PROCESS CHART	

Proposed Method ☐

SUBJECT CHARTED __Requisition for small tools__   DATE _____
Chart begins at supervisor's desk and ends at typist's desk in   CHART BY __J. C. H.__
purchasing department   CHART NO. __R 136__

DEPARTMENT __Research laboratory__   SHEET NO. __1__ OF __1__

DIST. IN FEET	TIME IN MINS.	CHART SYMBOLS	PROCESS DESCRIPTION
		●⇨☐D▽	Requisition written by supervisor (one copy)
		O⇨☐D▽	On supervisor's desk (awaiting messenger)
65		O⇨☐D▽	By messenger to superintendent's secretary
		O⇨☐D▽	On secretary's desk (awaiting typing)
		●⇨☐D▽	Requisition typed (original requisition copied)
15		O⇨☐D▽	By secretary to superintendent
		O⇨☐D▽	On superintendent's desk (awaiting approval)
		O⇨■D▽	Examined and approved by superintendent
		O⇨☐D▽	On superintendent's desk (awaiting messenger)
20		O⇨☐D▽	To purchasing department
		O⇨☐D▽	On purchasing agent's desk (awaiting approval)
		O⇨■D▽	Examined and approved
		O⇨☐D▽	On purchasing agent's desk (awaiting messenger)
5		O⇨☐D▽	To typist's desk
		O⇨☐D▽	On typist's desk (awaiting typing of purchase order)
		●⇨☐D▽	Purchase order typed
		O⇨☐D▽	On typist's desk (awaiting transfer to main office)
		O⇨☐D▽	
		O⇨☐D▽	
		O⇨☐D▽	
		O⇨☐D▽	
		O⇨☐D▽	
		O⇨☐D▽	
		O⇨☐D▽	
105		3 4 2 8	Total

Fig. 37. Process chart of an office procedure—present method.

For the cans, it would have been a waste of time to make a minute study of the cleaning and spraying operations with the idea of improving them, only to find later that all of them could be eliminated.

No matter how complicated or intricate the manufacturing process may be, a process chart can be constructed in the same manner and serves the same purpose as those in the examples given. It is some-

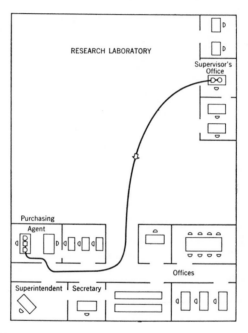

Fig. 38. Flow diagram of an office procedure—proposed method. Requisition is written in triplicate by supervisor and approved by purchasing agent.

times desirable to insert photographs of the work place or of a key set of motions at the appropriate point on the chart. Occasionally time values are set opposite each operation.

The Process Chart as an Aid in Plant Layout. The process chart is also a valuable aid in making a new layout or in rearranging equipment already in use. The following case will illustrate how B. C. Koch, while standards supervisor of the International Business Machines Corporation, used the process chart for this purpose. The process is that of making a magnet armature for a tabulator (Fig. 44).

As Figs. 42 and 43 show, 31 operations were originally required and the part traveled 5710 feet during the course of manufacture. The process charts in Figs. 42 and 46 do not show the temporary storage

Present Method ☐		PROCESS CHART	
Proposed Method ☒			

SUBJECT CHARTED ____Requisition for small tools_____ DATE _____
___Chart begins at supervisor's desk and ends at purchasing agent's desk_____ CHART BY _J. C. H._
_____ CHART NO. _R 149_
DEPARTMENT ____Research laboratory_____ SHEET NO._1_ OF _1_

DIST. IN FEET	TIME IN MINS.	CHART SYMBOLS	PROCESS DESCRIPTION
		●⇨☐D▽	Purchase order written in triplicate by supervisor
		O⇨☐D▽	On supervisor's desk (awaiting messenger)
75		O⇨☐D▽	By messenger to purchasing agent
		O⇨☐D▽	On purchasing agent's desk (awaiting approval)
		O⇨■D▽	Examined and approved by purchasing agent
		O⇨☐D▽	On purchasing agent's desk (awaiting transfer to main office)
		O⇨☐D▽	
		O⇨☐D▽	
		O⇨☐D▽	
		O⇨☐D▽	
		O⇨☐D▽	
		O⇨☐D▽	
		O⇨☐D▽	
		O⇨☐D▽	
		O⇨☐D▽	
		O⇨☐D▽	
		O⇨☐D▽	
		O⇨☐D▽	

			SUMMARY			
		O⇨☐D▽		PRESENT METHOD	PROPOSED METHOD	DIFFER-ENCE
		O⇨☐D▽				
		O⇨☐D▽	Operations O	3	1	2
		O⇨☐D▽	Transportations ⇨	4	1	3
		O⇨☐D▽	Inspections ☐	2	1	1
		O⇨☐D▽	Delays D	8	3	5
		O⇨☐D▽	Distance Traveled in Feet	105	75	30
75		1 1 1 3	Total			

Fig. 39. Process chart of an office procedure—proposed method.

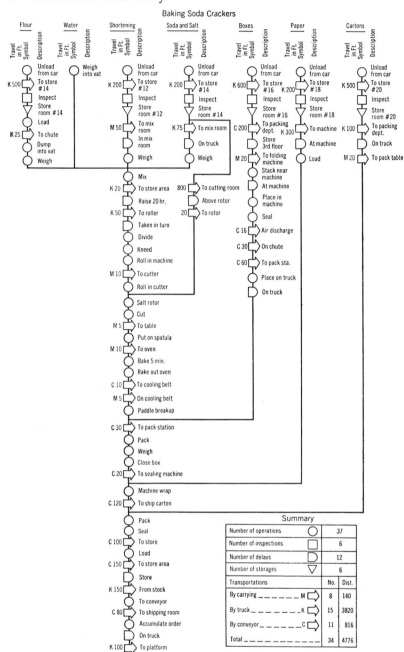

Fig. 40. Assembly process chart—baking soda crackers.

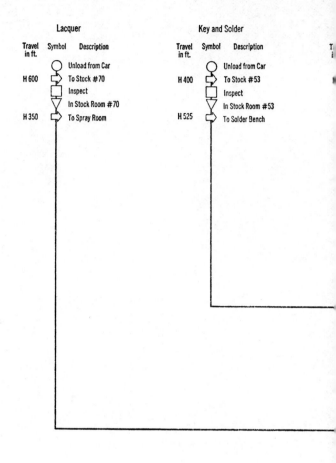

Lacquer			Key and Solder			T
Travel in ft.	**Symbol**	**Description**	**Travel in ft.**	**Symbol**	**Description**	**i**
	○	Unload from Car		○	Unload from Car	
H 600	⇨	To Stock #70	H 400	⇨	To Stock #53	
	☐	Inspect		☐	Inspect	
	▽	In Stock Room #70		▽	In Stock Room #53	
H 350	⇨	To Spray Room	H 525	⇨	To Solder Bench	

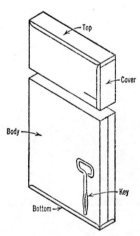

Fig. 41. Process chart for making, filling,

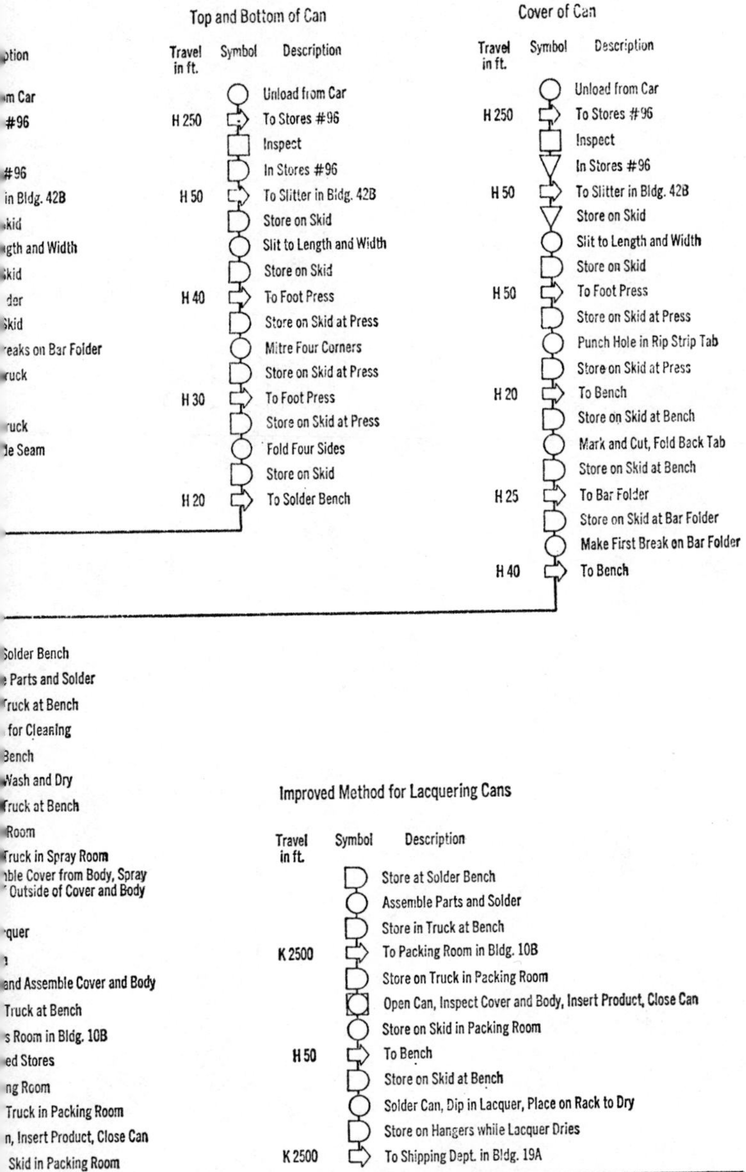

Top and Bottom of Can

Description	Travel in ft.	Symbol	Description
		○	Unload from Car
	H 250	▷	To Stores #96
		□	Inspect
		D	In Stores #96
	H 50	▷	To Slitter in Bldg. 42B
		D	Store on Skid
		○	Slit to Length and Width
		D	Store on Skid
	H 40	▷	To Foot Press
		D	Store on Skid at Press
		○	Mitre Four Corners
		D	Store on Skid at Press
	H 30	▷	To Foot Press
		D	Store on Skid at Press
		○	Fold Four Sides
		D	Store on Skid
	H 20	▷	To Solder Bench

Cover of Can

Travel in ft.	Symbol	Description
	○	Unload from Car
H 250	▷	To Stores #96
	□	Inspect
	▽	In Stores #96
H 50	▷	To Slitter in Bldg. 42B
	▽	Store on Skid
	○	Slit to Length and Width
	D	Store on Skid
H 50	▷	To Foot Press
	D	Store on Skid at Press
	○	Punch Hole in Rip Strip Tab
	D	Store on Skid at Press
H 20	▷	To Bench
	D	Store on Skid at Bench
	○	Mark and Cut, Fold Back Tab
	D	Store on Skid at Bench
H 25	▷	To Bar Folder
	D	Store on Skid at Bar Folder
	○	Make First Break on Bar Folder
H 40	▷	To Bench

(left column, partially cut off)

Solder Bench
Parts and Solder
Truck at Bench
for Cleaning
Bench
Wash and Dry
Truck at Bench
Room
Truck in Spray Room
ble Cover from Body, Spray
Outside of Cover and Body

quer

and Assemble Cover and Body
Truck at Bench
s Room in Bldg. 10B
ed Stores
ng Room
Truck in Packing Room
n, Insert Product, Close Can
Skid in Packing Room
h in Bldg. 13A
Skid at Bench
over to Body
Skid at Bench
Booth in Bldg. 31A
Skid at Booth
ver Soldered Seam
hile Lacquer Dries
ping Dept. in Bldg. 19A

Improved Method for Lacquering Cans

Travel in ft.	Symbol	Description
	D	Store at Solder Bench
	○	Assemble Parts and Solder
	D	Store in Truck at Bench
K 2500	▷	To Packing Room in Bldg. 10B
	D	Store on Truck in Packing Room
	○	Open Can, Inspect Cover and Body, Insert Product, Close Can
	D	Store on Skid in Packing Room
H 50	▷	To Bench
	D	Store on Skid at Bench
	○	Solder Can, Dip in Lacquer, Place on Rack to Dry
	D	Store on Hangers while Lacquer Dries
K 2500	▷	To Shipping Dept. in Bldg. 19A

SUMMARY		OLD METHOD		IMPROVED METHOD		DIFFERENCE	
Number of operations	○	8		3		5	
Number of inspections	□	2		1		1	
Number of delays	D	13		6		7	
Number of storages	▽	0		0		0	
Transportations		No.	Dist.	No.	Dist.	No.	Dist.
By motor truck — — — — K▷		4	11,300	2	5000	2	6300
By hand truck — — — — H▷		5	995	1	50	4	945
Total — — — — — — — —		9	12,295	3	5050	6	7245

rectangular can for export shipping of instruments.

symbols, because it is understood that in all the manufacturing operations in this factory there is a temporary storage immediately before and immediately after each operation.

A study of this process chart and of the flow lines on the floor plans in Fig. 43 led to a rearrangement of equipment. A small bench containing a disk grinder and some special filing fixtures was moved from the bench-work department to the drill-press department. This permitted an improvement in several operations. Figures 45 and 46 show those operations (numbers 8 to 16) affected by the change.

The improved layout resulted in the following:

1. All drilling operations were performed on one drill press.

2. Two inspection operations were combined.

3. Two straightening and two burring operations were combined.

4. Four moves were eliminated.

5. The total distance traveled was reduced from 5710 to 3750 feet, a reduction of 34%.

6. The total manufacturing time was reduced from 16.0 to 11.5 hours per hundred pieces, a reduction of 28%. (See Table 5.)

Utilization of Space in a Warehouse. A new addition to a factory warehouse was being proposed in order to provide additional storage space. Cases of product were stacked three skids high, and aisles were arranged as shown in *A* of Fig. 47. As the result of some hard thinking and imagination, utilization of

Travel in ft.	Symbol	Description
	1	Select Strip Stock in Raw Material Store
60		To Punch Press Dept.
	2	Blank
190		To Heat Treat Dept.
	3	Pack
	4	Anneal
	5	Unpack
190		To Punch Press Dept.
	6	Straighten
	7	Swage
180		To Elevator
		Up One Floor
370		To Straightening Dept.
	8	Straighten for Mill
	9	Straighten End Bearings
290		To Milling Dept.
	10	Straddle Mill
290		To Burring Dept.
	11	Burr Edges
370		To Elevator
		Down One Floor
180		To Punch Press Dept.
	12	Form Hook
180		To Elevator
		Up One Floor
370		To Straightening Dept.
	13	Straighten Post
	14	Straighten for Hollow Mill
430		To Large Drill Press Dept.
	15	Drill and Hollow Mill
80		To Small Drill Press Dept.
	16	Countersink and Ream
300		To Inspection Dept.
170	17	Inspect
170		To Burring Dept.
	18	Burr Ends
290		To Milling Dept.
	19	Mill Hook
290		To Straightening Dept.
	20	Grind and Burr Hook
	21	Straighten to Gauge
	22	File Hook to Gauge 1
	23	File Hook to Gauge 2
	24	Burr Hook
170		To Inspection Dept.
	25	Inspect
210		To Elevator
		Down One Floor
140		To Heat Treat Dept.
	26	Pack
	27	Carburize
	28	Unpack
	29	Harden Point
140		To Elevator
		Up One Floor
370		To Straightening Dept.
	30	Scrape off Lead
170		To Inspection Dept.
	31	Inspect
110		To Elevator
		Down One Floor
170		To Stock Room
	32	Store

Fig. 42. Process chart—old method of making armature.

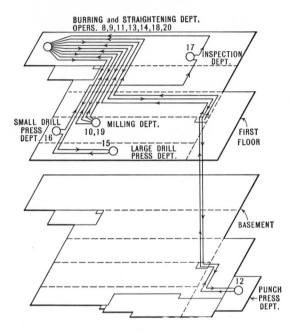

Fig. 43. Layout for making armature—old method.

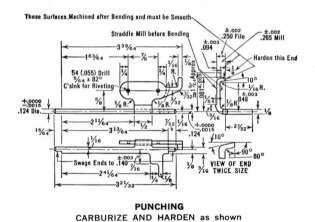

PUNCHING
CARBURIZE AND HARDEN as shown

Fig. 44. Magnet armature for IBM tabulating machine.

the present warehouse space was increased from 4.3 to 7.9 cases per square foot by the changes shown in *B, C,* and *D* of Fig. 47. This made the proposed addition to the warehouse unnecessary.

Gang Process Charts. The gang process chart is an aid in studying the activities of a group of people working together.[6] This chart is a composite of individual member process charts, arranged to permit

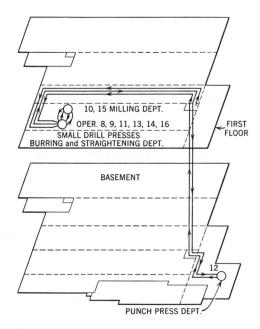

Fig. 45. Layout for making armature—improved method.

thorough analysis. Those operations which are performed simultaneously by gang members are indicated side by side. The basic purpose of the chart is to analyze the activities of the group and then compose the group so as to reduce to a minimum all waiting time and delays.

Construction

1. The same symbols are used as for ordinary process charts.

2. A process chart covers the cycle or routing followed by each member of the gang. Member charts are placed side by side, with steps

[6] The gang process chart was originated by John A. Aldridge, and the description of chart and illustrations presented here were developed by him. Also see "Gang Process Charts in Work Simplification," by John V. Valenteen, *Factory Management and Maintenance,* Vol. 104, pp. 125–127.

which are performed simultaneously shown on the same horizontal line. Figure 48 shows the form used for gang charts. The dots aid in chart construction, symbols being centered around the dots.

3. So that symbols of member charts may be placed close together, the various steps are given code numbers rather than entering de-

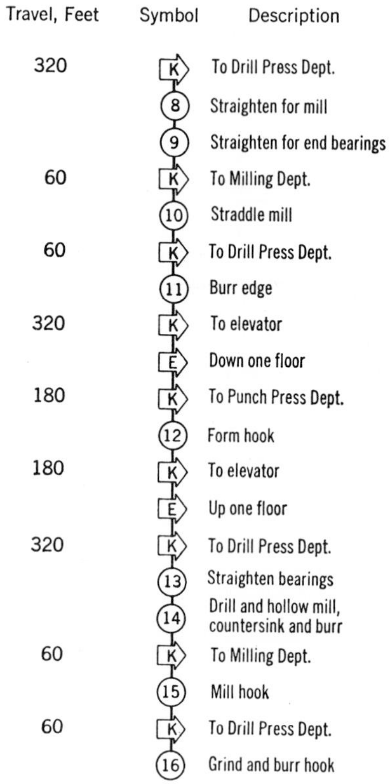

Travel, Feet	Symbol	Description
320	K	To Drill Press Dept.
	8	Straighten for mill
	9	Straighten for end bearings
60	K	To Milling Dept.
	10	Straddle mill
60	K	To Drill Press Dept.
	11	Burr edge
320	K	To elevator
	E	Down one floor
180	K	To Punch Press Dept.
	12	Form hook
180	K	To elevator
	E	Up one floor
320	K	To Drill Press Dept.
	13	Straighten bearings
	14	Drill and hollow mill, countersink and burr
60	K	To Milling Dept.
	15	Mill hook
60	K	To Drill Press Dept.
	16	Grind and burr hook

Fig. 46. Process chart—improved method of making armature.

scriptions beside each symbol. Numbers are entered in the center of each symbol and corresponding explanations are placed at the side of the chart. This eliminates repetition of the description when similar steps are repeated, and at the same time permits the member charts to be placed close together.

4. Attention must be paid to entering simultaneous steps side by side. It may be found that an operation performed by one member of the group continues while another is performing more than one operation. In such instances, the symbol is repeated at each step for

the operation which occupies the larger number of steps. On the chart in Fig. 48 it will be noted that the transportation distance was broken down to intervals of 20 feet, as movement over this distance was accomplished while one step of another worker was started and completed. Such divisions of transportation distances are approximate, but for the purpose of analysis are sufficient.

5. The chart should cover a complete cycle for the member performing the largest number of steps. Other gang members usually repeat their cycles during the largest member cycle.

6. Elements which do not occur in every cycle may be omitted from the chart. This includes preparatory work which is done before a cycle is started, such as obtaining supplies for an entire shop. On the other hand, if an operational step occurs at periodic intervals within the cycle, such as the moving of empty pallets as shown under operation 6 in Fig. 49, it should be included on the chart. If such an operation occurs every two or three cycles, enough cycles should be shown to include the operation.

7. The summary usually takes a different form from that described in the first part of this chapter. Steps per unit before and after study are used in gang summaries. This ratio is obtained by dividing the total steps on the chart by the total units handled for the cycles

Table 5. Summary of Savings Resulting from Improvement in Method of Making Magnet Armature

		Old Method		Improved Method		Difference	
Total number of operations	◯	31		28		3	
Transportations		No.	Dist.	No.	Dist.	No.	Dist.
By electric truck	K⟩	22	5250	19	3370	3	1880
By hand truck	H⟩	3	460	2	380	1	80
By elevator	E⟩	6	110	6	110	0	0
Total		31	5820	27	3860	4	1960

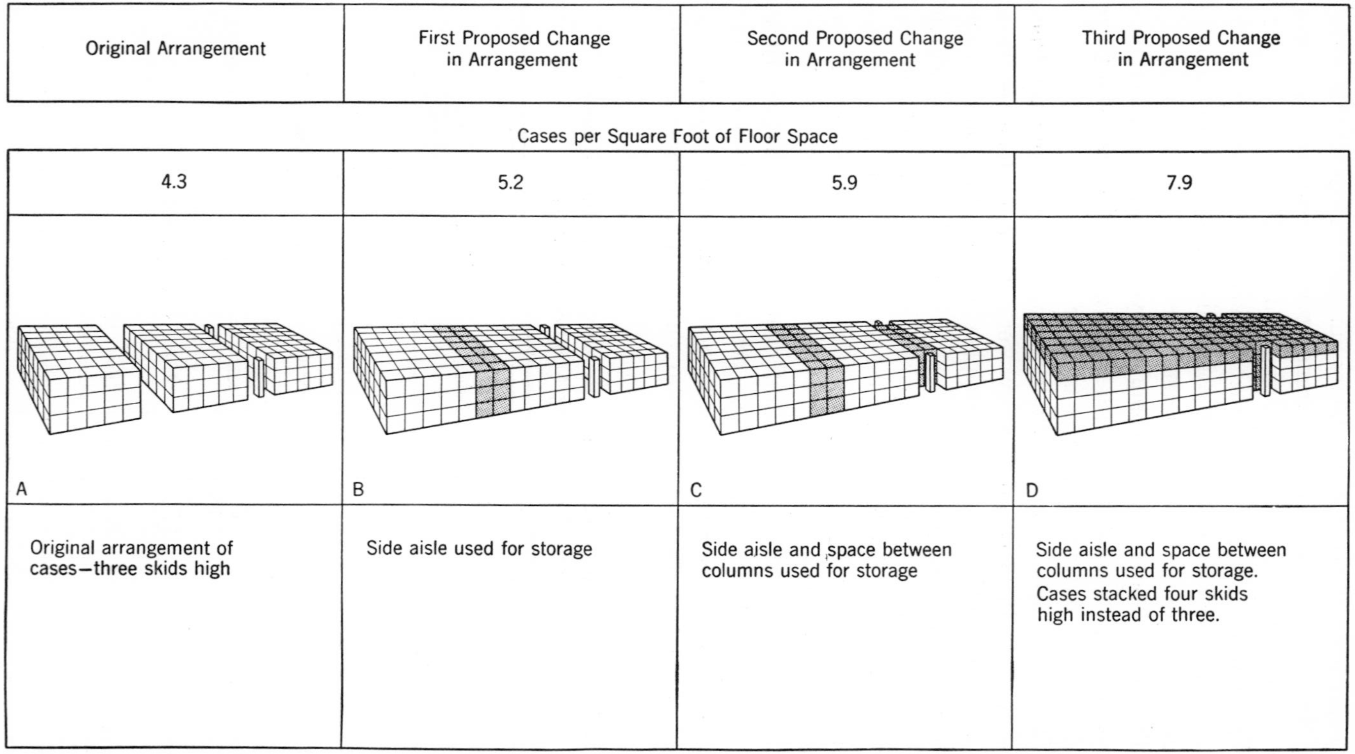

Original Arrangement	First Proposed Change in Arrangement	Second Proposed Change in Arrangement	Third Proposed Change in Arrangement
		Cases per Square Foot of Floor Space	
4.3	5.2	5.9	7.9

A	B	C	D
Original arrangement of cases—three skids high	Side aisle used for storage	Side aisle and space between columns used for storage	Side aisle and space between columns used for storage. Cases stacked four skids high instead of three.

Fig. 47. Four different methods of stacking cases of finished product in a factory warehouse.

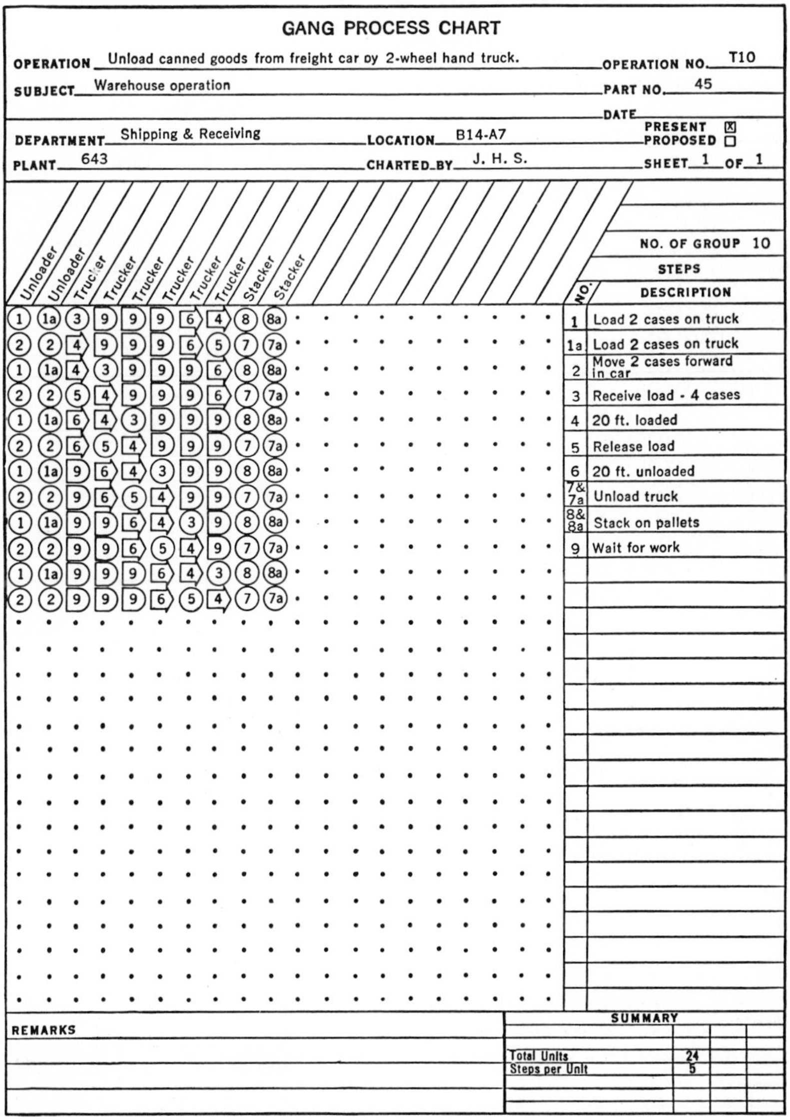

GANG PROCESS CHART

OPERATION __Unload canned goods from freight car by 2-wheel hand truck.__ OPERATION NO. __T10__

SUBJECT __Warehouse operation__ PART NO. __45__

DATE

DEPARTMENT __Shipping & Receiving__ LOCATION __B14-A7__ PRESENT ☒ PROPOSED ☐

PLANT __643__ CHARTED BY __J. H. S.__ SHEET __1__ OF __1__

Columns (worker roles): Unloader, Unloader, Trucker, Trucker, Trucker, Trucker, Trucker, Trucker, Stacker, Stacker

NO. OF GROUP 10 STEPS

NO.	DESCRIPTION
1	Load 2 cases on truck
1a	Load 2 cases on truck
2	Move 2 cases forward in car
3	Receive load - 4 cases
4	20 ft. loaded
5	Release load
6	20 ft. unloaded
7 & 7a	Unload truck
8 & 8a	Stack on pallets
9	Wait for work

REMARKS

SUMMARY

Total Units	24
Steps per Unit	5

Fig. 48. Gang process chart of unloading canned goods from freight car—present method.

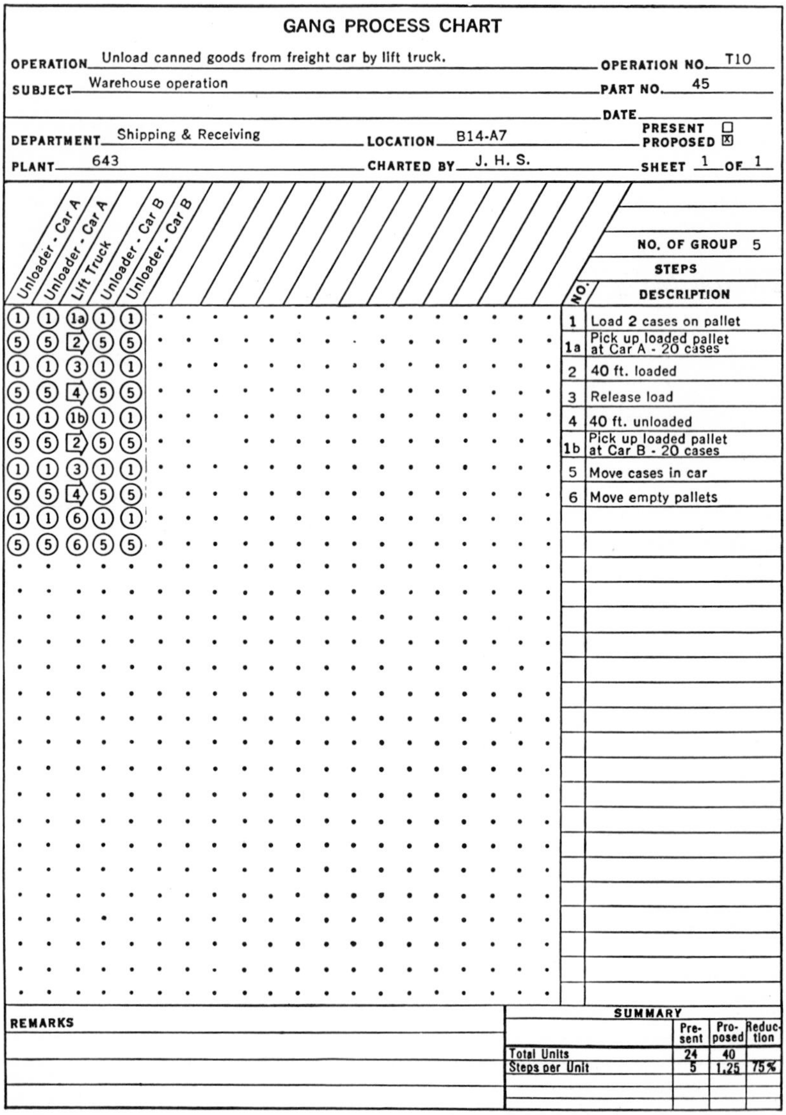

GANG PROCESS CHART

OPERATION___Unload canned goods from freight car by lift truck.___ OPERATION NO.___T10___

SUBJECT___Warehouse operation___ PART NO.___45___

DEPARTMENT___Shipping & Receiving___ LOCATION___B14-A7___ DATE_____

PLANT___643___ CHARTED BY___J. H. S.___

PRESENT ☐
PROPOSED ☒
SHEET _1_ OF _1_

						NO. OF GROUP STEPS 5
					NO.	DESCRIPTION
① ① ⑴ₐ ① ①	· · · · · · · · · · · · · · ·				1	Load 2 cases on pallet
⑤ ⑤ ② ⑤ ⑤	· · · · · · · · · · · · · · ·				1a	Pick up loaded pallet at Car A - 20 cases
① ① ③ ① ①	· · · · · · · · · · · · · · ·				2	40 ft. loaded
⑤ ⑤ ⑷ ⑤ ⑤	· · · · · · · · · · · · · · ·				3	Release load
① ① ⑴ᵦ ① ①	· · · · · · · · · · · · · · ·				4	40 ft. unloaded
⑤ ⑤ ② ⑤ ⑤	· · · · · · · · · · · · · · ·				1b	Pick up loaded pallet at Car B - 20 cases
① ① ③ ① ①	· · · · · · · · · · · · · · ·				5	Move cases in car
⑤ ⑤ ⑷ ⑤ ⑤	· · · · · · · · · · · · · · ·				6	Move empty pallets
① ① ⑥ ① ①	· · · · · · · · · · · · · · ·					
⑤ ⑤ ⑥ ⑤ ⑤	· · · · · · · · · · · · · · ·					

REMARKS	SUMMARY			
		Present	Proposed	Reduction
	Total Units	24	40	
	Steps per Unit	5	1.25	75%

Fig. 49. Gang process chart of unloading canned goods from freight car—proposed method.

represented on the chart. On the chart illustrated in Fig. 49 the total steps are 120 and the total units (cases) handled are 24. Four cases are loaded on a truck and 6 trucks are loaded in the cycle shown on the chart. One hundred and twenty divided by 24 equals 5 steps per unit.

8. A chart should not be constructed from observation of a single cycle. A number of cycles should be observed, as the amount of waiting time may vary from cycle to cycle. The average condition should be reflected by the chart.

Analysis. Four steps are followed in analyzing a gang process chart. First, the six questions what, who, where, when, how, and why are asked of the entire process. Next, each operation and inspection is analyzed by utilizing the same six questions. Third, the remaining transportations and storages are studied. These three steps are the same as those used in analyzing individual process charts. The fourth step consists in applying the "how" question in a new way after refinements have been completed under steps 1, 2, and 3. This question is asked: "How should the gang be composed to reduce waiting time to the minimum?" The following will assist the analyst to "balance" the gang under step 4:

1. Determine the class of operator having the largest amount of waiting time per cycle, and the class having the least.

2. Adjust the gang by decreasing number of operators least busy and increasing number of operators most busy. Generally, it is preferable to work towards a smaller rather than a larger gang.

A Specific Case. The activity to be considered is unloading a car of canned goods (Figs. 48 and 49). In answer to the "how" question it was decided that the work could be performed better if a lift truck were used to transport the material and if pallets were loaded in the car. It was determined that one lift truck could service two cars. These questions resulted in a radical change in the entire procedure. The substitution of the lift truck eliminated all truckers and stackers.

Arrangement of Lathes for Machining Ends of Casings. Figure 50 shows an installation of twelve Gisholt heavy-duty turret lathes laid out to machine taper seals concentric with threads for leakproof pressure joints on standard seamless line casings. Each pair of turret lathes is placed end to end, with a roller table between. The lathes machine the two ends of casings 30 to 40 feet long with a minimum of handling.

Rearrangement of Departments in a Hotel. The process chart has had widest use in the factory as an aid in eliminating operations, improving the layout of equipment, and reducing the amount of handling of materials. Because of the opportunity for large savings,

Fig. 50. Location of turret lathes on each side of table permits machining the ends of casings with a minimum of handling.

offices, banks, restaurants, and hotels are also using this approach in studying many of their processes.

Hotel Lowry has had notable success in this field. H. E. Stats, under whose direction this work was carried on, states that whereas the original objective was to study the plant layout for the purpose of improving space utilization and materials handling, this undertaking was so successful that the scope of the work was extended to include a well-rounded small-scale application of scientific management in:

1. Plant layout and materials handling.

2. Personnel, including training and improvement in individual and group productivity.[7]

3. Service functions, including cost accounting, analysis of printed forms, maintenance policies, etc.

4. Use of scientific administration in the over-all organization setup.[8]

Fig. 51. One container manufacturer placed scales in floor of elevator to eliminate unnecessary movement of folded cartons ready for shipment.

The hotel employed approximately 250 people and was able to carry on all engineering activities at a cost of less than one half of 1% of the total sales of the company.

The following paragraphs summarize some of the changes that were made:

[7] H. E. Stats, "Personnel Relations in Hotel Management," *Journal of Society for the Advancement of Management,* Vol. II, No. 4, p. 101.

[8] H. E. Stats, "Evolution of an Organization Plan," *Proceedings of the Minnesota Hotel Association.*

1. All stores were consolidated into one unit directly responsible to a newly created centralized purchasing department. Seven executives [9] who had spent part of their time buying under the old system, where each department did its own purchasing, were released from this responsibility and made available for additional supervisory duties.

2. The receiving department was combined with the central stores department, eliminating the receiving room entirely as a separate function.

3. A complete change in the location of the linen room, resulting in the consolidation of two other departments (the work-dispatching department and the service bureau), made it possible for the functions of these departments to be handled easily by the service bureau alone. Control on supplies and personnel was increased by the move.

4. The woodworking, paint, and repair shops, originally located in three separate rooms on different floor levels, were consolidated in a works department room, formerly used as a rubbish catch-all.

5. Relocation of the bottling department resulted in a large reduction of labor and equipment expenditures.

6. Removal of the butcher department from the stores on the basement level to the kitchen on the main floor eliminated the last necessity for direct departmental issues of raw materials. Cut meats are now more quickly available for preparation, and the butcher is now able to utilize his spare time on other kitchen work.

7. Perhaps the most revolutionary change in layout and materials handling was the centralization of all dishwashing in one department adjoining the main kitchen. This department is fed by an overhead chain conveyor which also runs to the coffee shop kitchen. The conveyor is also used for transporting food orders and raw materials between the kitchens.

Centralization of Dishwashing Department

a. Toweling of dishes and glasses is now completely automatic. New equipment dries dishes and glasses automatically.

b. Use of conveyor for transferring food orders allows complete shutdown of one kitchen when production falls below a certain point. Orders are transmitted by intercommunicating loud-speakers and delivered by conveyor.

c. Rhythm of moving conveyor paces other kitchen operations.[10]

d. The conveyor is used for storage of dirty dishes during production peaks, eliminating production bottleneck and breakage because of congestion in dishwashing department (Fig. 52).

---

[9] Manager, auditor, catering manager, chef, steward, housekeeper, and building superintendent.

[10] Kitchen department heads have reported an unusual psychological influence on employees in the department as a result of the constant, regular motion of the overhead chain conveyor. Apparently it assists the employees in maintaining a steady and smooth work pace, even when their activities are not directly connected with the conveyor.

e. The conveyor is used for temporary storage of clean dishes during production valleys, eliminating unnecessary handling and stacking.

f. Concentration of dishwashing personnel in one location simplifies supervision.

Fig. 52. Centralized dishwashing department, silver burnisher in foreground; two dishwashing machines and chain conveyor which carries dishes in background.

Installation of a Pipe Bridge in a Factory Building. The Procter and Gamble Company is making extensive use of methods design in the construction of its factory buildings. A careful analysis of the method of constructing a pipe bridge and of installing the pipe and conduit in it resulted in substantial savings in time and cost. The bridge shown in Figs. 53 and 54 was installed at the company's plant in Florida.[11] The bridge was built in a steel fabricating shop and delivered to the site in one piece. The normal procedure for erecting a bridge of this kind would be to pick it up with a crane, fasten it in place, and then install the pipe and conduit. A better method was developed which consisted of installing the pipe and conduit inside the

[11] Gunnar C. Carlson, "A Cost Reduction Program for Construction," *Proceedings Eighth Industrial Engineering Institute,* University of California, Los Angeles-Berkeley, p. 21, February, 1956.

Fig. 53. Pipe bridge, outside view.

Fig. 54. Pipe bridge, inside view.

bridge while it was still on the ground. In fact, the insulation and painting were also done in this position. Then the bridge, with the pipe in place, was raised into position with the same crane that would have been required to lift the empty bridge. This better method saved $2800 over the normal method.

Fig. 55. Model of mechanized production line at the Ford Motor Company. Careful analysis of the process was made, including the use of three-dimensional models of machines and operators, before the actual machinery and equipment were installed. (Courtesy of Ford Motor Company.)

Careful Process Analysis Is Required for Mechanized Production Lines. When a factory is laid out for the production of a specific product in quantity, the process of manufacture is studied with great care, and the machinery, equipment, and work stations are located so that the product will flow through the plant with the least amount of backtracking and lost motion. The path of travel for each part and subassembly is worked out before the equipment is installed in the plant.

The layout (Fig. 55) showing one department in the Ford plant illustrates this type of manufacture. Most factories, however, are not laid out in this manner. Rather, the material moves from work station to work station intermittently by truck, and in many cases little thought has been given to the sequence of operations or to the path of travel through the plant. Because of this fact there are usually many opportunities to save time and money through an analysis of the process.

Steps to Be Followed in Making a Process Chart and Flow Diagram

1. Determine the activity to be studied. Decide whether the subject to be followed is a person, product, part, material, or printed form. Do not change subjects during the construction of the process chart.

2. Choose a definite starting point and ending point in order to make certain that you will cover the activity that you want to study.

3. Draw the process chart on a sheet of paper of sufficient size to allow space for (a) the heading, (b) the description, and (c) the summary. The heading should identify the process being studied. The body of the process chart should contain a column for *Travel* (distance in feet), *Symbol, Description,* and possibly *Time.* The five process chart symbols should be used. Every step in the process should be shown if the analysis is to be of real value. Unnecessary steps and inefficiencies in the work must first be "seen" before they can be eliminated.

4. Include on the process chart a tabular summary showing the number of operations, number of moves of each kind, distance the part was moved, number of inspections, and number of storages and delays. After improvements have been made, a combined summary should be compiled giving this information for the old method, the proposed method, and the difference.

5. Obtain floor plans of the department or the plant, showing location of machines and equipment used in making the part. If these are not available, draw floor plans to scale. It is frequently desirable to mount the floor plans on a drawing board or table, cut out cardboard templates the size of the machines (scale ¼ inch = 1 foot), and use these when new arrangements for the equipment are suggested. Sometimes three-dimensional scale models of machines and equipment are used instead of templates (Fig. 55).

6. Draw on the floor plans in pencil the path of the part through the plant, noting the direction of travel by means of arrows. The flow diagram should be made on location and not from memory at a desk. Distances should be actually measured or paced off.

CHAPTER 8

Activity Charts; Man and Machine Charts

ACTIVITY CHARTS

Although the process chart and the flow diagram give a picture of the various steps in the process, it is often desirable to have a breakdown of the process or of a series of operations plotted against a time scale. Such a picture is called an activity chart. Figure 57 shows an activity chart for the operation of picking up castings from a tote box, carrying them 10 feet, and placing them in a sandblast. The sketch shown in Fig. 56 was made to emphasize the fact that the operator carried the castings 10 feet and returned empty handed the same distance.

The chart suggests the obvious fact that walking could be eliminated by placing the tote box beside the sandblast. This was not done originally because the sandblast was located on a 4-inch concrete platform. When an inclined plank runway was built, the power-lift truck was able to move the tote box of castings up to the sandblast, as shown in Fig. 58. Figure 59 shows how this eliminated the walking and enabled the operator to sandblast 75% more castings per hour. One man can now feed this sandblast, whereas it originally required two.

The activity chart is of special value for analyzing maintenance work, jobs involving people working in gangs, and operations where the work is unbalanced and where there is "necessary" idle time.[1]

MAN AND MACHINE CHARTS

The operator and the machine work intermittently on some types of work. That is, the machine is idle while the operator loads it and

[1] For use of activity charts in petroleum production operations see "Job Design," by H. G. Thuesen and M. R. Lohmann, *Oil and Gas Journal*, Vol. 41, pp. 115–118.

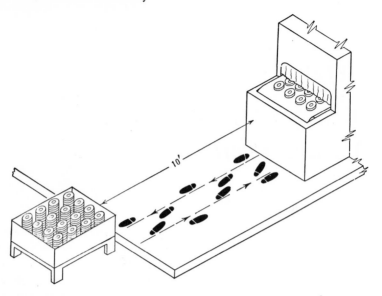

Fig. 56. Layout of work place for sandblasting castings—old method. Notice excessive walking.

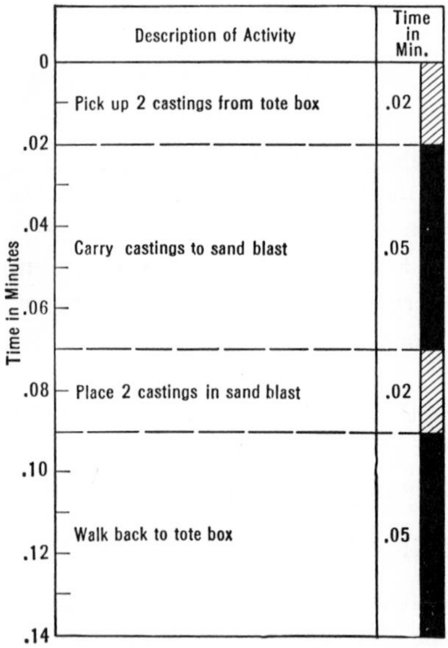

	Description of Activity	Time in Min.
0		
	Pick up 2 castings from tote box	.02
.02		
.04	Carry castings to sand blast	.05
.06		
.08	Place 2 castings in sand blast	.02
.10		
.12	Walk back to tote box	.05
.14		

(Vertical axis label: Time in Minutes)

Fig. 57. Activity chart for sandblasting castings—old method.

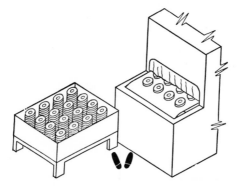

Fig. 58. Layout of work place for sandblasting castings—improved method. Unnecessary walking has been eliminated. One man now does the work of two.

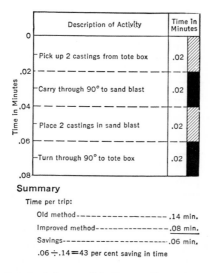

Fig. 59. Activity chart for sandblasting castings—improved method.

while he removes the finished work from it, and the worker is idle while the machine is in operation. It is desirable to eliminate idle time for the worker, but it is equally important that the machine be kept operating as near capacity as possible. In many instances an idle machine costs almost as much per hour as one in operation.

Drill Hole in Casting

Man	Machine
1. Pick up piece, place in jig, clamp, lower drill, throw in feed. Time, ½ minute. **(GET READY)**	Idle
Idle	2. Drill ½-inch hole in piece. Power feed. Time, 2.5 minutes. **(DO)**
3. Raise drill, remove piece, dispose, blow chips out of jig. Time, ¾ minute. **(PUT AWAY OR CLEAN UP)**	Idle

SUMMARY

	Man	Machine
Idle time	2.50 minutes	1.25 minutes
Working time	1.25	2.50
Total cycle time	3.75	3.75
Utilization in per cent	Operator utilization $= \dfrac{1.25}{3.75} = 33\%$	Machine utilization $= \dfrac{2.50}{3.75} = 67\%$

Fig. 60. Man and machine chart (simple form). It required a total of 3.75 minutes to drill the hole in the casting. During this time the operator worked 1¼ minutes and the machine was in operation 2½ minutes. The operator working time was 33% of the cycle, and the machine working time was 67% of the cycle.

The first step in eliminating unnecessary waiting time for the operator and for the machine is to record exactly when each works and what each does. Most operations consist of three main steps: (1) GET READY, such as putting material in the machine; (2) DO (doing the work), such as drilling a hole; and (3) PUT AWAY or clean up, such as removing the finished piece from the machine.

In Fig. 60, which shows the drilling of a hole in a steel casting with a power-feed drill, the steps performed by the man are listed on the left-hand side and the operation performed by the machine is listed on the right-hand side. This is a man and machine chart in its simplest form.

Very often a clearer picture of the relationship of the operator's working time and the machine time can be obtained by showing the information graphically to scale.

Purchasing Coffee. The simple task of purchasing a pound of coffee is used here to illustrate the operations performed by the customer, the clerk, and the coffee grinder (machine) in a grocery store. The customer walks into the coffee department and asks the clerk for one pound of coffee, specifying the brand and grind. The clerk gets the coffee, opens the package, sets the grinder, dumps the coffee into the grinder, and starts the machine. The customer and the clerk are idle during the 21 seconds the coffee is being ground.[2]

After the coffee is ground the clerk places it in the package and gives it to the customer. The customer then pays the clerk, who rings up the sale, gives the customer her change, and places the money in the register. The "work" or activity of the customer, clerk, and coffee grinder is shown graphically on the man and machine chart (see Fig. 61) and in tabular form at the bottom of the chart.

Possible Changes. The man and machine chart in Fig. 61 shows the excessive waiting time on the part of the customer and clerk while the coffee is being ground. This at once suggests that a supply of coffee be ground somewhat in advance so that the customer would not need to wait for her coffee to be ground. If this were done, the clerk could serve more than twice as many customers per hour, and the customer would spend less than half as much time at the coffee counter.

If the store were a large one employing a number of clerks and using a number of coffee grinders, the man and machine chart would indicate that the activities of the clerk be divided into two parts, one clerk selling the coffee and another grinding the coffee. Thus, in the hypothetical case, the coffee grinders would be kept in almost constant use, which would mean that fewer grinders would be needed. The clerks could work to better advantage inasmuch as there would be less idle

---

[2] Time is usually taken and recorded in decimal minutes or decimal hours, not in seconds. However, in describing certain motion and time study techniques to factory operators, as in the man and machine chart in Fig. 61, time may be expressed in seconds because most people are more familiar with this unit.

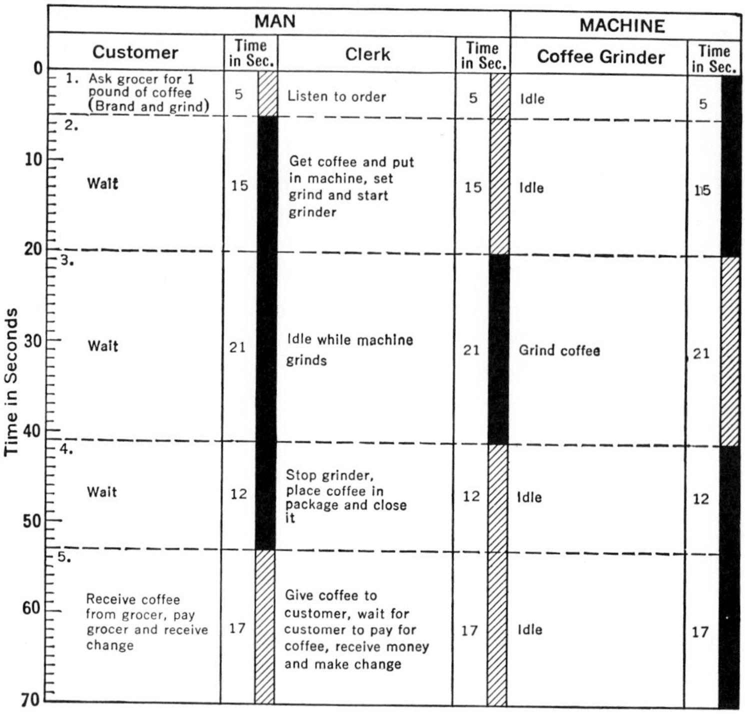

MAN				MACHINE	
Customer	Time in Sec.	Clerk	Time in Sec.	Coffee Grinder	Time in Sec.
1. Ask grocer for 1 pound of coffee (Brand and grind)	5	Listen to order	5	Idle	5
2. Wait	15	Get coffee and put in machine, set grind and start grinder	15	Idle	15
3. Wait	21	Idle while machine grinds	21	Grind coffee	21
4. Wait	12	Stop grinder, place coffee in package and close it	12	Idle	12
5. Receive coffee from grocer, pay grocer and receive change	17	Give coffee to customer, wait for customer to pay for coffee, receive money and make change	17	Idle	17

Summary

	Customer	Clerk	Coffee Grinder
Idle time	48 sec.	21 sec.	49 sec.
Working time	22	49	21
Total cycle time	70	70	70
Utilization in per cent	Customer utilization= $\frac{22}{70} = 31\%$	Clerk utilization= $\frac{49}{70} = 70\%$	Machine utilization= $\frac{21}{70} = 30\%$

Fig. 61. Man and machine chart showing activities involved in purchasing coffee in grocery store. The customer, the clerk, and the coffee grinder (machine) are involved in this operation. It required 1 minute and 10 seconds for the customer to purchase a pound of coffee in this particular store. During this time the customer spent 22 seconds, or 31% of the time, giving the clerk her order, receiving the ground coffee, and paying the clerk for it. She was idle during the remaining 69% of the time. The clerk worked 49 seconds, or 70% of the time, and was idle 21 seconds, or 30% of the time. The coffee grinder was in operation 21 seconds, or 30% of the time, and was idle 70% of the time.

time, and the customer would receive faster service. This would also tend to relieve congestion in the store during rush hours. Consequently, the store could handle more customers with a given floor area and with a given amount of equipment. However, it would be necessary to seal and date the bags containing the ground coffee, so that the customer would be assured of receiving freshly ground coffee.

Slitting Coated Fabric. Special fabric is coated with adhesive on continuous coating machines, and the finished material is taken off

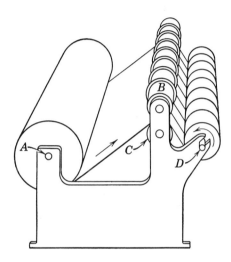

Fig. 62. Slitting machine. Coated fabric is drawn under slitting knives *B* onto the "wind up" shaft at *D*.

the drying racks in rolls approximately 3 feet wide and 2 feet in diameter. These rolls go to storage and later are removed and slit into narrower rolls to customers' orders.

Original Method. The material is slit on machines similar to the one shown in Fig. 62. The roll is placed on the shaft *A* at the back of the machine. The material is passed under rotating cutters *B*, which press against a rotating cylinder *C*, thus slitting the material into the desired width. The material is then rolled onto cardboard cores held in place on a shaft at *D*. After the desired length of fabric has been spooled, the machine is stopped and the cloth is cut parallel to shaft *D*. The operator, with the assistance of a helper, then places wrapping paper around the spooled material, attaches a label to each roll, and marks the grade, roll length, and other information on the label. The rolls are then removed from shaft *D* and placed on a skid. During this time the slitting machine is idle.

Improved Method. The following change was made in the method, increasing the capacity of the slitting machines 44%. A shaft was mounted on a pedestal shown in Fig. 63. After the desired length of

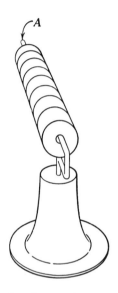

coated fabric had been slit, spooled, and cut off, the rolls were slid from shaft *D* of the slitting machine onto shaft *A* of the pedestal. This is a short and simple operation. The helper then wraps, labels, and marks the rolls while the machine operator immediately starts the slitting machine, eliminating much of the idle machine time. Because of the design of the machine it is necessary for the operator to manipulate the slitting machine controls while the fabric is being slit. The man and machine charts, Figs. 64 and 65, show the idle time and the working time before and after the new method was developed.

Results. The total cycle time was 5.2 minutes using the old method, or 11.5 cuts were made per hour. The new method reduced the cycle time to 3.6 minutes, which increased the output to 16.6 cuts per hour. This increase of 5.1 cuts per hour represents a gain of 44%. As the man and machine charts show, the machine utilization was increased from 42 to 61%. This was especially important in this case, as these slitting machines were operating 24 hours per day 7 days per week and were still unable to supply the demand for the product.

Fig. 63. Special pedestal. The rolls of fabric are transferred onto arm *A* to be wrapped and labeled.

Design of Machines and Equipment. Manufacturers of machines and equipment are confronted with the problem of designing machines that will do better work at a lower cost. In approaching this problem they should study the process and the individual operations from the point of view of the person who is doing the work, and design the machine or equipment to save his time and energy.

The fact that new equipment saves time by eliminating some operations is often used in advertising the equipment. Figure 66 is a reproduction of part of an advertisement used by a commercial laundry machinery manufacturer to show that an extractor of improved design

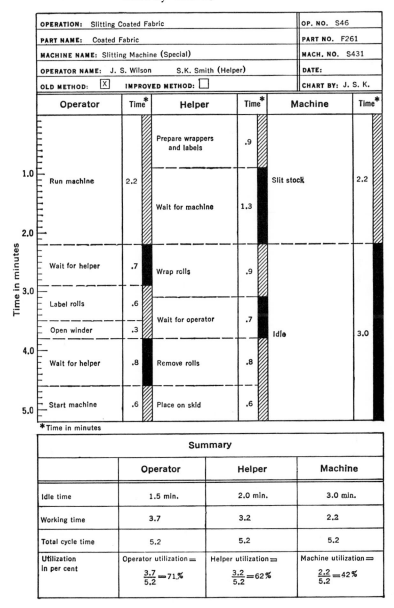

OPERATION: Slitting Coated Fabric			OP. NO. S46	
PART NAME: Coated Fabric			PART NO. F261	
MACHINE NAME: Slitting Machine (Special)			MACH. NO. S431	
OPERATOR NAME: J. S. Wilson S.K. Smith (Helper)			DATE:	
OLD METHOD: [X] IMPROVED METHOD: ☐			CHART BY: J. S. K.	

Operator	Time*	Helper	Time*	Machine	Time*
		Prepare wrappers and labels	.9		
Run machine	2.2			Slit stock	2.2
		Wait for machine	1.3		
Wait for helper	.7	Wrap rolls	.9		
Label rolls	.6				
Open winder	.3	Wait for operator	.7	Idle	3.0
Wait for helper	.8	Remove rolls	.8		
Start machine	.6	Place on skid	.6		

Time in minutes

Time in minutes: 1.0, 2.0, 3.0, 4.0, 5.0

*Time in minutes

Summary			
	Operator	Helper	Machine
Idle time	1.5 min.	2.0 min.	3.0 min.
Working time	3.7	3.2	2.2
Total cycle time	5.2	5.2	5.2
Utilization in per cent	Operator utilization = $\frac{3.7}{5.2} = 71\%$	Helper utilization = $\frac{3.2}{5.2} = 62\%$	Machine utilization = $\frac{2.2}{5.2} = 42\%$

Fig. 64. Man and machine chart for slitting coated fabric—old method. Total cycle time, 5.2 minutes. Total number of cuts per hour, 11.5.

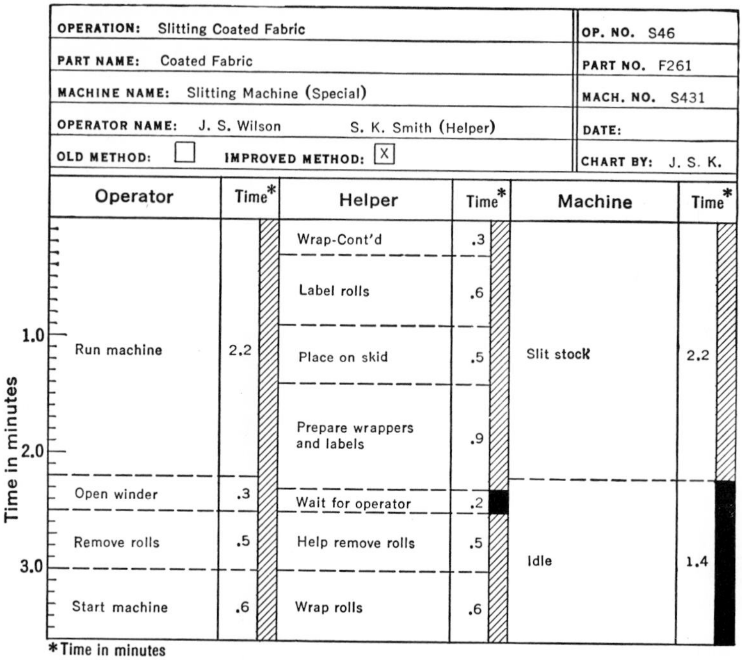

OPERATION: Slitting Coated Fabric				OP. NO. S46	
PART NAME: Coated Fabric				PART NO. F261	
MACHINE NAME: Slitting Machine (Special)				MACH. NO. S431	
OPERATOR NAME: J. S. Wilson S. K. Smith (Helper)				DATE:	
OLD METHOD: ☐ IMPROVED METHOD: ☒				CHART BY: J. S. K.	

Operator	Time*	Helper	Time*	Machine	Time*
		Wrap-Cont'd	.3		
		Label rolls	.6		
Run machine	2.2	Place on skid	.5	Slit stock	2.2
		Prepare wrappers and labels	.9		
Open winder	.3	Wait for operator	.2		
Remove rolls	.5	Help remove rolls	.5	Idle	1.4
Start machine	.6	Wrap rolls	.6		

(Time in minutes — vertical axis: 1.0, 2.0, 3.0)

*Time in minutes

Summary			
	Operator	Helper	Machine
Idle time	0.0 min.	0.2 min.	1.4 min.
Working time	3.6	3.4	2.2
Total cycle time	3.6	3.6	3.6
Utilization in per cent	Operator utilization = $\frac{3.6}{3.6} = 100\%$	Helper utilization = $\frac{3.4}{3.6} = 95\%$	Machine utilization = $\frac{2.2}{3.6} = 61\%$

Fig. 65. Man and machine chart for slitting coated fabric—improved method. Total cycle time, 3.6 minutes. Total number of cuts per hour, 16.6.

eliminates several hand operations and does in 8 minutes the work that formerly took 29½ minutes. A more complete description of this work is given here.

Extracting Water from Clothes in a Commercial Laundry—Ordinary Method. After clothes are washed in a commercial laundry, they are removed from the washing machine by hand, placed in a truck, moved to an extractor, and unloaded by hand from the truck into the

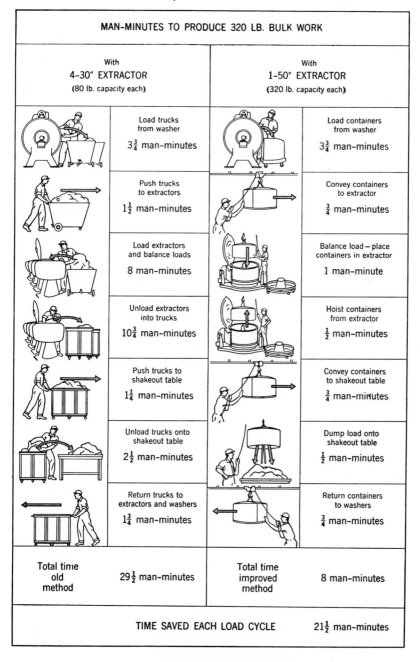

MAN-MINUTES TO PRODUCE 320 LB. BULK WORK

With 4-30″ EXTRACTOR (80 lb. capacity each)		With 1-50″ EXTRACTOR (320 lb. capacity each)	
	Load trucks from washer $3\frac{3}{4}$ man-minutes		Load containers from washer $3\frac{3}{4}$ man-minutes
	Push trucks to extractors $1\frac{1}{2}$ man-minutes		Convey containers to extractor $\frac{3}{4}$ man-minutes
	Load extractors and balance loads 8 man-minutes		Balance load – place containers in extractor 1 man-minute
	Unload extractors into trucks $10\frac{3}{4}$ man-minutes		Hoist containers from extractor $\frac{1}{2}$ man-minutes
	Push trucks to shakeout table $1\frac{1}{4}$ man-minutes		Convey containers to shakeout table $\frac{3}{4}$ man-minutes
	Unload trucks onto shakeout table $2\frac{1}{2}$ man-minutes		Dump load onto shakeout table $\frac{1}{2}$ man-minutes
	Return trucks to extractors and washers $1\frac{3}{4}$ man-minutes		Return containers to washers $\frac{3}{4}$ man-minutes
Total time old method	$29\frac{1}{2}$ man-minutes	Total time improved method	8 man-minutes
TIME SAVED EACH LOAD CYCLE			$21\frac{1}{2}$ man-minutes

Fig. 66. Chart used by a laundry machinery manufacturer to show how his extractor is designed to eliminate hand operations and save time.

extractor. The extractor lid is then closed, and the extractor is run at high speed 10 to 15 minutes, during which time the water is thrown out of the clothes by centrifugal force.

Fig. 67. Extractor used in commercial laundry to remove water from clothes by centrifugal force. The removable extractor container is made in halves, each of which is fitted with casters and a hinged bottom.

The extractor is then stopped, the lid opened, and the clothes removed by hand and placed in a truck. The truck is moved to a "shake-out" table, where the clothes are removed from the truck by hand and placed on the table.

Extractor with Removable Containers. An extractor (Fig. 67) is now being manufactured with a removable container or spinner basket, made in two parts or halves. Each of the two parts of the container is fitted with casters, and the bottom is hinged on one side and opens downward.

With this extractor the operation of extracting water from clothes is as follows. The halves of the container are moved to the washing machine, and the clothes are removed from the washing machine by hand and placed in them. The container halves are then shoved together to form a cylinder (Fig. 67). By means of a power hoist mounted on a monorail, the container is lifted up and moved over the extractor, balanced, and lowered in place. The extractor is run for 15 minutes. After the water is removed from the clothes, the extractor is stopped, the lid opened, and the container lifted out of the extractor with the hoist. It is then moved over the "shake-out" table, the hinged bottom of each half of the container opened downward, and the clothes are allowed to drop onto the "shake-out" table by gravity. The bottom of each container is then closed, and the extractor is returned to the washing machine for another load of clothes.

Mechanization and Automation. The use of the extractor with removable containers was a decided improvement over the previous method, and it is widely used in commercial laundries. However, a combination washer-extractor is now available in sizes from 135-pound (dry weight) to 350-pound capacity. Moreover, an automatic laundry machine will soon be on the market that will wash, rinse, dry, and polish flatwork at the rate of 100 feet per minute. This is equivalent to one dozen full-size sheets per minute.

CHAPTER 9

Operation Analysis

The over-all study of the process should result in a reduction in the amount of travel of the operator, materials, and tools, and should bring about orderly and systematic procedures. The man and machine chart often suggests ways of eliminating idle machine time and promotes a better balancing of the work of the operator and the machine.

After such studies have been completed, it is time to investigate specific operations in order to improve them. The purpose of motion study is to analyze the motions used by the worker in performing an operation, in order to find the preferred method. A systematic attempt is made to eliminate all unnecessary motions and to arrange the remaining necessary motions in the best sequence. It is when we come to the analysis of specific operations that motion study principles and techniques become most useful.

The extent to which motion study, as well as the other phases of motion and time study, should be carried will depend largely upon the anticipated savings in cost. As Table 1 shows, motion study may vary in extent from a cursory analysis followed by a general application of motion economy principles, to a detailed study of individual motions of each hand followed by a careful and extensive application of motion economy principles. The most elaborate analysis is possible, of course, only by means of full micromotion study, which will be explained in the chapters to follow.

Operation Charts. For those who are trained in the micromotion study technique—that is, those who are able to visualize work in terms of elemental motions of the hands—the operation chart, or the left- and right-hand chart, is a very simple and effective aid for analyzing an operation. No timing device is needed, and on most kinds of work the analyst is able to construct such a chart from observations of the operator at work. The principal purpose of such a chart is to

110

assist in finding a better way of performing the task, but this chart also has definite value in training operators.

Two symbols are commonly used in making operation charts. The

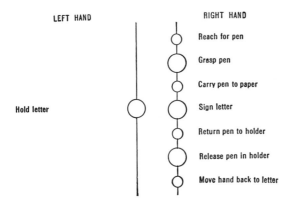

Fig. 68. Operation chart showing the movements of the two hands in signing a letter.

small circle indicates a transportation, such as moving the hand to grasp an article, and the large circle denotes such actions as grasping, positioning, using, or releasing the article. In signing a letter with a fountain pen the left hand holds the paper while the right hand performs the various movements indicated in Fig. 68.

The first step in making an operation chart or a left- and right-hand chart is to draw a sketch of the work place, indicating the contents of the bins and the location of tools and materials. Then watch the operator and make a mental note of his motions, observing one hand at a time. Record the motions or elements for the left hand on the left-hand side of a sheet of paper, and then in a similar manner record the motions for the right hand on the right-hand side of the sheet. As it is seldom possible to get the motions of the two hands in proper relationship on the first draft, it is usually necessary to redraw the chart.

Bolt and Washer Assembly. A left- and right-hand chart of the operation of assembling a lock washer, a steel washer, and a rubber washer onto a bolt is shown in Fig. 69. This operation is described fully on page 223. A glance at the chart shows that the left hand is holding the bolt while the right hand is doing useful work, assembling the washers. It is obvious that the motions of the two hands are unbalanced. The chart in Fig. 70 shows how the operation would appear if an assembly fixture were used and if the two hands worked together simultaneously.

When one has a detailed breakdown of the operation before him, he is in a much better position to question each element of the job and work out an easier and better method.

Assembling Rope Clips. The rope clip shown in Fig. 71 consists of three different parts: A, the U bolt; B, the casting; C, the hexagonal nuts. The rope clips were originally assembled in the following manner. The operator grasped a U bolt from bin 1 (Fig. 72) with her left hand and carried it up in front of her. Then she grasped a casting from bin 3 with her right hand and assembled it onto the bolt; and in a similar manner she grasped (from bin 2) and assembled in succession the two nuts onto the threaded ends of the bolt. She then disposed of the assembly with her right hand into bin 4 at her right. The operation chart for this operation is shown in Fig. 72.

Check Sheet for Operation Analysis. One approach to the problem of finding a better way of doing the work is to subject the operation to specific and detailed questions. If the several persons interested in the job consider these questions together, a more satisfactory solution is likely to result. In addition to studying the motions used in performing an operation, it is also desirable to give consideration to materials, tools, jigs, fixtures, handling equipment, working conditions, and other factors affecting the job. Finding the best way is not always easy, and considerable imagination, ingenuity, and inventive ability are required.

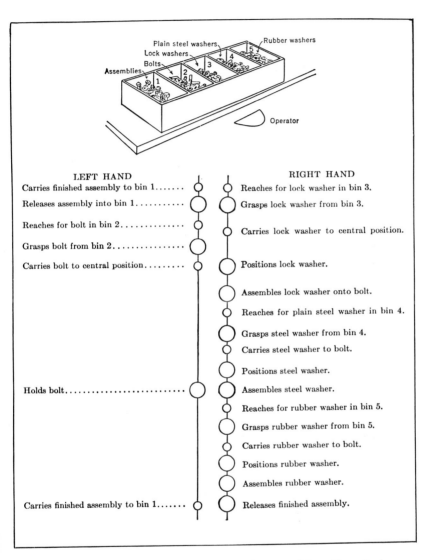

Fig. 69. Operation chart of bolt and washer assembly—old method.

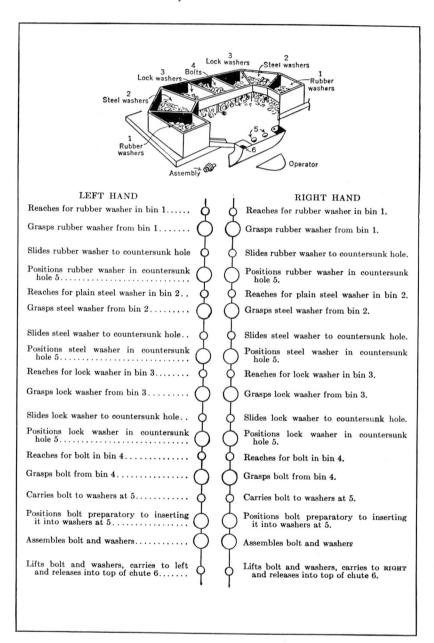

LEFT HAND			RIGHT HAND
Reaches for rubber washer in bin 1......			Reaches for rubber washer in bin 1.
Grasps rubber washer from bin 1.......			Grasps rubber washer from bin 1.
Slides rubber washer to countersunk hole			Slides rubber washer to countersunk hole.
Positions rubber washer in countersunk hole 5...........................			Positions rubber washer in countersunk hole 5.
Reaches for plain steel washer in bin 2..			Reaches for plain steel washer in bin 2.
Grasps steel washer from bin 2........			Grasps steel washer from bin 2.
Slides steel washer to countersunk hole..			Slides steel washer to countersunk hole.
Positions steel washer in countersunk hole 5...........................			Positions steel washer in countersunk hole 5.
Reaches for lock washer in bin 3........			Reaches for lock washer in bin 3.
Grasps lock washer from bin 3.........			Grasps lock washer from bin 3.
Slides lock washer to countersunk hole..			Slides lock washer to countersunk hole.
Positions lock washer in countersunk hole 5...........................			Positions lock washer in countersunk hole 5.
Reaches for bolt in bin 4..............			Reaches for bolt in bin 4.
Grasps bolt from bin 4...............			Grasps bolt from bin 4.
Carries bolt to washers at 5...........			Carries bolt to washers at 5.
Positions bolt preparatory to inserting it into washers at 5................			Positions bolt preparatory to inserting it into washers at 5.
Assembles bolt and washers...........			Assembles bolt and washers
Lifts bolt and washers, carries to left and releases into top of chute 6.......			Lifts bolt and washers, carries to RIGHT and releases into top of chute 6.

Fig. 70. Operation chart of bolt and washer assembly—improved method.

Therefore, the cooperation of such persons as the foreman, the tool
designer, and the operator is often of decided value to the analyst.

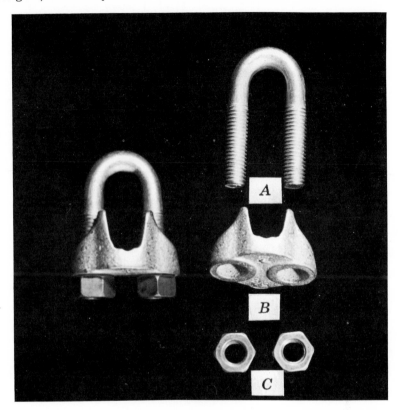

Fig. 71. Rope clip assembly: *A*, U bolt; *B*, casting; *C*, nuts.

After recording all that is known about the job, the various phases
of the operation should be considered:

I. *Materials*

  1. Can cheaper material be substituted?
  2. Is the material uniform and in proper condition when brought to the
     operator?
  3. Is the material of proper size, weight, and finish for most economical
     use?
  4. Is the material utilized to the fullest extent?
  5. Can some use be found for scrap and rejected parts?
  6. Can the number of storages of material and of parts in process be
     reduced?

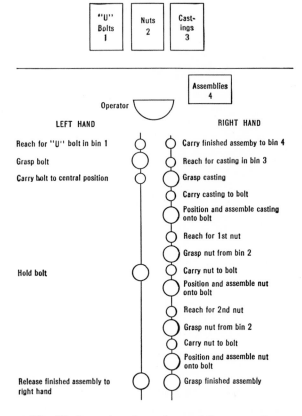

Fig. 72. Operation chart of assembling rope clips.

II. *Materials Handling*

1. Can the number of times the material is handled be reduced?
2. Can the distance moved be shortened?
3. Is the material received, moved, and stored in suitable containers? Are the containers kept clean?
4. Are there delays in the delivery of material to the operator?
5. Can the operator be relieved of handling materials by the use of conveyors?
6. Can backtracking be reduced or eliminated?
7. Will a rearrangement of the layout or combining of operations make it unnecessary to move the material?

III. *Tools, Jigs, and Fixtures*

1. Are the tools the best kind for this work?
2. Are the tools in good condition?

3. If metal-cutting tools, are the cutting angles of the tools correct, and are they ground in a centralized tool-grinding department?
4. Can tools or fixtures be changed so that less skill is required to perform the operation?
5. Are both hands occupied by productive work in using the tools or fixtures?
6. Can slide feeds, ejectors, holding devices, etc., be used?
7. Can an engineering change be made to simplify the design?

IV. *Machine*

A. Setup
1. Should the operator set up his own machine?
2. Can the number of setups be reduced by proper lot sizes?
3. Are drawings, tools, and gauges obtained without delay?
4. Are there delays in making inspection of first pieces produced?

B. Operation
1. Can the operation be eliminated?
2. Can the work be done in multiple?
3. Can the machine speed or feed be increased?
4. Can an automatic feed be used?
5. Can the operation be divided into two or more short operations?
6. Can two or more operations be combined into one? Consider the effect of combinations on the training period.
7. Can the sequence of the operation be changed?
8. Can the amount of scrap and spoiled work be reduced?
9. Can the part be pre-positioned for the next operation?
10. Can interruptions be reduced or eliminated?
11. Can an inspection be combined with an operation?
12. Is the machine in good condition?

V. *Operator*

1. Is the operator qualified mentally and physically to perform this operation?
2. Can unnecessary fatigue be eliminated by a change in tools, fixtures, layout, or working conditions?
3. Is the base wage correct for this kind of work?
4. Is supervision satisfactory?
5. Can the operator's performance be improved by further instruction?

VI. *Working Conditions*

1. Are the light, heat, and ventilation satisfactory on the job?
2. Are washrooms, lockers, restrooms, and dressing facilities adequate?
3. Are there any unnecessary hazards involved in the operation?
4. Is provision made for the operator to work in either a sitting or a standing position?
5. Are the length of the working day and the rest periods set for maximum economy?
6. Is good housekeeping maintained throughout the plant?

This list of questions, although by no means complete, shows some of the elements that enter into a thorough consideration of the problem of finding the best way of doing work. The list is typical of a check sheet that can be prepared for use in a specific plant.

Another approach to the problem is to divide the job into the three phases: (1) get ready; (2) do the work (or use); and (3) put away or clean up, as has already been mentioned. The second phase is the primary object of the work, and the first and the third phases are

Fig. 73. Metal box: *A*, box cover; *B*, box bottom.

auxiliary to it. Often the get-ready and the cleanup can be shortened and simplified without impairing the do or use phase of the operation.

Spray Inside and Outside of Metal Box Covers and Bottoms. This example shows the steps that were taken to improve the method of spray-painting black enamel on the two parts of a small metal box. Of the questions listed in the preceding section, the one that seemed to give the greatest promise in this case was IV-B-6—"Can two or more operations be combined into one?"—referring to the possibility of spraying the inside and the outside of the container in a single operation.

When a systematic attempt is made to find a better method, it is seldom that the first one tried proves to be the best. Finding the most economical method for doing a given task is usually a process of development and invention. The following case illustrates this in an excellent manner.

The boxes (Fig. 73), made in slightly different sizes and shapes, are used for such products as surgical instruments and sewing machine attachments. The container is composed of a cover and a bottom, which fit together (Figs. 74 and 75). The containers are manufactured in lots of 5000 to 10,000.

Fig. 74. Clamps for holding box covers and bottoms for spraying by the original method: *A*, for spraying inside; *B*, for spraying outside.

Original Method. The operator, standing in front of the spray booth, procured an unsprayed box cover or bottom with the right hand from a tote box at her right and placed it on the metal fixture *A* shown in Fig. 74, which she held in her left hand. Grasping the spray gun in

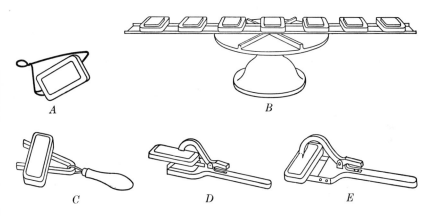

Fig. 75. Devices for use in spray-painting box covers and bottoms by "improved" methods: *A*, steel spring hook for dipping; *B*, rack and turntable; *C*, magnetic fixture; *D*, mechanical fixture; *E*, improved mechanical fixture—the one finally adopted.

her right hand and holding the part inside the spray booth, she sprayed the inside surface, then disposed of the part on a screen tray. When the screen tray was filled (35 covers or bottoms) it was placed on an oven rack, and an empty screen was positioned at the left of the spray booth.

When an oven rack was full, the oven man moved it into the oven on the other side of the room, where it was baked for 1½ hours. The rack was then removed and cooled, and the outside of the parts were sprayed, using fixture *B* shown in Fig. 74. The sequence of motions used in spraying the outside was similar to that for spraying the inside. The box parts were again baked in the oven for 1½ hours. When removed and cooled, they were ready for the final inspection.

Improved Methods. The following methods were tried in the order indicated.

1. *Steel Spring Hooks.* It was apparent that considerable savings would result if a way could be devised that would permit the operator to spray both the inside and the outside of the box cover or bottom in a single operation.

Several designs of spring hooks were tried, similar to those used for another type of container, which held the piece from the inside.

Results. It was found that the blast from the air gun would blow the piece from the hook. Hooks made from stiffer spring made hooking too difficult for girl operators. This method was discarded as impractical.

2. *Dipping in Enamel.* Since some products were being satisfactorily dipped in enamel and baked in a continuous oven, it was suggested that an attempt be made to dip the boxes. Wire hangers, *A* in Fig. 75, were made and dipping was tried.

Results. An air pocket formed in the upper corner of the box covers, which prevented the enamel from making contact with the metal. Also, the enamel failed to drain out of the lower corner properly. The device for dipping the boxes was discarded.

3. *Spray Outside on Turntable.* The box covers were sprayed on the inside in the old manner and placed on a narrow tray, *B* in Fig. 75. When seven box covers had been placed on the tray the outsides of all of them were sprayed and the trayful then sent to the oven for baking.

Results. The air from the spray gun blew the boxes off the rack. If heavy corrugations or teeth were cut in the edges of the tray, they disfigured the enamel finish on the boxes. This method was discarded.

4. *Magnetic Fixture.* A permanent magnet, *C* in Fig. 75, was used to hold the box cover while the operator sprayed both the inside and the outside in one operation.

Results. This proved to be a satisfactory method for holding the cover, but it was difficult to get the sprayed pieces off the magnet to the screen tray. The suggestion was not used.

5. *Mechanical Fixture.* A fixture, *D* in Fig. 75, was made so that the cover rested on two knife-edges and was held mechanically in place with a needle point.

Results. This device was satisfactory in that it permitted the inside and outside of the cover to be sprayed at one operation and it was easy to release the piece and dispose of it on the screen tray. However, the two knife-edges tended to scrape the enamel off the edges of the cover as it slid off the holder and onto the screen tray.

6. *Improved Mechanical Fixture.* A holder, *E* in Fig. 75, was built with two parallel knife-edges that did not scrape off the enamel in disposing of the sprayed part.

Results. This fixture proved to be entirely satisfactory, and several were made of aluminum and immediately put into use on production work. Each operator is supplied with two fixtures, allowing one to soak in solvent while the other is being used.

The improved method, using this fixture, proved to be superior to the old method in the following ways:

1. The operator now sprays both the inside and the outside of the box cover or bottom at one operation. This effects a saving of approximately 25% in direct labor.

2. The covers and bottoms are baked only once instead of twice. This reduces the use of the baking ovens 50%, and also reduces the indirect labor for handling racks and trays 50%.

3. An additional saving results since the investigation showed that the insides of the box covers and bottoms were being sprayed with a dull-finish enamel and the outsides with a glossy-finish enamel. There is no need for the dull finish on the inside, and it is more expensive than glossy. Use of the dull has been discontinued; the entire box is now sprayed with glossy enamel. This alone has saved in one year more than enough to pay for all the experimental fixtures that were used in the development work.

Cleanup Work. Janitor or cleanup work represents a sizable part of the office and factory payroll. For example, the wages paid for cleanup work in the Ford Motor Company run into millions of dollars per year, with more than 5000 men employed on this kind of work. In some organizations such work accounts for as much as 10 to 15% of the total wages.

In discussing this subject Lawrence A. Flagler, formerly of the Procter and Gamble Company, states, "A survey of our factory clean-up costs revealed the fact that clean-up represented one of the largest single classifications of wage expense. It showed that there were more than 700 people in the company engaged in this kind of work. . . . It is my estimate that there are at least 150,000 full-time factory clean-up men employed in this country."[1]

Some of the results of a careful study of cleaning tools and equipment and of cleanup methods made by one organization are given here. They show what may be accomplished by setting out to answer such a single question as "Are the tools the best kind for this work?" (III-1, page 116). Although these findings apply to conditions in this particular concern, many of the results of this investigation are basic and have wide application.

The first step was to find the best equipment. Since the tools that the janitor uses cost but a few dollars per year, and since they represent less than two tenths of 1% of the total cleanup costs, it is false economy to purchase any but the most efficient tools.

Cleaning with Mop. Mopping of floors is one of the important classes of janitor work. Of the 700 people on cleanup work at Procter and Gamble, for example, the equivalent of 215 spend their full time mopping floors.

An analysis of the operations used in mopping floors indicated that the following factors were most important in the selection of a mop:

1. High ratio of water absorption to give maximum transfer of water to and from the floor for each stroke of the mop.
2. Minimum retention of water in the mop after wringing, in order to reduce the dead weight and its corresponding higher fatigue allowances.
3. Proper shape of the mop to provide maximum surface contact between the mop and the floor.
4. Minimum weight of handle, hardware, or fixtures.
5. The wearing qualities of the mop.

Specifications for a Mop. Factory tests made of more than 40 different styles and kinds of mops resulted in the following specifications for a good mop:

1. Mops should be of wide tape type to be used with detachable handles.
2. The mop should be made of a good grade of 4-ply, soft roving, long staple yarn free from linters and foreign material.
3. The length of the mop strands should be 38 to 42 inches, taped in the middle with good cotton duck at least 5 inches wide. The completed mop

---

[1] Lawrence A. Flagler, "Motion Study Applied to Factory Clean-Up," *Abstract of Papers Presented at the Management Conference,* University of Iowa, *Extension Bulletin,* **458,** p. **9.**

should be 6¼ to 6¾ inches in width after sewing on the tape with at least three rows of double stitching. The mop is not to be sewed in the folded shape, in order that both sides may be used to equalize wear.

4. The average dry weight of the cotton should be 23½ to 24½ ounces for wet mopping and 31½ to 32½ ounces for dry mopping.

5. The mop handle should be 60 inches long, 1¼ inches in diameter, and should have an aluminum knob at the end.[2]

6. The mop attachment device should be of the claw or clamp type, wherein the mop is folded, placed in the open clamp, and the wing nut tightened. The hardware on the mop should be light in weight and made of rust-resisting material.

Figure 76 shows a good mop and a poor one. The "ferrule" mop on the right is unsatisfactory for factory work. The mop is too small, the handle is too short, and the ferrule where the mop is attached to the handle prevents the mop from lying flat on the floor.

The "head" mop on the left is well designed. The handle is long, with a knob on the end. Because the head mop lies flat on the floor, there is 30% more cotton in contact with the floor than with a ferrule mop of equal weight. In addition, the head mop fits the wringer better and 10% more water can be removed, making for faster pickup of dirty water from the floor, less dead weight for the janitor to handle, and fewer wringing operations.

Recommended Method of Mopping. The recommended method for mopping is the use of the "side to side" stroke rather than the "push or pull" stroke. The janitor positions himself in the middle of the stroke length, with his feet spread well apart and at right angles to the direction of the stroke (Fig. 77). The mop handle is grasped over the end with one hand and approximately 15 inches down the handle with the other hand. The mop is placed flat on the floor and passed from side to side in front of the janitor, in the form of an arc. The arc should be slight, as too wide an arc will greatly increase the effort required in that the arms are extended in front of the body at a lower muscular efficiency. The mop should pass in front of the janitor and within about 3 inches of his feet. At the ends of the stroke the mop is slightly looped to reverse the direction. Centrifugal force in describing the arc spreads the mop strands to increase the area covered in the stroke. Periodically, depending on the floor condition, the mop is flopped over to give an equal distribution of water and to use both

[2] When mopping in open unobstructed areas, the mop stroke can be lengthened from 12 feet 1½ inches to 12 feet 10½ inches, an increase of 6.9%, by the use of a knob on the end of the regular handle.

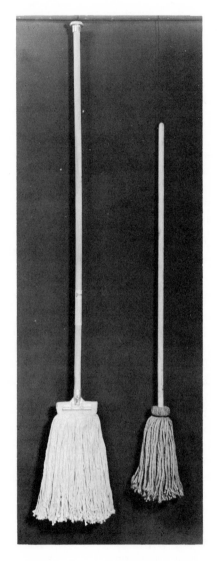

Fig. 76. Two common types of mops: "head" mop on the left; "ferrule" mop on the right.

sides of the mop effectively. As the boundary is approached, the janitor reverses his position 180 degrees at the end of the stroke; with proper timing this motion can be accomplished with only a momentary loss of time. The optimum length of stroke for a janitor of average

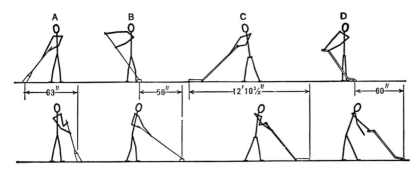

A. CLEANING WITH SWEEP BROOM

Length of Broom Handle, 54 Inches

At start of stroke: Shoulder is normal, feet normal and in a position for a forward step. Right arm is straight.

At end of stroke: Shoulder is turned 45 degrees, broom is just past vertical.

B. CLEANING WITH PUSH BRUSH

Length of Brush Handle, 68 Inches

At start of stroke: Shoulder is turned 45 degrees, right arm is horizontal.

At end of stroke: Shoulder is normal, feet positioned for normal step forward on return stroke. Left arm is straight at 45 degrees with vertical. Back is turned 45 degrees at hip.

C. CLEANING WITH MOP

Length of Mop Handle, 60 Inches

At start of stroke: Shoulder is turned 90 degrees to right, with right arm at 45 degrees and down 14 inches on the handle. Weight is shifted to right foot, with back inclined 4 inches to right. Left hand is grasping end of handle.

At end of stroke: Shoulder is turned 90 degrees to left, with right arm at 45 degrees and down 14 inches on the handle. Weight is shifted to left foot, with back inclined 4 inches to left. Left hand is grasping end of handle.

D. CLEANING WITH VACUUM TOOL

Height of Tool above Floor, 29 Inches
Length of Tool, 56 Inches

At start of stroke: Shoulder is turned 90 degrees, right arm is at 45 degrees to back, and feet are together.

At end of stroke: Shoulder is normal, right arm is 45 degrees to front, and a forward step is taken.

Fig. 77. Description of recommended methods for using broom, brush, mop, and vacuum cleaner.

height is 12 feet, which with an effective width of 0.70 foot will result in a coverage of 8.4 square feet per stroke.

In planning the work, the direction of mopping should be arranged so that a full stroke can be used. For example, an 11½-foot by 16-foot storage bay should be mopped with an 11½-foot stroke perpendicular to the 16-foot dimension. As most factories have a uniform size of storage bay, it is possible for the janitor to standardize his starting position so as to use the optimum length of stroke. In mopping aisles the direction of the stroke should parallel the aisle. For example, a 32% saving in time is possible when a 5-foot by 120-foot aisle is mopped with lengthwise instead of crosswise strokes. Another reason for parallel mopping of aisles is that splashing of the mop against the mop boards or materials in storage is minimized.

Much time is lost in transporting water in small buckets. A specially designed mop truck has been developed, with three large water compartments having a 42-gallon capacity for clean water and a 37-gallon capacity for dirty water. The temperature of the clean water should not fall below 130° F. for effective use.

By wearing nonskid sandals the janitor is able to keep his feet dry and is in less danger of slipping. This brings an increase in output of 5%.

With all the improvements of mopping methods and equipment, the coverage is now 2000 square feet per man-hour, in comparison with slightly less than 1000 square feet per man-hour before the installation of the improvements.

Although space does not permit detailed analysis and recommendations for each of the other tools that the janitor uses, brief reference will be made to a few of them.

Cleaning with Push Brush. The most effective tools and methods for sweeping floors will depend upon such factors as kind and amount of dirt, kind of floor, kind and amount of obstructions, and the desired cleanliness of the floor. In general the following conclusions have been reached:

1. Push brushes made of Russian bristles are recommended for dry and light dirt.
2. Push brushes made of fiber are recommended for wet and heavy dirt.
3. Depending on the amount of obstruction, widths of brushes should vary from 18 to 36 inches.
4. Brush handles should be at least 68 inches long.
5. Corn brooms should never be used except for very special cases.

Cleaning with Vacuum Tool. One concern found most efficient for its particular conditions a vacuum cleaner with a high-speed motor mounted over a dust-collecting can on casters, with the filter exposed on the discharge side of the pump. Studies showed that a cleaning tool 12 inches wide is most effective for areas with an average degree of obstruction. An aluminum handle with a double bend and a swivel at the point where the hose is fastened to the vacuum cleaner is the best. The most efficient stroke was found to be looping the tool across the floor at the end of the stroke (Fig. 78) rather than making an abrupt change in direction.

Start

TYPE A

Length of stroke.... 60 in.
Width of stroke..... 12 in.
Coverage.......... 200%
Area per stroke...... 5 sq. ft.
Std. time per stroke
            0.03198 min.
Std. time per 100 sq. ft.
            0.639 min.

Start

TYPE B

Length of stroke.... 60 in.
Width of stroke..... 12 in.
Coverage.......... 100%
Area per stroke..... 10 sq. ft.
Std. time per stroke
            0.05880 min.
Std. time per 100 sq. ft.
            0.588 min.

Start

TYPE C

Length of stroke.... 60 in.
Width of stroke..... 12 in.
Coverage.......... 103%
Area per stroke...... 8 sq. ft.
Std. time per stroke
            0.04362 min.
Std. time per 100 sq. ft.
            0.545 min.

Fig. 78. Three types of vacuum-cleaner strokes. From the theoretical calculations shown here and from factory practice, Type C was found to be most effective.

Washing Windows. In one plant windows were washed with wet rags, dried with chamois, and polished with a dry cloth. The method was changed to washing the windows with a wet sponge, drying with a squeegee, and cleaning the edges at the sash with a dry rag. The increase in production was from 316 panes measuring 13 by 10 inches, to 910 panes of the same size, in a given length of time.

CHAPTER 10

Micromotion Study

Micromotion study provides a technique for recording and timing an activity. It consists of taking motion pictures of the operation with a clock in the picture or with a motion picture camera operating at a constant and known speed. The film becomes a permanent record of both method and time and may be re-examined whenever desired.

Purposes of Micromotion Study. Micromotion study was originally employed for job analysis work, but in recent years new uses have been found for this valuable tool. Micromotion study may be used for the following purposes: as an aid in studying the activities of two or more persons on group work, as an aid in studying the relationship of the activities of the operator and the machine, as a means of timing operations (instead of using stop-watch time study), as an aid in obtaining motion-time data for synthetic time standards, as a permanent record of the method and time of activities of the operator and the machine, and for research in the field of motion and time study.

As valuable as micromotion study is for these purposes, however, its two most important uses are: (1) to assist in finding the most efficient method of doing work, and (2) to assist in training individuals to understand the meaning of motion study and, when the training is carried out with sufficient thoroughness, to enable them to become proficient in applying motion economy principles.

Micromotion Study as an Aid in Improving Methods. Micromotion study provides a technique that is unique for making a minute analysis of an operation. As will be explained in detail later, the procedure consists of (1) filming the operation to be studied, (2) analyzing the film, (3) charting the results of the analysis, and (4) developing an improved method through the problem-solving process.

Micromotion study is usually associated with camera speeds of

128

960 or 1000 frames per minute, but faster speeds may be used to study very fast hand motions or complex operations. When the film is projected on the screen, the pictures are enlarged many times to facilitate the analysis of the motions. Each movement of the worker can be timed to any degree of accuracy desired.

Although micromotion study provides a convenient, accurate and positive means of studying work, it is used to only a limited extent for improving methods. In fact, micromotion analysis is not necessary for studying a large majority of the operations to be improved. One who understands the technique and the principles of motion study can, in most cases, visualize the operation completely and, by applying the principles that go to make good motion economy, determine methods that should be used. Motion study may usually be carried out without taking a motion picture and making the full analysis that micromotion study requires. Moreover, a micromotion study, although not prohibitive in cost, does require special motion picture equipment, film, and considerable time for the analysis. Micromotion study for determining methods of doing work has a place in industry, but not so large a place as some maintain. This is the less valuable of the two main purposes of micromotion study.

Micromotion study should be treated like any tool—as something to be used when it is profitable to do so. It might, for example, profitably be utilized in the investigation of short-cycle operations that are highly repetitive or largely manual in character, of work produced in large volume, or of operations performed by large numbers of workers. These factors alone do not always determine whether a micromotion study should be made. In fact, a micromotion study is often the last resort. Sometimes in a complex operation it is difficult to get the motions of the two hands balanced without the aid of the simo chart, which is a graphic picture of the motions on paper.

Micromotion Study as an Aid in Teaching. Industry has been slow to realize the fact that micromotion study is of greatest value in aiding one to understand motion study. From its definition motion study would appear to be very simple and easily understood. However, there is a knack to getting at the real meaning of it, and to being able to understand it in its entirety.

It is essential for the observer to become proficient in detecting and following the motions used by the worker in performing his task. He must *see* the motions made by the operator's right hand, and by his left hand, even noting what the fingers of each hand do. It is necessary to be able to detect where one motion ends and another begins. As the Gilbreths state, ". . . one must have studied motions and measured them until his eye can follow paths of motions and judge lengths of motions, and his timing sense, aided by silent rhythmic counting, can estimate times of motion with surprising accuracy. Sight, hearing, touch, and kinesthetic sensations must all be keenly developed."[1]

The term *motion-minded* has been used to describe this ability of the person who has trained himself to follow unconsciously the motions of the worker and check them against the principles of motion economy with which he is familiar. Micromotion study is of great assistance in training individuals to become motion-minded.

R. M. Blakelock once said:

> . . . the greatest value of micromotion training comes through the ability to visualize industrial operations in terms of motions . . . the ability to visualize the motions that are necessary to perform each step of an operation, and to recognize which are and which are not good motion practice, rather than think in such terms as describe steps in the operation itself.
>
> Most time study observers, as they record steps in the operation, think in terms of elemental operations, such as "drills one hole," "faces off side," "rivets end," or "assembles part 2 to part 3," making no analysis of the motions of the operator, and giving little thought to them unless there is a glaring case of bad motions that is quite obvious.[2]

Blakelock, while in charge of the motion study division at the Schenectady plant of the General Electric Company, seldom found it necessary to make a micromotion study to determine proper methods for doing work. He applied the principles of motion study without needing to resort to the motion picture camera. However, he did make extensive use of this technique for training members of the organization.

For information concerning the use of micromotion study by the Fort Wayne works of the General Electric Company and by other companies, see Chapter 35.

Memomotion Study. Motion pictures must be made and projected at approximately normal speed if one wants a fairly accurate reproduction of motion of people and objects. But for certain types of man and machine activities, motion pictures made at 50 or 100 frames per

---

[1] F. B. and L. M. Gilbreth, *Applied Motion Study,* Sturgis & Walton, New York, 1917, p. 61.

[2] R. M. Blakelock, "Micromotion Study Applied to the Manufacture of Small Parts," *Factory and Industrial Management,* Vol. 80, No. 4, pp. 730–732.

minute are quite satisfactory. The term *memomotion study* [3] has been suggested to designate this form of micromotion study.

The Gilbreths, using a hand-cranked camera, took pictures at very slow speeds to study group activities,[4] and time-lapse photography using a motor-driven camera has long been employed for studying the growth of plants and flowers. In recent years many new uses have been found for this valuable technique. In addition to its applications in the factory and office, memomotion study serves in studying such activities as check-in operations at airline counters, the manner in which customers select items in a self-serve store, and the flow of traffic on highways and in stores and banks.

The major advantage of slow-speed pictures over those made at normal speed is the great savings in film cost and in the time required for film analysis. With the film exposed at 50 frames per minute instead of 960, film cost is only about 6% as great.

CYCLEGRAPHIC AND CHRONOCYCLEGRAPHIC ANALYSIS

The cyclegraphic and the chronocyclegraphic methods of analysis developed and used by the Gilbreths are described on page 18. These techniques have had limited use in this country as a means of improving methods, although they seem to have met with greater favor in Great Britain.[5] We have used the cyclegraph to some extent for training purposes, as an aid in describing a motion pattern used in performing a task,[6] and to dramatize the superiority of one method or motion pattern over another.

[3] M. E. Mundel, *Motion and Time Study*, 3rd ed., Prentice-Hall, Englewood Cliffs, N. J., 1960, p. 301.

[4] "Our methods and devices have been criticized as being specially adapted to problems involving the minutia of motions, but too expensive for the general time study purposes. A moment's consideration will show that the turning of the crank of the cinematograph may be done as slowly as the requirements of the particular case of time study demand. In fact, we have made films that were taken at the rate of one picture every ten minutes. With the sixteen pictures to the foot, a foot will last 160 minutes, or two hours and forty minutes, at a total maximum cost of six cents." From *Fatigue Study*, by Frank B. Gilbreth, Sturgis and Walton Co., New York, 1916, p. 126.

[5] A. G. Shaw, *The Purpose and Practice of Motion Study*, Columbine Press, London, 1960, pp. 92–121.

[6] G. E. Clark, "A Chronocyclegraph That Will Help You Improve Methods," *Factory Management and Maintenance*, Vol. 112, No. 5, pp. 124–125, May, 1954. See also "Catching Waste Motions with the Camera," *Supervisory Management*, Vol. 1, No. 1, pp. 53–56, December, 1955.

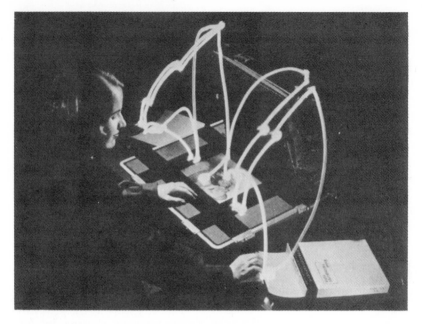

Fig. 79. Motions needed to make a print with hand-operated photoprinter.

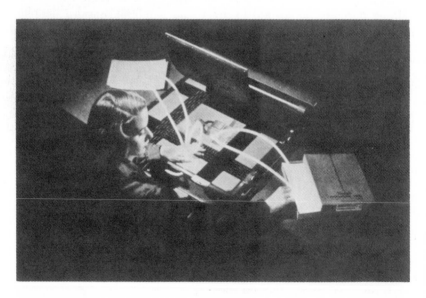

Fig. 80. Motions needed to make a print with power printer of improved design.

Fig. 81. Motions needed to wrap rolls of cellophane—old method.

Fig. 82. Motions needed to wrap rolls of cellophane—improved method.

The improved power printer developed by Eastman Kodak Company shown in Fig. 80 requires many fewer motions than the hand-operated printer in Fig. 79. These two illustrations were included in the annual report of this company, under the caption "An Improved Professional Product Resulting from Development Work."

The illustrations in Figs. 81 and 82 show the old method and the improved method of wrapping rolls of cellophane. Originally the rolls were wrapped in two layers of cellophane and placed in an inner box and an outer box; then the 52-pound package was lifted onto a pallet for shipment. The method shown in Fig. 82 has been greatly simplified, as the motion pattern shows. Now rolls are wrapped in a single layer of cellophane and shipped on a pallet that is loaded with the aid of a small elevator. These two illustrations are from a publication, *This is Du Pont*, of the E. I. du Pont Company. Under the caption "Modern Technology Spurs New Advances" the following statement is made:

The contrasting pictures [Figs. 81 and 82] show a simple improvement in the method of doing a day-to-day job. Hundreds of these are effected in du Pont operations every year. In sum total, such improvements are as significant a part of modern technology as the discovery of a new product, the development of a new process, or the creation of intricate and costly new tools.

When Slit Roll Packer Irvin Coleman can handle cellophane more quickly and easily, then he can accomplish more in a day's time. Economists call that "increased productivity." And there has been no more important element in the rise of America's industrial economy than the steady rise in the individual's productivity due to technological improvements.

Three things happen when an employee of industry can turn out more in less time. Cost of product drops, enabling more people to buy it. The operator does not have to work as long to make his product, which in time leads to shorter hours and higher earnings. Finally, as the nation, which is the sum of its people, produces more, its material well-being is enhanced.

CHAPTER 11

Fundamental Hand Motions

Most work is done with the two hands, and all manual work consists of a relatively few fundamental motions which are performed over and over again. *Get* or *pick up* and *place* or *put down* are two of the most frequently used groups of motions. In most cases *get* is followed by some use or process element, such as driving a nail with a hammer, using a wrench to tighten a bolt, or writing with a pen. In using a fountain pen the sequence of motions would be *get* pen, write—that is, *use* pen, *place* pen in holder. Although *get* and *place* represent two very common groups of motions, they are not fundamental motions in themselves.

Frank B. Gilbreth, in his early work in motion study, developed certain subdivisions or events which he thought common to all kinds of manual work. He coined the word *therblig* (Gilbreth spelled backwards) in order to have a short word with which to refer to any of these seventeen elementary subdivisions of a cycle of motions.[1] Not all of these seventeen therbligs are pure or fundamental elements in the sense that they cannot be further subdivided, but they are the best classification of hand motions that we have. The experienced analyst has no difficulty in using the therbligs in industrial applications.

The term *therblig* is more convenient to use than "hand motion" or "motion element," and perhaps carries a more precise meaning than "motion." Although the word *therblig* is familiar to industrial engineers, the term *motion* or *hand motion* is preferred when discussing the subject of micromotion study with factory and office personnel. Uncommon terms and symbols (such as the mnemonic therblig symbols) may be a handicap in a training program and are to be avoided whenever possible.

The seventeen fundamental hand motions, together with their letter

[1] F. B. and L. M. Gilbreth, "Classifying the Elements of Work," *Management and Administration*, Vol. 8, No. 2, p. 151, August, 1924.

symbols, mnemonic symbols, and color designations,[2] are shown in Fig. 83. The definitions of these motions are given on the following pages.

Name of Symbol	Therblig Symbol		Explanation-suggested by	Color	Color Symbol	Dixon Pencil Number	Eagle Pencil Number
Search	Sh	⟋O	Eye turned as if searching	Black		331	747
Select	St	→	Reaching for object	Gray, light		399	734½
Grasp	G	∩	Hand open for grasping object	Lake red		369	744
Transport empty	T E	⌣	Empty hand	Olive green		391	739½
Transport loaded	T L	⌣	A hand with something in it	Green		375	738
Hold	H	⌂	Magnet holding iron bar	Gold ochre		388	736½
Release load	RL	⌢	Dropping content out of hand	Carmine red		370	745
Position	P	9	Object being placed by hand	Blue		376	741
Pre-position	P P	⏀	A nine-pin which is set up in a bowling alley	Sky-blue		394	740½
Inspect	I	◯	Magnifying lens	Burnt ochre		398	745½
Assemble	A	#	Several things put together	Violet, heavy		377	742
Disassemble	D A	⫲	One part of an assembly removed	Violet, light		377	742
Use	U	U	Word "Use"	Purple		396	742½
Unavoidable delay	U D	⌝o	Man bumping his nose, unintentionally	Yellow ochre		373	736
Avoidable delay	A D	⌞o	Man lying down on job voluntarily	Lemon yellow		374	735
Plan	Pn	℘	Man with his fingers at his brow thinking	Brown		378	746
Rest for overcoming fatigue	R	ℇ	Man seated as if resting	Orange		372	737

Fig. 83. Standard symbols and colors for fundamental hand motions.

Definitions of Fundamental Hand Motions

1. *Search* (Sh): that part of the cycle during which the eyes or the hands are hunting or groping for the object. Search begins when the

[2] The color symbols are included in order to indicate color on the printed simo charts in this book. These color symbols should *not* be used in the actual construction of simo charts. Colored pencils should be used instead.

eyes or hands begin to hunt for the object, and ends when the object has been found.

The original list of the Gilbreth motions contained the therblig *find.* Since find occurs at the end of the therblig search, and since it is a mental reaction rather than a physical movement, it is seldom used in micromotion analysis work. Therefore find is omitted from the list of fundamental hand motions here.

2. *Select* (St): the choice of one object from among several. In many cases it is difficult if not impossible to determine where the boundaries lie between search and select. For this reason it is often the practice to combine them, referring to both as the one therblig *select.*

Using this broader definition, select then refers to the hunting and locating of one object from among several. Select begins when the eyes or hands begin to hunt for the object, and ends when the desired object has been located.

EXAMPLE. Locating a particular pencil in a box containing pencils, pens, and miscellaneous articles.

3. *Grasp* (G): taking hold of an object, closing the fingers around it preparatory to picking it up, holding it or manipulating it. Grasp begins when the hand or fingers first make contact with the object, and ends when the hand has obtained control of it.

EXAMPLE. Closing the fingers around the pen on the desk.

4. *Transport empty* (TE): moving the empty hand in reaching for an object. It is assumed that the hand moves without resistance toward or away from the object. Transport empty begins when the hand begins to move without load or resistance, and ends when the hand stops moving.

EXAMPLE. Moving the empty hand to grasp a pen on the desk.

5. *Transport loaded* (TL): moving an object from one place to another. The object may be carried in the hands or fingers, or it may be moved from one place to another by sliding, dragging, or pushing it along. Transport loaded also refers to moving the empty hand against resistance. Transport loaded begins when the hand begins to move an object or encounter resistance, and ends when the hand stops moving.

EXAMPLE. Carrying the pen from the desk set to the letter to be signed.

6. *Hold* (H): retention of an object after it has been grapsed, no movement of the object taking place.[3] Hold begins when the movement of the object stops, and ends with the start of the next therblig.

EXAMPLE. Holding bolt in one hand while assembling a washer onto it with the other.

7. *Release load* (RL): letting go of the object. Release load begins when the object starts to leave the hand, and ends when the object has been completely separated from the hand or fingers.

EXAMPLE. Letting go of the pen after it has been placed on the desk.

8. *Position* (P): turning or locating an object in such a way that it will be properly oriented to fit into the location for which it is intended. It is possible to position an object during the motion *transport loaded*. The carpenter, for example, may turn the nail into position for using while he is carrying it to the board into which it will be driven. Position begins when the hand begins to turn or locate the object, and ends when the object has been placed in the desired position or location.

EXAMPLE. Lining up a door key preparatory to inserting it in the keyhole.

9. *Pre-position* (PP): locating an object in a predetermined place, or locating it in the correct position for some subsequent motion. Pre-position is the same as *position* except that the object is located in the approximate position that will be needed later. Usually a holder, bracket, or special container of some kind holds the object in a way that permits it to be grasped easily in the position in which it will be used. Pre-position is the abbreviated term used for *pre-position for the next operation*.

EXAMPLE. Locating or lining up the pen above the desk-set holder before releasing it. The pen may then be grasped in approximately the correct position for writing. This eliminates the therblig position that would be required to turn the pen to the correct writing position if it were resting flat on the desk when grasped.

10. *Inspect* (I): examining an object to determine whether or not it complies with standard size, shape, color, or other qualities previously determined. The inspection may employ sight, hearing, touch, odor, or taste. Inspect is predominantly a mental reaction and may occur

---

[3] Gilbreth did not classify hold as a separate therblig but considered it a form of grasp.

simultaneously with other therbligs. Inspect begins when the eyes or other parts of the body begin to examine the object, and ends when the examination has been completed.

EXAMPLE. Visual examination of pearl buttons in the final sorting operation.

11. *Assemble* (A): placing one object into or on another object with which it becomes an integral part. Assemble begins as the hand starts to move the part into its place in the assembly, and ends when the hand has completed the assembly.

EXAMPLE. Placing cap on mechanical pencil.

12. *Disassemble* (DA): separating one object from another object of which it is an integral part. Disassemble begins when the hand starts to remove one part from the assembly, and ends when the hand has separated the part completely from the remainder of the assembly.

EXAMPLE. Removing cap from mechanical pencil.

13. *Use* (U): manipulating a tool, device, or piece of apparatus for the purpose for which it was intended. Use may refer to an almost infinite number of particular cases. It represents the motion for which the preceding motions have been more or less preparatory and for which the ones that follow are supplementary. Use begins when the hand starts to manipulate the tool or device, and ends when the hand ceases the application.

EXAMPLE. Writing one's signature in signing a letter (use pen), or painting an object with spray gun (use spray gun).

14. *Unavoidable delay* (UD): a delay beyond the control of the operator. Unavoidable delay may result from either of the following causes: (*a*) a failure or interruption in the process; (*b*) an arrangement of the operation that prevents one part of the body from working while other body members are busy. Unavoidable delay begins when the hand stops its activity, and ends when activity is resumed.

EXAMPLE. If the left hand made a long transport motion to the left and the right hand simultaneously made a very short transport motion to the right, an unavoidable delay would occur at the end of the right-hand transport in order to bring the two hands into balance.

15. *Avoidable delay* (AD): any delay of the operator for which he is responsible and over which he has control. It refers to delays which the operator may avoid if he wishes. Avoidable delay begins when

	Name and Definition of Motion	Symbol	Description of Motion	Illustration
1	**TRANSPORT EMPTY** (Transport Empty refers to moving the empty hand in reaching for an object. It is assumed that the hand moves without resistance toward or away from the object. Transport empty begins when the hand begins to move without load or resistance and ends when the hand stops moving.)	TE	Reach for pen.	
2	**GRASP** (Grasp refers to taking hold of an object, closing the fingers around it preparatory to picking it up, holding it or manipulating it. Grasp begins when the hand or fingers first make contact with the object and ends when the hand has obtained control of it.)	G	Take hold of pen – close thumb and fingers around pen.	
3	**TRANSPORT LOADED** (Transport Loaded refers to moving an object from one place to another. The object may be carried in the hands or fingers or it may be moved from one place to another by sliding, dragging, or pushing it along. Transport loaded also refers to moving the empty hand against resistance. Transport loaded begins when the hand begins to move an object or encounter resistance and ends when the hand stops moving.)	TL	Carry pen to paper.	

	Name and Definition of Motion	Symbol	Description of Motion	Illustration
4	**POSITION** (Position consists of turning or locating an object in such a way that it will be properly oriented to fit into the location for which it is intended. It is possible to position an object during the motion transport loaded. The carpenter, for example, may turn the nail into position for using while he is carrying it to the board into which it will be driven. Position begins when the hand begins to turn or locate the object and ends when the object has been placed in the desired position or location.)	P	Position pen on paper for writing.	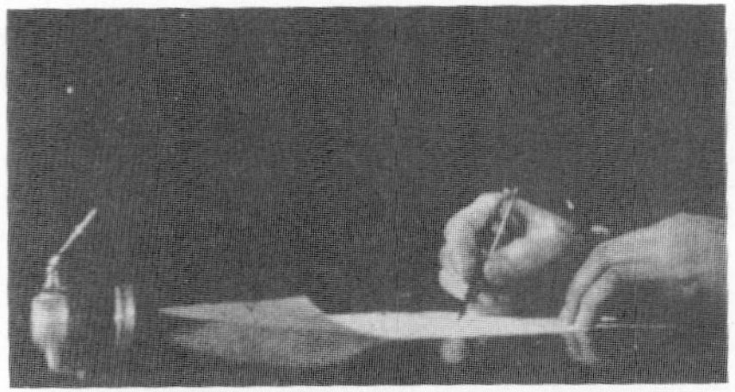
5	**USE** (Use consists of manipulating a tool, device, or piece of apparatus for the purpose for which it was intended. Use may refer to an almost infinite number of particular cases. It represents the motion for which preceding motions have been more or less preparatory and for which the ones that follow are supplementary. Use begins when the hand starts to manipulate the tool or device and ends when the hand ceases the application.)	U	Sign letter.	
6	**TRANSPORT LOADED**	TL	Return pen to holder.	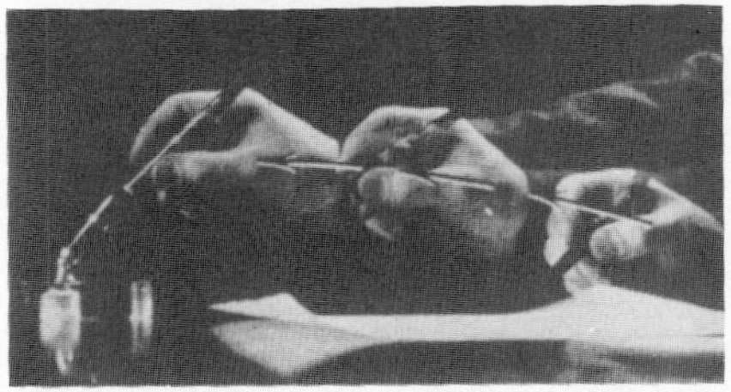

Fig. 84. Fundamental motions of the right hand in signing a letter.

(*Continued*)

	Name and Definition of Motion	Symbol	Description of Motion	Illustration
7	**PRE - POSITION** (Pre-position refers to locating an object in a predetermined place or locating it in the correct position for some subsequent motion. Pre-position is the same as position except that the object is located in the approximate position that it will be needed later. Usually a holder, bracket, or special container of some kind is used for holding the object in a way that permits it to be grasped easily in the position in which it will be used. Pre-position is the abbreviated term used for pre-position for the next operation.)	PP	Position pen in holder.	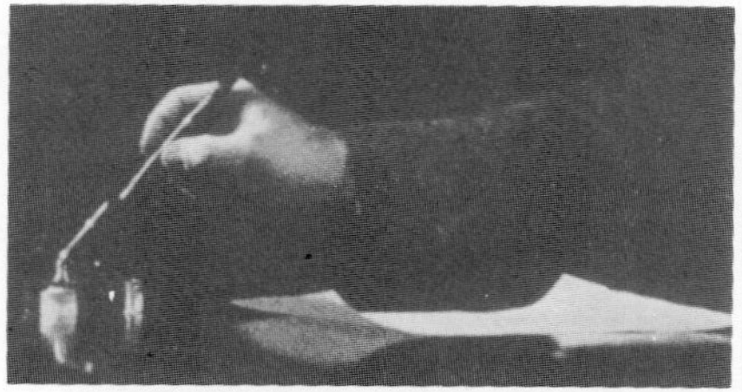
8	**RELEASE LOAD** (Release Load refers to letting go of the object. Release load begins when the object starts to leave the hand and ends when the object has been completely separated from the hand or fingers.)	RL	Let go of pen.	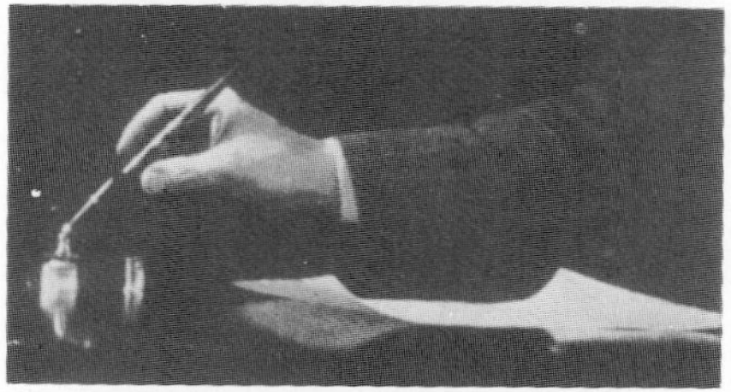
9	**TRANSPORT EMPTY**	TE	Move hand back to letter.	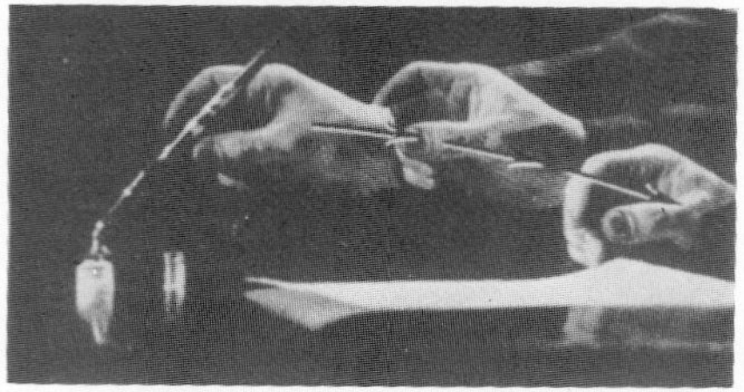

Fig. 84 (*Continued*)

Motions of the Left Hand

Illustration	Name of Motion	Symbol
	TRANSPORT EMPTY Reach for pencil in tray.	TE
	SELECT Select the automatic pencil from among the other pencils in the tray. The eyes aid the hand in searching for and selecting the automatic pencil.	St
	GRASP Close thumb and fingers around barrel of pencil.	G
	TRANSPORT LOADED Carry pencil from tray to vertical position in front of body. Also: **POSITION (in transit)** Pencil is in horizontal position when grasped. It is turned to vertical position in transit.	TL P

Motions of the Right Hand

Symbol	Name of Motion	Illustration
UD	**UNAVOIDABLE DELAY** The right hand is idle – there is nothing for it to do. Therefore this delay is called unavoidable.	
TE	**TRANSPORT EMPTY** Right hand moves empty to pencil cap.	

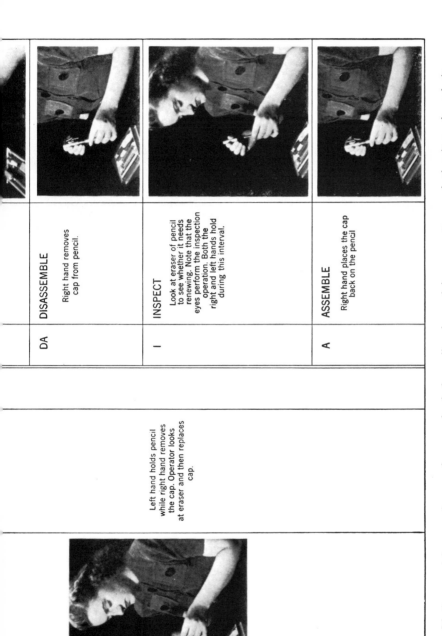

DA	**DISASSEMBLE** Right hand removes cap from pencil.	
I	**INSPECT** Look at eraser of pencil to see whether it needs renewing. Note that the eyes perform the inspection operation. Both the right and left hands hold during this interval.	
A	**ASSEMBLE** Right hand places the cap back on the pencil	

Left hand holds pencil while right hand removes the cap. Operator looks at eraser and then replaces cap.

Fig. 85. Motions used in removing cap from mechanical pencil and examining eraser to see whether it needs replacing.

the prescribed sequence of motions is interrupted, and ends when the standard work method is resumed.

EXAMPLE. The operator stops all hand motions.

16. *Plan* (Pn): a mental reaction which precedes the physical movement, that is, deciding how to proceed with the job. Plan begins at the point where the operator begins to work out the next step of the operation, and ends when the procedure to be followed has been determined.

EXAMPLE. An operator assembling a complex mechanism, deciding which part should be assembled next.

17. *Rest for overcoming fatigue* (R): a fatigue or delay factor or allowance provided to permit the worker to recover from the fatigue incurred by his work. Rest begins when the operator stops working, and ends when work is resumed.

Motions Used in Signing a Letter. It is a relatively easy matter to learn the names of these fundamental motions. For example, in signing a letter, the sequence of motions is *transport empty* (reach for pen), *grasp* (take hold of pen), *transport loaded* (carry pen to paper), *position* (place pen on paper at correct position for writing), *use* (sign letter), *transport loaded* (return pen to holder), *pre-position* (position pen in holder), *release load* (let go of pen), and *transport empty* (move hand back to letter). These motions are fully defined and illustrated in Fig. 84.

Motions Used in Removing the Cap from a Mechanical Pencil. To provide further practice in learning the fundamental motions, Fig. 85 shows the motions used in picking up a mechanical pencil from a tray, removing the cap from the pencil, and examining the pencil to see whether the eraser needs renewing.

Notice that in this case the left hand has a *select* following the transport empty and preceding the grasp. *Select* refers to the choice of one object from among several. In removing the fountain pen from the holder (see page 111), only one pen was present; consequently, no selection was required. In the second case, the mechanical pencil is in a box with other pencils, and the particular pencil desired must be *selected* from among the others.

Pinboard. It seems natural for most people, when observing another person at work, to notice the material being handled or the tools being used rather than the motions made in performing the task. After one becomes "motion-minded," that is, after one has learned the

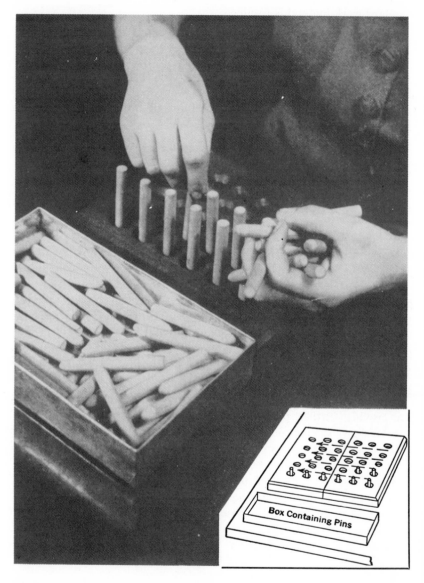

Fig. 86. Inserting pins in board, using the one-handed method. Left hand *holds* pins; right hand works productively. It takes 0.62 minute to fill the board.

Fig. 87. Inserting pins in board, using simultaneous motions, both hands working together. It takes only 0.41 minute to fill the board using this method.

classification of hand motions, this situation is changed. The observer then notices the motions made with the right hand and those made with the left hand, and then proceeds to use those motions which are easy and effective and to discard awkward, fatiguing, and ineffective motions. People who accomplish the most do not necessarily work hardest. Rather, they make every motion count—they use good work methods. We are not at all interested in the "speed-up" or "stretch-

Illustration	Name of Motion	Symbol
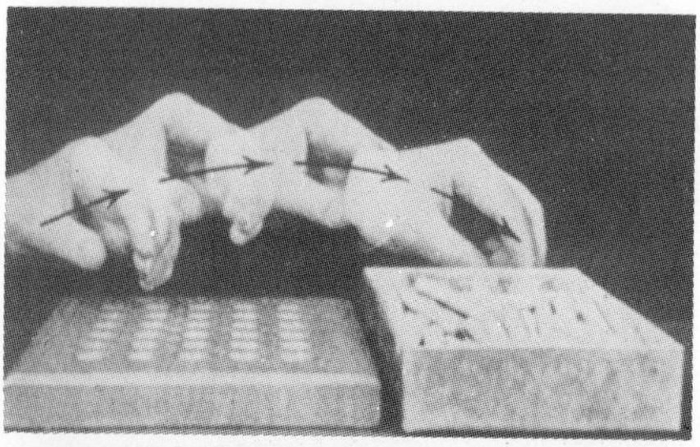	**TRANSPORT EMPTY** Reach for pin.	TE
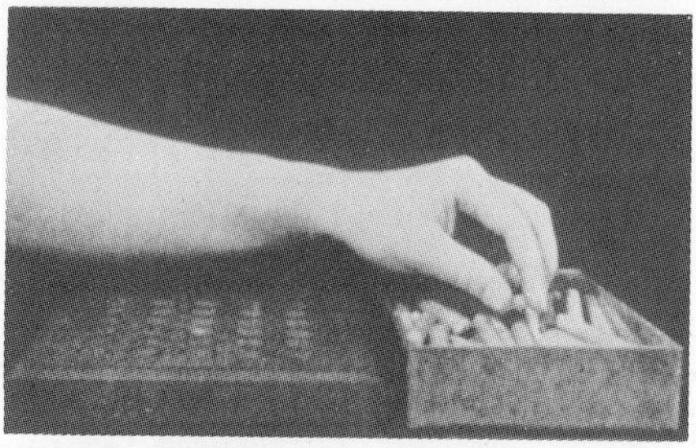	**SELECT** Select one pin from among those in box. The eyes aid the hand in searching for a particular pin. This searching and then spotting or finding a particular pin is called <u>select.</u>	St
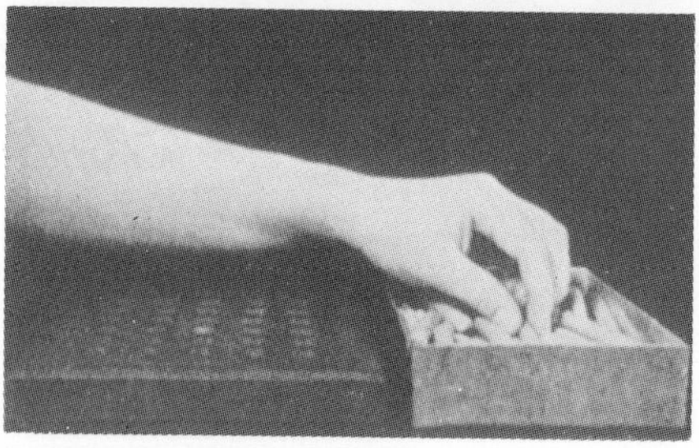	**GRASP** Close thumb and fingers around the pin selected.	G

Fig. 88. Fundamental motions used in inserting pin in pinboard. The operator is using simultaneous symmetrical motions in filling the pinboard (Fig. 87). Since the motions of the left hand and the right hand are the same, those of the left hand only are shown here.

Illustration	Name of Motion	Symbol
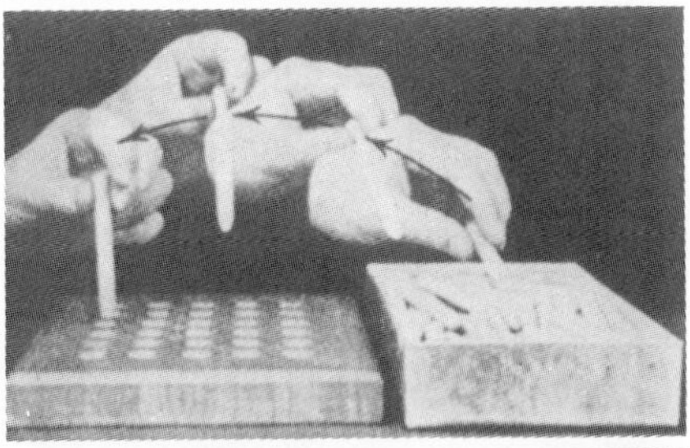	**TRANSPORT LOADED** Carry pin from tray to hole in board into which it will be inserted. **Also:** **POSITION (in transit)** Pin is turned into vertical position as it is transported to board.	TL P
	**POSITION** Pin is lined up directly over the hole in the board into which it is to be inserted.	P
	**ASSEMBLE** Insert pin into hole in board.	A
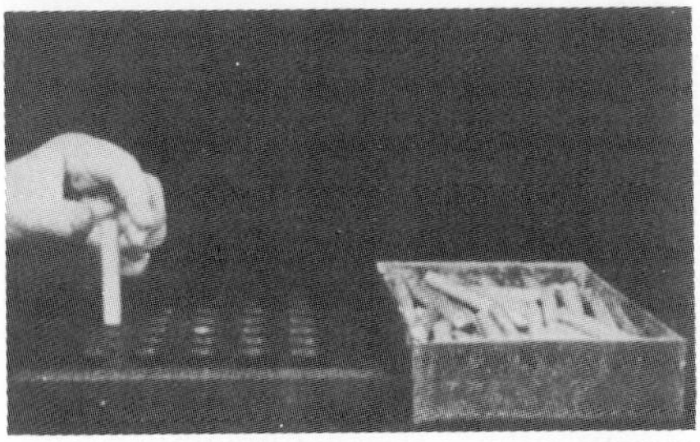	**RELEASE LOAD** Open fingers – let go of pin.	RL

Fig. 88 (*Continued*)

out." We are interested in getting more quality work done with less expenditure of energy. Excessive speed is no substitute for good work methods.

To illustrate what is meant by developing a better method through the analysis of hand motions and the application of principles of motion economy, let us consider the task of filling a board containing thirty holes with thirty wooden pins (Fig. 86). You will notice that there are five rows of six holes to a row in the board. The pins are square on one end and bullet-shaped on the other. The job is to fill the board with the pins as quickly as possible, inserting the pin in the hole with the bullet nose down.

Ninety-five people out of a hundred would fill the board by the method shown in Fig. 86. The left hand grasps a handful of pins from the box and holds them while the right hand gets pins from the left, one at a time, and places them in the board. The right hand is working in a very effective manner inasmuch as it is performing the desired task, that is, filling the board with pins. Notice that the left hand is doing very little productive work; most of the time it is merely holding the pins.

If both hands were to work simultaneously at getting and placing the pins in the holes, the operator's efforts would be much more effective. We are now applying one of the "principles of motion economy" which will be presented later. We are having a preview of one of these principles now.

Using this improved method (Fig. 87) it is obvious that the left-hand *hold* has been eliminated, and the left hand, like the right, now performs useful motions. The two hands work together in a symmetrical manner, getting the pins and placing them in the holes in the board.

Results. One study showed that it requires 0.62 minute to fill the board using the one-handed method, whereas but 0.41 minute is required with the two-handed method. This is a saving of 34% in time.

The fundamental motions of the left hand in filling the pinboard are shown in Fig. 88.

Pinboard Demonstration. It is suggested that the pinboard be used as a demonstration in introducing the subject of motion study to an individual or a group. Let each person try various methods of filling the pinboard, timing himself on each method. It is convincing indeed to discover that there is an easier and a far faster way to fill the board than the conventional one-handed method. Specifications for the pinboard and pins are given in Fig. 360.

CHAPTER 12

Motion Study and Micromotion Study Equipment

The motion picture camera is perhaps the most important piece of equipment used in motion study and micromotion study work. The first cameras were of the hand-cranked type, using film 35 mm. wide. The camera, mounted on a tripod, was fitted to take either single exposures or pictures at varying speeds up to 100 frames per second or faster.

Today the professional camera using 35-mm. film has entirely given way to the amateur motion picture camera using 16-mm. or 8-mm. film for motion study work. Although 8-mm. equipment is used to a limited extent for motion study work, there is a definite trend toward the standardization of 16-mm. equipment in this field.

Figure 89 shows six motion picture cameras, arranged according to the speed at which they normally operate. Since the spring-driven camera (*B* of Fig. 89) is most widely used, it will be described first.

Spring-Driven Camera. The typical amateur motion picture camera is very compact and light in weight. It is operated by a spring-driven motor which runs approximately ½ minute with one winding. The speed at which the film passes through the camera is regulated by a governor that maintains a constant speed (within ±10%) until the spring motor runs down and stops. Some camera motors slow down sharply near the end of their "run" and need more frequent windings if an approach to a uniform speed is to be maintained.

The typical camera can be loaded or unloaded in the daylight, and takes either a 50-foot or a 100-foot roll of film. The normal speed is 16 exposures per second. A 100-foot roll of film exposed at normal speed will last approximately 4 minutes. Although the amateur spring-driven camera may be operated satisfactorily for most kinds of outdoor work without a tripod, for motion study work a tripod is recommended. Since there are a number of excellent cameras on the market

149

A. SLOW SPEED—50 or 100 pictures or frames per minute. Time-lapse attachment for camera—electric-motor driven or solenoid operated

B. NORMAL SPEED—960 frames per minute. Spring-driven or electric motor-driven camera

C. MODIFIED NORMAL SPEED—1000 frames per minute. Electric motor-driven camera

Fig. 89. Six motion picture cameras arranged according

D. SOUND SPEED—1440 frames per minute. Spring-driven or electric motor-driven camera

E. HIGH SPEED—64 to 128 frames per second. Spring-driven or electric motor-driven camera

F. VERY HIGH SPEED—1000 to 3000 frames per second. Electric motor-driven camera

to the speed at which they normally operate.

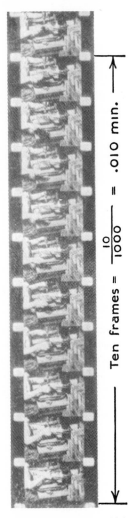

Ten frames = $\dfrac{10}{1000}$ = .010 min.

(Film 16 mm. in width, enlarged one half.)

Fig. 90. Motion picture film made with constant-speed motor-driven camera at 1000 frames per minute. This strip of film shows part of a counterboring operation on a drill press.

suitable for this work, no attempt will be made to describe them here.

A motion picture camera should have at least the following features if it is to be used satisfactorily for micromotion study work:

1. Lens—f.2.4 or faster; f.1.9 is preferred.
2. Focus—adjustable from 4 feet or closer to infinity (a camera with a fixed-focus lens is not satisfactory).
3. Film capacity—100 feet.
4. Accurate film meter.

The following additional features are desirable but not absolutely necessary:

5. Variable-speed spring motor which operates from one-half normal speed (8 frames per second) to four times normal speed (64 frames per second).
6. Interchangeable lenses or zoom lens.
7. Electric motor drive attachment.

Electric Motor-Driven Camera. Considerable use is being made of motion picture cameras driven by a constant-speed electric motor. The most common speed for an electric motor-driven camera is 1000 frames per minute (C of Fig. 89). This speed is slightly faster than normal speed, which is 16 frames per second or 960 per minute. Camera speeds other than 1000 frames per minute may be obtained by changing the gear ratio between the motor and the camera or by a separate motor drive. The camera shown in A of Fig. 89 is operated by a solenoid at 50 or 100 frames per minute. D of Fig. 89 shows a motor-driven camera for taking pictures at sound speed, or 1440 frames per minute. The cameras shown in E and F of Fig. 89 are for high-speed work.

An enlarged print from a strip of film made at a speed of 1000 frames per minute is shown in Fig. 90. The time interval that elapsed from one frame to the next on this film was exactly $\frac{1}{1000}$ of a minute. The motions of the hand shown as taking place during the exposure of these 10 frames required $1\frac{0}{1000}$ of a minute (0.010).

Since the camera operates at a constant and known speed, the microchronometer (Fig. 91) is not needed to indicate time on the film. Consequently, it does not occupy valuable space in the picture or shut out motions of the operator being studied. It is easy to assign time values to the motions, since no study need be made of the position of the clock's hands. If the film is projected on a screen, it is possible to know the exact projection speed by means of a tachometer attached to the projector (Fig. 321). In other words, one can project the film on the screen at exactly the same speed at which it was made, or at a faster or slower speed of known value.

Although the electric motor-driven camera has certain advantages, a good amateur motion picture camera of the regular spring-driven type is perfectly satisfactory for most ordinary motion study and micromotion study work.

Camera Speeds. The amateur motion picture camera operates in such a way that one frame, or the film for one exposure, is suddenly "pulled down" or jerked in front of the lens of the camera during an instant when the camera shutter has closed the lens. After the film is in place, the rotating shutter opens and permits the subject to be photographed. The shutter then closes and the next frame is pulled down for the next exposure, and so on. The shutter is closed one third to one half of the time that the camera is in action, depending on the design of the shutter. The ratio of the size of the open segment in the shutter to that of the closed segment determines the exposure time for one rotation of the shutter. The shutter makes one complete revolution each time an exposure is made. Therefore, if the camera is operating at the normal speed of 16 exposures per second, and has a shutter with an open segment of 180 degrees, the time that the lens will be open during one revolution is $\frac{1}{16} \times \frac{180}{360}$ or $\frac{1}{32}$ of a second.

The motion picture camera photographs intermittent scenes. In photographing moving subjects there is an instant ($\frac{1}{32}$ of a second in the preceding case) between successive exposures during which no record of action that has been taking place is made on the film. It is for this reason that successive frames on the film show the moving object at different points along its line of motion (see enlarged print in Fig. 99). The hand reaching for an object is shown first a foot away from the object, then 10 inches, then 8 inches, etc. Where the movement of the subject is relatively rapid, the moving object appears to be blurred. The right hand in Fig. 105 appears blurred in exposures 2, 7, and 8; during the short instant when the shutter was open, the hand moved a sufficient distance to cause the blur. These blurs are eliminated by exposing the film at a more rapid rate. Had the picture been made at 32 instead of 16 exposures per second, the time during which the shutter remained open would have been $\frac{1}{64}$ of a second, and the hand would have moved but one half the distance. This would have reduced or entirely eliminated the blur.

With the camera operating at normal speed, it frequently happens that the hand, for example, changes direction entirely while the shutter is closed. If an operator should reach for an object, the hand might be shown moving to the right on one frame of the film; on the next frame it might be shown moving to the left. During the instant that the shutter was closed the hand had actually continued to move to the right, grasped a piece of material, and was on its return movement when the next exposure was made. For very accurate studies such hidden motions are undesirable. To prevent them it is necessary to operate the camera at higher speeds.

Although the camera normally operates at a speed of 16 exposures per second, amateur spring-driven cameras are available which operate at speeds as high as eight times normal.[1]

For ordinary micromotion study work the normal camera speed is satisfactory. For studying rapid hand motions it may become necessary to use twice normal speed, and in evaluating very short and fast motions such as a "sliding grasp" under laboratory conditions, speeds of 5000 exposures per minute or higher may be required.

Motion Pictures Are Easy to Make. The amateur motion picture camera is designed so that the average person is able to make satisfactory pictures without practice. However, pictures for motion study work inside the factory are more difficult to make than outdoor pictures. Most people are able to make very satisfactory pictures by following the directions which come with the camera. Even though a person may be able to make successful pictures of ordinary factory operations, it will be to his advantage to learn as much about photography as he can.[2]

[1] The camera shown in F of Fig. 89 will take motion pictures at speeds up to 3000 frames per second. Professor H. E. Edgerton has taken still photographs with an exposure of a millionth of a second, and motion pictures at 6000 exposures a second. See H. E. Edgerton, J. K. Germeshausen, and H. E. Grier, "High Speed Photographic Methods of Measurement," *Journal of Applied Physics*, Vol. 8, No. 1, p. 1.

[2] *How to Make Good Movies*, Eastman Kodak Co., Rochester, N. Y.

A motion picture data sheet similar to that shown in Fig. 96 is of real assistance in improving one's ability to take good pictures under widely varying conditions. The data sheet provides a permanent record of all important factors connected with the taking of the pictures. The information on this sheet may be used as a check if pictures do not turn out satisfactorily, or as a reference if pictures are to be made

Fig. 91. Electric motor-driven microchronometer. The large hand makes 20 revolutions per minute, and the small hand makes 2.

under new conditions. The mere necessity of recording the various items on the data sheet, such as diaphragm opening, focus, distance of the subject from the camera, and number of spotlights used, automatically prevents the beginner from making the picture before he has adjusted his camera and completed his setup. The greatest value of using such a data sheet, as will be explained more fully in Chapter 13, is that together with the film number it forms a permanent record for completely identifying any piece of film.

Microchronometer. Since the number of exposures made on the film in any given time interval will depend upon the speed of the camera, and since the speed of a spring-driven camera is not constant, it is necessary to place some accurate timing device in the picture so that the time interval from the exposure of one frame to the exposure of the next will be indicated on the film.

Gilbreth developed a spring-driven, fast-moving clock called a microchronometer, capable of indicating time to $\frac{1}{2000}$ of a minute. The dial of the clock was divided into 100 equal spaces, and the hand made 20 revolutions per minute. Synchronous motor-driven clocks are now used. Such clocks are very accurate, and the hands may be geared to indicate time intervals of any desired length. The clock shown in Fig. 91 is driven by a small synchronous motor. It has 100 equal divisions on the dial. The large hand makes 20 revolutions per minute, and the small hand 2 revolutions per minute. Each division on the dial indicates $\frac{1}{2000}$ of a minute. The clock reading in Fig. 91 is 652. By changing the gear ratio inside the clock, the large hand will make 50 revolutions per minute and the small hand 5. By using this latter arrangement it is possible to read time intervals of $\frac{1}{5000}$ of a minute without interpolation. The clock is operated at this fast speed only when the film is exposed at 2000 frames per minute or faster.

When the electric motor-driven camera is used, the microchronometer is not needed unless it is wanted for the purpose of quickly identifying particular motions or places in a cycle. It is sometimes so used.

Illumination. Insofar as possible, motion pictures should be made by daylight. Frequently some additional illumination is required. Such illumination is easily provided by portable spotlights or "photoflood" lamps fitted with a suitable reflector and supported by a tripod. With either the standard spotlight or the photoflood lamp, the problem of lighting the subject to be photographed is easily solved. Ordinarily two light sources should be used for best results. The lamps should be placed so that the work place and the particular motions to be studied are properly lighted, without deep shadows. If the person making the picture remembers that he is later going to study the motions in detail, he will be more likely to see that they are properly lighted.

With the development of TRI-X motion picture film the need for artificial lighting has been greatly reduced.

Laboratory. Some insist that wherever possible the motion pictures should be made in a special laboratory apart from the main production floor. This requires that the regular tools and equipment be moved into the laboratory and the regular operators be brought in from the factory. Studies of the operation can then be made without disturbing production in the factory. Although this procedure has many advantages, it is now common practice to take the pictures at the work place in the factory. This practice is less costly, aids in securing the cooperation of the workers, and tends to remove some of the mystery of motion and time study. Where the work is of such a nature that the laboratory setup is possible, and where an extended study is warranted, it is not unusual to carry the investigation into the laboratory.

There are other uses for the laboratory which alone would justify its existence. A laboratory is indispensable for storing motion picture equipment, for analyzing the film, for constructing simo charts, and for showing film to those concerned with improving methods. The motion study laboratory may be used as a classroom by members of the organization who are interested in learning the micromotion study technique. It frequently serves as a conference room for foreman and supervisor training programs. Figure 19 shows the motion study laboratory at the Fort Wayne works of the General Electric Company as it appeared in 1929. Figure 20 shows the methods laboratory at the Packard Electric Division of General Motors Corporation. The Armstrong Cork Company Industrial Engineering Center is shown in Fig. 315.

Motion Picture Film. Professional motion pictures are made on negative film. This film is developed, and from it positive prints are made for use in theater projectors. With the development of the amateur camera and 16-mm. film, a reversal process was perfected. The amateur 16-mm. reversal film now most commonly used is coated with a photographic emulsion which permits the film to be exposed in the camera in the usual way. It is then sent to the laboratory, where it is processed in such a way as to produce a positive directly on the original film base. The user receives his original roll of film from the processing station as a positive, ready for projection on the screen. Duplicates of the original 16-mm. film may be obtained when desired.

TRI-X film is most commonly used for motion study work, but colored motion pictures also are finding wide application in this field. All amateur motion picture film is made from an acetate base, is not flammable, and is known as "safety" film.

Indexing and Storing Film. If motion pictures are used extensively, adequate provision should be made for indexing the film and caring for it. One method that has been found successful is to assign a number to each picture and place a small card bearing this number in the picture at the time the film is exposed.

A motion picture data sheet similar to that shown in Fig. 96 is filled out at the time the picture is made, and this sheet is kept as a permanent record. The roll of film or the film loop is then placed in a box, properly labeled, and filed by number in a drawer in a metal filing cabinet. The film may be cross-indexed as to kind of operation (drilling, spray painting, inspection, etc.), and also as to department, kind of product, or in any other way that seems desirable.

Motion Picture Projector. The motion picture projector is indispensable for analyzing film, as it must be studied frame by frame in minute detail. Frequently the motions of several members of the

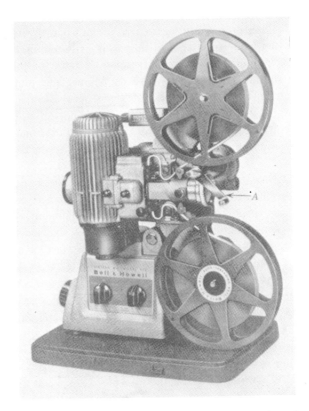

Fig. 92. Motion picture projector (16-mm.) designed for general motion and time study work. The hand crank *A* may be used for film analysis.

body, such as the fingers, arms, and feet, must be studied. This study requires that the same film be analyzed a number of times, once for each member of the body studied.

The most suitable projector for this purpose is a small one of light weight, which can be moved around easily on the desk or table. The projector should have a low-power bulb or special filters, so that the heat developed by it will not buckle or warp the film when it remains stationary in front of the lens for a long period of time. The projector may be fitted with a lens of short focal length (1 inch), to project a relatively large picture on a screen placed near the projector. The projector should have a hand crank geared to it so that one turn of this crank advances the film one frame in front of the lens for film analysis (A in Fig. 92). By giving this crank a quick turn the frame of film is pulled down in front of the projection lens so quickly that the movement of the subject may be noted on the screen. This aids in finding the points where motions begin and end, or where change of direction occurs. This projector is also equipped with a mechanical counter which counts the frames passing by the lens. This counter

Fig. 93. Motion picture projector (16-mm.) adapted for film analysis. The special control box A makes it possible to operate this projector from a distance. The frame counter B permits the analysis of films made with an electric motor-driven camera.

is especially useful when analyzing film made with a constant-speed motor-driven camera. The projector shown in Fig. 93 is controlled by the switches on box A.[3] This control box makes it possible to operate the projector from a distance.

[3] For a detailed description of this projector see D. B. Porter and L. P. Granath, "How to Convert Projectors for Motion Study," *Factory Management and Maintenance,* Vol. 97, No. 7, pp. 49–50.

The projector shown in Fig. 94 has been designed for film analysis demonstrations before a large audience. Special filters and an auxiliary blower protect the film even when a 1000-watt lamp is used in the projector.

Fig. 94. Motion picture projector (16-mm.) adapted for film analysis demonstrations before a large audience. The special control box *A* makes it possible to operate this projector from a distance. The projector can be stopped without damage to film, even with a 1000-watt lamp.

A projector with a high-power bulb is needed when pictures are to be shown to a large group of people. If this projector (Fig. 321) is fitted with a tachometer and a variable-speed motor, it can be used to show pictures made with a constant-speed camera. The pictures may be shown at the exact speed at which the operator was working when the pictures were made, or at faster or slower speeds of known value.[4]

[4] See Chapter 24 for a discussion of the use of motion pictures for performance rating of operators.

Fig. 95. Projection booth for film analysis.

List of Equipment for Motion Study Work. To summarize, the following equipment is recommended where a fairly extensive program is to be carried on:

1. One motion picture camera, preferably with f.1.9 lens, adjustable focus from 4 feet to infinity, or zoom lens, and film capacity of 100 feet.
2. One metal tripod with tilting and panoraming head.
3. One exposure meter.
4. Three or four photoflood lamps with reflectors.
5. Two tripods for photoflood lamp reflectors.
6. One microchronometer.
7. One motion picture projector with low-power bulb and hand crank, for film analysis.
8. One motion picture projector with high-power bulb, for showing pictures to large audiences.
9. One portable screen.
10. Suitable cabinets for storage of film.
11. One titling outfit.
12. One rewind, editor, and film splicer.
13. One steel cabinet for storage of equipment.

CHAPTER 13

Making the Motion Pictures

Motion pictures can be used for numerous purposes in motion and time study work. They are frequently made (1) for micromotion study and memomotion study, (2) for obtaining work sampling data, (3) for training factory operators, (4) for showing the current method of doing a particular job,[1] (5) for performance rating in time study work, and (6) for motion study research.

Since films for micromotion study are perhaps the most difficult to make, an explanation is given here of the procedure to be followed. It will be assumed that a particular operation has been selected for study.

Operator to Be Studied. The first step is to select one or more operators as subjects for making the motion picture. It is of greatest value to make the pictures of those operators who are the most highly skilled and who perform the work in the most satisfactory manner. Every operator who gives promise of contributing something to the establishment of the improved method should be studied. It is often desirable from a psychological standpoint to make motion pictures of everyone performing the operation. It is unlikely that information of much value will be obtained from the inexperienced workers; consequently only a few feet of film need be made of these. Occasionally it has been found that the "lazy worker" is using better methods than some of the more energetic operators. This, of course, is true because he attempts to get his work done with the least expenditure of energy.

It is very important and necessary that the workers and the supervisor are told just what is going to be done. Their cooperation should be sought from the very beginning, and is seldom difficult to obtain.

---

[1] This fourth use may be supplemented by a picture taken after an improvement in method has been worked out. Such pictures are sometimes referred to as "before and after" films.

In most cases the workers will give their very best performance while the motion pictures are being made, because they know that a permanent record is being made and that their fellow workers as well as the executives may review their work on the screen.

It should be emphasized that motion study makes no effort to force the worker to "move faster," but studies his motions to find the shortest and best ones to use. Motion study aids in finding the easiest and least fatiguing way of doing the work. If the best operators obtainable are used as subjects for the study, the analyst is likely to progress more rapidly on the solution of his problem than if he selects the inexperienced workers. The motions that the operators use are the things being studied, not the speed that they exhibit.

As stated at the beginning of Chapter 1 of this book, motion and time study has several objectives. It is the purpose of time study, for example, to determine a time value in minutes or hours which permits the qualified operator to work day after day and week after week without harm or undue fatigue to himself, always being able to perform the task in this standard or specified time. In making micromotion studies, however, it is expected that the superior workers who act as subjects may perform at a faster speed than the "standard" calls for. No one can object to this, for in motion study the discovery of the very best possible *way* of doing the work is the first and foremost object. Those operators who will aid most in determining this method are the ones who should be studied.

Placing the Camera. Assuming that the operator or operators to be studied have been selected and understand that a micromotion study is to be made, the motion study analyst is ready to set up his equipment and make the picture.

Although it is not necessary to have pictures of a quality equal to that of professional movies, it is essential that the pictures should be sufficiently clear when projected to give all necessary details. They should be sharply focused, and they should be taken from such an angle as to give a satisfactory picture of all motions of the operator.

The camera should be placed as close to the subject as possible without omitting anything necessary from the picture. Both the work place and the actions of the operator should be considered in positioning the camera. The motions of the operator may occur in two directions—those made perpendicular to the line of sight and those made parallel to the line of sight. The camera should be placed at such an angle, relative to the operator and the work place, that a majority of the motions will be perpendicular to the line of sight.

Not only does such an arrangement tend to permit a sharp focus throughout the cycle, but it also makes the analysis of the film easier. It is less difficult to judge the nature and extent of movements made at right angles to the line of sight than it is to judge movements made toward or away from the line of sight. The view finder on most motion picture cameras is sufficiently accurate even at close range to show what will be included in the picture.

The camera should preferably be placed to include the entire range of the worker's motions for the cycle. Seldom is it desirable to follow the movements of the operator by moving the camera as the cycle progresses. It is difficult to anticipate the movements of the operator and almost impossible to keep all his motions in the picture at all times.

In some cases motion pictures may profitably be made from more than one position, although this is by no means required on every operation. It is desirable, however, to make a few pictures of the operator and the work place from a distance in order to have a complete record of the job, and incidentally a good picture of the operator.

It is sometimes advantageous to place behind the operator a cross-sectioned screen made with white lines drawn on a black background, forming 4-inch squares. In certain cases the workbench or the floor may be marked off in a similar manner. This is done to assist in determining the location and the extent of motions when the film is being analyzed. Everything should be done that will assist in making the analysis of the film easy. Such small details as the color of the operator's clothes have an important effect upon the ease with which the motions may be analyzed.

On some occasions Gilbreth used what he called a "penetrating screen" because it gave further assistance in studying and measuring motions.[2] The penetrating screen resulted from a double exposure of the film. The first exposure photographed a screen of black material marked off into small squares with white lines. This screen was located on the work place, at a position where the motions of the operator's hands normally occurred. Having been photographed, the screen was removed, the film was rewound in the darkroom on the original reel, and then the motion picture of the operator was made in the usual way. After the film was processed and projected, the operator appeared to be working with this transparent cross-sectioned screen across the work place.

[2] F. B. and L. M. Gilbreth, *Applied Motion Study,* Sturgis & Walton Co., New York, 1917, p. 86.

The camera should be mounted on a tripod, which should be placed securely on the floor or on top of a solid table or bench so that the camera will be free of vibration while it is in operation.

Lighting. Daylight is preferable to artificial light for making pictures; however, indoors it is often necessary to supplement daylight with some artificial light. Photoflood lamps with suitable reflectors usually supply this additional illumination. These lamps should be located to light properly the darkest places in the picture. It is better to have too much illumination on the subject and stop the diaphragm opening of the camera down than it is to have too little illumination and get a dark picture.

In placing the lighting units it should be remembered that the intensity of illumination falling on an object from the lamp varies inversely as the square of the distance between the source and the object. If the lamp that is 10 feet from the object is moved up to a distance of 5 feet, the intensity of illumination on the object is increased four times. The surest way to know whether the object is sufficiently illuminated is to use an exposure meter.

Making the Motion Picture. If a microchronometer is used, it should be placed so that its entire face will appear in the picture and yet not hide any of the motions of the operator or any part of the work place that should be included in the picture. Neither should motions of the operator interfere with the clock's being in full view at all times. The microchronometer should be in focus if it is to be easily read when photographed.

The camera is loaded with film, the film-footage meter set to zero, and the diaphragm opening adjusted for the lighting conditions present and for the speed at which the camera is to operate if the speed is different from normal. The distance from the center of action of the operator to the lens of the camera should be carefully measured with a tape measure (estimating distance is not satisfactory) and the camera focused accurately. This is particularly important when the camera is placed near the work to be photographed.

To identify the film a card bearing such information as operation name, part number, date of study, department number, and film number is often placed in front of the camera and photographed on the first few frames of film. Another method is to place a single number or symbol so that it shows in the picture during the entire "run." A different number is used for each setup or run. This number is referred to as the "film number." A special motion picture data sheet bearing the film number, such as the one shown in Fig. 96, is used to list all

MOTION - PICTURE DATA SHEET

PLACE L. C. Smith & Corona Typewriters, Inc. Groton, N.Y.	**FILM NO.** C1
OPERATION Form Links	**OP. NO.** 15
PART NAME Link for Portable Typewriter	**PART NO.** 357

MACHINE NAME Special Bench Fixture	**MACHINE NUMBER** Fixture No. 1364	**DEPT.** No. 9	
OPERATOR NAME & NO. M. S. Fost A1		**DATE**	
EXPERIENCE ON JOB An average operator, often on other work	**MATERIAL** M. S. wire cut to length		

BEGIN 11:00 AM	**FINISH** 11:50 AM	**ELAPSED**	**UNITS FINISHED**	**RATING**
CAMERA MAKE EK Co.	**CAMERA SERIAL NO.** 03221	**CAMERA SPEED IN FRAMES/SEC.** 16	**LENS** f.1.9	
DISTANCE OF CAMERA FROM SUBJECT 4' 2"	**FOCUS SETTING** 4	**EXPOSURE METER READING**	**DIAPHRAGM OPENING** f.4.0	
KIND OF FILM E. K. Pan.	**SENSITIVITY RATING**	**NO. OF SPOTS** Two	**WATTAGE**	**A** 500 **C** **B** 500 **D**

Material was placed on the top of the bench at a distance of 10 inches from the center of the fixture

Operator

X

Special Bench Fixture

Finished parts were disposed of by dropping on bench top at the left of the operator

⊕ A ⊕ B

Work Bench Top 27" from Floor

Fixture 2¾" above Bench Top

Link before and after forming

⊕ SPOT LIGHT

▶ CAMERA

‖ WINDOW

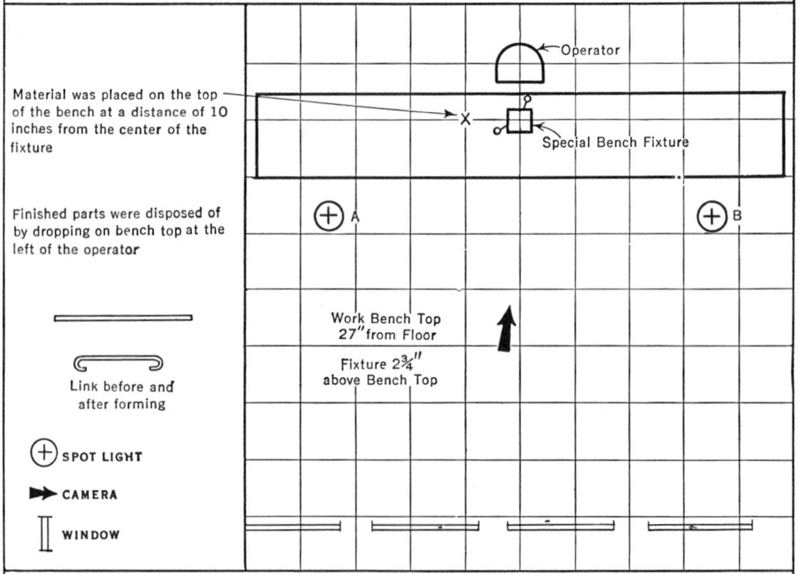

TOOLS, JIGS, GAUGES:— Special fixture operated with both the right and the left hands.

MADE BY

Fig. 96. Motion picture data sheet.

data pertaining to the particular study. This sheet forms a permanent record of the information about the work being filmed, as well as of data pertaining to the mechanics of making the picture. Since the same symbol appears on each frame of the film in a given run, it is always possible to refer to the data sheet in order to identify any piece of film.

The analyst should estimate or measure with a watch the time required for a cycle if he has not already obtained this information from previous time standards or from the production department. There should be plenty of material ahead of the operator, and everything should be in readiness so that there will be no unnecessary interruptions while the pictures are being made. The operator should be allowed to work some time after the lights are turned on before the camera is set in motion. Some operators require time to become accustomed to the new surroundings and to work off nervousness. Most workers present no serious problem on this last score.

The film is then exposed, making pictures of as many cycles of the operation as desired. It is impossible to give rules as to the number of cycles of an operation to be photographed. That depends upon the circumstances surrounding each case, but a sufficient number of cycles should be taken to give a representative record of the job. It is better to make too many pictures of an operation than too few.

Outline of Procedure for Making Motion Pictures

1. Secure the cooperation of the foreman and the operators before attempting to make the motion picture. Explain why the picture is being made.

2. Determine whether electricity is available for the photoflood lamps, microchronometer, and camera if an electric motor-driven camera is to be used.

3. Locate the camera to give the best picture of the cycle of the operation. Use the view finder to ascertain whether the entire cycle is covered.

4. Locate the photoflood lamps to give adequate intensity of illumination without deep shadows. See that the darkest places are properly lighted.

5. Place the microchronometer so that it will be in the picture and in focus. See that it does not obscure any part of the operation.

6. Place the card bearing the film number or other identification in the picture, preferably near the microchronometer.

7. Have sufficient film in the camera for the number of cycles to be photographed.

8. Determine the proper diaphragm opening by means of an exposure meter and adjust the diaphragm setting on the camera.

9. Measure the distance of the subject from the camera lens and adjust the focus setting on the camera to correspond.

10. Fill in the motion picture data sheet.

11. Turn on the lights, start the microchronometer, and make the picture.

CHAPTER 14

Film Analysis

After the motion picture has been made and the film processed, it is placed in the projector and shown on the screen, where it may be examined. Since the film contains an exact record of the activities photographed, a process chart, activity chart, man and machine chart, or an operation chart can be made from the film as well as from the actual activity. If the film is projected at the same speed at which it was made, a time study can also be made from the film. In this chapter, however, an explanation will be made of the method of analyzing the film for simo (simultaneous-motion-cycle) chart construction. Before starting the analysis it is customary for the analyst to run the films through the projector several times in order to familiarize himself with the entire operation. A particular cycle is then selected to be analyzed in detail.

The extent to which the movements of the hands, arms, legs, head, and trunk will be analyzed depends largely upon the nature of the work. Most operations selected for micromotion study analysis involve either benchwork or short-cycle machine work requiring motions of the hands only. It is usually satisfactory to consider the hand as a unit in making the analysis. That is, it is not necessary to analyze the motions of each finger independently. Occasionally, however, an operation will be studied in which several body movements take place. When such detailed analysis is required, it is entirely possible to adapt the technique to this use, although much more time is required when all the members of the body must be considered separately.

In the bolt and washer assembly, which is to be an example, the simplest form of analysis will be used, that is, analysis of hand motions. When the thumb and index finger of the right hand grasp a washer, it will be assumed that the right hand grasps the washer, and so on.

Forms for Recording Motion Analysis Data. As the film is analyzed, the data are transferred to a data sheet, commonly called an analysis sheet. Various forms have been devised for this, and the one used will depend upon the type of work studied and the extent to which the analysis is to be carried. The forms in Figs. 97 and 100 are very satisfactory for right-hand and left-hand analysis. The extra column on the form in Fig. 97 provides space for the analysis of a third member of the body, such as the foot in punch-press work or the knee in knee-controlled sewing machine operation. The analysis sheet in Fig. 98 was designed for use by Macy's Department Store when a complete analysis was to be made.

Film Analysis of the Bolt and Washer Assembly. The enlarged print of one cycle of the film showing the operation "Assemble Three Washers on Bolt" is reproduced in Fig. 99. The pictures were taken at normal speed of 16 exposures per second, and the microchronometer speed was 20 revolutions per minute. The operation is described in detail on page 223. The enlarged print in Fig. 99 will be analyzed in the same manner as if the analyst had the actual film before him. In that case, however, the task would be easier, since he would have the film in a projector and could greatly enlarge it on the screen.

Analysis of the Left-Hand Motions. The motions of the left hand of the operator are usually analyzed first. The film is then run back to the beginning of the cycle and analyzed for the motions of the right hand.

To begin the analysis, the film is run through the projector until the beginning of a cycle is found. This is usually the point where the hand begins its first *transport empty* therblig. Sometimes it is best to begin the analysis at a point where both the right and the left hands begin or end their therbligs together. If the enlarged print of film in Fig. 99 is examined, it will be noted that the microchronometer is located at the left of the operator, the card bearing the film number (B21) appears above the clock, and the material to be assembled is located in small bins directly in front of the operator. The exact arrangement of the material is shown in Fig. 69. The large clock hand makes 20 revolutions per minute; the small hand makes 2. There are 100 equal divisions on the dial of the clock; therefore, time is indicated directly in $\frac{1}{2000}$ of a minute by the large hand. This time interval of $\frac{1}{2000}$ of a minute was called a "wink" by Gilbreth.

The first frame of film in the upper left-hand corner of Fig. 99 shows the operator holding the head of the bolt with her left hand and completing the assembly of the last washer on the bolt with her

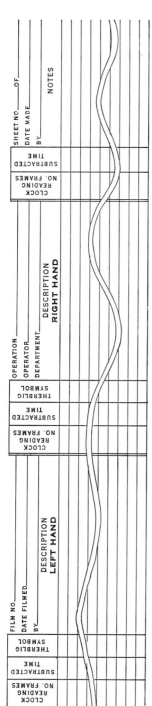

Fig. 97. Micromotion study analysis sheet.

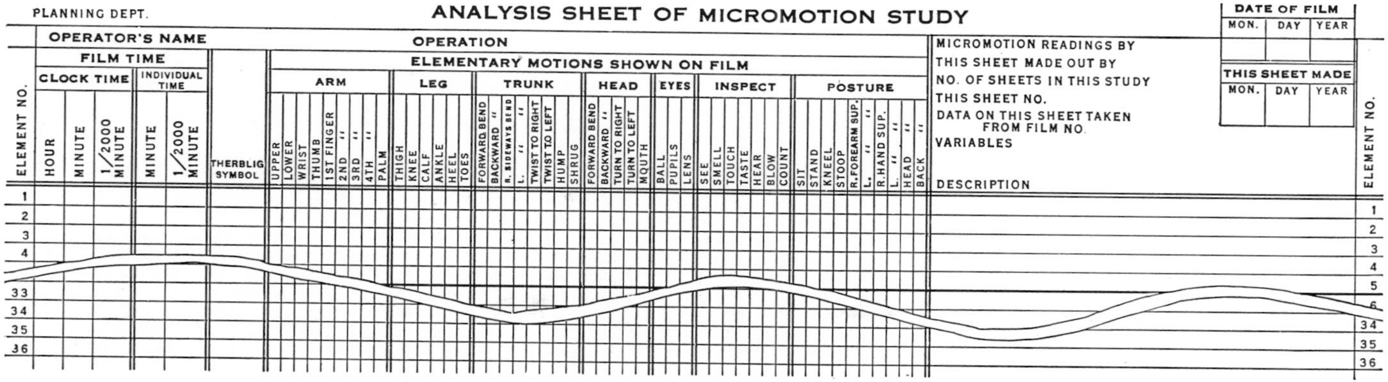

Fig. 98. Form for complete micromotion study analysis.

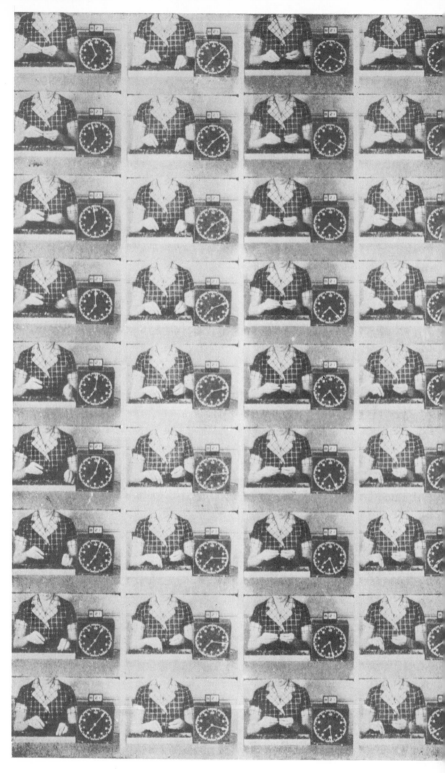

Fig. 99. Print of motion picture film showing one com

le of the bolt and washer assembly—old method.

right hand. The second frame shows the operator in the act of beginning to carry the finished assembly (with her left hand) to the bin nearest the clock, where she will dispose of it. This frame is an excellent place to begin the analysis, as it shows the two hands at the instant they are beginning to separate. The clock reads 595, meaning $^{595}\!/_{2000}$ of a minute from zero.

MICROMOTION STUDY
ANALYSIS SHEET

PART   Bolt and washer assembly - Old Method   DEPARTMENT   AY16   FILM NO.   B21

OPERATION   Assemble 3 washers on bolt   OP. NO.   A32

OPERATOR   M. Smith 1C634   DATE   ANALYSED BY   M.E.R.   SHEET NO.   1   OF   1

CLOCK READING	SUBTRACTED TIME	THERBLIG SYMBOL	DESCRIPTION LEFT HAND	CLOCK READING	SUBTRACTED TIME	THERBLIG SYMBOL	DESCRIPTION RIGHT HAND
595	7	TL	Carries assembly to bin	595	26	TE	Reaches for lock washer
602	2	RL	Releases assembly	621	6	St+G	Selects and grasps washer
604	4	TE	Reaches for bolt	627	7	TL	Carries washer to bolt
608	2	St+G	Selects and grasps bolt	634	6	P	Positions washer
610	17	TL	Carries bolt to working position	640	12	A+RL	Assembles washer onto bolt and releases
627	5	P	Positions bolt	652	8	TE	Reaches for steel washer
632	104	H	Holds bolt	660	8	St+G	Selects and grasps washer
736	7	TL	Carries assembly to bin	668	9	TL	Carries washer to bolt
743	2	RL	Releases assembly	677	3	P	Positions washer
745				680	10	A+RL	Assembles steel washer and releases
				690	6	TE	Reaches for rubber washer
				696	10	St+G	Selects and grasps rubber washer
				706	9	TL	Carries washer to bolt
				715	5	P	Positions washer
				720	16	A+RL	Assembles washer and releases
				736			
		Time in 2000ths of a minute					

Fig. 100. Analysis sheet for bolt and washer assembly—old method.

The motions of the left hand are recorded on the analysis sheet (Fig. 100) in the column "Description Left Hand." The clock reading is recorded in the first column, 595 being the time at which the motion *transport loaded* begins. The symbol for this motion is placed in the third column and the motion is described, "Carries assembly to bin." [1] The film is then examined frame by frame until this therblig (for the left hand) ends. The frame of film showing the left hand in the act

---

[1] The best practice is to let the therblig symbol indicate the action, thus making it unnecessary to include the verb in the description. The description of the first therblig would therefore have been "Assembly to bin" instead of "Carries assembly to bin." The verbs have been included on the analysis sheets and simo charts in this book in order to aid the reader in learning the meanings of the therblig symbols.

of releasing the assembly also shows the clock to read 602. Therefore, 602 is recorded in the second horizontal line and in the first vertical column. Since the operator's left hand is now beginning the therblig *release load*, the symbol for this therblig is placed in the third vertical column, and the description of the therblig, "Releases assembly," is recorded in the fourth column. The analyst now turns to the film and examines it further, looking for the end of the *release load* and for the beginning of the next motion, *transport empty*. The very next frame of film shows the operator's left hand in the act of moving empty to the bin of bolts; consequently the *release load* therblig has ended, and the *transport empty* motion has begun. The clock is read 604, and this reading is recorded in the third horizontal line and in the first vertical column. The symbol for *transport empty* is placed in the third column, and the description of the therblig is noted in the fourth column. In a like manner the film is examined through the entire cycle, the analyst noting where one motion ends and the next one begins and recording the data on the analysis sheet. After the analysis has been made for both hands, the clock readings are subtracted to get the elasped time for each motion. These subtracted times are recorded in the second vertical column.

Analysis of the Right-Hand Motions. After the motions of the left hand have been analyzed, the film is run back to the starting place and the motions made by the right hand are analyzed and recorded on the right side of the analysis sheet in Fig. 100. In the second frame of film in the upper left-hand corner of Fig. 99, the operator's right hand is beginning the motion *transport empty*, the hand moving to the bin of lock washers. Therefore, the clock reading 595 is recorded in the first vertical column, "Clock Reading," for the right hand in Fig. 100. The therblig symbol for *transport empty* is placed in the third vertical column, and the description of the motion is recorded. The film is then studied frame by frame until the point is found where the motion *transport empty* for the right hand ends and the next one begins. This motion is a long one because the hand moves very slowly in order to allow time for the left hand to dispose of the assembly and procure a bolt. It is not until the frame in the middle of the second row of pictures of Fig. 99 that the operator begins to *select* and *grasp* a lock washer from the bin on the bench. The clock is read 621, and the data are recorded on the analysis sheet; the analysis is continued in this manner for the remainder of the cycle. After the analysis has been made for both hands and the subtracted time obtained, it is possible to picture the entire cycle easily and accurately. The left hand is analyzed independently of the right hand except that the cycle must begin and end at approximately the same point for the two hands. It must be remembered that the subtracted time shown in Fig. 100 is in 2000ths of a minute.

As many cycles of the operation may be analyzed as seem necessary. Usually one or two are all that are required if care is used in their selection.

CONSTRUCTION OF SIMULTANEOUS-MOTION-CYCLE CHARTS

The time for each therblig recorded on the analysis sheet may be shown to scale by means of a simultaneous-motion-cycle chart, commonly called a *simo chart*. Either the analysis sheet or the simo chart may be made independently, or the simo chart may be constructed from the data on the analysis sheet.

When a full simo chart showing every moving member of the body is made, it is customary to use a sheet of cross-section paper 22 inches wide with lines ruled 10 to the inch. For operations longer than one-half minute some analysts use paper with lines ruled 10 to the half inch, or paper ruled in millimeters in order to condense the chart. However, since the divisions are closer together, this paper is more difficult to use than the decimal-inch paper. Headings containing information such as that shown at the top of Fig. 98 are often printed in quantities and are pasted across the top of the cross-section paper.

For many operations, however, it is not necessary to construct a complete chart of all the moving members of the body. A simo chart of the two hands for the bolt and washer assembly is shown in Fig. 101. Exactly the same procedure would be used to construct a chart showing the motions of the arms, legs, head, trunk, and other parts of the body.

The vertical scale shown in the center of the chart in Fig. 101 represents time in 2000ths of a minute. The therblig description, symbol, color, and relative position in the cycle all appear on the chart. The time required for each motion is drawn to scale in the vertical column and colored to represent the particular motion. The sheet is arranged much like the analysis sheet. The clock read 595 at the beginning of the motion *transport loaded* for the left hand; therefore this point is located on the vertical scale by a heavy black horizontal line at the top of the column. The first motion, *transport loaded*, required seven winks ($\frac{7}{2000}$ of a minute); therefore, seven divisions are marked off on the vertical column for the left hand and a heavy black line is drawn in. The space above this horizontal black line is then colored

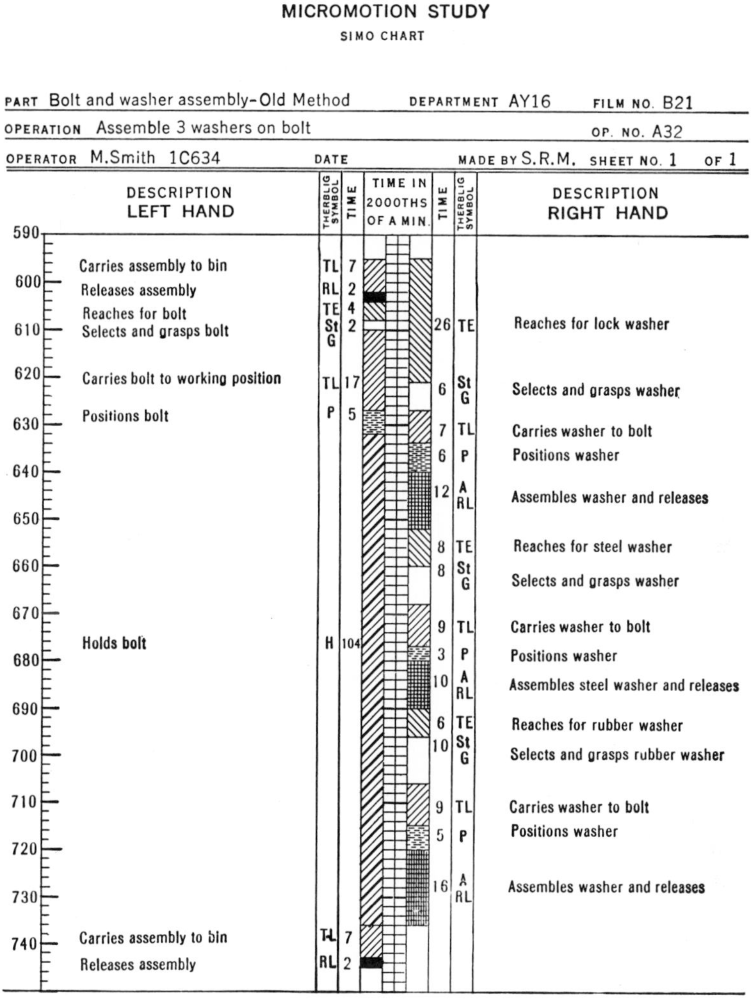

MICROMOTION STUDY
SIMO CHART

PART Bolt and washer assembly–Old Method   DEPARTMENT AY16   FILM NO. B21

OPERATION   Assemble 3 washers on bolt   OP. NO. A32

OPERATOR M.Smith 1C634   DATE   MADE BY S.R.M. SHEET NO. 1   OF 1

DESCRIPTION LEFT HAND	THERBLIG SYMBOL	TIME	TIME IN 2000THS OF A MIN.	TIME	THERBLIG SYMBOL	DESCRIPTION RIGHT HAND
590						
600 Carries assembly to bin	TL	7				
Releases assembly	RL	2				
Reaches for bolt	TE	4				
610 Selects and grasps bolt	St G	2		26	TE	Reaches for lock washer
620 Carries bolt to working position	TL	17		6	St G	Selects and grasps washer
630 Positions bolt	P	5		7	TL	Carries washer to bolt
				6	P	Positions washer
640				12	A RL	Assembles washer and releases
650						
				8	TE	Reaches for steel washer
660				8	St G	Selects and grasps washer
670				9	TL	Carries washer to bolt
680 Holds bolt	H	104		3	P	Positions washer
				10	A RL	Assembles steel washer and releases
690				6	TE	Reaches for rubber washer
700				10	St G	Selects and grasps rubber washer
710				9	TL	Carries washer to bolt
720				5	P	Positions washer
730				16	A RL	Assembles washer and releases
740 Carries assembly to bin	TL	7				
Releases assembly	RL	2				

Fig. 101. Simo chart for bolt and washer assembly—old method.

solidly in green [2] with a pencil, No. 375. The next therblig for the left hand is *release load,* which required two winks. In a similar manner this elapsed time is marked off on the vertical scale immediately below the heavy black line, and another horizontal line is drawn in. The area for this therblig is colored red. And so for the

[2] Since color cannot be reproduced in this book, color symbols are used instead. For standard therblig colors see Fig. 83.

MICROMOTION STUDY

SIMO CHART

PART Bolt and Washer Assembly–Improved Method DEPARTMENT AY16 FILM NO. X75

OPERATION Assemble 3 washers on bolt   OP. NO. A32

OPERATOR M.S. Bowen   1C4327   DATE   MADE BY S.R.M. SHEET NO. 1   OF 1

DESCRIPTION LEFT HAND	THERBLIG SYMBOL	TIME	TIME IN 2000THS OF A MIN.	TIME	THERBLIG SYMBOL	DESCRIPTION RIGHT HAND
Reaches for rubber washer	TE	10		10	TE	Reaches for rubber washer
Selects and grasps washer	St G	1		1	St G	Selects and grasps washer
Slides washer to fixture	TL	13	20	13	TL	Slides washer to fixture
Positions washer in fixture and releases	P RL	14		14	P RL	Positions washer in fixture and releases
Reaches for steel washer	TE	12	40	12	TE	Reaches for steel washer
Selects and grasps washer	St G	1		1	St G	Selects and grasps washer
Slides washer to fixture	TL	17	60	17	TL	Slides washer to fixture
Positions washer in fixture and releases	P RL	13	80	13	P RL	Positions washer in fixture and releases
Reaches for lock washer	TE	12		12	TE	Reaches for lock washer
Selects and grasps washer	St G	1	100	1	St G	Selects and grasps washer
Slides washer to fixture	TL	14		14	TL	Slides washer to fixture
Positions washer in fixture and releases	P RL	8	120	8	P RL	Positions washer in fixture and releases
Reaches for bolt	TE	10		10	TE	Reaches for bolt
Selects and grasps bolt	St G	10		10	St G	Selects and grasps bolt
Carries bolt to fixture	TL	12	140	12	TL	Carries bolt to fixture
Positions bolt	P	8		8	P	Positions bolt
			160			
Inserts bolt through washer	A	48	180	48	A	Inserts bolt through washer
			200			
Withdraws assembly	DA	3		3	DA	Withdraws assembly
Carries assembly to top of chute	TL	10		10	TL	Carries assembly to top of chute
Releases assembly	RL	1	220	1	RL	Releases assembly

(Left margin scale markings: 600, 620, 640, 660, 680, 700, 720, 740, 760, 780, 800, 820)

Fig. 102. Simo chart for bolt and washer assembly—improved method.

remainder of the cycle the motions are marked off to scale and each area is colored with the standard therblig color. The motions made by the right hand are charted on the right-hand side of the sheet in exactly the same way as those for the left hand.

Figure 102 shows the simo chart for one cycle of the *improved method* of bolt and washer assembly as described on page 223.

Analysis of the Link-Forming Operation. The simple operation of bending a "hook" on each end of a short piece of wire to form a link (Fig. 103) for a portable typewriter was the subject of a number of studies. Because this operation involves the use of a fixture and because it is short in length, it will serve as an example.

The camera used for making the motion pictures of this operation was driven by a synchronous motor and operated at a constant speed of 1000 exposures per minute. Therefore, no microchronometer was needed, and the time interval from one frame of film to the next was exactly $\frac{1}{1000}$ of a minute. (See Fig. 105.)

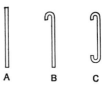

A   B   C

Fig. 103. Link for portable typewriter: *A*, soft steel wire; *B*, link with one end formed; *C*, finished link.

Description of the Link-Forming Operation. The material from which the link was formed consisted of soft steel wire cut from wire stock 0.045 inch in diameter to uniform length of $1\frac{1}{4}$ inches. The material was supplied in metal containers to the operator, who emptied the stock of cut wire on the linoleum-topped bench at the right of the fixture as she needed it. The link was formed in the following manner.

The pictures in Fig. 105 give a reproduction of each element of one

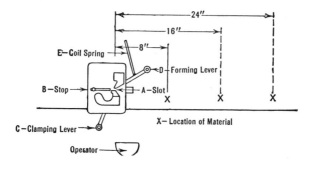

Fig. 104. Fixture and layout of the work place for forming link.

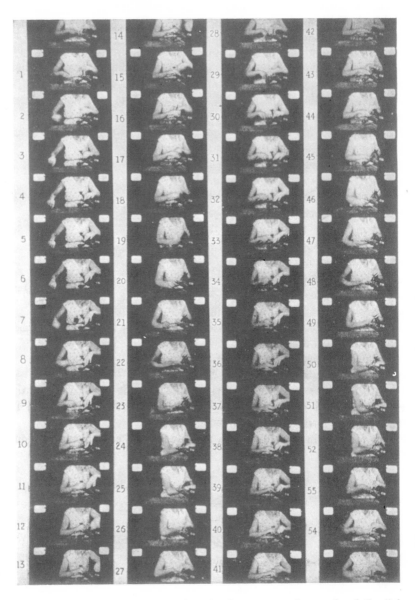

Fig. 105. Print of motion picture film showing one complete cycle of the link-forming operation.

complete cycle. The fixture was mounted securely on the bench so that its top surface was 2¾ inches above the top of the bench. The top of the bench was 27 inches above the floor. The material was spread out over the surface of the bench top so that it could be grasped more easily. The operator, seated behind the bench, picked up one piece of material with the thumb and index finger of her right hand, carried it to the left, and inserted it into the slot A in the fixture (Fig. 104). The operator pressed the piece of material against a stop B in the fixture, and at the same time she clamped the piece into place by moving lever C to the left with her left hand. Then with the right hand she grasped the knob of the forming lever D, which extended to the right of the fixture and was about 3 inches above the top of the bench. The right hand rotated the forming lever in the clockwise direction about the center of the fixture as an axis, through approximately 180 degrees, the radius of rotation being 8 inches. The lever was moved in a plane parallel to the top of the bench. This movement of the lever formed the "hook" on one end of the link. The operator then returned the lever in the counterclockwise direction toward its original position. A coil spring E, fastened to the forming lever and to the bench, assisted the operator in returning the lever to its original position. This spring made it possible for the operator to release the forming lever after she had returned it through about one-half its return travel, the spring pulling the lever back the remainder of the distance.

After the operator had released the knob of the lever, she moved her hand slightly to her right and into a position about 4 inches in front of her, and waited an instant while her left hand removed the half-formed link from the slot in the fixture. Then the two hands together turned the link end for end and placed it back into the slot. Care was taken to ensure that the "hook" was turned in the proper direction so that after the link had been completely formed the two hooks would be on the same side. As the right hand held the link in place in the fixture, the left hand moved the lever C to the left, clamping the link in the slot as in the first part of the cycle. The right hand then grasped the knob of the forming lever and, as before, moved it through 180 degrees in the clockwise direction, forming the second end of the link. While the right hand was forming the end of the link, the left hand continued to hold lever C in its position as far to the left as it would go, clamping the link in the fixture while it was being formed. After the link was formed, the right hand returned the forming lever toward its original position, releasing the knob of this lever directly in front of the operator. She then moved her hand to her right to pick up a piece of material from the bench for the beginning of the next cycle. In the meantime the left hand released the knob of lever C and reached forward to remove the fin-ished formed link from the slot in the fixture. The left hand then carried the link to the left, where it was dropped on top of the bench. During this time the operator was looking to her right, where the right hand was grasping a piece of material from the top of the bench for the next cycle.

Figure 106 shows the simo chart for the link-forming operation.

Complete Analysis of Hand Motions. Confusion sometimes occurs when making a left- and right-hand motion analysis because some members of the arm are performing certain motions while other mem-bers are performing other motions. The first motion on the simo chart in Fig. 107, for example, shows the thumb and the first and second fingers of the right hand performing a grasp while the palm and the third and fourth fingers of the same hand are holding a smooth piece of bone. The operation is folding and creasing sheets of paper by the improved method described on page 242. The right hand carries the bone at all times, although it is used during only a small part of the cycle.

When a complete analysis of an operation is made, each member of the arm and hand is analyzed separately—the upper arm, lower arm, wrist, first finger, second finger, and so forth. The film is run back to the starting place after the analysis of each member, and a separate vertical column on the simo chart is required for recording the motions of each. Although Fig. 107 does not show the movements of the head, trunk, and legs, it does show all motions of the arms, hands, and fingers.

Had a simple left- and right-hand simo chart been made of the paper-folding operation, it would have shown only those therbligs performed by the thumb and the first and second fingers. A note would have been included on the chart to show that the bone was carried in the right hand throughout the entire cycle.

Using the Simo Chart. When the simo chart of the operation has been made, the task of finding a better way of doing the work has just begun. A thorough study of the chart is ordinarily the first step in this task.

The simo chart aids one in grasping a picture of the complete cycle in all its details, and assists in working out better combinations of the most desirable motions. The simo chart of Fig. 101 shows in a

MICROMOTION STUDY
SIMO CHART

PART Link for typewriter .... DEPARTMENT 9 .... FILM NO. C18

OPERATION Form link for typewriter .... OP. NO. G11

OPERATOR A.S.Sanders   A2 .... DATE .... MADE BY S.A.R. SHEET NO. 1 OF 1

DESCRIPTION LEFT HAND	THERBLIG SYMBOL	TIME	TIME IN 1000THS OF A MIN.	TIME	THERBLIG SYMBOL	DESCRIPTION RIGHT HAND
Returns clamping lever and releases it	TL RL	2				
Moves hand to fixture	TE	3		4	TE	Reaches for material
				2	St G	Selects and grasps one piece
Grasps formed link in fixture	G	6				
				6	TL	Carries piece to fixture
Carries formed link to left and releases it	TL RL	3				
Moves to clamping lever and grasps knob	TE G	3		6	P RL	Inserts piece in fixture and releases it
Moves lever to extreme left	TL	3		2	TE G	Reaches for forming lever and grasps knob
Holds lever in this position	H	6		5	U	Forms 1st end of link
Returns lever to original position and releases it	TL RL	3		3	TL RL	Returns forming lever and releases it
Moves hand to fixture	TE	3		4	TE UD	Moves hand toward fixture and waits for left hand
Grasps piece, turns it end for end in fixture and releases it	G P RL	8		4	TE	Moves hand to fixture
Moves to clamping lever and grasps knob	TE G	3		8	P	Assists left hand in turning piece end for end in fixture
Moves lever to extreme left	TL	3		2	TE G	Reaches for forming lever and grasps knob
Holds lever in this position	H	8		5	U	Forms 2nd end of link
Returns clamping lever and releases it				3	TL RL	Returns forming lever and releases it

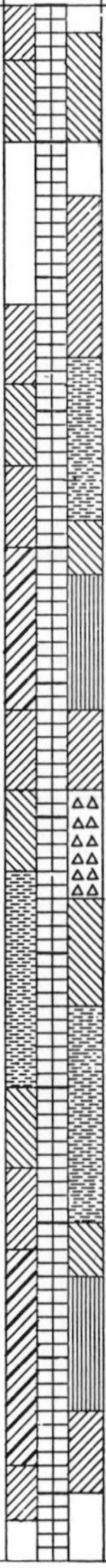

Fig. 106. Simo chart of link-forming operation.

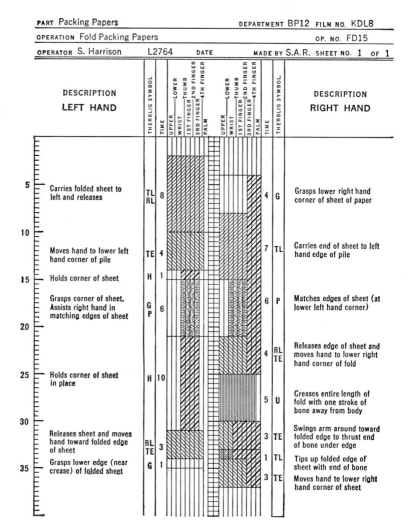

Fig. 107. Simo chart of folding paper.

very clear way that the left hand is used during most of the cycle for holding the bolt. This at once suggests that some mechanical device be substituted which will permit the left hand as well as the right to do more useful work.

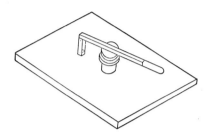

Fig. 108. Fixture used for pressing cheese in cup—old method.

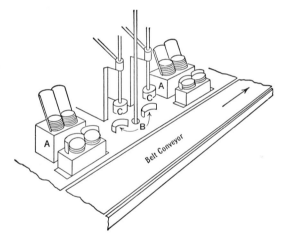

Fig. 109. Fixture used for pressing cheese in cups—improved method: *A*, supply of tinfoil liners, covers, etc.; *B*, nests for positioning cups under plungers; *C*, plungers (operated by foot) for pressing cheese into cups.

It is often found that the sequence of motions in one kind of work may be used in other kinds, or a particularly good sequence in one operation may suggest a more efficient sequence in another. The simo chart shows very distinctly where delays occur in the cycle, and it aids in finding an effective way of eliminating these delays.

The next five chapters give some methods of attack that may be useful in this problem of improving the way of doing a given task.

MICROMOTION STUDY
SIMO CHART

PART: ½ lb. and 1 lb. Cheddar Cheese   FILM NO.: S1486

OPERATION: Crimp Foil Liner over Top, Press Cheese in Cup,   OP. NO.:

Code Date, Move to Wrapper   DATE:

OPERATOR: M. Sanderson   MADE BY: M.G.S.   SHEET NO.: 1 of 1

NO. UNITS PER CYCLE: 2-Cup Cycle   ORIGINAL METHOD: X   IMPROVED METHOD:

FOOT	LEFT HAND	1000THS OF A MINUTE		RIGHT HAND
	T.E. to cup and aid right hand	10	10	T.E. to cup and move to position
	Crimp foil liner over cheese	28	26	Crimp foil liner over cheese
		— 25 —		
	Aid right hand	5	7	Aside cup
	T.E. to cup and grasp	8	17	Hold lever up
	Place cup in position to press	15	— 50 —	
			17	Lower lever pressing cheese to cup, raise
	Hold cup	11	— 75 —	
	Aside cup to table	7	7	Hold lever up
	T.E. pick up inserts, hold while right hand places, aside (6 at a time)	18	18	T.E. to instruction inserts in left hand, grasp and place on top of cheese in cup
		— 100 —		
	Aid place cover	13	32	Place cover and T.E. for next cover
	Shove cup aside	5		
—	Move next cup in position	14	— 125 —	
	Turn cups upside down (1 in each hand)	10	10	Same as left
	T.E. for stamp pad, hold while right stamps, aside pad	13	— 150 — 13	T.E. for stamp-stamp cover-aside stamp- avg. 12 per time
	Shove to wrapper (1 each hand)	5	5	Same as left
	T.E. to cup and aid right hand	10	10	T.E. to cup and move to position
	Crimp foil liner over cheese	28	— 175 — 26	Crimp foil liner over cheese
	Aid right hand	5	— 200 — 7	T.L. and R.L. cup
	T.E. to cup and grasp	8	17	Hold lever up
	T.L. cup in position to press	15		
		— 225 —	17	Lower lever pressing cheese in cup, raise
	Hold cup	11		
	T.L. cup to table	7	7	Hold lever up
	T.E. pick up inserts, hold while right hand places, aside (6 at a time)	18	— 250 — 18	T.E. to instruction inserts in left hand, grasp and place on top of cheese in cup
	Aid place cover	13	32	Place cover and T.E. for next cover
		— 275 —		
	Shove cup aside	5		
	Move next cup in position	14		
	Turn cups upside down (1 in each hand)	10	— 300 — 10	Same as left
	Reach for stamp pad, hold while right stamps, aside pad	13	13	T.E. for stamp-stamp cover-aside stamp- avg. 12 per time
	Shove to wrapper (1 each hand)	5	**324** 5	Same as left
	NOTE: In order to compare Old and New Methods a one-cup cycle of Old Method has been shown twice.			————Indicates that element probably could be completely eliminated. - - - -Indicates element probably could be reduced.

Fig. 110. Modified simo chart of one operation in the process of packing Cheddar cheese in paper cups by the old method. Size of chart, 8½ × 17 inches.

MICROMOTION STUDY
SIMO CHART

PART: ½ lb. and 1 lb. Cheddar Cheese	**FILM NO.:** S1347
OPERATION: Crimp Foil Liner over Top, Press Cheese in Cup,	**OP. NO.:** c-27
Code Date, Move to Wrapper	**DATE:**
OPERATOR: M. Sanderson **MADE BY:** M.G.S.	**SHEET NO.** 1 OF 1
NO. UNITS PER CYCLE: 2 Cup Cycle **ORIGINAL METHOD:**	**IMPROVED METHOD:** X

FOOT	LEFT HAND	1000THS OF A MINUTE		RIGHT HAND
	T.E. and G. cup	7	8	T.E. and G. cup
	T.L. and R.L. cup	9	10	T.L. and R.L. cup
	Fold over tinfoil liner on first cup	28 —25	26	Aid left hand to fold over tinfoil liner on first cup
	R.L. and T.E.	6 —50	6	Slide forward into depression
	G. and T.L. to position	6	6	T.E. to next cup
	Fold over tinfoil liner on second cup	28 —75	28	Aid left hand to fold over tinfoil liner on second cup
	Grasp cup	5	4	T.E. and G. first cup
	Slide into position to press	9	10	Idle
	Hold while press	7 —100	7	Hold while press
	T.E. to insert	4	5	Idle
	Grasp insert	22	7	T.E. to insert
		—125	14	Grasp insert
	T.L. to cover	5	8	T.L. to cover
	Position insert and G. cover	7	8	Position insert and G. cover
	T.L. to cup	10 —150	6	T.L. to cup
	Position	6	6	Position
	Slide out 3″-4″	4	4	Slide out 3″-4″
	PRESS COVER DOWN	3	3	PRESS COVER DOWN
	Grasp	4	4	Grasp
	Turn over and R.L.	11 —175	11	Turn over and R.L.
			3	T.E. TO STAMP
	Idle	29	5	G. and T.L. stamp to cup
			13	Stamp 2 cups
		—200		
		—210	8	T.L. and R.L. stamp
		—225		
		—250		Net reduction 35%
		—275		
		—300		
		—324		Last 40 frames of this operation could be eliminated by stamping from bottom at the time of pressing.

Fig. 111. Modified simo chart of one operation in the process of packing Cheddar cheese in paper cups by the improved method. Size of chart, 8½ × 17 inches.

Possibility Charts. After the suggestions for the improvement of the method have been secured and when worthwhile changes seem practicable, a possibility chart may be constructed. This is a simultaneous-motion-cycle chart showing the proposed method. A competent analyst will be able to draw up such a chart listing the necessary motions in order and indicating the time for each. Possibility charts can be made with a surprising degree of accuracy by one trained in the micromotion study technique and experienced in making them.

Modified Simo Chart. Some organizations find it satisfactory to list the motions and the motion times on the simo chart, plotting the motion time values on a graduated scale, as shown in Figs. 110 and 111. The colored therblig identifications are not used. Moreover, such charts may be prepared with a typewriter, using hectograph carbon paper on forms printed with hectograph ink. With such a master, a number of copies can be made. The simo charts for one operation in the process of packing Cheddar cheese in paper cups, shown in Figs. 110 and 111, were made in this manner.

In this case the company had several branch plants packing cheese, and the charts together with films of the original method and of the improved method were sent to each plant so that the plants could benefit from this work.

Original Method of Packing Cheese in Cups. The original method for "Crimp Foil Liner over Top, Press Cheese in Cup, Code Date, and Move to Wrapper" was as follows. The operator, standing at a table, picked up a cup from the table at her extreme left, placed it on the table directly in front of her, folded the tinfoil liner down on the cheese, placed it in a hand fixture (Fig. 108), and lowered the lever which pressed the cheese down in the cup. Then the cup was removed from the fixture, an insert and a cover were placed on top of the cup, and the cover was pressed down. The cup was then turned over and a code date stamped on the bottom, using a hand stamp and an ink pad.

In investigating operations that seem to lend themselves to the use of duplicate fixtures, it is often desirable to plot on the simo chart the motions used in producing two units, as it is easier to compare the old with the improved method. This was done in Fig. 110.

Having this new fixture in mind, the engineer drew heavy solid vertical lines beside the motions that he thought could be eliminated entirely, and heavy dotted vertical lines beside the motions that he thought probably could be reduced in time. This procedure gives a fairly definite idea of the potential savings that can be expected after the proposed fixture has been built and put into use.

Improved Method of Packing Cheese in Cups. The improved method (Fig. 109), after it was worked out and put into effect in the factory, was filmed and a simo chart made of it (Fig. 111). In the new method the left and right hands each picked up from the conveyor belt a cup filled with cheese and placed the two cups on the front edge of the table. Both hands then folded the tinfoil liner down on the first cup and placed it in a nest in the fixture. The foil was then folded down on the second cup, which was placed in the other nest in the fixture. The two hands held the two cups while the foot operated the two plungers which came down vertically, pressing the cheese into the cups. The operator then grasped an insert and then a cover in each hand and placed them on the cups. The cups were removed a few inches from the fixtures, the covers were pressed into place, and the cups were turned over and the code date stamped on the bottom of each.

The two simo charts, when placed side by side, make it easy to visualize the changes that were made in the method. Originally 0.324 minute was required for two cups, whereas the improved method requires only 0.210 minute, a reduction in time of 35%.

CHAPTER 15

The Use of Fundamental Hand Motions

Although the definition of each therblig was given in Chapter 11, further explanation is needed in certain cases. Also, since each motion requires, for its performance, time and energy on the part of the worker, the elimination of motions or the better arrangement of such as are indispensable constitutes part of the regular technique of improving methods of work. Information that will aid in making better use of the therbligs is included in this chapter, and a check list follows the discussion of each therblig.

Select. The time for select is frequently so short in duration that it is impossible to measure it with the camera at ordinary speeds. When this is the case it is advisable to combine it with either the preceding or the following motion. Since select usually precedes grasp, it is good practice to combine these two. The symbols for both motions should be included on the analysis sheet, and the color for the more important or the predominating motion should be used in making the simo chart. Usually it will be the motion other than select.

Color can be seen more quickly than shape; therefore color should be used to aid in selecting or sorting whenever possible. For example, in sorting photographic snapshot prints into batches after the printing, developing, and drying operations, it was found desirable to use inks of different colors for stamping identification numbers on the back of the paper before printing. Sorting these prints by color of the ink is much easier and faster than sorting by key letters or figures.

Also, painting a tool the same color as the place where it is to be kept in the drawer saves time when putting it away and finding it the next time.[1]

[1] F. B. Gilbreth, *Motion Study,* Van Nostrand Co., Princeton, N. J., 1911, p. 47.

Check List for Select

1. Is the layout such as to eliminate searching for articles?
2. Can tools and materials be standardized?
3. Are parts and materials properly labeled?
4. Can better arrangements be made to facilitate or eliminate select
   —such as a bin with a long lip, a tray that pre-positions parts,
   and a transparent container?
5. Are common parts interchangeable?
6. Are parts and materials mixed?
7. Is the lighting satisfactory?
8. Can parts be pre-positioned during preceding operation?
9. Can color be used to facilitate selecting parts?

Grasp. There are two main types of grasp: (1) *pressure grasp*,
as in grasping a pencil lying flat on a table top; and (2) *full-hook
grasp*, as in grasping a pencil lying on a table top with one end raised
an inch or so, so that the thumb and fingers are able to grasp by
reaching around it (hook) instead of grasping by pinching.[2]

An investigation [3] of grasping small pieces of wire used in making
a link for a portable typewriter (page 179) showed that it required
twice as long to grasp a piece using a pressure grasp as it did using
a full-hook grasp. The same investigation revealed that the time for
the grasp was not greatly affected by the distance through which the
hand moved in either the motion preceding or the one following the
grasp, other conditions being constant.

The results of a study [4] of the time required to grasp washers from
a flat surface using a hook grasp and a pressure (pinch) grasp are
given in Table 6. The operator merely picked up a washer from one
flat surface, carried it through a distance of 5 inches, and disposed
of it onto another flat surface. The time for the motions grasp, transport loaded and position, and release load and transport empty was
measured very accurately. Circular washers ½ inch in diameter, with

[2] For a classification of "get" see Fig. 269.
[3] Ralph M. Barnes, "An Investigation of Some Hand Motions Used in Factory
Work," *University of Iowa Studies in Engineering, Bulletin* 6, p. 29. For results
of other similar motion study investigations see Ralph M. Barnes and M. E.
Mundel, "Studies of Hand Motions and Rhythm Appearing in Factory Work,"
Bulletin 12; "A Study of Hand Motions Used in Small Assembly Work," *Bulletin*
16; "A Study of Simultaneous Symmetrical Hand Motions," *Bulletin* 17; and
Ralph M. Barnes, M. E. Mundel, and John M. MacKenzie, "Studies of One- and
Two-Handed Work," *Bulletin* 21.
[4] *University of Iowa Studies in Engineering, Bulletin* 16, p. 10.

Table 6. Time Required to Grasp, Carry, and Dispose of Washers from a Flat Surface, Using a Hook Grasp and a Pressure Grasp

Thickness in Inches	1/32		1/8		1/4		1/2	
	Hook	Pressure	Hook	Pressure	Hook	Pressure	Hook	Pressure
Time in Minutes	0.01527	0.01960	0.01524	0.01590	0.01630	0.01450	0.01750	0.01428
Time in Per Cent (Shortest Time = 100%)	100	138	100	112	107	102	115	100

a $\frac{1}{8}$-inch hole in the center, and $\frac{1}{32}$, $\frac{1}{8}$, $\frac{1}{4}$, and $\frac{1}{2}$ inch thick were used.

The time for grasp using the hook grasp tended to increase slightly as the washer thickness increased, whereas the time for grasp using the pressure grasp decreased markedly as the washer thickness increased. The time for grasping the thinnest washer ($\frac{1}{32}$ inch thick) using a pressure grasp was 297% greater than that for grasping the thickest washer ($\frac{1}{2}$ inch thick).

It is usually quicker and easier to transport small objects by sliding than by carrying. In grasping a small object such as a coin or a washer preparatory to transporting it by sliding, the grasp consists merely of touching the ball of the index finger to the top surface of the object; whereas in grasping the same object preparatory to transporting it by carrying, the grasp consists of closing the thumb and index finger around the piece. A study [5] showed that the grasp preceding the slide required as little as $\frac{1}{30}$ as long as the grasp preceding the carry.

Check List for Grasp

1. Is it possible to grasp more than one object at a time?
2. Can objects be slid instead of carried?
3. Will a lip on the front of the bin simplify grasp of small parts?
4. Can tools or parts be pre-positioned for easy grasp?

[5] *University of Iowa Studies in Engineering, Bulletin* 6, p. 32.

5. Can a special screwdriver, socket wrench, or combination tool be used?
6. Can a vacuum, magnet, rubber finger tip, or other device be used to advantage?
7. Is the article transferred from one hand to another?
8. Does the design of the jig or fixture permit an easy grasp in removing the part?

Transport Empty and Transport Loaded. Investigations show (1) that it requires a greater period of time to move the hand through a long distance than through a short distance, other conditions being constant; (2) that the average velocity of the hand is greater for long motions than for short ones; and (3) that in such motions as transport empty and transport loaded an experienced operator moves the hand through almost identically the same path in going from one point to another in consecutive cycles of a repetitive operation. The particular study relating to this last point was made by projecting the film, one frame at a time, on a sheet of paper and marking the position of the tip of the index finger. Connecting these points by pencil lines gave the path of the motion in two dimensions. By placing the camera perpendicular to the path of motion it was possible to secure a close approach to a true record of the motion path.

A movement of the hand, such as transport empty or transport loaded, is ordinarily composed of three phases: (1) starting from a still position the hand accelerates until it reaches a maximum velocity; (2) it then proceeds at a uniform velocity; and (3) finally it slows down until it comes to a dead stop. If the hand changes direction and returns over the same path, as in making a mark back and forth across a sheet of paper, there will be an appreciable length of time at the end of the stroke during which the hand is motionless, that is, while it is changing direction.[6]

For example, in a simple hand motion 10 inches in length, one study [7] showed the distribution of these events to be as follows: 38% of the cycle time for acceleration, 18% for movement at uniform velocity, 27% for retardation, and 17% for stop and change direction (see Fig. 140).

The time required to move the hand is affected by the nature of the motions which precede and which follow the transport. For example, when a delicate or fragile object is transported and placed carefully

[6] *Ibid.,* pp. 37–51.
[7] *Ibid.,* p. 48.

in a small receptacle, the time for the transport will be longer than when the transport is followed by an ordinary disposal such as tossing a bolt into a box. The manner in which an object is grasped and the way that it must be carried and positioned may also affect the time for the transport.[8]

The path of the hand in reaching for a small washer is shown in Figs. 112 and 113. The time to move the hand from A to B is less than that required to move the same distance from C to D because of the change in direction of the hand in the latter case. Barriers and ob-

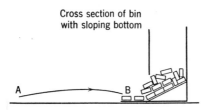

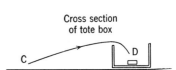

Fig. 112. Straight-line motion of the hand. The hand moves from A to B to pick up a washer which rests on the lip of the bin. The cross section of the bin shows sloping bottom which feeds washers forward.

Fig. 113. A longer hand motion is required because the hand must reach down into box to get washer. The side of the box forms a barrier, making it necessary for the hand to change direction in going from C to D.

structions that retard free hand motion or that require a change in direction should be eliminated whenever possible.

Gauging Hard-Rubber Washers. Mention has already been made of the fact that it is usually quicker and easier to transport small objects by sliding than by carrying. That a "grasp and slide" is definitely faster than a "grasp and carry" apparently results from the shorter grasp rather than from a saving in time in the transportation.

The inspection of small hard-rubber washers for thickness is another illustration of the use of a sliding transport. The purpose of this operation is to reject all washers that are too thick or too thin, as well as those having burrs on the edges. The washers have the following dimensions: outside diameter 0.280 ± 0.002 inch, inside diameter 0.188 ± 0.002 inch, and thickness 0.085 ± 0.005 inch. The gauge used for this operation (Fig. 114) was developed by W. R. Mullee while at the American Hard Rubber Company.

[8] *University of Iowa Studies in Engineering, Bulletin* 16, p. 20.

The metal bar A forms a "go" gauge and the bar B a "no-go" gauge with the base C, which is a heavy metal plate set at an angle with the bench top. The washers to be inspected are drawn from the hopper D by hand into the upper section of the inclined go gauge. Those washers that do not slide underneath the bar A are too thick

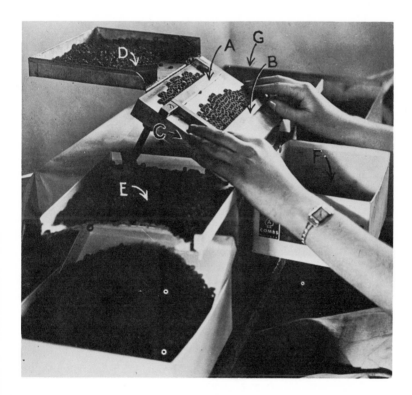

Fig. 114. Special gauge for inspecting hard-rubber washers for thickness: A, go gauge; B, no-go gauge; C, base plate of gauge; D, supply of washers; E, rejected washers (oversize); F, rejected washers (undersize); G, good washers.

and are slid in multiple to the chute E at the left of the gauge. The pieces that go through the gauge A drop down into the middle compartment. If they are too small they slide under the gauge B and drop into the box F directly in front of the operator. Washers that are the correct size are slid off into the chute G at the right.

All movements of the washers in this operation are sliding transports. The washers are not picked up at any place in the cycle. They are not handled individually but are shuffled back and forth in groups across the metal plate and against the bar gauges so that gravity is able to act as the force which tends to pull them through the gauge. The height and angle at which the gauge is mounted above the bench are such as to make the task as easy and comfortable as possible. With this arrangement one operator inspects 30,000 washers per day.

Effect of Eye Movements on Transport Time. In any activity where the eyes must direct the hands, the eye movements and eye fixations often control the operation. In such work it is necessary to study the relationship of the eye movements to the hand motions.

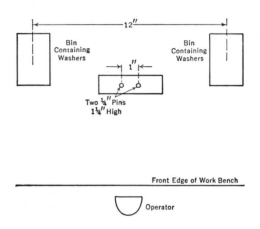

Fig. 115. Layout of work place for the assembly of washers onto pins.

A study [9] was made to obtain information bearing on the question "How are eye and hand movements coordinated when simultaneous symmetrical motions are performed?" A simple operation of picking up a washer in each hand and placing them, bright side up, on two vertical pins in the center of the work place (Fig. 115) was used for the investigation. Nine different operators did the job, and careful records were made of eye and hand movements by means of an eye-movement camera. Figure 116 is an eye-hand simo chart of one cycle of the operation. The results of this study showed that, in reaching for the washers in the two bins, the eyes went first to the right bin, then to the left bin, and finally to the center pins. In most cases the eyes led the hands to the bins and also to the pins. In this operation it seemed to make no difference whether the eyes were focused on the

---

[9] Study made by Dr. D. U. Greenwald in the University of Iowa Industrial Engineering Laboratory.

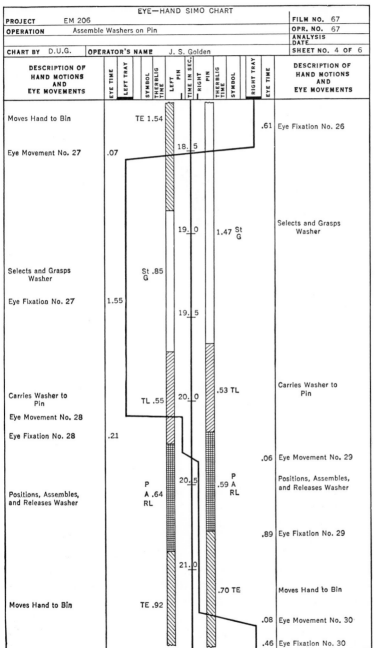

Fig. 116. Eye-hand simo chart for one cycle of an assembly operation.

right pin or the left pin. Apparently the eyes could direct the hands equally well when focused on either pin.

Check List for Transport Empty and Transport Loaded

1. Can either of these motions be eliminated entirely?
2. Is the distance traveled the best one?
3. Are the proper means used—hand, tweezers, conveyors, etc.?
4. Are the correct members (and muscles) of the body used—fingers, forearm, shoulder, etc.?
5. Can a chute or conveyor be used?
6. Can "transports" be effected more satisfactorily in larger units?
7. Can transport be performed with foot-operated devices?
8. Is transport slowed up because of a delicate position following it?
9. Can transports be eliminated by providing additional small tools and locating them near the point of use?
10. Are parts that are used most frequently located near the point of use?
11. Are proper trays or bins used, and is the operation laid out correctly?
12. Are the preceding and following operations properly related to this one?
13. Is it possible to eliminate abrupt changes in direction? Can barriers be eliminated?
14. For the weight of material moved, is the fastest member of the body used?
15. Are there any body movements that can be eliminated?
16. Can arm movements be made simultaneously, symmetrically, and in opposite directions?
17. Can the object be slid instead of carried?
18. Are the eye movements properly coordinated with the hand motions?

Hold. Hold is a therblig that frequently occurs in assembly work and in hand-manipulated machine operations. It is one of the easiest therbligs to eliminate, and disposing of it often leads to substantial increases in output. The elimination of the hold therblig in the bolt and washer assembly (page 223), for example, was largely responsible for the 50% increase in output.

The hand should not be used for a "vise"—a mechanical device of some kind is usually much more economical for holding. In fact, when one hand is used for holding, the operator has reduced his capacity for productive handwork by 50%. Although not every hold therblig can be eliminated, certainly all such therbligs in a cycle are vulnerable points of attack for improving the method.

Check List for Hold

1. Can a vise, clamp, clip, vacuum, hook, rack, fixture, or other mechanical device be used?
2. Can an adhesive or friction be used?
3. Can a stop be used to eliminate hold?
4. When hold cannot be eliminated, can arm rests be provided?

Release Load. Although release load is often of very short duration, it should always be included in the analysis. In the operation "Assemble Three Washers on Bolt" (page 223) the operator released the washer after having assembled it onto the bolt. This motion required such a short time that it could not be measured with the camera at ordinary speeds, and consequently this motion was combined with the preceding one, as shown in Fig. 101.

Release load should be short. If it is of long duration some change should be made in the operation to shorten it. The discussion of drop delivery on page 271 suggests some possible changes.

Check List for Release Load

1. Can this motion be eliminated?
2. Can a drop delivery be used?
3. Can the release be made in transit?
4. Is a careful release load necessary? Can this be avoided?
5. Can an ejector (mechanical, air, gravity) be used?
6. Are the material bins of proper design?
7. At the end of the release load, is the hand or the transportation means in the most advantageous position for the next motion?
8. Can a conveyor be used?

Position and Pre-position. The difference between position and pre-position may be illustrated by the simple operation of picking up a pen, writing, and returning it to its holder.[10] The motions involved in this operation are shown in Table 7.

After the pen is carried to the paper it is necessary to *position* it, that is, to bring the pen down on the sheet of paper at the correct

---

[10] This refers to the usual form of desk-set pen.

Table 7. Motions Used in Writing

Steps Used in Writing	Name of Motion		Time in Thousandths of a Minute
1. Reaches for pen	Transport empty	TE	10
2. Grasps pen	Grasp	G	3
3. Carries pen to paper	Transport loaded	TL	8
4. Positions pen for writing	Position	P	3
5. Writes	Use	U	44
6. Returns pen to holder	Transport loaded	TL	9
7. Inserts pen in holder	Pre-position	PP	6
8. Lets go of pen	Release	RL	1
9. Moves hand to paper	Transport empty	TE	9

place on the line to begin writing. This is a *position* motion. The writing completed, the pen is returned to the holder. The motion *transport loaded* is followed by *pre-position* (rather than by position), because the pen rests in the holder in such a way that it can be grasped in the position in which it will be used the next time. Had the pen been placed in a horizontal pen holder on the desk top, the motion sequence would then have been *transport loaded* and *position* (rather than pre-position), because the pen would have been resting in such a way that it could not have been grasped in the correct position for using. However, had the pen merely been dropped on the desk top, the motion sequence would have been *transport loaded* and *release load,* since no positioning or pre-positioning would have occurred.

Positioning Pins in Bushings with Beveled Holes. Beveled holes in bushings, funnel-shaped openings in fixtures, and bullet-nosed pins all tend to reduce positioning time.

The results of a study [11] of the time required to position and insert pins in bushings with beveled holes are given in Table 8. The operation consisted of grasping a quarter-inch brass pin 1¼ inches long from a magazine, carrying it through a distance of 5 inches, position-

---

[11] *University of Iowa Studies in Engineering, Bulletin* 12, p. 19.

ing and inserting it in the hole in the bushing, withdrawing the pin, and disposing of it in a tray on the table top. The time for the motions transport loaded, position, and assemble and disassemble was accurately measured.

The study was conducted in two parts, one where the clearance between the pin and the hole in the bushing was 0.002 inch, and the other where the clearance was 0.010 inch.

Least time was required to position the pin in the bushing with the 45-degree bevel, (1) in Table 8. Seventy-three per cent more time was required to position the pin in the bushing with no bevel (5) when the clearance was 0.002 inch.

Check List for Position

1. Is positioning necessary?
2. Can tolerances be increased?
3. Can square edges be eliminated?
4. Can a guide, funnel, bushing, gauge, stop, swinging bracket, locating pin, spring, drift, recess, key, pilot on screw, or chamfer be used?
5. Can arm rests be used to steady the hands and reduce the positioning time?
6. Has the object been grasped for easiest positioning?
7. Can a foot-operated collet be used?

Table 8. Time Required to Position Pins in Bushings with Beveled Holes

		1		2		3		4		5
	Clearance Between the Pin and the Hole in the Bushing in Inches									
	0.002	0.010	**0.002**	0.010	**0.002**	0.010	**0.002**	0.010	**0.002**	0.010
Time in Minutes	**0.0047**	0.0027	**0.0054**	0.0029	**0.0048**	0.0039	**0.0049**	0.0038	**0.0081**	0.0067
Time in Per Cent (Shortest Time = 100%)	**100**	100	**115**	107	**102**	144	**104**	141	**172**	248

Check List for Pre-position [12]

1. Can the object be pre-positioned in transit?
2. Can tool be balanced so as to keep handle in upright position?
3. Can a holding device be made to keep tool handle in proper position?
4. Can tools be suspended?
5. Can tools be stored in proper location to work?
6. Can a guide be used?
7. Can design of article be made so that all sides are alike?
8. Can a magazine feed be used?
9. Can a stacking device be used?
10. Can a rotating fixture be used?

Inspect. In inspection work [13] the time for the therblig inspect is usually proportional to the reaction time of the individual and the type of the stimulus used. Only an individual with fast reaction time should be employed on inspection operations. Good eyesight is a second essential requirement for success on this kind of work.

As to the type of stimulus, the data in Table 9 show that, other conditions being equal, a person reacts more quickly to sound than to light, the time being 0.185 second for the former and 0.225 second for the latter. Reaction to touch is the quickest of all, being 0.175 second.[14]

Inspection of Printed Labels. The manufacturer of pharmaceutical products has the important task of making certain that the proper label is applied to the bottle or container of the products he produces. For most products the bottles are filled, capped, labeled, and inserted in cartons on automatic machines. Some labels are printed in two or three different colors, and each label must be correct and complete in every respect. Since labels are printed on offset presses and there must be a separate run for each color, there is the possibility that the press may misprint a sheet and consequently produce a label that is imperfect. In order to make certain that all labels are acceptable, they are individually inspected.

[12] Pre-position is discussed on page 297.

[13] See also pages 277 to 283.

[14] A slight difference in reaction time results from different attitudes of mind on the part of the operator. For example, if the operator's mind is concentrated primarily on the *stimulus,* the reaction times are likely to be a little slower than those indicated. However, if his attention is directed primarily to the *muscular sensations* involved in reacting, the reactions will be a little faster.

Table 9. Average Speed of Reaction

(Thousandths of a second)

Type of Stimulus	Reaction Time
Simple reaction—visual stimulus. Subject was instructed to press telegraph key as quickly as possible after light flashed.	225
Simple reaction—auditory stimulus. Subject was instructed to press telegraph key as quickly as possible after electric buzzer sounded.	185
Simple reaction—touch stimulus. Subject was instructed to press telegraph key as quickly as possible after feeling bar touch hand.	175
Simple reaction—electric shock stimulus. Subject was instructed to press telegraph key as quickly as possible after receiving electric shock on hand.	140
Choice reaction—visual stimulus. Subject could react to two lights. If the right light flashed, the subject pressed the right key. If the left light flashed, the subject pressed the left key.	325
Timed action stimulus—touch stimulus. Subject is given notice of the approaching stimulus. The subject watched the operator's descending hand and was instructed to react as soon as the operator's hand touched the key.	50

To facilitate the first step in the inspection of labels, Eli Lilly and Company has small rectangular bars printed on one edge of the sheet at the time the label is printed (Fig. 117). There is a bar for each color, and the bars are located side by side on one edge of the sheet. The sheets are riffled or fanned out and placed flat on the inspection table. If all sheets are properly printed, there will be an uninterrupted line across the edges of the sheets for each color as shown in Fig. 118. If there is a blank sheet or if there is a sheet with one color printing missing, it is easy to detect this, remove the sheet, and scrap it. This procedure has not only been a great time-saver but has also made it possible to detect and remove imperfect labels at the source rather than later in the process after the labels have been cut to size.

Check List for Inspect

1. Can inspect be eliminated or overlapped with another **operation**?
2. Can multiple gauges or tests be used?

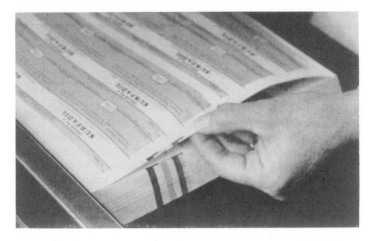

Fig. 117. Small bars are printed on the edge of the sheet at the time the label is printed. There is a bar for each color, and these bars are located side by side to facilitate inspection.

Fig. 118. The printed sheets of labels are fanned out for inspection. Blank sheets or sheets with one color missing will show up as a break in the line and can be removed and scrapped.

3. Can a pressure, vibration, hardness, or flash test be used?
4. Can the intensity of illumination be increased or the light sources rearranged to reduce the inspection time?
5. Can a machine inspection replace a visual inspection?
6. Can the operator use spectacles to advantage?

Assemble, Disassemble, and Use. The following explanation is included here to clarify the meaning of assemble and use. *Use* always refers to the use of a tool or device for the purpose for which it was intended. Thus, in Table 7 the actual writing was a use therblig. Similarly, painting, drilling, and sawing are all use therbligs. If a nut is assembled onto a bolt by hand, this motion is *assemble;* whereas if a wrench is used for this operation, the sequence is assemble (fit wrench to nut), use (turn nut down), and disassemble (remove wrench from nut).

Frequently a tool will be held in the palm of the hand when not in use. For example, the clerk checking boxes in a shipping department may place a crayon mark on certain items as they pass by on a conveyor. The use therblig would not include the entire cycle, but only that part during which the crayon is actually used for marking. The use of the bone in folding paper (Fig. 141) presents another example of this.

Some analysts advocate limiting use to ultimate objectives and restricting assemble to such temporary acts as fitting a tool to its work. Thus, any permanent assembly of two or more parts would be use even when no tool is involved. Since this interpretation is likely to result in some confusion to the beginner, and in view of the fact that the former interpretation is more widely accepted, use will, in this book, always refer to the use of a tool or device for the purpose for which it was intended; and assemble will be understood to consist of placing one object into or onto another object with which it becomes an integral part.

Painting with Spray Gun. The scope of the *use* therblig is so wide that it is impossible to cite representative cases. However, one illustration will be included because it gives an interpretation of this therblig that is often overlooked. The operation is painting with a spray gun the motor unit of an electric refrigerator.

From observation of the operation it was apparent that the operator was wasting paint, because he was missing the surface, "spraying air" by spraying past corners, and making sweeping flourishes during which little or none of the paint was being directed at the motor unit. In this operation the use motion involved not only time but material as well. Therefore, shortening the use motion meant saving both time and paint.

A micromotion study of this operation, made of the best operator in the plant, showed that during 23% of the time the spray gun was in use the paint was not hitting the surface of the unit being sprayed, but was being wasted "spraying air."

By careful training of the operator, and by some changes in the work place, including a power-driven, foot-controlled turntable for the work and three fixed spray guns mounted above the turntable, the following results were obtained:

Savings in time	50%
Reduction in rejects	60%
Direct labor savings per year	$3750.00
Savings in paint per year	$5940.00
Cost to develop and install new method	$1040.00

Not only did the total savings in direct labor and paint resulting from the improved method amount to a substantial sum, but also of importance was the great reduction in rejects.

Check List for Assemble, Disassemble, and Use

1. Can a jig or fixture be used?
2. Can an automatic device or machine be used?
3. Can the assembly be made in multiple? Or can the processing be done in multiple?
4. Can a more efficient tool be used?
5. Can stops be used?
6. Can other work be done while machine is making cut?
7. Should a power tool be used?
8. Can a cam or air-operated fixture be used?

ACCURATE MEASUREMENT OF FUNDAMENTAL MOTION TIME

The motion-time values for several of the investigations referred to in this chapter were given in hundred-thousandths of a minute. An electrical recording kymograph was used for making these time measurements. Paper tape, similar to adding-machine tape, is drawn across the kymograph table by means of the two rollers at the front of the machine. A synchronous motor drives the rollers, drawing the paper through at a uniform velocity of 2000 inches per minute.

Solenoid-operated pens are mounted above the paper tape so that each of the pens makes contact with it and draws a straight line on the tape as it passes through. Each pen is connected to a solenoid in such a way that closing the solenoid circuit moves the pen toward the solenoid and perpendicular to the motion of the paper through the kymograph, thus putting a jog in the ink line on the moving tape.

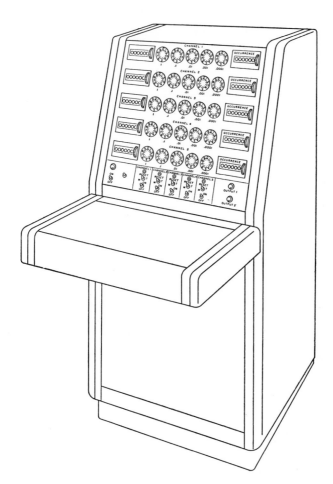

Fig. 119. Automatic electronic timer measures time in $\frac{1}{10.000}$ of a second.

By means of photoelectric cells and other devices it is possible to measure various parts of an operation without interfering in any way with the natural movements of the operator.

Automatic Electronic Timer. Although the kymograph is a satisfactory device for measuring short time intervals for motion study, it is a tedious and time-consuming task to measure the distance between the jogs in the lines on the paper tape. An electronic timer (Fig. 119) has been designed and built in the University of California laboratories, and is now being used by R. B. Andrews for measuring time for fundamental hand motions. The instrument is so designed that when the hand makes contact with an object, a pulse is started which causes an electronic switch to turn on, and then at the end of the contact the switch is turned off. The counter can be read directly, in ten-thousandths of a second, at the start of the motion and again at the end. It is possible to obtain mass data by timing 10 or 100 consecutive cycles. Likewise, different hand motions can be grouped together and timed as a unit. Information on individual cycles can be obtained by the use of a digital printer which rounds off the time to hundredths of a second. Because of the ease with which short time intervals can be measured, this apparatus greatly facilitates motion study research in the laboratory.[15]

[15] For other similar equipment see Stanley M. Block, "Semtar, Automatic Electronic Motion Timer," *Journal of Industrial Engineering,* Vol. 12, No. 4, pp. 276–288, July–August, 1961.

CHAPTER 16

Human Engineering

Work methods design seeks to determine the most effective combination of the man, the machine, and the working environment. In attempting to find this most effective combination, it is necessary to determine which functions can be performed better by man and which by machine. Man has certain inherent abilities which surpass existing machines, whereas existing machines surpass man in certain ways. Also, the matter of economy enters into the determination of the man-machine combination. For example, as the volume of a manufactured product increases, greater mechanization usually becomes profitable. More and more of the activities are performed by the machine and fewer by the man. The ultimate is, of course, a completely automatic process with no direct labor required.

However, most activities include some labor requirements, and the work methods analyst begins his study by defining the problem and by proceeding to design a man-machine system that best meets the specified objectives or goals. A knowledge of the inherent capacities and abilities of the human being is of vital importance in designing the process, the equipment, the work method, and the environment to best suit the people who will do the work. This body of knowledge pertaining to work methods and equipment design is the accumulation of certain rules and principles which have been developed over many years, and from the results of carefully designed research experiments. Many of the "rules for efficiency and fatigue reduction" and "principles of motion economy" are not universal in their application, and some of them have not been verified by controlled experiments; nevertheless they do serve a useful purpose as a check list to aid in finding preferred work methods and in designing equipment.

Although engineers, physiologists, and psychologists have for many years conducted experiments dealing with work design problems, a rather highly specialized group of these people were called upon during World War II to aid in solving man-machine problems as they pertained to the design, operation, and maintenance of military equipment. For example, the controls on planes, ships, and submarines became so complex that many failures resulted because the operators could not do all the things expected of them. The term *human engineering* was used to refer to this area of activity. Human engineering is reported to have as its goal "the adaptation of human tasks and working environment to the sensory, perceptual, mental, physical, and other attributes of people. This adaptation for human use applies to such functions as the design of equipment, instruments, man-machine systems, and consumer products, and to the development of optimum work methods and work environment."[1]

Much of the work in the field of human engineering has been in the form of carefully controlled research experiments and has been supported by various military and governmental agencies. The literature[2] in the field of human engineering takes the form of detailed research reports, abstracts of research studies which have been assembled into logical and systematic presentations, and handbooks and tables which provide a ready reference to such groups as work methods designers, equipment designers, and product designers and industrial stylists, as well as those groups working in the field of military, missile, and space vehicles. This growing body of knowledge is useful to all those interested in work methods design.

A person ordinarily does three things (Fig. 120) in performing any task:

1. Receives information—through the sense organs, eyes, ears, touch, etc.

[1] Ernest J. McCormick, *Human Engineering,* McGraw-Hill Book Co., New York, 1957, p. 1.

[2] Alphonse Chapanis, *Research Techniques in Human Engineering,* The John Hopkins Press, Baltimore, Md., 1959. Alphonse Chapanis, W. R. Garner, and C. T. Morgan, *Applied Experimental Psychology: Human Factors in Engineering Design,* John Wiley & Sons, New York, 1949. Henry Dreyfuss, *The Measure of Man—Human Factors in Design,* Whitney Library of Design, New York, 1960. W. F. Floyd and A. T. Welford (editors), *Symposium on Human Factors in Equipment Design,* H. K. Lewis & Co., Ltd., London, 1954. W. F. Floyd and A. T. Welford (editors), *Symposium on Fatigue,* H. K. Lewis & Co., Ltd., London, 1953. *Handbook of Human Engineering Data,* 2nd Ed., Tufts College, Medford, Mass., 1952. Wesley E. Woodson, *Human Engineering Guide for Equipment Designers,* University of California Press, Berkeley-Los Angeles, 1964.

2. Makes decisions—acts on the information obtained and on the basis of his own knowledge.
3. Takes action—action resulting from the decision that has been made. The action may be purely physical, such as operating a machine, or it may involve communication, such as giving oral or written instructions.

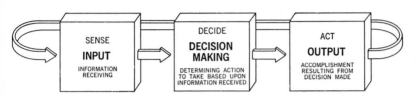

Fig. 120. The basic control cycle consists of three parts: sense, decide, and act.

The designer of machines, equipment, the work method, and the work environment must have an understanding of the way the human being functions, his body dimensions, his physical limitations, and the conditions under which he performs most effectively. In designing any process or operation, the question arises as to just which activities should be performed by the man and which by the machine. The following summary [3] may be useful in this connection:

Human beings appear to surpass existing machines in their ability to:

1. Detect small amounts of light or sound.
2. Receive and organize patterns of light or sound.
3. Improvise and use flexible procedures.
4. Store large amounts of information for long periods and recall relevant facts at the appropriate time.
5. Reason inductively.
6. Exercise judgment.
7. Develop concepts and create methods.

Existing machines appear to surpass human beings in their ability to:

1. Respond quickly to control signals.
2. Apply great force smoothly and precisely.
3. Perform repetitive routine tasks.
4. Store information briefly and then erase it completely.
5. Perform rapid computations.
6. Perform many different functions simultaneously.

[3] Ernest J. McCormick, *op. cit.*, p. 421.

On important jobs it may be desirable to tabulate the several ways of performing each part of the task, beginning with the manual method and progressing step by step to the fully mechanized method. This may make it easier to find the preferred method—the one with lowest cost. This procedure can aid in solving the man or machine question.

Human engineering today is concerned primarily with problems in connection with complicated military equipment and space hardware. However, it behooves all those responsible for work methods design to master the information coming out of research activities in the area of human engineering.

WORK METHODS DESIGN

General Statements

The work should be organized so that the operator receives only essential information through the appropriate sensory channels and at the time and place needed. The information should be presented in a way that permits the operator to react to it in the optimum manner.

In the decision-making phase of a task, the work should be arranged so that interpretations and decisions will be as nearly automatic as possible. The number of choices which the operator must make at a given time should be as few as possible.[4]

The work method should be designed so as to enable the operator to perform the task in the shortest possible time and with the greatest ease and satisfaction. The number of body members and the number of motions should be as few as possible, and the length of motions should be as short as possible. The job should be designed so that it results in the lowest energy expenditure and the least physiological stress, as measured by calories per minute and heart beats per minute.[5]

[4] *Ibid.*, p. 439.
[5] See Chapter 33, "Measuring Work by Physiological Methods."

HUMAN MEASUREMENTS
OF THE ADULT MALE AND FEMALE,
AND BASIC DISPLAY DATA AND BASIC CONTROL DATA

The information shown on the drawings [6] in Figs. 121–124 was developed after many years of research by Henry Dreyfuss and his associates. Such information is of great value in designing the machine, the work place, and the environment to best suit the person.

---

[6] Chart reproduced with permission from *The Measure of Man,* by Henry Dreyfuss, published by Whitney Library of Design, New York, 1960.

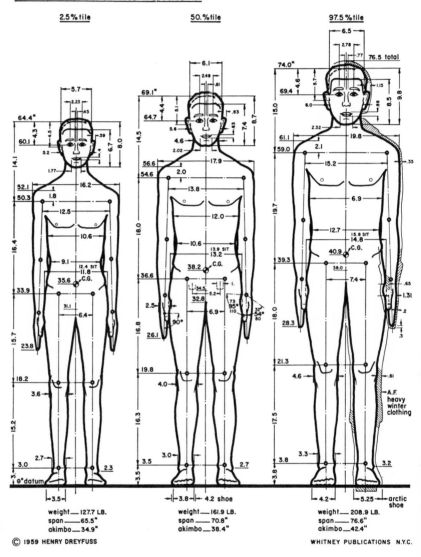

Fig. 121. Human dimensions of the standing adult male. (Chart from *The Measure of Man* by Henry Dreyfuss, published by Whitney Library of Design, New York.)

ANTHROPOMETRIC DATA — STANDING ADULT MALE

ACCOMMODATING 95% OF U.S. ADULT MALE POPULATION

2.5 % tile   50. % tile   97.5 % tile

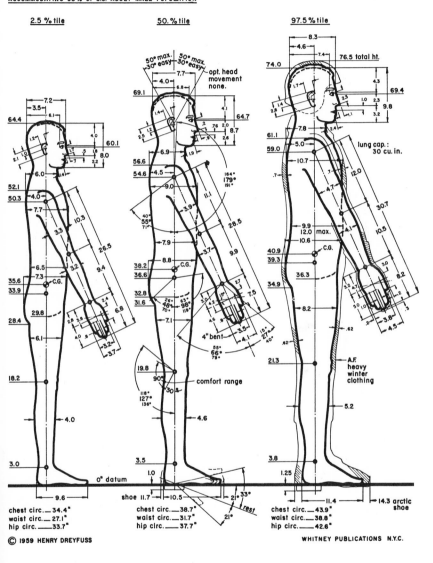

chest circ.— 34.4"
waist circ.— 27.1"
hip circ.— 33.7"

chest circ.— 38.7"
waist circ.— 31.7"
hip circ.— 37.7"

chest circ.— 43.9"
waist circ.— 38.8"
hip circ.— 42.6"

© 1959 HENRY DREYFUSS

WHITNEY PUBLICATIONS N.Y.C.

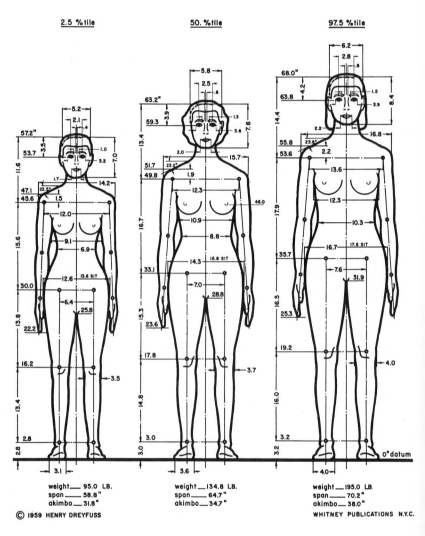

Fig. 122. Human dimensions of the standing adult female. (Chart from *The Measure of Man* by Henry Dreyfuss, published by Whitney Library of Design, New York.)

ANTHROPOMETRIC DATA — STANDING ADULT FEMALE
ACCOMMODATING 95% OF U.S. ADULT FEMALE POPULATION

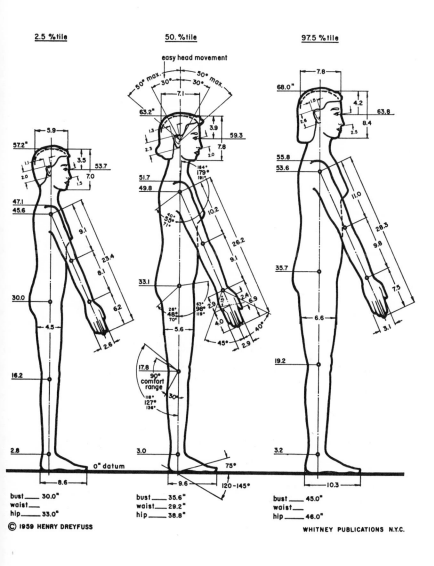

bust___ 30.0"
waist___
hip___ 33.0"

bust___ 35.6"
waist___ 29.2"
hip___ 38.8"

bust___ 45.0"
waist___
hip___ 46.0"

© 1959 HENRY DREYFUSS

WHITNEY PUBLICATIONS N.Y.C.

BASIC DISPLAY DATA

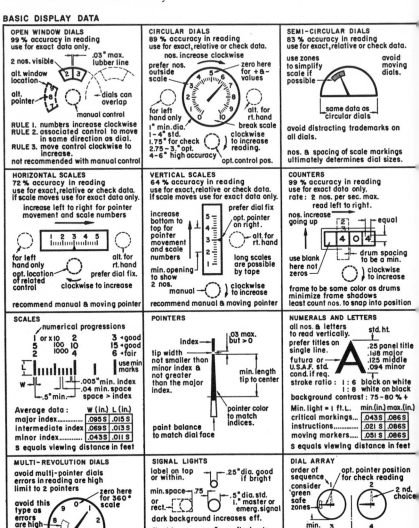

© 1960 HENRY DREYFUSS   WHITNEY PUBLICATIONS N.Y.

Fig. 123. Basic display data. (From *The Measure of Man* by Henry Dreyfuss, published by Whitney Library of Design, New York.)

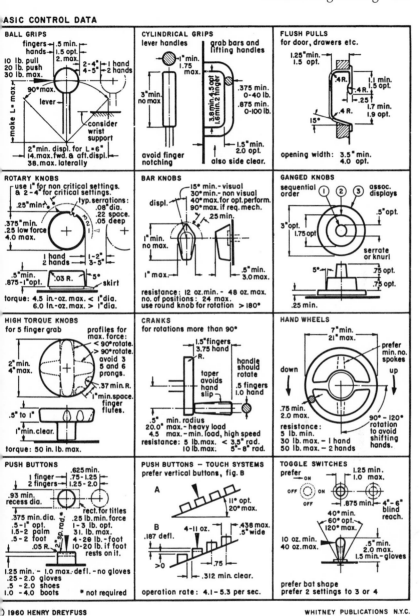

Fig. 124. Basic control data. (From *The Measure of Man* by Henry Dreyfuss, published by Whitney Library of Design, New York.)

Principles of Motion Economy

A Check Sheet for Motion Economy and Fatigue Reduction

These twenty-two rules or principles of motion economy may be profitably applied to shop and office work alike. Although not all are applicable to every operation, they do form a basis or a code for improving the efficiency and reducing fatigue in manual work.

Use of the Human Body

1. The two hands should begin as well as complete their motions at the same time. (Page 222.)
2. The two hands should not be idle at the same time except during rest periods. (Page 222.)
3. Motions of the arms should be made in opposite and symmetrical directions, and should be made simultaneously. (Page 222.)
4. Hand and body motions should be confined to the lowest classification with which it is possible to perform the work satisfactorily. (Page 235.)
5. Momentum should be employed to assist the worker wherever possible, and it should be reduced to a minimum if it must be overcome by muscular effort. (Page 237.)
6. Smooth continuous curved motions of the hands are preferable to straight-line motions involving sudden and sharp changes in direction. (Page 241.)
7. Ballistic movements are faster, easier, and more accurate than restricted (fixation) or "controlled" movements. (Page 245.)
8. Work should be arranged to permit easy and natural rhythm wherever possible. (Page 247.)
9. Eye fixations should be as few and as close together as possible. (Page 249.)

Arrangement of the Work Place

10. There should be a definite and fixed place for all tools and materials. (Page 256.)
11. Tools, materials, and controls should be located close to the point of use. (Page 258.)
12. Gravity feed bins and containers should be used to deliver material close to the point of use. (Page 268.)
13. Drop deliveries should be used wherever possible. (Page 271.)
14. Materials and tools should be located to permit the best sequence of motions. (Page 273.)
15. Provisions should be made for adequate conditions for seeing. Good illumination is the first requirement for satisfactory visual perception. (Page 273.)
16. The height of the work place and the chair should preferably be arranged so that alternate sitting and standing at work are easily possible. (Page 283.)
17. A chair of the type and height to permit good posture should be provided for every worker. (Page 286.)

Design of Tools and Equipment

18. The hands should be relieved of all work that can be done more advantageously by a jig, a fixture, or a foot-operated device. (Page 289.)
19. Two or more tools should be combined wherever possible. (Page 295.)
20. Tools and materials should be pre-positioned whenever possible. (Page 297.)
21. Where each finger performs some specific movement, such as in typewriting, the load should be distributed in accordance with the inherent capacities of the fingers. (Page 298.)
22. Levers, crossbars, and hand wheels should be located in such positions that the operator can manipulate them with the least change in body position and with the greatest mechanical advantage. (Page 301.)

CHAPTER 17

Principles of Motion Economy as Related to the Use of the Human Body

As valuable as the general statements at the end of Chapter 16 (page 212) may be, experience shows that the use of "check lists," "rules for fatigue reduction," and "principles of motion economy" can be helpful in work methods design. On several occasions Gilbreth listed certain "rules for motion economy and efficiency"[1] which govern hand motions, and from time to time other investigators in this field have added to the list.

Additional research which will enlarge our knowledge of the inherent capacities of the various members of the human body is greatly needed. There is much yet to be done in determining the fundamental laws which permit the maximum amount of productive effort with a minimum of fatigue. Although the material in this chapter is discussed under the heading "principles of motion economy," it might perhaps have been more accurately designated as "some rules for motion economy and fatigue reduction."

In attempting to collect and codify the information which is already available as a guide in determining methods of greatest economy, one is confronted with a number of difficulties. If general principles are stated, they are likely to be abstract and of little practical use; whereas if narrower rules with specific illustrations are presented, they may lack universality of application. In the past it has been customary to make general statements of the principles without including additional information or practical applications. This has been unsatisfactory and has retarded the use of motion and time study.

[1] F. B. and L. M. Gilbreth, "A Fourth Dimension for Measuring Skill for Obtaining the One Best Way," *Society of Industrial Engineering Bulletin*, Vol. 5, No. 11, November, 1923.

It is the purpose of this and the following two chapters to interpret by means of specific illustrations some of the general rules or principles of motion economy which have been and are now being successfully used. Not all the principles presented in these chapters are of equal importance, nor does this discussion include all the factors which enter into the determination of better methods for doing work. These principles do, however, form a basis—a code or a body of rules —which, if applied by one trained in the technique of motion study, will make it possible to increase greatly the output of manual labor with a minimum of fatigue.

These principles will be presented under the following three subdivisions:

I. Principles of motion economy as related to the use of the human body.
II. Principles of motion economy as related to the arrangement of the work place.
III. Principles of motion economy as related to the design of tools and equipment.

PRINCIPLES OF MOTION ECONOMY AS RELATED TO THE USE OF THE HUMAN BODY

1. The two hands should begin as well as complete their motions at the same time.

2. The two hands should not be idle at the same time except during rest periods.

3. Motions of the arms should be made in opposite and symmetrical directions and should be made simultaneously.

These three principles are closely related and can best be considered together.[2] It seems natural for most people to work productively with one hand while holding the object being worked on with the other hand. This is usually undesirable. The two hands should work together, each beginning a motion and completing a motion at the same time. Motions of the two hands should be simultaneous and symmetrical.

It is obvious that in many kinds of work more can be accomplished by using both hands than by using one hand. For most people it is advantageous to arrange similar work on the left- and right-hand sides of the work place, thus enabling the left and right hands to move together, each performing the same motions. The symmetrical movements of the arms tend to balance each other, reducing the shock

[2] *Ibid.*, p. 6.

and jar on the body and enabling the worker to perform his task with less mental and physical effort. There is apparently less body strain when the hands move symmetrically than when they make nonsymmetrical motions, because of this matter of balance.

Some examples will be cited to show how better methods were developed through the analysis of the hand motions with which you

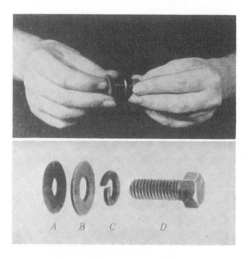

Fig. 125. Bolt and washer assembly: *A*, special rubber washer; *B*, flat steel washer; *C*, lock washer; *D*, ⅜-inch × 1-inch bolt.

are now familiar, and through the application of the first three principles of motion economy.

Bolt and Washer Assembly. A manufacturing concern uses eight bolts ⅜ inch by 1 inch, fitted with three washers each (see Figs. 125 and 126), in the final assembly of one of its products. This operation was facilitated by having the three washers previously assembled onto the bolt; consequently the bolt and washers were assembled by girls at benches in another department.

Old Method. The bolt and washer assembly was originally made in the following manner. Containers with the bolts, lock washers, steel washers, and rubber washers were arranged on the top of the bench as shown in Fig. 69. The operator reached over to the container of bolts, picked up a bolt with her left hand, and brought it up to position in front of her. Then with the right hand she picked up a lock washer from the container on the bench and placed it on the bolt, then a flat steel washer, and then a rubber washer. This completed the assembly, and with the left hand the operator disposed

Fig. 126. The hole in the rubber washer is slightly smaller than the outside diameter of the bolt so that when the bolt is forced through the hole it is gripped, thus preventing the washers from falling off the bolt.

Fig. 127. Bins, fixture, and chute for bolt and washer assembly.

of it in the container to her left. Figure 100 gives the analysis sheet for this operation, and Fig. 99 shows the pictures of one cycle.

It is readily seen that every one of the three principles named above was violated when the operation was performed in this way, although it is the customary method of doing such work. The left hand *held* the bolt during most of the time while the right hand worked productively. The motions of the two hands were neither simultaneous nor symmetrical.

Improved Method. A simple fixture was made of wood and surrounded by metal bins of the gravity-feed type, as shown in Figs. 127, 128, and 129. The bins containing the washers are arranged in duplicate so that both hands can move simultaneously, assembling washers for two bolts at the same time. As seen from Fig. 127, bins 1 contain the rubber washers, bins 2 the flat steel washers, bins 3 the lock washers, and bin 4, located in the center of the fixture, contains the bolts. The bottoms of the bins slope toward the front at a 30-degree angle so that the materials are fed out onto the fixture board by gravity as the parts are used in assembly.

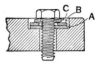

Fig. 128. Enlarged view of recess in wood fixture for assembling bolt and washers: *A,* rubber washer; *B,* steel washer; *C,* lock washer.

Two countersunk holes or recesses were made in the front of the fixture (Fig. 128) into which the three washers fitted loosely, the rubber washer on the bottom, the flat steel washer next, and the lock washer on top. A hole slightly larger than the diameter of the bolt went through the fixture, as shown in Fig. 128. A metal chute was placed around the front of the wood fixture, with openings to the right and to the left of the two recesses so that assembled bolts and washers might be dropped into the top of this chute and carried down under the bench to a container (Fig. 129).

In assembling the bolt and washers, as the chart in Fig. 102 shows, the two hands move simultaneously toward the duplicate bins 1, grasp rubber washers which rest on the wood fixture in front of the bins, and slide the rubber washers into place in the two recesses in the fixture. The two hands then, in a similar way, slide the steel washers into place on top of the rubber washers, and then the lock washers are slid into place on top of these. Each hand then grasps a bolt and slips them through the washers, which are

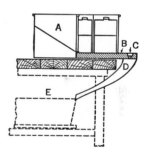

Fig. 129. Cross section of bins showing chute for drop delivery: *A,* bins with sloping bottom; *B,* top of fixture; *C,* countersunk holes in top of fixture; *D,* chute; *E,* container for finished assemblies.

lined up so that the holes are concentric. The hole in the rubber washer is slightly smaller than the outside diameter of the threads on the bolt so that when the bolt is forced through the hole it is gripped and thus permitted, with the three washers, to be withdrawn vertically upward wihout losing the washers (Fig. 126). The two hands release the assemblies simultaneously over the metal chute. As the operator begins on the next cycle with the hands in this position, the first and second fingers of each hand are in position to grasp the rubber washer, which is almost at the tip of the fingers.

A detailed study of the old and the improved methods of assembling the bolt and washers shows:

Average time per assembly, old method	0.084 minute
Average time per assembly, improved method	0.055 minute
Time saved	0.029 minute
Increase in output = 53% [3]	

[3] The results of an improved method are sometimes expressed in "increase in output in per cent," and sometimes in "time saved in per cent." These two percentages do not mean the same thing. Perhaps the following computations may serve to clarify this point.

Increase in Output in Per Cent

$$\frac{\begin{bmatrix} \text{Pieces produced} \\ \text{per minute,} \\ \text{new method} \end{bmatrix} - \begin{bmatrix} \text{Pieces produced} \\ \text{per minute,} \\ \text{old method} \end{bmatrix}}{\begin{bmatrix} \text{Pieces produced per minute,} \\ \text{old method} \end{bmatrix}} \times 100 = \text{Increase in output in per cent}$$

EXAMPLE

Time for assembly, old method = 0.084 minute
Number of assemblies per minute, old method = 1 ÷ 0.084 = 11.9
Time per assembly, new method = 0.055 minute
Number of assemblies per minute, new method = 1 ÷ 0.055 = 18.2

$$\frac{18.2 - 11.9}{11.9} \times 100 = 53\% \text{ increase in output}$$

Savings in Time in Per Cent

$$\frac{\begin{bmatrix} \text{Time per piece,} \\ \text{old method} \end{bmatrix} - \begin{bmatrix} \text{Time per piece,} \\ \text{new method} \end{bmatrix}}{[\text{Time per piece, old method}]} \times 100 = \text{Savings in time in per cent}$$

EXAMPLE

Time per piece, old method = 0.084 minute
Time per piece, improved method = 0.055 minute

$$\frac{0.084 - 0.055}{0.084} \times 100 = 35\% \text{ savings in time}$$

The improved method as opposed to the old method of assembling the bolt and washers conforms to each of the three principles of motion economy already mentioned. The two hands begin and end their motions at the same instant, and they move simultaneously in opposite directions. There is no idle time, and neither hand is used as a vise for holding material while the other one does the work, as under the old method.

Fig. 130. Arrangement of work place—old method. Filling mailing envelope with four sheets of advertising material.

Filling Mailing Envelope with Advertising Material. This operation involved inserting four sheets of advertising material in a mailing envelope and tucking in the envelope flap. The job consisted of picking up the sheets one at a time with the right hand, transferring them to the left hand, jogging them, and then inserting them in the envelope (Fig. 130). It is obvious that the *left* hand was idle part of the time, held the sheets part of the time, and worked in an inefficient manner during the rest of the cycle. Also, the *right* hand was idle part of the time.

Improved Method. Two small triangular pieces made from cardboard and tape were fastened to flat sheets of cardboard (Fig. 131). The advertising material was stacked against the two sides of the triangular pieces, which served as fixtures, enabling the operator to pick up two sheets at a time with each hand. Rubber finger stalls facilitated the grasping. Since sheets of the particular size shown in Fig. 132 were mailed out at frequent intervals, the work place to handle this job was set up permanently. Triangular wood blocks are shown at *A*. The operation now consists of grasping two sheets of paper at a time with each hand, drawing them together, jogging them on block *B*, and inserting them in the envelope.

When the operator used the old method of filling envelopes, picking up the sheets one at a time, her production was 350 per hour. Using the improved work-place layout and the better method, she was able

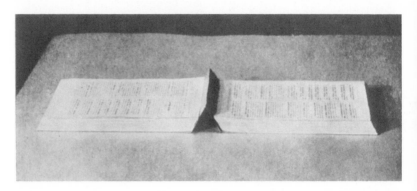

Fig. 131. Temporary fixture for assembling two sheets of advertising material.

to fill 750 envelopes per hour. The new method is so much easier than the old that she has more than doubled her output.

Folding Paper Cartons. Frankfurters are usually packed in cardboard cartons for shipment to the retail store. The cartons are delivered to the packing house in flat bundles, and these flat cartons must be formed and the end flaps folded over and locked together before they can be filled with frankfurters. A cover slightly larger than the bottom is placed over the filled carton bottom, telescope fashion (Fig. 133). The shape and design of the cover and bottom of the carton are alike; that is, both the cover and the bottom are folded the same way.

Old Method. The operator walked 10 feet, got bundle of 50 carton "flats," and carried them to carton-folding table. Using both hands, she grasped a group of 8 flat cartons, broke all seams or scored lines to facilitate forming, and placed flats on table with ends toward her. She then grasped sides of carton flap and simultaneously bent bottom and side flaps toward center of carton (Fig. 134). Holding left side flap in position, she inserted tongue of right flap into retaining groove

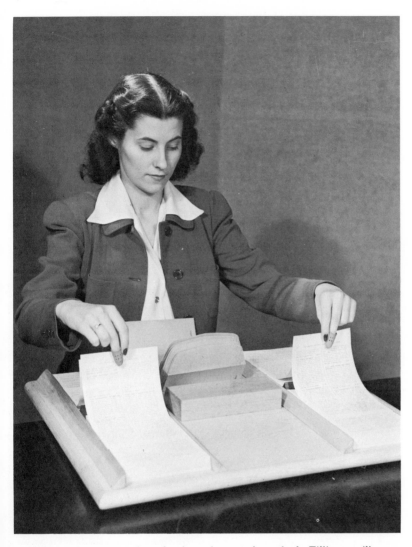

Fig. 132. Arrangement of work place—improved method. Filling mailing envelope with four sheets of advertising material. *A*, triangular blocks; *B*, block on which sheets are jogged.

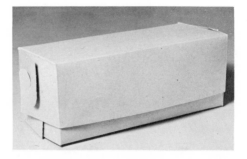

Fig. 133. Carton for packing frankfurters.

Fig. 134. Old method of folding cartons. Left hand holds one carton flap while right hand locks the other flap to the first.

of left flap. She then pushed the partly formed carton forward approximately 4 inches to nest. The above procedure was repeated until four cartons were folded at one end.

The operator took the group of partly formed cartons from position on table and turned them end for end so the unfolded ends were toward

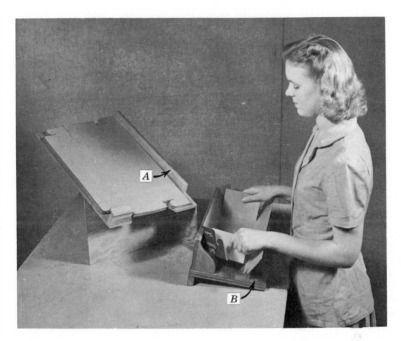

Fig. 135. Improved method of folding cartons. Fixture aids in forming carton and holds it while both hands lock the flaps. Flaps on both ends of carton are assembled at the same time. Operator doubled her output.

her. She then repeated the folding operation on the second end of the carton flat and placed the completely formed carton on the conveyor, to be filled with frankfurters.

Improved Method. A simple wood fixture, shown in Fig. 135, was designed and built.[4] Bundles of flat cartons are now delivered by truck near the packing table. The operator gets a bundle of 50 flats, carries them to the table, and places them in holder *A* in Fig. 135. With her left hand she reaches to lower end of pile of carton flats, grasps end of middle flap, brings flap to position over forming fixture *B*. With her right hand she disposes of previously formed carton to

[4] Improved method developed by Eugene J. Smith.

conveyor. Then with her right hand she grasps middle flap on right end of carton, which is already positioned in fixture. Holding both middle end flaps, one in each hand, she bends flaps upward and pushes carton into fixture. The fixture forces rear side flap up 90 degrees, folds two rear end flaps forward 90 degrees, and folds front side and end flaps up approximately 45 degrees. With both hands simultaneously, the operator then reaches to front end flaps and folds end flaps to rear and toward center of carton until ends of tongues on front flaps are inside notches of rear end flaps. While holding end flaps in position with fingers, the operator reaches with thumbs to front corners of carton, pushes front of carton to rear in order to lock tongues of front end flaps into notches of rear end flaps, and thus completes forming carton. The operator then disposes of finished carton onto conveyor.

Results. The improved method enabled the operator to double her output. The fixture cost approximately $10 to make. The improved method was superior to the old for two reasons: (1) elimination of the operation of breaking "seams" or scored lines of fold on carton flats before forming; (2) elimination of holding one flap in position with one hand while assembling the second flap with the other.

Nonsymmetrical Motions. Frequently the nature of the work prevents the operator from moving his arms simultaneously in opposite and symmetrical directions. When this is the case, it may be that the work can be arranged so that the operator can move his arms simultaneously in directions perpendicular to each other. An example of this type of movement is shown in Fig. 136. The operation is wrapping and boxing electric switches. The old method was to place the product to be wrapped on one end of a sheet of wrapping paper, and then to finish the operation by a folding and rolling process. The product was then placed in a fiber box and the lid was put on. This method of wrapping and boxing was wasteful of time and effort, as well as of paper.

In the improved method two narrow strips of paper are drawn from supply boxes (*A* and *B* of Fig. 136) across the top of the fiber box by perpendicular motions of the two arms. The switch is then placed on top of the paper and pushed down into the box, both ends of the paper being folded over the switch with simultaneous motions of the two hands. Finally the lid is placed on the box. The new method of wrapping and boxing the electric switch requires 40% less time than the old method.

There is a certain balance and ease of muscular control to these motions performed at right angles which make them definitely superior to motions of the arms in the same direction. However, they are not so easy as simultaneous motions of the arms in opposite directions, and should be used only when the former motions are impossible.

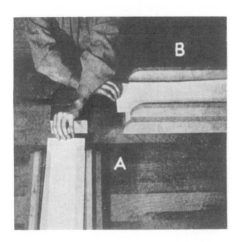

Fig. 136. Simultaneous motions of the arms perpendicular to each other. The operation is that of wrapping and boxing electric switches.

One- and Two-Handed Work. The results of a study [5] of the time to select and grasp, transport, and dispose of machine-screw nuts from two types of bins with the right hand alone, with the left hand alone, and with both hands working together are shown in Table 10.

The operation consisted of selecting and grasping machine-screw nuts (Nos. 2 and 8) from a bin, carrying them through a distance of 5 inches, and disposing of them in a hole in the table top. The study was made using a rectangular bin, and was then repeated using a bin with tray. These bins are shown in Table 12 in the next chapter. The operator worked first with the right hand alone, then with the left hand alone, and finally with both hands.

Least time was required for a total cycle when only the right hand was used. A cycle for the left hand required 6% more time with the rectangular bin, and 12% more time with the bin with tray; and a cycle with both hands required 30 to 40% more time. However, since two cycles were performed simultaneously when the two hands were used, the time chargeable to each cycle was considerably less than when only the right hand was used.

[5] *University of Iowa Studies in Engineering, Bulletin* **21.**

Table 10. Study of One- and Two-Handed Work

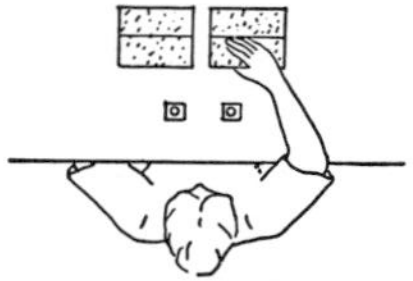

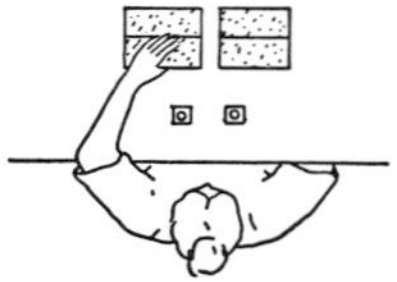

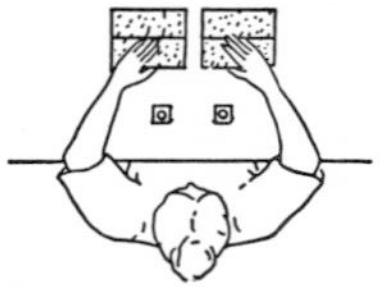

		Right Hand Working Alone		Left Hand Working Alone		Both Hands Working Together	
		Rectangular Bin	Bin with Tray	Rectangular Bin	Bin with Tray	Rectangular Bin	Bin with Tray
SELECT AND GRASP Nut from bin at A (see figure above).	Time in Minutes	0.00723	0.00438	0.00822	0.00520	0.01307	0.00674
	Time in Per Cent (Shortest Time = 100%)	100	100	114	118	181	154
TRANSPORT LOADED Carry nut through distance of 5 inches —from A to B.	Time in Minutes	0.00292	0.00235	0.00347	0.00234	0.00380	0.00270
	Time in Per Cent (Shortest Time = 100%)	100	100	119	100	130	115
RELEASE LOAD Drop nut into 1-inch hole in table top at B.	Time in Minutes	0.00403	0.00403	0.00380	0.00453	0.00463	0.00500
	Time in Per Cent (Shortest Time = 100%)	106	100	100	112	122	124
TRANSPORT EMPTY Move hand to bin at A for nut.	Time in Minutes	0.00314	0.00277	0.00282	0.00304	0.00308	0.00337
	Time in Per Cent (Shortest Time = 100%)	111	100	100	110	110	122
TOTAL CYCLE	Time in Minutes	0.01730	0.01351	0.01832	0.01510	0.02459	0.01778
	Time in Per Cent (Shortest Time = 100%)	100	100	106	112	142	131

Under the conditions observed in this investigation and with the operators studied, there was considerable evidence to indicate that a good one-handed operator was also a good two-handed operator, and a relatively poor one-handed operator was also a relatively poor two-handed operator. This suggests that the introduction of two-handed simultaneous work in place of less efficient one-handed work will not inconvenience any one operator very much more than another operator.

4. Hand and body motions should be confined to the lowest classification with which it is possible to perform the work satisfactorily.

The five general classes of hand motions are listed here because they emphasize that material and tools should be located as close as possible to the point of use, and that motions of the hands should be as short as the work permits. The lowest classification, which is shown first, usually requires the least amount of time and effort and probably produces the least fatigue.

General Classification of Hand Motions

1. Finger motions.
2. Motions involving fingers and wrist.
3. Motions involving fingers, wrist, and forearm.
4. Motions involving fingers, wrist, forearm, and upper arm.
5. Motions involving fingers, wrist, forearm, upper arm, and shoulder. This class necessitates disturbance of the posture.

The operator shown in Fig. 137 is manually operating a swing saw for rough-cutting lumber to length in a furniture factory. A better method was devised (Fig. 138), and the operator now operates the saw by means of a motor control switch at his right hand while feeding the stock with his left hand. A guard directly in front of the saw protects the operator. In addition to this saving of time and effort on the part of the operator, belt conveyors were installed which carry the lumber to the off-bearer and the scrap directly to the incinerator. Originally each operator needed a helper; now one helper serves three operators.[6]

As desirable as it may be to keep hand motions as short as possible, it is incorrect to assume that finger motions are less fatiguing than

[6] Martin S. Meyers, "Evaluation of the Industrial Engineering Program in Small Plant Management," *Proceedings Sixth Industrial Engineering Institute,* University of California, Los Angeles-Berkeley, p. 37.

Fig. 137. Manual method of operating swing saw in furniture factory—old method.

Fig. 138. Swing saw is now operated by a control button at the operator's right hand—improved method.

motions of the forearm. One has only to remember his early instruction in writing to know that free, loose forearm and wrist movements are easier, faster, and more uniform than finger motions. In telegraphy the substitution of a telegraph key which moved in the lateral direction instead of vertically was the result of the observation that lateral movement permitted the operator to work with a freer and looser wrist.[7]

In another investigation of movements, it was found that finger motions were more fatiguing, less accurate, and slower than motions of the forearm.[8] All evidence seems to show that the forearm is the most desirable member to use for light work, and that in highly repetitive work motions about the wrist and elbow are in all respects superior to those of the fingers or shoulders.

Physiological Cost of Body Bending. Body movements are time-consuming and result in high physiological costs as well. We recently made a systematic study of an operation consisting of picking up 5-pound bricks under various conditions.[9] Change in energy expenditure and change in heart rate were measured, and the results pertaining to a part of the study are shown in Table 11 and Fig. 139. In Method A, the operator picked up the brick from a platform 5 inches above the floor and placed it on a bench 33 inches high; this involved major body bending. In Method B the bricks were moved from a platform 37 inches high to a bench 33 inches high, which called for minor body movements. The operators worked at four different speeds. At the slowest speed, there was an increase in heart rate from 94 beats per minute to 105, and at the highest speed an increase from 106 to 148 beats per minute. Likewise, the energy expenditure increased from 2.5 calories (abbreviation for kilocalories) per minute to 5.4 at the low speed and from 4.5 to 10.1 at the high speed. These data give a quantitative evaluation of the physiological cost of work involving extreme body bending.

5. *Momentum should be employed to assist the worker wherever possible, and it should be reduced to a minimum if it must be overcome by muscular effort.*

The momentum of an object is its mass multiplied by its velocity. In most kinds of factory work the total weight moved by the operator

[7] M. Smith, M. Culpin, and E. Farmer, "A Study of Telegrapher's Cramp," Industrial Fatigue Research Board, *Report* 43, 1927.

[8] R. H. Stetson and J. A. McDill, "Mechanisms of the Different Types of Movement," *Psychological Monograph,* Vol. 32, No. 3, Whole No. 145, p. 37, 1923.

[9] Study made in the UCLA Work Physiology Laboratory by Ralph M. Barnes, Robert B. Andrews, James I. Williams, and B. J. Hamilton.

Table 11. Physiological Costs of Two Different Methods of Handling Brick

Operator	METHOD A — Major Body Bending								METHOD B — Minor Body Movements							
	Energy Expenditure in Calories per Minute				Heart Rate in Beats per Minute				Energy Expenditure in Calories per Minute				Heart Rate in Beats per Minute			
	Number of bricks moved per minute				Number of bricks moved per minute				Number of bricks moved per minute				Number of bricks moved per minute			
	16	22	28	34	16	22	28	34	16	22	28	34	16	22	28	34
1	5.4	5.7	6.8	8.5	102	104	109	131	2.8	3.1	3.3	5.8	92	92	92	113
2	5.4	6.8	7.9	10.2	110	126	134	155	2.3	2.6	3.3	3.8	100	97	107	109
3	5.3	6.8	8.5	11.7	102	113	126	159	2.5	2.7	3.0	3.8	90	97	97	95
Average	5.4	6.4	7.7	10.1	105	114	123	148	2.5	2.8	3.2	4.5	94	95	99	106

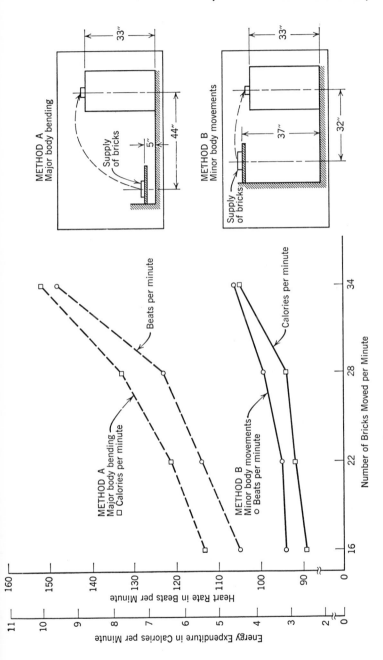

Fig. 139. Physiological cost of two different methods of handling brick. Method *A.* major body bending; Method *B,* minor body movements.

may consist of three components: the weight of the material moved, the weight of the tools or devices moved, and the weight of the part of the body moved.[10] It is often possible to employ momentum of the hand, the material, or the tool to do useful work. When a forcible stroke is required, the motions of the worker should be so arranged that the stroke is delivered when it reaches its greatest momentum.[11] In laying a brick wall, for example, "If the bricks are conveyed from the stock platform to the wall with no stops, the momentum can be made to do valuable work by assisting to shove the joints full of mortar. If, instead of being utilized, the momentum must be overcome by the muscles of the bricklayer, fatigue . . . will result.

"The ideal case is to move the brick in a straight path and make the contact with the wall overcome the momentum." [12]

The improved method of candy dipping explained on page 245 is another illustration of the utilization of momentum for the performance of useful work. The piece to be dipped was submerged under the surface of the melted sugar by the right hand at the end of a long return stroke of the hand. The momentum developed in this movement of the hand and the empty dipping fork was employed in doing useful work instead of being dissipated by the muscles of the dipper's arm.

There are many times when momentum has no productive value, and its presence is undesirable in that the muscles must always counteract the momentum developed. When such is the case, the three classes of weight or mass named previously should be studied for the purpose of reducing each to the minimum. In addition, the velocity of the motions should be kept low by using the shortest motions possible. A number of tools are most effective when they are made as light in weight as possible. Such tools do not depend upon momentum or the use of a blow to function properly. For many kinds of work a heavy shovel or a heavy trowel is more fatiguing to use than a light one of the same dimensions and rigidity.

Many additional considerations enter into the determination of the proper size and weight of materials and tools to produce maximum efficiency. Unfortunately, the accumulated data are of little value here. Each case, as a rule, is surrounded by circumstances and conditions peculiar to itself. Consequently each problem must be the subject for special investigation.

[10] F. B. Gilbreth, *Motion Study*, Van Nostrand Co., Princeton, N. J., 1911, p. 63.
[11] C. S. Myers, *Industrial Psychology in Great Britain*, Jonathan Cape, London, 1926, p. 88.
[12] F. B. Gilbreth, *op. cit.*, p. 78.

6. Smooth continuous curved motions of the hands are preferable to straight-line motions involving sudden and sharp changes in direction.

The simple operation of moving a pencil back and forth across a sheet of paper consists of two phases, the movement and the stop and change direction. The results of a study [13] of the simple hand motions *transport loaded* (away from the body), *stop* and *change direction,* and

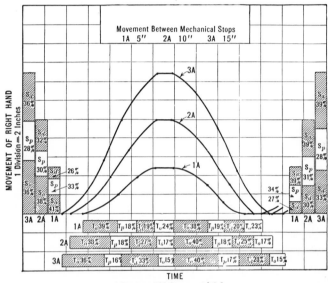

Fig. 140. Curves showing movement of right hand through varying distances between mechanical stops.

transport loaded (toward the body) are given in Fig. 140. This figure shows that 75 to 85% of the time to make a complete back and forth movement is used in actually moving the hand, and the remaining 15 to 25% of the time in changing direction of the hand; that is, during 15 to 25% of the time the hand and the pencil are motionless. Further studies [14] show that continuous curved motions are preferable to straight-line motions involving sudden and sharp changes in direction. Such abrupt changes in direction are not only time consuming but also fatiguing to the operator.

[13] Investigation by J. Wayne Deegan, *University of Iowa Studies in Engineering, Bulletin* 6, pp. 37–51. Also see results of study by A. B. Cummins, *University of Iowa Studies in Engineering, Bulletin* 12, pp. 8–18.

[14] See footnote 13.

In many jobs in the shop and office it is possible to use these smooth curved motions. Some examples will be given here.

Folding Paper. The first illustration is folding rectangular sheets of paper used in packing X-ray films. The sheets vary in size from 3 inches by 5 inches to 12 inches by 15 inches folded. Although several

Fig. 141. Path of hand in creasing folded sheet of paper—old method. There is an abrupt change in direction at *D* and also at *E*. Two strokes of the bone are used to crease the fold.

million of these sheets of paper are folded per year, it was found to be more economical to fold them by hand than by machine because of the many different sizes used.

Old Method. The worker, holding a smooth piece of bone in the palm of her right hand (Fig. 141), grasped the lower right-hand corner *A* of the sheet of paper to be folded. She folded this end of the sheet over to point *B*, where the two hands matched or lined up the two corners of the sheet of paper. Then, swinging the right hand away from the body and using the bone as a creasing tool, she struck the folded sheet of paper about midpoint at *C*, creasing the fold from

C to *D*. At *D* she stopped, changed direction abruptly, and doubled back, creasing the entire length of the fold from *D* to *E*. At *E* the hand again changed direction and swung around to *F*, where the end of the bone was inserted under the edge of the creased sheet to assist the left hand in disposing of it on the pile of folded sheets at *G*.

Fig. 142. Path of hand in creasing folded sheet of paper—improved method. The hand makes a smooth S curve, creasing the fold with one stroke of the bone. Output was increased 43%.

Improved Method. In the improved method (Fig. 142) the worker grasps the lower right-hand corner *A* of the sheet of paper to be folded. She folds this end of the sheet over to point *B*, where the two hands match or line up the two corners of the sheet of paper. She then moves the right hand through a smooth S curve, the bone striking the paper and beginning to crease at *X* and ending at *Y*. The entire crease is completed with the single stroke of the bone. The hand then swings around in a curved motion from *Y* to *Z*, where, as in the old method, the end of the bone is inserted under the creased sheet to assist the left hand in disposing of it on the pile of folded sheets at *G*.

Results. By the improved method only one creasing motion was required to complete the cycle instead of the two (one short and one long one) in the old method. Moreover, in the improved method two curved motions of the hand were used instead of two complete change directions and one 90-degree change direction.

A micromotion study of these two methods shows that 0.009 minute was required to crease the fold by the old method and 0.005 minute by the improved method. The improved method of creasing the fold, plus some other changes in the cycle, reduced the total time from 0.058 to 0.033 minute per cycle, enabling the operator to increase her output 43%.

Dipping Candy. Another illustration of the value of curved motions over straight-line motions with sudden changes in direction is a candy-dipping operation.[15]

Old Method. The dipping process was carried out in the following manner. A "center" (an almond, walnut, Brazil nut, or caramel) was placed in a pot of melted sugar by using the left hand, and was covered with the melted sugar by working it with a fork held in the right hand. The finished piece of candy was then placed with the right hand on the tray to the right of the operator. Approximately 2 seconds were required to dip each piece.

Although the lines in Fig. 143 do not show the exact movements of the right hand, they give a picture of the principal motions used. While the left hand was placing a center in the container of melted sugar, the right hand carried the empty fork from the tray *A* to the container *B*, and took up some of the thick melted sugar and pulled it over the center at *X*. When the hand reached *C* it moved to the left side of the container, the center being carried along with the end of the fork under it. The center was picked up at *D* and carried to the tray where it was deposited. The objections to this method of dipping were that the hand stopped at *B* and changed direction sharply, and then at *C* the direction was almost reversed. This stopping and sudden changing of direction placed unnecessary strain on the muscles of the arm.

Improved Method. The improved method of dipping is shown diagrammatically in Fig. 144. The center is dipped by a smooth sweeping motion of the hand instead of by a number of short zigzagging motions as in the old method. In the improved method the hand, after

[15] E. Farmer, "Time and Motion Study," Industrial Fatigue Research Board, *Report* 14, pp. 36–41. Figures 143 and 144 reproduced by permission of the Controller of H. M. Stationery Office, London.

disposing of the finished piece of candy on the tray, moves from A to B as before, but reaches B in the middle of an inward and downward curve with the hand in its strongest position for doing work. This makes it possible to utilize the momentum developed in the movement $A–B$ in doing the most fatiguing part of the work, the dipping being the part of the process that offers the greatest resistance to the hand. In the old process this dipping motion was made by a short backward movement just after the hand had stopped and changed its direction. Furthermore, the momentum developed during the motion $A–B$ was wasted in the old method, since the hand motion was checked at B in order to change its direction. By using the downward motion of the

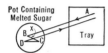

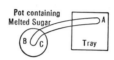

Fig. 143. Old method of dipping candy.

Fig. 144. Improved method of dipping candy.

hand in the improved method, the melted sugar is swept over the center, and going under the surface in the second part of the curve, it comes up at C. The piece of candy is then deposited on the tray with a circular motion to "finish off" the candy. In the new method the hand takes easy smooth movements with all changes in direction effected by curves.

Results. The improved method was taught to a group of workers in the factory, and after a short period of training an average increase in production of 27% resulted. However, since many workers had used the old method for years, it was difficult to persuade some of them to give the new method a fair trial. A new dipping room equipped with new-style tables and trays was started, and new operators were trained in the proper method of dipping. After 3 months' work in this new room, these new workers were producing an average of 88% more than the workers of the same standing in the original room.

7. Ballistic movements are faster, easier, and more accurate than restricted (fixation) or "controlled" movements.

Voluntary movements of the members of the human body may be divided into two general classes or groups. In the *fixation* or controlled movements, opposing groups of muscles are contracted, one group against the other. For example, in bringing the pencil down to the paper preparatory to writing, two or more sets of muscles are in action. The positive sets of muscles propel the hand, and the antagonistic sets oppose the movement. When the two sets of muscles act in an uneven or unbalanced manner, motion of the hand results. When the two sets of muscles exactly balance each other, the hand remains in a fixed position, although it is ready to act in any direction at any instant. The finger-and-thumb method of writing is an excellent illustration of fixation movements.

The *ballistic* movement is a fast, easy motion caused by a single contraction of a positive muscle group with no antagonistic muscle group contracting to oppose it. The contraction of the muscles throws the member of the body into motion, and since these muscles act only through the first part of the movement, the member sweeps through the remainder of the movement with its muscles relaxed. The ballistic movement is controlled by the initial impulse, and once under way its course cannot be changed.[16] A ballistic stroke may terminate (1) by the contraction of the opposing muscles, (2) by an obstacle, or (3) by dissipation of the momentum of the movement, as in swinging a golf club.

The ballistic movement is preferable to the fixation movement and should be used whenever possible. It is less fatiguing, for the muscles contract only at the beginning of the movement and are relaxed during the remainder of the movement. The ballistic movement is more powerful, faster, more accurate, and less likely to cause muscle cramp. It is smoother than the fixation movement, which is caused by the contraction of two sets of muscles, one acting against the other continuously. The skilled carpenter swinging his hammer in driving a nail illustrates a ballistic movement. He aims his hammer, then throws or swings it. The muscles are contracted only during the first part of the movement; they idle along the rest of the way. The swinging curves of an orchestra conductor's baton are another illustration of ballistic movement. P. R. Spencer understood the value of ballistic movements; the "free-hand writing" which he taught is known to everyone to produce greater speed and accuracy with less fatigue than is possible with the finger-and-thumb method of writing, where the muscles of the hand are tightly drawn. The ballistic movement is the one taught to telegraph operators, piano players, violin players,

---

[16] L. D. Hartson, "Analysis of Skilled Movements," *Personnel Journal,* Vol. 11, No. 1, pp. 28–43, June, 1932.

and athletes, all of whom must use fast and accurate motions or movements.

It is not difficult to develop the free, loose, easy movements of the wrist and forearm. The hand should move about the wrist for the shorter motions, and the forearm about the elbow for the longer motions. Experiments show that wrist and elbow movements are faster than finger or shoulder movements.[17]

8. Work should be arranged to permit an easy and natural rhythm wherever possible.

Rhythm is essential to the smooth and automatic performance of an operation. Rhythm may be interpreted in two different ways. Perhaps it is most frequently understood to mean the speed or the rapidity with which repeated motions are made. Reference is commonly made to the rhythm of walking or breathing. The operator feeding material into a machine is said to work with a rhythm depending upon the speed of the machine. Rhythm, then, in this sense, refers to the regular repetition of a certain cycle of motions by an individual.

Rhythm may be interpreted in a second way:

A movement may be perfectly regular, uniform, and recurrent and yet not give the impression of rhythm. If one moves the hand or the arm in a circle, the hand may be made to pass a point in a circle much oftener per second than the tempo of the slower rhythms requires, and yet there will be no feeling of rhythm *so long as the hand moves uniformly and in a circle*. In order to become rhythmic in the psychological sense, the following change in the movement is necessary. The path of the hand must be elongated to an ellipse; the velocity of the movement in a part of the orbit must be much faster than in the rest of the orbit; just as the hand comes to the end of the arc through which it passes with increased velocity, there is a feeling of tension, of muscular strain; at this point the movement is retarded, almost stopped; then the hand goes on more slowly until it reaches the arc of increased velocity. The rapid movement through the arc of velocity and the sudden feeling of strain and retarding at the end of this rapid movement constitute the beat. In consciousness they represent one event, and a series of such events connected in such a movement cycle may be said provisionally to constitute a rhythm. Every rhythmic beat is a *blow*. . . . In all forms of activity where a rhythm is required, the stroke, the blow, the impact, is the thing; all the rest is but connection and preparation.[18]

[17] Wm. L. Bryan, "On the Development of Voluntary Motor Ability," *American Journal of Psychology,* Vol. 5, No. 2, p. 171.

[18] R. H. Stetson, "A Motor Theory of Rhythm and Discrete Succession," *Psychology Review,* Vol. 12, No. 4, p. 258.

Rhythm, either in the sense of a regular sequence of uniform motions, or in the sense of a regular sequence of accented motions, is of value to the worker. Uniformity, ease, and even speed of work are promoted by the proper arrangement of the work place, tools, and materials. The proper sequence of motions enables the worker to establish a rhythm which assists in making the operation practically an automatic performance—the operator does the work without mental effort.

In many kinds of work there is an opportunity for the operator to accent certain points in a cycle of motions. For example, every punch-press operator, feeding the press by hand, tends to feed the sheet of material forward with a sudden thrust which constitutes an accented point in the cycle. Where the work permits, it is most natural for the worker to fall into a rhythm in this second sense.

Individual Rhythm. Some have suggested that each individual has a "natural" rhythm or speed of movement that permits him to work with least effort. Some have urged that individuals should be permitted to work at this natural speed and that no outside force, such as a wage incentive, should be exerted to cause the individual to work faster than his natural rhythm.[19] Since it seems difficult to determine what the natural rhythm is for any person and since most workers can be taught to change their rhythm in performing the same work (to work at different speeds or use different sets of motions), it seems that too much emphasis should not be placed on this so-called natural rhythm. Habit acts in a powerful way to affect the speed and the sequence of motions which a worker uses in performing a task. Once the habit is formed, it *does* require real effort on the part of the worker to change or modify this habit.

To illustrate this point, a typewriter company had several polishers of long experience who had for several years been polishing a particular part of the typewriter. These polishers had been accustomed to take a definite number of strokes across the polishing wheel, and they knew the finish that the piece should have to pass inspection. In a new design of the typewriter this particular piece was located in a more obscure position than formerly and did not need such a high polish. The polishers were told just how the piece was to be polished for the new typewriter, and they were carefully instructed as to the finish that would now be required to pass inspection. The operators, however, found it difficult to change their habits. They "forgot" to

[19] E. Farmer, "Time and Motion Study," *Engineering and Industrial Management,* Vol. 7 (N.S.), No. 5, p. 138.

take fewer strokes; as a result they were turning out work that was of higher quality than needed and their output was lower than it should have been. With constant and persistent attention, after 4 days these polishers were able to produce parts having just the finish desired and at a proportionately faster speed in pieces per hour.

Nearly every worker finds that a conscious effort and some persistence are required to do a new task or to perform an old one in a new way. For most people, change is by no means impossible and usually can be readily made. There are cases where a certain sequence of motions has been made by a person for such a long period of time that it is unwise to try to change it. This can, perhaps, also be said about the speed at which some people work.

When a worker becomes fatigued or when he is distracted or voluntarily wishes to produce less, he may either slow down his speed and maintain a slower rhythm, or he may introduce delays or interruptions into the cycle, in the form of extra motions.

Effect of Fatigue on Rhythm. In a study of polishing in a silverware factory, it was found that during the morning the polishers worked at a uniform rate and the units were finished at regular intervals.[20] In the afternoon, however, the pressure used in holding the knife or the spoon against the polishing wheel increased, more strokes were used, and the time for polishing each piece was greater than in the morning when a regular rhythm was maintained. Fatigue, then, seems to break up the rhythm and disturb the coordination that makes for rapid and easy work. "The tired worker is, therefore, not only working slower than when she is fresh, but is also expending her energy extravagantly." [21]

9. *Eye fixations should be as few and as close together as possible.*

Eye Movements. Although some kinds of work can be performed with little or no eye direction, where visual perception is required it is desirable to arrange the task so that the eyes can direct the work effectively; that is, the work place should be so laid out that the eye fixations are as few and as close together as possible.

Figure 145 shows head, eye, and hand motions of the operator performing a simple assembly operation. Small steel washers enameled green on one side and black on the other were to be assembled with green side up in the fixture directly in front of the operator. Dupli-

[20] E. Farmer and R. S. Brooke, "Motion Study in Metal Polishing," Industrial Fatigue Research Board, *Report* 15, pp. 1–65.

[21] *Ibid.,* p. 51.

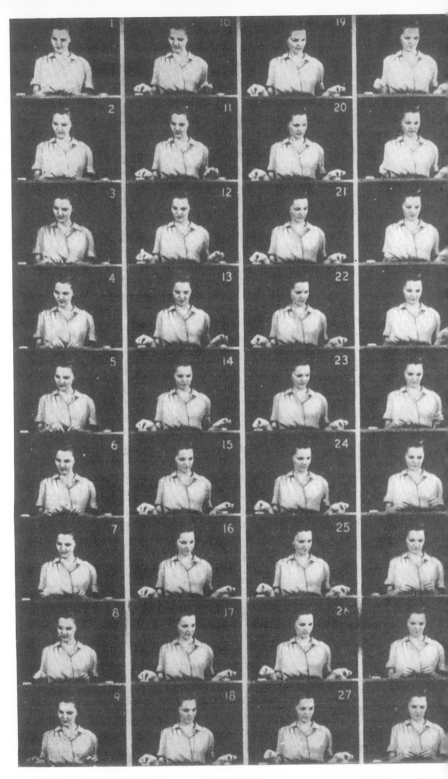

Fig. 145. Print of motion picture film showing eye and hand motions of the operator assembling small parts.

cate bins containing the washers were located on each side of the fixture. As the figure shows, it was necessary for the operator to look first to the right and then to the left before grasping the washers. The first strip of film in Fig. 145 shows the operator looking to her right preparatory to grasping a washer from the bin at her right. The second and third strips of film show her looking to her left and grasping a washer from the bin at her left. The fourth strip of film shows the two hands moving simultaneously, carrying washers to the fixture. The 36 consecutive frames of film were made at 1000 exposures per minute.

The distance that the eyes and the hands have to move and the nature of the operation will determine whether the hands must wait for the eyes, thus increasing the time to perform the task. In this case, had the containers been placed directly in front of the operator, the head movements would have been eliminated entirely and the eye movements would have been greatly reduced.

Packaging Small Parts. A study [22] of various methods of packaging small parts was made in order to find the most effective one. The operation consisted of placing seven small screws of four different sizes in a small envelope and then sealing it. Figure 146 shows the layout of the work place for the old method. By arranging the materials as shown in Fig. 147 a substantial saving in time resulted. However, a second improvement was made later (Fig. 148) which further reduced the time for the operation. Some visual direction was required in this operation, and the work place as finally arranged enabled the operator to reduce the extent of the head and eye movements and also to shorten the hand motions. This further illustrates that eye motions should always be considered in determining the best method of doing a task.

Eye-Hand Coordination. In a study [23] of the effect of practice on individual motions of a punch-press operation, one of the observations involved eye movements.

The operation was the forming of a relay contact bar. The fixture and work-place arrangements shown in Figs. 149 and 150 were designed to duplicate the mechanical movements and hand motions of the actual factory operation.

[22] This study was made by Bert H. Norem and John M. MacKenzie.
[23] Ralph M. Barnes, James S. Perkins, and J. M. Juran, "A Study of the Effect of Practice on the Elements of a Factory Operation," *University of Iowa Studies in Engineering, Bulletin* 22.

Fig. 146. Layout of work place for packaging wood screws—old method: *A*, envelopes with gummed flap; *B*, ½-inch No. 5 wood screws; *C*, ¾-inch No. 5 wood screws; *D*, 1-inch No. 7 wood screws; *E*, 1-inch No. 9 wood screws; *F*, moistener; *G*, filled envelopes.

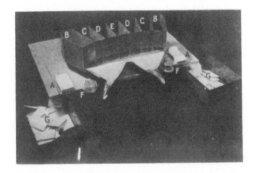

Fig. 147. Layout of work place for packaging wood screws—first improvement.

Fig. 148. Layout of work place for packaging wood screws—second improvement.

The eye movements and the hand motions of the beginner (Figs. 149 and 150-I) are as follows.

As the tweezers start to open when releasing the part in the die, the eyes shift to the part in the left hand to direct the tweezer grasping of the next part. The first fixation of the eyes occurs at *A* in Fig. 149-I.

Before the right hand releases the part into the tweezers, the eyes

I. As the tweezers start to open when releasing the part in the die, the eyes shift to the part in the left hand to direct the tweezer-grasping of the next part. First fixation at A.

II. Before the right hand releases the part in the tweezers, the eyes shift to the supply tray to select the next part. Second fixation at B.

III. After the left hand is sufficiently well directed towards the part on the supply tray, the eyes shift to the die to direct the right hand in locating the part over the pilot pins. Third fixation at C.

IV. The eyes remain fixed on the die until the part is properly located. The part is ejected by a foot pedal as the right reaches for the next part.

Fig. 149. Punch-press operation, showing eye fixations and hand motions of a beginner. Three fixations were used per cycle.

shift to the supply tray to select the next part. The second fixation occurs at *B* in Fig. 149-II.

If the left hand is sufficiently well directed toward the part on the supply tray, the eyes shift to the die to direct the right hand in locating the part over the pilot pins. The third fixation occurs at *C* in Fig. 149-III.

The eyes remain fixed on the die until the part is properly located. The part is ejected by a foot pedal as the right hand reaches for the next part.

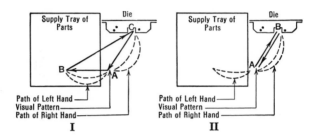

Fig. 150. Punch-press operation. Schematic drawings, showing eye fixations and hand motions. I, three eye fixations; II, two eye fixations.

After 10,000 cycles of practice, however, 56% of the cycles had three fixations and the remaining 44% had two fixations. At first the cycle time averaged 0.0584 minute; after 10,000 cycles of practice the average time was 0.0258 minute. When only two fixations occurred, the hand movements were the same but the eyes did not fixate on the supply of parts. The eyes would fixate on the part as it was transferred from the left hand to the right hand at *A* in Fig. 150-II, and then would move to the fixture to direct in locating the part over the pilot pins at *B* in Fig. 150-II. Although at first it was necessary to look at the parts in the tray to facilitate the grasping, after practicing a less-defined picture was required (Fig. 151). It is believed that attention was directed to the parts and to the hand in grasping them, but that it was not essential for the eyes to see the parts so clearly.

It seems that the better coordination resulting from practice not only enabled the operator to perform each of the motions in less time (although they were not all affected in the same way with practice) but also reduced the number of fixations required.

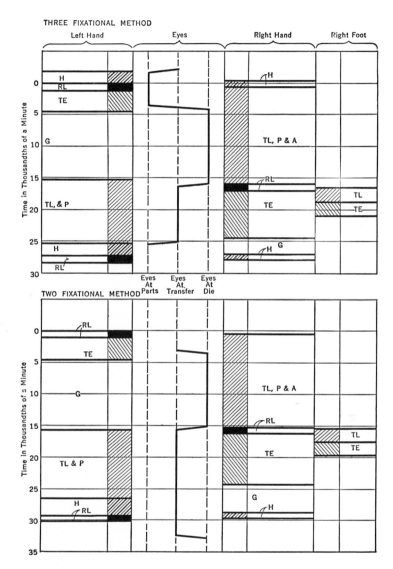

Fig. 151. Eye-hand simo chart of punch-press operation, showing the three-fixation method and the two-fixation method.

CHAPTER 18

Principles of Motion Economy as Related to the Work Place

✓ **10. There should be a definite and fixed place for all tools and materials.**

The operator should always be able to find the tools and materials in the same location. Likewise, finished parts and assembled units should be disposed of in fixed places. For example, in the assembly of the bolt and washers, the hand should move without mental direction to the bin containing the rubber washers, then to the bin containing the steel washers, then to the lock washers, and finally to the bolts. It should be unnecessary for the operator to have to think where the materials are located.

Definite stations for materials and tools aid the worker in habit formation, permitting the rapid development of automaticity. It cannot be emphasized too strongly that it is greatly to the worker's advantage to be able to perform the operation with the least conscious mental direction. Frequently, materials and tools are scattered over the work place in such a disorderly fashion that the operator must not only exert mental effort, but must also hunt around in order to locate the part or tool needed at a given instant. The workers are very much in favor of having definite stations for materials and tools, since this reduces fatigue and saves time. There can be no virtue in requiring the worker to exert the unnecessary effort of deciding just what tool to pick up next or what part to assemble next, when by simply arranging the materials and tools properly, with a little practice the operator will automatically perform the work in the proper sequence, at a rapid rate, and with a minimum expenditure of effort.

When the eyes must direct the hand in reaching for an object, the eyes ordinarily precede the hand. However, if materials or tools are located in a definite place and if they are always grasped from the

256

same place, the hand automatically finds the right location and in many cases the eyes may be kept fixed on the point where the tools or materials are used.

Shipping Room Table. Motion study principles have been successfully applied in many nonmanufacturing establishments, such as offices, restaurants, hotels, department stores, and mail-order houses. Figure

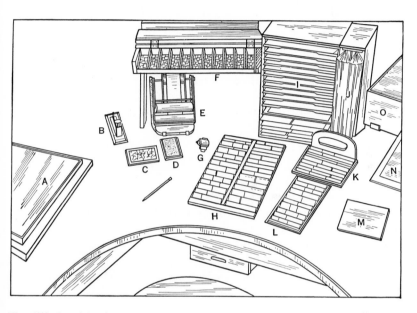

Fig. 152. Special table for weighing, stamping, and billing parcel-post packages for shipment in mail-order house: A, scales; B, stapler; C, pin-disposal box; D, stamp pad; E, gum tape dispenser; F, postage stamps; G, counter; H, refund vouchers; I, supply of form letters; J, re-order blanks; K, "due us" forms; L, postal collection forms; M, scratch pad; N, adding machine; O, irregular billing papers.

152 shows a special semicircular work table designed for weighing, stamping, and billing parcel post packages in the shipping room of a large mail-order house.[1] The packages to be shipped come down to the work table on a slide at the extreme left of the table, are weighed, stamped, billed, and pushed off onto a belt conveyor at a point adjacent to the incoming slide. It is not necessary to lift the package. Note that the table has "cutouts" for scales, pins, forms, stamp pad, adding machine, etc. A drawer under the table provides a place for the

[1] Illustration and data courtesy of John A. Aldridge.

Fig. 153. Special tool-chest trucks provided for maintenance electricians at Douglas Aircraft Company plants. Service cribs or shops are centrally located in the production areas, thus enabling electricians to answer trouble calls by traveling short distances with all the tools they will need. (From "Decentralized Maintenance for Continuous Output," by A. T. Kuehner, *Factory Management and Maintenance*, Vol. 101, No. 3, pp. 123–128.)

operator's personal belongings. This work-place layout is typical of the care with which every activity in this organization has been studied and indicates how the work has been made easier.

11. *Tools, materials, and controls should be located close to the point of use.*

Very frequently the work place, such as a bench, machine, desk, or table, is laid out with tools and materials in straight lines. This is incorrect, for a person naturally works in areas bounded by lines which are arcs of circles.

Normal Working Area. Considering the horizontal plane, there is a very definite and limited area which the worker can use with a normal expenditure of effort. There is a normal working area for the right hand and for the left hand, working separately, and for both hands working

Fig. 154. Normal and maximum working areas in the horizontal plane.

together (Figs. 154 and 155). The normal working area for the right hand is determined by an arc drawn with a sweep of the right hand across the table. The forearm only is extended, and the upper arm hangs at the side of the body in a natural position until it tends to swing away as the hand moves toward the outer part of the work place. The normal working area for the left hand is determined in a similar manner. The normal arcs drawn with the right and left hands

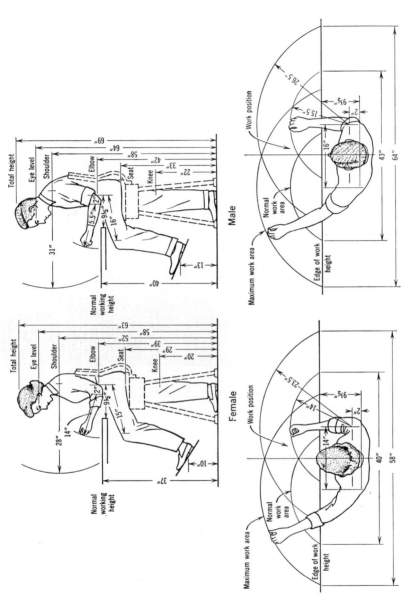

Fig. 155. Dimensions of normal and maximum working areas in the horizontal and vertical planes as developed and used by the Process Development Section of the General Motors Manufacturing Staff. (From Richard R. Farley, "Some Principles of Methods and Motion Study as Used in Development Work." *General Motors Engineering Journal*, Vol. 2, No. 6, pp. 20–25.)

will cross each other at a point in front of the worker. The overlapping area constitutes a zone in which two-handed work may be done most conveniently.

Maximum Working Area. There is a maximum working area for the right hand and for the left hand, working separately, and for both hands working together (Figs. 154 and 155). The maximum working area for the right hand is determined by an arc drawn with a sweep of the right hand across the table, with the arm pivoted at the right

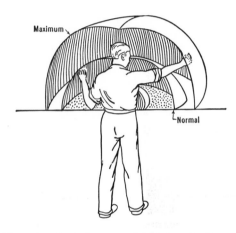

Fig. 156. Normal and maximum working space in three dimensions.

shoulder. The maximum working area for the left hand is determined in a similar manner by an arc drawn with a sweep of the left hand. The overlapping area formed by these two maximum arcs constitutes a zone beyond which two-handed work cannot be performed without causing considerable disturbance of posture, accompanied by excessive fatigue.

Each hand has its normal working space in the vertical plane as well as in the horizontal plane, in which work may be done with the least time and effort (Fig. 156). A maximum work space in the vertical plane may also be determined, beyond which work cannot be performed without disturbing the posture. In locating materials or tools above the work place, consideration should be given to these facts.

Figures 157 and 158 emphasize the importance of arranging the material *around* the work place and as close in as possible. In Fig. 157 the five bins containing materials are outside the maximum working

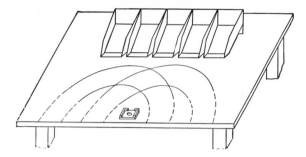

Fig. 157. Incorrect work-place layout. Bins are located too far from the assembly fixture. The operator must bend forward to get parts from bins.

area, necessitating bending the body to reach them. In Fig. 158 the bins have been located within the normal working area, permitting a third-class motion which requires no movement of the body. The use of a duplicate fixture and duplicate bins arranged symmetrically on each side of the fixture permits the two hands to make simultaneous motions in opposite directions in performing the operation. Such an arrangement facilitates natural, easy, rhythmical movements of the arms.

Those tools and parts that must be handled several times during an operation should be located closer to the fixture or working position than tools or parts that are handled but once. For example, if an

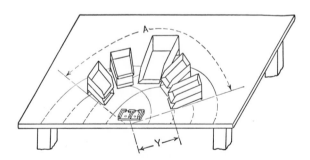

Fig. 158. Correct work-place layout. Bins are located close to the fixture, enabling the operator to get parts from any of the bins with easy, quick forearm motions. In many kinds of work the eyes must direct the hands. In such cases the work area should be located directly in front of the operator so that eye fixations will be as few and as close together as possible. In other words, angle A should be as small as possible, and distance Y should be as short as the nature of the work will permit.

operation consists of assembling a number of screws into a metal switch plate, the containers for the screws should be placed closer to the fixture than the containers for the plates. This is done because only one plate must be transported from the container to the fixture per cycle, whereas several screws have to be transported from their containers to the fixture.

In considering this point it is equally important to remember that the parts must be arranged in such a way as to permit the shortest eye movements, the fewest eye fixations, and the best sequence of motions, and to aid the operator in rapidly developing automatic and rhythmical movements.

Results of Moving Parts Closer to Fixture. The production of one model of a radio requires the assembly of 260 separate parts or sub-assemblies. Two hand movements are required to pick up each part from the supply bin and process or assemble it—one movement of the hand to the bin and one from the bin. By shortening the distance 6 inches for reaching each of these parts, there is a saving in time of 34,000 hours per year.

Number of parts moved	260
Movements (motion of hand to and from bin)	2
Average saving in time to move hand 6 inches shorter distance, minute	0.002

or

$$260 \times \frac{2 \times 0.002}{60} = 0.017 \text{ hour per radio set}$$

This saving of 0.017 hour or 62 seconds per radio set per day is extremely small. However, since this company makes 8000 sets per day, the savings per day are

$$8000 \times 0.017 = 136 \text{ hours per day}$$

Consider this production to run 250 working days per year:

$$250 \text{ days} \times 136 \text{ hours per day} = 34,000 \text{ hours saved per year}$$

Another way to look at this is in total distance saved. If 6 inches is saved in the movement of the hand to the bin and another 6 inches from the bin, the total savings are 12 inches, or 1 foot per piece.

$$260 \text{ pieces} \times 1 \text{ foot} = 260 \text{ feet saved per set}$$

8000 sets $\times$ 260 feet per set = 2,080,000 feet or 394 miles saved per day. 250 working days $\times$ 394 miles per day = 98,500 miles saved per year.[2]

Arrangement of Machines. The following statement might be considered as a corollary to rule 11: *In the continuous or progressive type of manufacturing, machines, process apparatus, and equipment should be arranged so as to require the least possible movement on the part of the operator.*

The machines in Fig. 159 are laid out in a straight line along a trucking aisle. Space is provided between the machines for a skid platform on which material is placed before and after being processed.

Fig. 159. Machines laid out in the conventional way. Material is moved to and from the machines on skid platforms by lift trucks. An aisle permits access to each machine.

Fig. 160. Machines laid out parallel to a belt or roller conveyor. Material is moved to and from the machines by conveyor, and no trucking aisle is needed. The operator turns through 180 degrees to use the conveyor.

When one man operates several machines, it is necessary for him to walk a considerable distance because the machines are spread out over so much floor space.

The trucking aisle is unnecessary and walking is reduced when the machines are located along a conveyor. Machines are frequently placed parallel to the conveyor as in Fig. 160. Such an arrangement, although better than the one illustrated in Fig. 159, still requires the operator to turn completely around in transporting material from the machine to the conveyor, and vice versa. A better arrangement is shown in Fig. 161, where the machines are placed perpendicular to the conveyor and close to it. This arrangement permits the operator to move material to and from the conveyor with less movement of the body.

There is still a fourth method of laying out machines, which can often be used to advantage (Fig. 162). The machines that can be operated by one man are grouped close together, so that the minimum

[2] This case developed for use in RCA training course by G. A. Godwin while industrial engineer for RCA Victor Division of Radio Corporation of America.

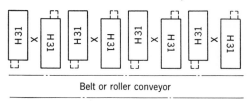

Belt or roller conveyor

Fig. 161. Machines laid out perpendicular to a belt or roller conveyor. The operator turns through 90 degrees to use the conveyor.

time is required for the operator to move from loading one machine to removing the finished piece from the next and loading it. Often machines used to perform successive operations on a part can be grouped together so that the part, in the form of a casting or forging, might begin the process at A (skid platform in Fig. 162), the first operation being performed by machine H31, the next by machine L12, and the third by H31B. The machine time and the handling time would have to be so balanced that the operator could keep the machines in operation without too much loss of machine time. From the third machine the part is sent, if necessary, to the next group of machines by means of a chute shown at D.

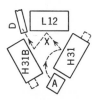

Fig. 162. Machines arranged in groups. Machine time and handling time are so balanced that one man can keep the entire group in operation.

Shipping Department Operations. The application of motion economy principles to shipping department operations may result in substantial savings in time and labor costs.

Old Packing Bench. A packing bench designed by C. H. Cox for use in the shipping department of Merck & Company shows how the principle of locating materials and tools close in front of the operator made it possible for the operator to do his work more easily and faster.[3]

The packing operation consisted mainly of packing for shipment bottles and boxes containing chemicals. Figure 163 shows the original 9-foot-long flat-top packing bench containing a tape machine, glue pots, nail boxes, hammer, stencil brush, knife, scissors, etc. Hoods, wrappers, and pads for individual container protection and special box labels were stored along the back half of the bench and in the cabinets below the working surface. None of the equipment or material

[3] C. H. Cox, "Work Simplification Applied to the Shipping Department," *American Management Association, Production Series* 115, p. 3.

Fig. 163. Packing bench—old design.

was pre-positioned in definite locations. Since the actual packing was not done on this bench, all equipment and material were outside the maximum working area. Boxes were set up and packed on the 2-foot by 3-foot packing "buck" set perpendicular to the packing bench. A loose bale of excelsior was placed on the floor opposite the bench and on the left side of the buck.

The operator carried each piece of stock from shelf truck to bench, hooded or wrapped each item, carried it to the box, placed it in position, stepped to the left for excelsior, stepped back to the box, and placed the excelsior. Each use of glue, labels, or stencil brush meant

Fig. 164. Packing bench—improved design.

several steps along the packing-bench area. The operator in Fig. 163 is cutting tape for sealing the packed box on the buck.

New Packing Bench. The new packing bench (Fig. 164) combines all three former units into one fixture. All equipment and material are conveniently positioned within the maximum working area. On the left is a tin-lined excelsior bin, and on the right is a packing buck. A compartment for hammer, stencil brush, knife, etc., is located above the buck; above this is a drawer for nails and tacks; to the left of this drawer is a holder for the operator's pencil and a slide for his production record. On the extreme right side of the bench a shelf holds the tape dispenser; a large compartment houses a new-style glue dispenser; then come four compartments for commonly used corrugated separators, and a small pigeonhole slot for special stencils. The extreme left-hand side of the fixture contains material infrequently used, such as six sizes of special labels, asbestos pads, large-size hoods, and long strips of corrugated wrappers.

Fig. 165. Standard bins of the gravity-feed type.

If the operator desires to do so, he may stand in one position to select and make up the box, select stock to be packed from the monorail carrier, reach glue and tape, select all internal packing material, reach necessary stencils and stencil brush, and record the work on his production record. Although it is not recommended that the operator stay in one fixed location, this packing bench has eliminated thousands of unnecessary steps for each packer every day.

12. *Gravity feed bins and containers should be used to deliver material close to the point of use.*

A bin with sloping bottom permits the material to be fed to the front by gravity and so relieves the operator of having to dip down into the container to grasp parts (Figs. 127 and 129). However, it is not always possible to slide material into position as in the bolt and washer assembly. More frequently bins like those shown in Fig. 165

Fig. 166. Standard work-place equipment: *A*, lip tray, length (from back to lip) 5½ inches, width 2⅛ inches, 4¼ inches, or 8½ inches; *B*, edge tray, length (from back to lip) 4¼ inches or 8½ inches, width 5½ inches; *C*, open bin—bench type, length (from back to front) 8 inches, width 5 inches, 8 inches, or 10 inches, depth 8 inches; *D*, open bin—rack type, length (from back to front) 8 inches, width at back 8½ inches, width at front 5½ inches, depth 3 inches; *E*, curved rack to support trays, depth 5⅜ inches, height 4⅞ inches; *F*, universal brackets for mounting fixtures; *G*, tote-box dolly designed to hold tote boxes of material or finished work.

are used. Where many different parts are required, as in the assembly of an electric switch, it becomes necessary to nest the bins one above the other in order to have the material within convenient reach of the operator.

Bins of standard sizes, such as those shown in Fig. 165, are standard equipment in many plants. The bins are interchangeable and are made in three heights and three widths. By the use of these standard-unit·

bins, any combination can be made to suit a particular job. It is difficult to give a general rule as to the proper size of bins for a particular operation. Some companies try to have their bins large enough to hold material for 4 hours' work, which probably is an economical size for many kinds of material.

Figure 166 shows standard work-place equipment used by the RCA Manufacturing Methods Division.[4] Bins, tool holders, flat trays, solder iron holders, etc., are interchangeable and may be mounted with equal facility on a workbench, drill press, or riveting machine, or hung on any standard rack in any position. This standard equipment is entirely

Fig. 167. Bin with tray attached to facilitate *select* and *grasp* of parts.

flexible and can be readily adapted for the manufacture of new radio apparatus. When a new type of radio is to be put into production, it is a simple matter to disassemble the standard bins and equipment and set them up again for the new job. The workbench itself is made in standard sections and is fitted with pipe to carry compressed air and conduit for electric power. When a long bench is needed, several standard bench sections are bolted together, electric lines being coupled together and plugged into the main power circuit. The regular setup man is able to complete the job, making it unnecessary to have an electrician or a pipe fitter.

Figure 167 shows a bin with a long spout or tray attached. This tray facilitates the grasping of very small parts. A number of parts are drawn from the bin to the tray. It is then easy to select and grasp individual parts. The ordinary gravity bin (Fig. 165) may have a tray spout attached to it. This type of bin is superior to that shown in Fig. 167, since it does not need to be refilled so often.

[4] Illustration and data courtesy of RCA Victor Division of Radio Corporation of America.

Table 12. Time Required to Grasp, Carry, and Dispose of Machine-Screw Nuts and Machine Screws from Various Types of Bins

	1—Hopper Type Bin		2—Rectangular Bin		3—Bin with Tray	
	Nuts	Screws	Nuts	Screws	Nuts	Screws
Time in Minutes	**0.0138**	0.0157	**0.0148**	0.0161	**0.0116**	0.0143
Time in Per Cent (Shortest Time = 100%)	**119**	110	**128**	113	**100**	100

A Study of Three Types of Bins. The results of a study [5] of the time to grasp machine screws and machine-screw nuts from various types of bins are shown in Table 12.

The operation consisted of selecting and grasping with the right hand a machine screw or nut from a bin, carrying it through a distance of 5 inches, and releasing it into a hole in the table top. The time for each of the motions select and grasp, transport loaded, release load, and transport empty was accurately measured. The bin with tray (3) required the least time; the hopper-type bin (1) required 19% more time; and the rectangular bin (2) required 28% more time, than did bin 3.

13. *Drop deliveries should be used wherever possible.*

The work should be arranged so that the finished units may be disposed of by releasing them in the position in which they are completed, thus delivering them to their destination by gravity. This saves time, and moreover the disposal of the objects by simply releasing them frees the two hands so that they may begin the next cycle simultaneously without breaking the rhythm. If a chute is used to carry the finished parts away, it should be located so that the parts can be released in the position in which they are finished, or as close to this point as possible.

[5] *University of Iowa Studies in Engineering*, Bulletin **16**, p. 28; also, *Iron Age*, Vol. 19, No. 13, pp. 32–37.

A perfect example of this is shown in Fig. 168. The operation is burring a hole in the end of a small angle plate. The drill is fed by means of a foot pedal, and the angle plate is held in position for burring by means of a fixture. The fixture is mounted on the drill-press table and extends up through a plywood board mounted 6 inches above the table. This board serves as an auxiliary work place, making it unnecessary to cut disposal holes through the drill-press table

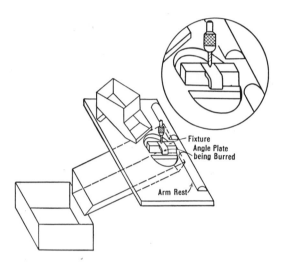

Fixture
Angle Plate
being Burred

Arm Rest

Fig. 168. Foot-operated drill press for burring small parts. Finished parts drop out of the fixture into the disposal chute by gravity.

itself. Holes cut in the board on either side of the fixture lead to a disposal chute underneath.

The part to be burred is placed in the fixture, and the drill is brought down against it. This holds the part in position while it is being burred, and when the burring is completed and the drill is raised, the burred plate drops out of the jig by gravity into the top of the disposal chute. It was economical to equip the drill press as described here because of the large quantity of burring to be done.

In the bolt and washer assembly (Fig. 127) it was necessary to lift the finished assemblies out of the fixture and move them a few inches to one side before releasing them into the chute. A still better arrangement would have been to have the assemblies drop through the fixture by moving some sort of a trip on the bottom of the fixture which could have been actuated by a foot pedal. This arrangement, however, would have added to the cost of the fixture and was not justified in the factory where this fixture was used.

Many people do not appreciate the amount of time that may be used in disposing of finished parts. A study was recently made of gauging small pins in a fixture mounted on the front edge of the table and disposing of them by tossing them into a tote box located first at a distance of 3 inches behind the fixture, then at a distance of 10 inches, and finally at a distance of 20 inches. The time required for the motions transport loaded and release load was least when the pins were tossed into the bin nearest the fixture. Eighteen per cent more time was required for the bin at 10 inches and 34% more at 20 inches.

14. *Materials and tools should be located to permit the best sequence of motions.*

The material required at the beginning of a cycle should be placed next to the point of release of the finished piece in the preceding cycle. In the assembly of the bolt and washers (Fig. 127) the rubber washers were in bins located next to the chute into which the assemblies were disposed as the last motion of the previous cycle. This arrangement permitted the use of the two hands to best advantage at the beginning of the new cycle.

The position of the motion in the cycle may affect the time for its performance. For example, the time for the motion transport empty is likely to be longer when it is followed by the motion select than when it is followed by a well-defined motion such as a grasp of a pre-positioned part. The reason for this is that the mind begins to select during the transport empty. When the motion transport loaded is followed by a position motion, it is slowed down by the mental preparation for the position. The time for the motion grasp is affected by the hand velocity preceding the grasp. A satisfactory sequence of motions in one kind of work may aid in determining the proper sequence in other types of work.

15. *Provisions should be made for adequate conditions for seeing. Good illumination is the first requirement for satisfactory visual perception.*

Visual perception may take place under such widely varying conditions that adequate provisions for seeing in one kind of work are not always most suitable for another. For example, the provisions for seeing on such very fine work as watch making would be different from those recommended for inspecting "leather cloth" or tin plate for surface defects. However, if adequate illumination is provided, seeing is made easier in every case, although this may not be the complete solution of the problem. By adequate illumination is meant (1) light of sufficient intensity for the particular task, (2) light of the proper color and without glare, and (3) light coming from the right direction.

It should be borne in mind that the visibility of an object is determined by the following variables: [6] brightness of the object, its contrast with its background, the size of the object, the time available for seeing, the distance of the object from the eye, and other factors such as distractions, fatigue, reaction time, and glare. These variables are so related that a deficiency in one may be compensated by an augmentation of one or more of the others, provided all factors are above certain limiting values.[7]

The intensity of illumination falling on an object and the reflection factor of the object or that of its background should be considered together in providing adequate illumination. For example, the pages of a telephone directory are dark in color, and the contrast between the printed letter and the page is not so great as that of printing on good book paper. The paper of the directory reflects only 57% of the incident light, whereas book paper reflects about 80%. Two to three times as much light is required to read a telephone directory as is required to read with equal facility the same critical details of names and numbers printed with blacker ink on white book paper.[8] The task of sewing on very dark cloth is difficult even under the best conditions of lighting. For example, dark cloth of 4% reflection factor would require 200 foot-candles to produce the same brightness as 10 foot-candles on white cloth.[9] A knowledge of this point suggests the use of greater intensity of illumination or lighter background for work with objects with a low reflection factor or for very fine work. The size of the image of the object falling on the retina of the eye must be sufficiently large to allow adequate discrimination of the details. This factor requires greatest consideration in very fine work. An in-

[6] M. Luckiesh and F. K. Moss, "The Applied Science of Seeing," *Transactions of the Illuminating Engineering Society,* Vol. 28, p. 846.

[7] M. Luckiesh and F. K. Moss, "The Human Seeing-Machine," *Journal of the Franklin Institute,* Vol. 215, No. 6, p. 647.

[8] M. Luckiesh, *Seeing and Human Welfare,* Williams & Wilkins, Baltimore, p. 85.

[9] M. Luckiesh and F. K. Moss, "The Applied Science of Seeing," *Transactions of the Illuminating Engineering Society,* Vol. 28, p. 854.

crease in the illumination on the object, or an increase in the contrast between the object and its surroundings, produces the same effect, within limits, as a decrease in the distance between the eye and the object.

Relief of Eyestrain on Fine Assembly Work. The following case [10] shows the changes that may be made to improve the seeing on fine assembly work. The operation was assembling and adjusting the parts

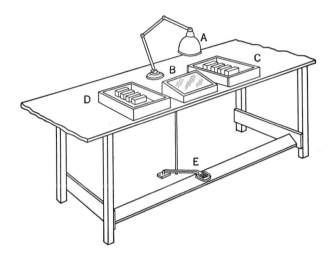

Fig. 169. Improved lighting unit: *A*, adjustable lamp for direct light; *B*, background light; *C*, finished work; *D*, work awaiting adjustment; *E*, foot-operated switch for adjustable lamp.

of a delicate electric meter mechanism. The task was performed by men, and about three quarters of an hour was required for each unit. Eyestrain and fatigue were excessive, owing to the fact that on certain parts of the operation the illumination was so inadequate in relation to the smallness of the parts that the work had to be held close to the eyes.

To remedy this condition a rest period was introduced and improvements were made in the illumination of the work place. Figure 169 shows the improved lighting units. Because certain parts of the operation could be done best by silhouetting the mechanism against an illuminated background, a background light was placed on the work bench and was kept "on" all the time. When it was necessary to view

[10] J. H. Mitchell, "The Relief of Eyestrain on a Fine Assembly Process," *The Human Factor*, Vol. 10, No. 10, p. 341.

the assembly under direct light, the foot pedal was depressed, turning on the upper lamp. Tests showed that the best color for the background light was white or pale yellow, and that it should be free of glare.

The effects of the rest period and of the improved illumination on six men in the experimental group over the period of the test were an improvement in the quality of the work and also an increase in output of 20%. The rest period was included as working time in calculating hourly output.

Use of Special Spectacles for Very Fine Work. On certain kinds of very fine work the eye must be kept very near the object, however high the intensity of illumination may be. The constant use of the eyes on objects at such close range imposes a serious strain [11] on the muscles of convergence and accommodation.[12] Experiments show that special spectacles are advisable to permit the eyes to assume their normal condition. An increase in output approximating 12% has been found to result from the use of glasses on such work as mounting lamp filaments, "linking" in hosiery making, and "drawing-in" in weaving processes.[13]

Time for Seeing. Seeing can take place only after the eyes come to a stop and are focused on the object. In the process of reading a printed page, for example, the eyes do not make a continuous movement along the line, but rather move in a series of jumps or leaps. The eyes begin at the left-hand end of the printed line and progress from one fixation to the next along the line to the right-hand end of the line. They then move back to the left-hand end of the next line with a single smooth sweep, during which movement they see nothing. The movements of the two eyes are coordinated, and one cannot move voluntarily without the other. The number of movements and pauses which the eyes make in reading a line of print will vary, usually from three to seven, depending upon the length of the line, the visibility of the print, the skill of the reader, and other factors.

It is generally agreed that the optimum length of line is 3 to 4

[11] H. C. Weston and S. Adams, "On the Relief of Eyestrain among Persons Performing Very Fine Work," Industrial Fatigue Research Board, *Report* 49, p. iii, 1928.

[12] "When a near object is viewed, two muscular actions take place simultaneously, the one causing a slight rotation of the eyes inwards toward each other, thus allowing the image to fall on the same point of the retina in each eye, the other a change in the curvature of the lenses of the eyes, the object being thereby kept in focus. The former of these is known as *convergence,* the latter as *accommodation. . . .*" *Ibid.,* p. iii.

[13] *Ibid.,* p. 5.

inches, and that it should not exceed 4 inches by very much. Ten-point type such as you are reading now seems to be the optimum size, although there is some evidence to show that the optimum may cover a considerable range.[14]

Fixation pauses require on the average 0.17 second. Tests show that the shortest interval of time possible for a person to gain an adequate visual impression of an object varies from 0.07 to 0.30 second, the average being 0.17 second.[15] The intensity of illumination affects the time required for seeing. "If an object of 50 per cent contrast can just be seen under a certain intensity of illumination when the time available is 0.30 second, the intensity of illumination must be trebled if it is to be visible when the time is reduced to 0.07 second." [16]

INSPECTION WORK

The provision for adequate conditions for seeing is of paramount importance in inspection work. Such work is usually highly repetitive, exacting in nature, and predominantly mental in its demands. Constant attention and almost continuous use of the eyes are required in many kinds of inspection work. Perception of a defect must be followed by instant action on the part of the inspector to reject the defective part. Some individuals are able to see smaller differences than others and to perceive the same differences with greater speed. Since reaction time and visual acuity are important elements in most inspection work, it is essential that persons be selected by means of suitable tests before being employed for such work.[17]

Inspection of Metal Spools. Some practical applications are included here to show how provisions were made for adequate conditions for seeing. The first case is the inspection of metal bobbin spools for dents, scratches, heavy paint, light paint, and bent flanges. Since the improved method of inspection employs a number of principles of motion economy in addition to those for adequate seeing, this operation is presented in some detail.

Original Method. The inspector was seated at a table, as shown in Fig. 170. The spools to be inspected were placed at her left in a large steel tote box *A*. The good spools were arranged in order in the small metal tray *B* at her right. Defective spools were tossed into trays at

[14] M. D. Vernon, *The Experimental Study of Reading,* Cambridge University Press, London, pp. 165–166.

[15] M. Luckiesh, *Seeing and Human Welfare,* Williams & Wilkins, Baltimore, p. 96.

[16] *Ibid.,* p. 96.

[17] S. Wyatt and J. N. Langdon, "Inspection Processes in Industry," Industrial Health Research Board, *Report* 63, p. 46.

the back of the table and directly in front of the inspector. They were classified as *C*, bent ends; *D*, light paint; *E*, overlap barrels; *F*, off-center flanges; *G*, culls; *H*, heavy paint.

Elements of the Operation. The inspector, turning to the tote box (previously positioned by the supply man) at her left, grasped spools

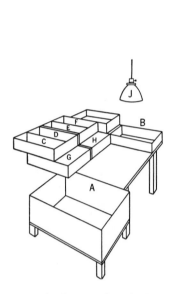

Fig. 170. Layout of work place for inspection of metal spools—old method: *A*, supply of spools to be inspected; *B*, good spools; *C-D-E-F-G-H*, rejected spools. Light was supplied from lamp at *J*.

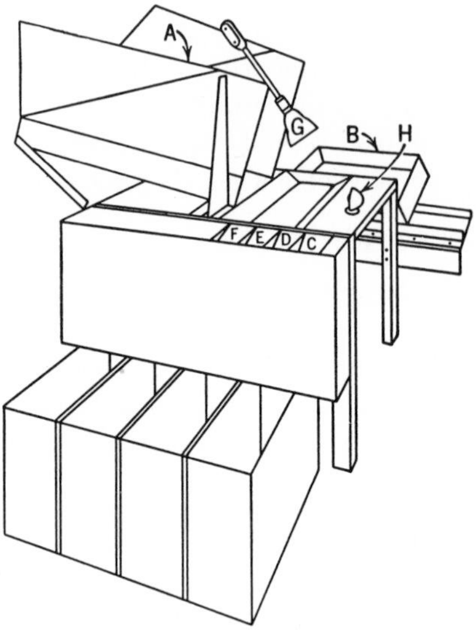

Fig. 171. Layout of work place for inspection of metal spools—first improved method: *A*, supply hopper, spools to be inspected; *B*, good spools; *C-D-E-F*, rejected spools. Note the location of lamps at *G* and *H*.

with both hands and carried them to the table in front of her, where she deposited them. This was repeated until a pile had been accumulated. The inspector procured an empty tray for the good spools from a pile at her right. She also positioned empty trays for various kinds of defects.

The inspector then proceeded with the inspection of the spools in the following manner.

1. She picked up one spool from the pile with the thumb and index finger of each hand, inspected the outside of flanges by looking straight

down on them, tipped spools slightly, and then by turning spools inspected them for bent ends. She turned spools end-for-end and repeated the above elements for the other flanges. Then she tipped spools back horizontally, and by turning spools around inspected for defects on the inside of flanges. If the spools were good, she flipped them back into the palm of her hand; if a defect was found, she disposed of the spool in the proper reject tray. These elements were repeated until three or four spools (depending upon the size of the spools) had been accumulated in each hand.

2. The inspector placed the spools held in her right hand in the tray of good spools at her right. She then transferred the spools accumulated in the left hand to the right hand, and placed these in the tray with her right hand. During this time the left hand was idle. The inspector then moved both hands to the pile in front of her and repeated the elements in 1.

3. As tiers of good spools were built up in the tray, the operator jogged the spools into position, pushing the tier over against the preceding one; or if it was the first tier, she pushed it over against the side of the tray.

4. When a tray was filled, the inspector made out a ticket and placed it in the end of the tray. She then placed the tray on the back of the table, where it was picked up by the supply man.

First Improved Method. The inspector was seated at a table, as shown in Fig. 171. The spools to be inspected were placed in the hopper A, Fig. 171, by the supply man and were fed by gravity down on the inspection table. The good spools were placed in order in the tray B at the inspector's right. This tray was tipped up at an angle and was placed at the correct height for disposing of the spools with least effort. When a defective spool was found, it was placed by the left hand into one of the four openings in the top of the table at the inspector's left. These spools went by chute to trays on the floor. Defective spools were classified as C, bent ends; D, light paint; E, heavy paint; or F, culls.

Elements of the Operation

ELEMENTS FOR GOOD SPOOLS

Left Hand	Right Hand
1. Pick up two spools.	1. Pack good spools in tray.
2. Transfer one spool to right hand.	2. Receive one spool from left hand.

3. Inspect upper flange under upper light.
4. Turn spool 60 degrees.
5. Inspect other flange in front of lower light.
6. Inspect barrel while spool is rotated between thumb and index finger, under upper light.
7. Flip spool to palm of hand.
8. Pick up one spool.

3. Inspect upper flange under upper light.
4. Turn spool 60 degrees.
5. Inspect other flange in front of lower light.
6. Inspect barrel while spool is rotated between thumb and index finger, under upper light.
7. Flip spool to palm of hand.
8. Pick up one spool.

Repeat elements 3, 4, 5, 6, 7, and 8 until there are three or four spools in each hand.

9. Transfer spools to right hand.

9. Grasp spools from left hand.

ELEMENTS FOR DEFECTIVE SPOOLS

1. When heavy paint, bent ends, or "jams" are found in elements 3, 5, or 6, reject spools to disposal chute. When light paint is found in element 3, inspect spool (elements 5 and 6) for other defects before rejecting.

1. Pick up another spool.

2. Grasp spool from right hand and reject it.

2. When a defective spool is found in right hand, transfer it to left hand and get new spool.

AUXILIARY ELEMENTS

1. Procure empty tray from pile behind inspector and position tray on table at right.
2. When tray is full make out ticket and place in end of tray.
3. Push finished tray of work to back of table ready for collection by the supply man.

Comparison of the Two Methods of Inspection. The first improved method of inspection was superior to the old method in the following ways:

1. Two lights on the new table furnished illumination for inspection, so that it was necessary only to turn the spools 60 degrees to inspect both ends. In the old method, using but one light, it was necessary to turn the spools end-for-end or 180 degrees. In the first improved method the intensity of illumination was greatly increased so that at the point of inspection there was 150 foot-candles. The bulbs were completely shielded to prevent glare.

2. The work of the two hands was so arranged that there was practically no idle time during the cycle.

3. The supply of spools was placed in the hopper by the supply man, and they were fed by gravity (occasionally pulled by the inspector with a hook) down to the inspection table. This saved the time of lifting the spools from the tote box to the table as required in the old method.

4. The rejected spools were dropped in openings located conveniently near the working position of the hands. In the old method the inspector had to toss the spools into trays piled in front of her.

5. The tray for receiving good spools was located at the proper height and was tipped up at a convenient angle.

6. The tray of finished work rested on a metal track and could be easily shoved to the back of the table, from which the supply man removed it. The inspector was not required to lift full trays of work.

7. Inspectors were given a 5-minute rest period at the end of each hour, and they were enthusiastic about this. Formerly one 5-minute rest period was provided in the morning and one in the afternoon.

8. Arm rests on the front of the table tended to steady the hands and reduce fatigue. Chairs were carefully adjusted to fit the individual inspector.

Training Inspectors. Considerable study was required in designing the new table and in determining the proper procedure for the inspection elements themselves. After the most satisfactory method was worked out, the inspectors were carefully trained. Slow-motion pictures were used to show the sequence of motions, and only after very careful and persistent training were the inspectors able to do the work in the proper manner and thus accomplish the expected amount of work per day.

Savings. Using the first improved method, the inspectors were able to inspect *twice* as many spools per day as formerly, and apparently with less eyestrain and fatigue. Less than half the floor space was required for the inspection work, and the department had a neater appearance than formerly. The quality of inspection did not suffer by the increased output per inspector. The new tables cost less than $25 each.

Second Improved Method. The first improvement in the method of inspecting spools was put into effect without requiring any change in the design of the spool itself and without the use of mechanical equipment, precision gauges, or other apparatus beyond a worktable of special design equipped with two ordinary 60-watt lamps located at a definite place and angle on the worktable. However, with the increasing volume of spools manufactured and with the increasing hourly wages paid to the operators, the whole matter of inspection of the spools was again carefully studied. It was concluded from this study that equipment could eventually be designed that would perform

Fig. 172. Layout of work place for inspection of metal spools—second improved method: *A*, supply of spools to be inspected; *B*, holes for feeding spools to be inspected to gauges; *C*, lead-in segment of gauge; *D*, inclined track gauge; *E*, good spools; *F*, rejected spools.

the inspection operation automatically. Rather than attempt to design a fully automatic machine at the outset, it was decided to make improvements in the method one step at a time. First, the spools were redesigned so that both ends were identical. This made it unnecessary to stack the good spools in order in the trays. Since the limiting factor of the inspection operation was the operator's ability to handle the spools, a track gauge was designed. As the spools rolled down the two inclined track gauges (*D* in Fig. 172), the spools were gauged to their inside and outside tolerances simultaneously. If the spool rolled through the gauge, it passed inspection; if not, it was removed by the operator as a reject. In order to reduce positioning time in feeding spools to the gauges, two elliptical holes were cut in the table top (*B* in Fig. 172). A funnel of special shape placed below the holes allowed the operator to drop the spools with only minor alignment. The funnels or guides deposited the spools onto the lead-in segments (*C* in Fig. 172) of the gauge.

This second improvement in method eliminated visual inspection entirely and resulted in an increase of 108% over the first improved method.

With further increase in volume, it became economical to design a vibratory bowl feeder to supply reels to the track gauges, and the operator then needed only to remove rejects from the four track gauges which made up one working unit. This method resulted in a further increase of 125% over the second improved method. For a summary of the several improvements on this operation see Fig. 197.

Inspection by Transmitted Light. Products made of transparent or translucent material may be inspected by transmitting the light through the product. Broken fibers, knots, and other defects in cloth are easily detected; bubbles, cracks, and foreign material in glass and cellulose show up when transmitted light is used for inspection.

In one plant transmitted light was used for inspecting milk bottles for dirt, cracks, grease, and pieces of broken glass. In a trough just above the moving belt on which the washed bottles passed on their way to the bottling machines, 200-watt bulbs were installed base to base. Approximately 150 foot-candles of light was present on the belt surface. The back portion of the inspection surface was painted white to reveal black defects, and the conveyor belt was black to aid in detecting pieces of broken glass which might be resting on the bottom of the bottle. An operator could inspect bottles at the rate of 128 per minute as they passed by on the conveyor.

16. *The height of the work place and the chair should preferably be arranged so that alternate sitting and standing at work are easily possible.*

The worker should be permitted to vary his position by either sitting or standing as he prefers.[18] Such an arrangement enables the individual to rest certain sets of muscles, and a change of position always tends to improve the circulation. Either sitting or standing for long periods of time produces more fatigue than alternately sitting or standing at will. In many kinds of work provision can easily be made for

[18] "First Principles of Industrial Posture and Seating," New York Department of Labor, *Special Bulletin* 141, p. 2.

this sitting-standing combination. So important is this from the point of view of health that some states have laws requiring that the work place be arranged to permit either sitting or standing.

SEATS AND WORKTABLES. As far as, and to whatever extent, in the judgment of the commission, the nature of the work permits, the following provisions shall be effective. Seats shall be provided at worktables or machines for each and every woman or minor employed, and such seats shall be capable of such adjustment and shall be kept so adjusted to the worktables or machines that the position of the worker relative to the work shall be substantially the same whether seated or standing. Worktables, including cutting and canning tables and sorting belts, shall be of such dimensions and design that there are no physical impediments to efficient work in either a sitting or a standing position, and individually adjustable foot rests shall be provided. New installations to be approved by the commission.[19]

Although it would be preferable to have the height of the work place and the chair fit the particular operator who has to use them, this cannot always be done. It may be necessary in many cases to make the benches of such height that they will be most suitable for the average worker.

The height of the worker's elbow above the standing surface is commonly taken as the starting place for determining the proper height of the work place and the chair. The Industrial Welfare Commission of California has found that the height of the average worker's elbow above the standing surface is 40 inches for women (for men this would be 2 or 3 inches higher), and that a large percentage of the workers will not vary 1½ inches from this measurement.[20]

With 40 inches taken as the average elbow height of the female workers (the range being from 34 to 45 inches) and with the hand allowed to work 1 to 4 inches lower than the elbow, the average height of the working surface should be 36 to 39 inches. The chair should be 25 to 31 inches high, depending upon the proportions of the individual. Such table and chair heights permit the worker to either stand or sit at work, with the elbow and the hand maintained at the same position relative to the work place.

Space between Top of Seat and Undersurface of Bench Top. The work place should be constructed to permit plenty of leg room for the worker. Braces, shafts, and other obstructions under the work place often interfere with the natural position of the worker, causing

---

[19] "Seating of Women and Minors in the Fruit and Vegetable Canning Industry," California Industrial Welfare Commission, *Bulletin* 2a, p. 3.
[20] *Ibid.*, p. 3.

poor posture and discomfort. Such obstructions should not be permitted. The workbench should preferably be not over 2 inches thick, and there should be 6 to 10 inches of space between the top of the chair seat and the undersurface of the bench.

A bench 37 inches high will be too high for the short person, but this can be corrected by placing a rack of the proper height on the floor for the worker to stand on. For the tall worker a small rack or platform can sometimes be placed on top of the bench to raise the

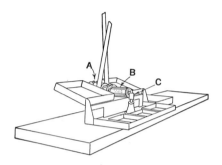

Fig. 173. Bench lathe cut in half to permit sitting or standing at work. This lathe is used for the assembly of hard-rubber syringe parts.

height of the work place. Where this cannot be done, the tall worker is handicapped while standing at work, but this is not necessarily true when she is seated.

In some kinds of work it is necessary to have equipment or material containers mounted on top of the workbench. This has the effect of adding to the "thickness" of the bench. A work place more than 5 inches thick cannot ordinarily provide a comfortable sitting-standing position for the worker.

The minimum table height for a comfortable position is determined by another limiting factor. A distance of not more than 8 inches between the elbow height and the underside of the table can be permitted if a restful position is to be maintained. A distance much greater than this interferes with the natural position of the knee. Using this distance of 8 inches as the limiting factor and allowing 1 inch for the thickness of the bench, we find the minimum height of the top of the bench to be 33 inches.

In some plants bench lathes have been cut in half (Fig. 173) and mounted on the bench with the axis of the spindle perpendicular to the front edge of the bench. This arrangement permits sitting-standing and facilitates working with both hands.[21]

Arm Rest. Often the work is of such a nature that it is desirable to provide arm rests at the work place. Arm rests are most effective on work that requires little movement of the forearms, with the hands working at approximately the same position, often at some distance from the body, for long periods of time. Light drilling, tapping, and reaming operations are frequently of this type. On such work it is restful to have padded metal or wood arm rests placed on top of or at the edge of the workbench, in a position to support the forearm. The arm rests need not interfere with the necessary working movements of the arms or hands. Figure 168 shows such an arm rest.

Foot Rest. When high chairs are used, a foot rest should be provided. The foot rest should preferably be attached to the floor or the bench; though this is less desirable, it may be fastened to the chair. The foot rest should be of ample width and depth to permit the entire bottoms of both feet to rest on it and allow for some movement. It usually requires a depth of 12 inches or more.

17. A chair of the type and height to permit good posture should be provided for every worker.

The following statements explain clearly what is meant by good posture.[22]

Good Standing Posture. When a person is standing properly, the different segments of the body—head, neck, chest, and abdomen—are balanced vertically one upon the other so that the weight is borne mainly by the bony framework and a minimum of effort and strain is placed upon the muscles and ligaments. In this posture, under

---

[21] W. R. Mullee, "Motion Study Is Safety's Partner," *National Safety News,* Vol. 34, No. 5, p. 23.

[22] New York State Labor Law, Section 150: "A sufficient number of suitable seats, with backs where practicable, shall be provided and maintained in every factory, mercantile establishment, freight or passenger elevator, hotel and restaurant, for female employees who shall be allowed to use the seats to such an extent as may be reasonable for the preservation of their health. In factories, female employees shall be allowed to use such seats whenever they are engaged in work which can be properly performed in a sitting posture. In mercantile establishments, at least one seat shall be provided for every three female employees, and if the duties of such employees are to be performed principally in front of a counter, table, desk or fixture, such seats shall be placed in front thereof, or if such duties are to be performed principally behind such counter, table, desk or fixture, they shall be placed behind the same."

normal conditions the organic functions—respiration, circulation, digestion, etc.—are performed with least mechanical obstruction and with greatest efficiency.

Good Sitting Posture. The thing which should always be insisted upon in the use of the body in any way is that the body should be kept straight from the hips to the neck and should not be allowed to flex or bend at the waistline. Any position which allows this bending

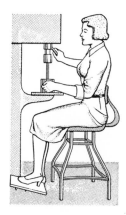

Fig. 174. A well-designed and properly adjusted back rest helps to hold the spine erect by relieving the strain on the back muscles.

Fig. 175. The chair seat should be form-fitting and the front edges should be well rounded.

lowers the vitality of the individual, leads to strain of the back, and naturally lessens efficiency.

The most frequent violation of the good sitting posture occurs when the individual slumps in his chair or assumes a sideways slouch, both of which are fatiguing and impair the health.

When the worker is seated, the chair should aid and not hinder him in maintaining good posture. A good chair should have the following features.

1. The chair should be adjustable in height so that it may be readily fitted to the particular individual who is to use it. Instead, nonadjustable chairs may be obtained in different sizes and issued to the workers according to their height. Such chairs, however, are not generally considered as practical as the adjustable type. The chair should be adjusted to a height that permits

the worker to sit comfortably with both feet resting on the floor or the foot rest (Fig. 174).

2. The chair should be rigidly built, preferably of steel frame with wood or padded seat and back. The edges of the seat and back should be rounded so that no sharp edges can cause discomfort and impede the circulation. Swivel chairs and chairs with casters are not recommended for factory work unless absolutely necessary. The easy movement of such chairs tends to cause unsteadiness while being used. This is particularly noticeable if the work requires some muscular effort. The chair may be provided with smooth metal "sliders" which permit the operator to shove it back out of the way without disturbing his work when he wishes to work standing.

3. The chair seat should be form-fitting. A saddle seat permits the weight of the body to be evenly distributed and so promotes comfort. The front edges of the seat should be well rounded (Fig. 175). For normal work the front edge of the chair should be approximately 1 inch higher than the back edge. When the person works leaning forward, the seat of the chair should be approximately flat. The seat should be of sufficient width to accommodate the body—16 to 17 inches is none too wide. However, the seat should not be over 14 to 16 inches in depth. The shallow seat permits the body to bend at the hip when leaning forward, whereas a deep seat tends to prevent this and to cause the body to bend at the waistline, putting a curve in the spine and disturbing the posture. The deep seat also tends to cut off the circulation of the blood through the underside of the thighs near the knee.

4. A back rest should be provided to support the lower part of the spine (Fig. 174). To do this the chair should not have a horizontal cross slat or bar lower than 6 inches above the seat. The body should sit well back on the seat so that the back rest can support the small of the back. The lower edge of the back should be 6 to 7 inches above the seat, depending upon the individual. The back rest may be 3 to 4 inches wide and 10 to 12 inches broad. It may be small and yet give satisfactory support. It can be so designed that it will not interfere with the movements of the individual's arms while working. It is important that the back rest be adjustable so that it may be fitted to the worker's body. When the worker leans forward while working, the chair back is of no use; however, the worker can use it while resting and it serves a very valuable purpose in being there for momentary relaxation.

CHAPTER 19

Principles of Motion Economy as Related to the Design of Tools and Equipment

18. *The hands should be relieved of all work that can be done more advantageously by a jig, a fixture, or a foot-operated device.*

From observation of the tools and fixtures usually found in the factory, it is obvious that many tool designers do not give much thought to the principles of motion economy when they design them. In most cases the fixtures are made for hand operation only, whereas foot-operated equipment would permit the operator to have both hands free to perform other motions.

Foot-Operated Tools and Fixtures. A hand tool can often be attached to or incorporated with a simple foot press or a modified arbor press in such a way that the tool is manipulated entirely by the foot. The electric soldering iron A in Fig. 176 is raised and lowered by the foot pedal B. After the soldered joint is made and as the iron is raised, valve C on the compressed air line opens and a stream of air cools the soldered joint. One company saved 50% in time on the operation of soldering a wire to the end of a flat metal electric static shield by the use of this foot-operated soldering iron.

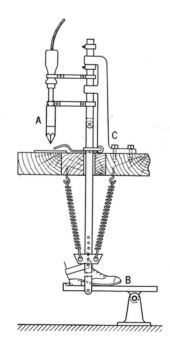

Fig. 176. Foot-operated soldering iron.

The operation of cutting and welding pipe is an important activity in the maintenance and construction department of many process industries. To make this job easier and to permit the welder to work more effectively, the Procter and Gamble Company designed and built

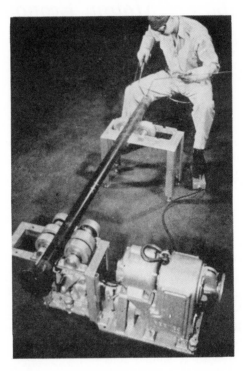

Fig. 177. Foot-controlled motor-driven unit rotates pipe for welder.

a foot-controlled unit (Fig. 177) which rotates the pipe or tube while the welder works in a comfortable position.

It is sometimes possible to use two foot pedals to actuate different parts of a jig, fixture, or machine. Such a setup should cause no difficulty for the operator. We are all familiar with the fact that the automobile has several pedals which the driver manipulates with ease, often traveling at high speed.

Opening a Shipping Carton. The operation shown in Fig. 178 consists of opening a flat shipping carton and folding over the bottom flaps preparatory to filling it with boxes of breakfast cereal in the packing room of the cereal factory. Cartons are delivered to the packing table and are stacked horizontally as shown in Fig. 178.

Improved Method. This operation is the same as the one described above. However, the cartons are stacked on the table vertically (Fig. 179) instead of horizontally. A simple fixture made of heavy wire, designed by E. H. Hollen, is used to aid the operator in folding in the two end flaps and the two side flaps.

Since the time required to open the carton and fold in bottom flaps is so short with this improved method, the same operator who does this also fills the carton with boxes of cereal (D of Fig. 179). The carton is then moved onto the automatic sealing machine, which applies glue and seals both ends.

Notice that the fixture contains no moving parts. It was made from 20 feet of No. 9 gauge wire and a piece of board, at a total cost of a few dollars. Using this fixture, the operator can open cartons in less than half the time required by the original method. The actual increase in output was 112%. The fixture saves many hand motions. Also as a result of this study, the carton was redesigned, saving over $20,000 per year in carton cost.

Rethreading Machine. The rethreading machine shown in Fig. 180, built at a cost of $786, produces 1100 parts per hour.[1] It replaced a standard two-spindle threading machine which cost $1356 and produced 600 parts per hour. The improved machine, incorporating a number of principles of motion economy, was built under the direction of O. W. Habel, Factory Manager of Saginaw Steering Gear Division, General Motors Corporation.

Some of the reasons why the new machine enabled the operator to double his output are the following:

1. Hand motions were replaced by foot and mechanical movements. The operator, seated at the machine, picks up a blank with each hand from a convenient position on the table and places them in the two-station fixture. Pressure of the right foot on the air valve operates the clamping fixture. Pressure on the left foot pedal brings the die head down.

2. Work does not pass from one hand to the other, and the finished work is dropped down a chute into a tote box.

3. Hand and eye motions are kept within the normal working space.

[1] O. W. Habel and G. G. Kearful, "Machine Design and Motion Economy," *Mechanical Engineering*, Vol. 61, No. 12, p. 897.

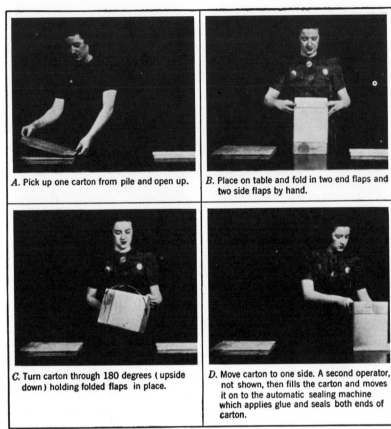

A. Pick up one carton from pile and open up.

B. Place on table and fold in two end flaps and two side flaps by hand.

C. Turn carton through 180 degrees (upside down) holding folded flaps in place.

D. Move carton to one side. A second operator, not shown, then fills the carton and moves it on to the automatic sealing machine which applies glue and seals both ends of carton.

Fig. 178. Opening shipping cartons—old method.

4. The hands are not used for holding or for manipulating any machine parts. There are no small controls to hunt for and manipulate.

5. Clamping fixtures are provided with bell mouths to facilitate the positioning of the blanks.

It is scarcely necessary to cite further illustrations of foot-operated apparatus, for their use is common. In fact, the question might properly be asked about almost any kind of bench or machine work, "Can a foot-operated device of some kind be used to facilitate the work?"

Design of Foot Pedals. Although the foot pedal is one of the most common devices for freeing the hands for productive work, most pedals are poorly designed.

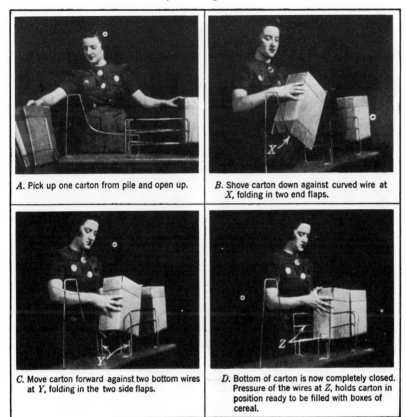

A. Pick up one carton from pile and open up.

B. Shove carton down against curved wire at X, folding in two end flaps.

C. Move carton forward against two bottom wires at Y, folding in the two side flaps.

D. Bottom of carton is now completely closed. Pressure of the wires at Z, holds carton in position ready to be filled with boxes of cereal.

Fig. 179. Opening shipping cartons—improved method. Output was increased 112%. Also as a result of this study the carton was redesigned, saving over $20,000 per year in carton cost.

Pedals might be classified as (1) those requiring considerable effort for manipulation, and (2) those requiring little effort. The first class is well illustrated by the garment press and by certain foot-operated punch presses and shears. The second class is illustrated by the trip on the power punch press, the control on the electric sewing machine, and the pedals shown in Figs. 176 and 177.

Particularly in the first class, where considerable force is required for manipulation, the pedal should be of sufficient width to permit its operation by either foot. Some pedals are placed across the entire front of the machine to facilitate this. The pedal should also be designed so that the operating foot can carry part of the body weight.

Fig. 180. Machine for rethreading ball studs, production rate 1100 per hour.

Poorly arranged pedals, such as the one shown in Fig. 181, tend to put all the body weight on one foot, throwing the body out of its normal position and resulting in excessive strain and fatigue for the operator.[2]

Study of Five Types of Pedals. Figure 182 shows the results of a study [3] made to determine the relative effectiveness of five different types of pedals. Each pedal was depressed against a tension spring requiring 20 inch-pounds for one complete stroke. For example, pedal 1 had the fulcrum under the heel, and the ball of the foot moved through a distance of 2 inches against a resistance of 10 pounds. All the pedals were operated as trip type, such as would be found on a punch press. That is, the operator was asked to depress the pedal as rapidly as possible, and the time for each pedal stroke was measured with a kymograph. The results of the study (Fig. 182) show that the operator using pedal 1 took the least time per stroke. Pedal 4 required the longest time— 34% more than pedal 1.

Fig. 181. Poorly designed pedal.

19. Two or more tools should be combined wherever possible.

It is usually quicker to turn a small two-ended tool end-for-end than it is to lay one tool down and pick up another. There are many examples of two-tool combinations—tack hammer and tack puller, two-ended wrench, pencil and eraser—and the designer of the "handset" telephone used this idea when he incorporated the transmitter and the receiver in one unit.

Two very convenient tools which have been developed at a midwestern electrical equipment company are illustrated in Figs. 183 and 184. The first one replaces the screwdriver and tweezers—it holds the screw while it is being assembled. The second tool replaces a wrench

[2] L. A. Legros and H. C. Weston, "On the Design of Machinery in Relation to the Operator," Industrial Fatigue Research Board, *Report* 36, p. 13. Figure 181 reproduced by permission of the Controller, H. M. Stationery Office, London.

[3] Ralph M. Barnes, Henry Hardaway, and Odif Podolsky, "Which Pedal Is Best?" *Factory Management and Maintenance*, Vol. 100, No. 1, p. 98.

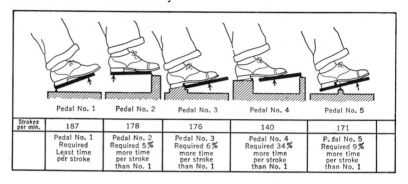

Strokes per min.	187	178	176	140	171	
	Pedal No. 1 Required Least time per stroke	Pedal No. 2 Required 5% more time per stroke than No. 1	Pedal No. 3 Required 6% more time per stroke than No. 1	Pedal No. 4 Required 34% more time per stroke than No. 1	Pedal No. 5 Required 9% more time per stroke than No. 1	

Fig. 182. Results of a study of five types of pedals.

and a screwdriver. This device permits the bolt to be set to the proper position and at the same time allows the operator to lock the nut in place by means of the "sleeve wrench" which slips over the screwdriver.

The practice of using only the thumb and first and second fingers is so common that attention is called to the fact that the third and fourth fingers and the palm should also be employed wherever possible. The combination screwdriver and wrench shown in Fig. 184, for example, permits the entire hand to be used. The thumb and first and second fingers manipulate the wrench while the palm and third and fourth fingers manipulate the screwdriver.

Fig. 183. Combination screwdriver and tweezers.

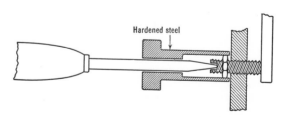

Fig. 184. Combination screwdriver and wrench.

The multiple-spindle air-operated nut runner shown in Fig. 185 is used to tighten all five wheel nuts at once.[4] The cable suspends the wrench in a convenient position and makes the job easier.

Fig. 185. Multiple-spindle air-operated nut runner can tighten all five wheel nuts at once.

20. *Tools and materials should be pre-positioned whenever possible.*

Pre-positioning refers to placing an object in a predetermined place. in such a way that when next needed it may be grasped in the position in which it will be used. For pre-positioning tools, a holder in the form of a socket, compartment, bracket, or hanger should be provided, into which or by which the tool may be returned after it is used, and where it remains in position for the next operation (Fig. 186). The tool is always returned to the same place. The holder should be of such design that the tool may be quickly released into its place from the hand. Moreover, the holder should permit the tool to be grasped in the same manner in which it will be held while being used. The most familiar example of pre-positioning is the pen desk set which

[4] San Jose Assembly Plant of the Ford Motor Company.

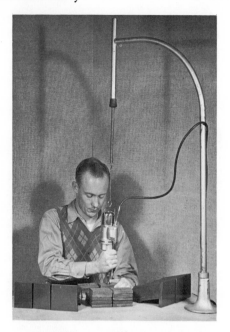

Fig. 186. Electric motor-driven wrench.

holds the pen in writing position even when not in use, and from or to which it may be easily and quickly removed or returned.

21. *Where each finger performs some specific movement, such as in typewriting, the load should be distributed in accordance with the inherent capacities of the fingers.*

The normally right-handed person performs work with less fatigue and greater dexterity with the right hand than with the left. Although most people can be trained to work equally well with either hand on most factory operations, the fingers have unequal inherent capacities for doing work. The first and second fingers of the two hands are ordinarily superior in their performance to the third and fourth fingers.

Arrangement of Typewriter Keys. A study made to determine the ideal arrangement of the keys of the typewriter for maximum efficiency [5] also illustrates this difference in the capacities of the fingers (Fig. 187). That part of the study which is of most interest here

[5] R. E. Hoke, "Improvement of Speed and Accuracy in Typewriting," *Johns Hopkins Studies in Education*, No. 7, pp. 1–42.

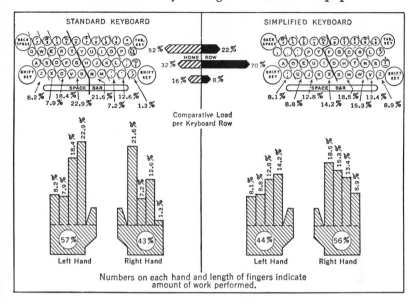

Fig. 187. Comparison of present typewriter keyboard with new simplified keyboard. Figures indicate comparative loads for each row, hand, and finger. The new keyboard at the right has the letters rearranged so that the right hand carries its share of the load. Seventy per cent of the commonly used words are written on the "home row" where the fingers are placed.

revealed that the ability of the right hand as compared to that of the left was as 100 to 88.87, or approximately as 10 to 9. This agrees with the findings of another investigator already cited.[6] The data in Table 13 show the ideal load in strokes, based on the abilities of the fingers.

Table 13. Relative Finger Loads on the "Ideal" and on the Present Keyboard

	Left Hand				Right Hand			
Finger	4	3	2	1	1	2	3	4
Ideal load	855	900	975	1028	1097	1096	991	968
Present keyboard load	803	658	1492	1535	1490	640	996	296

[6] Wm. L. Bryan, "On the Development of Voluntary Motor Ability," *American Journal of Psychology,* Vol. 5, No. 2, p. 123.

The finger loads required by the present typewriter keyboard are shown for comparison.

These data indicate that the first and second fingers of the right hand should carry the greatest load, whereas the fourth finger of the left hand should carry the smallest. They also show that the total load of the right hand using the present typewriter is 3422, and of the left hand 4488, or a ratio of 100 to 131.25, when it should be as 100 to 88.87. Thus there is an overload of the left hand of 47.7% as compared with the load of the right hand.

It was suggested from the analysis of the data secured in this investigation that the keys of the typewriter should be so arranged that the letters occuring most frequently would be typed with the fingers capable of carrying the greatest load. In fact, many investigators in this field have proposed new keyboards. The one shown on the right in Fig. 187 is the Dvorak-Dealey "simplified" keyboard [7] tested over a period of years at the University of Washington. Researches conducted under a grant from the Carnegie Foundation for the Advancement of Teaching seem to indicate that the simplified typewriter keyboard eliminates many of the defects of the standard keyboard, and that it is:

1. Easier to master in that it requires less time to attain any particular level of typing speed.

2. Faster, since it makes higher net rates possible for average typists.

3. More accurate, since fewer typing errors are made.

4. Less fatiguing through simplifying the stroking patterns and through adapting the hand and finger loads to the relative hand and finger abilities.[8]

Despite the advantages listed above, two practical problems have prevented any wide use of the simplified keyboard thus far. First, typewriters with the simplified keyboard are not generally available in offices and schools, and people learning to type hesitate to use a keyboard that is not in general use. Second, tests [9] seem to show that

[7] A. Dvorak, N. I. Merrick, W. L. Dealey, and G. C. Ford, *Typewriting Behavior,* American Book Co., New York, 1936, p. 219. Reproduced by permission.

[8] Dwight D. W. Davis, "An Evaluation of the Simplified Typewriter Keyboard," *Journal of Business Education,* Vol. 11, No. 2, p. 21, October, 1935; A. Dvorak, "There Is a Better Typewriter Keyboard," *National Business Education Quarterly,* Vol. 12, No. 2, pp. 51–58, December, 1943.

[9] Earl P. Strong, *A Comparative Experiment in Simplified Keyboard Retraining and Standard Keyboard Supplementary Training,* General Services Administration, Washington, D. C., 1956.

Fig. 188. Friden perforator with Dvorak simplified keyboard.

it is not economical to retrain experienced typists to use the simplified keyboard. General use of this basically superior simplified keyboard must wait until ways can be found to solve these problems. However, the new keyboard may find use in another area. A revolution is taking place in the printing industry, and it may be that the simplified keyboard will be used for the tape perforators in composing rooms and for direct input to computers (Fig. 188).

22. Levers, crossbars, and hand wheels should be located in such positions that the operator can manipulate them with the least change in body position and with the greatest mechanical advantage.

Some machine-tool manufacturers understand that it is possible to build a machine that will perform its functions satisfactorily and at the same time will be easy to operate. Unless a machine is fully automatic, the amount of work that it will produce depends to some extent upon the performance of the operator. The more convenient the machine is to operate, the greater the production is likely to be.

The Gisholt Machine Company, for example, has incorporated a speed selector in its universal turret lathes. This device enables the operator to obtain easily and quickly any one of the several available spindle speeds. It is power-operated; the operator simply sets a dial, and the machine automatically makes the shift to give the correct spindle speed. Figure 189 shows the automatic and manual control center for the Gisholt automatic chucking lathe.

The operator should not be required to leave his normal working position to operate his machine. The levers should be placed in such

Fig. 189. Automatic and manual control center for automatic chucking turret lathe. All controls are located in front of the headstock so that all working units can be easily observed and controlled from the normal operating position.

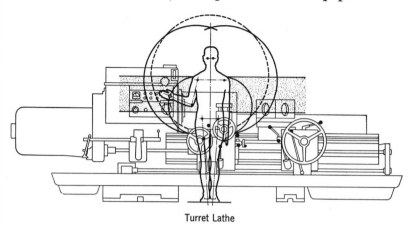

Turret Lathe

Fig. 190. Controls are grouped near normal operating area. Major control panel is within easy reach of operator's left hand. (From Robert H. Hose, "Designing the Product to Suit Human Dimensions," *Product Engineering*, Vol. 26, No. 9, p. 171.)

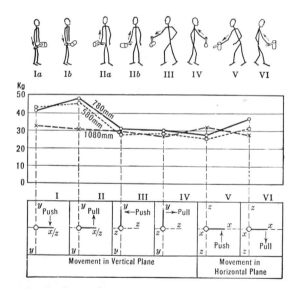

Fig. 191. Results of the study of levers.

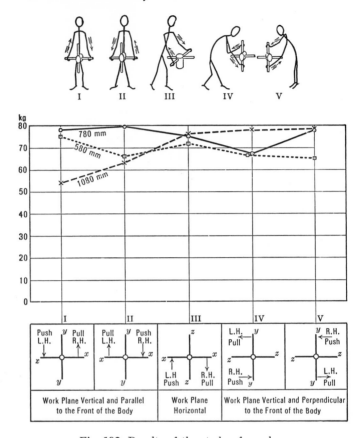

Fig. 192. Results of the study of crossbars.

a way that he need not bend over or twist his body in an uncomfortable manner when manipulating them (Fig. 190). Where this ideal condition cannot be provided, the nearest approach to it should be adopted.

It is well known that levers can be operated more effectively in certain positions and at certain heights than at others.[10] A very ex-

[10] A. Chapanis, W. R. Garner, and C. T. Morgan, *Applied Experimental Psychology: Human Factors in Engineering Design,* John Wiley & Sons, New York, 1949; *Handbook of Human Engineering Data,* 2nd ed., Tufts College Institute for Applied Experimental Psychology, 1951; Wesley E. Woodson, *Human Engineering Guide for Equipment Designers,* University of California Press, Berkeley, 1964, L. E. Davis, "Human Factors in Design of Manual Machine Controls," *Mechanical Engineering,* Vol. 71, pp. 811–816.

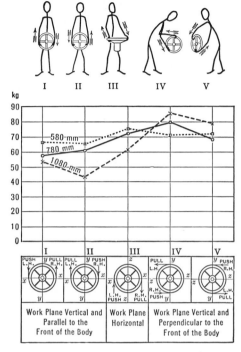

Fig. 193. Results of the study of hand wheels.

haustive study was made to determine the effectiveness of levers, crossbars, and hand wheels located in both horizontal and vertical planes, and at three different heights from the floor.[11] These devices were arranged so that the force of the push or pull was indicated in kilograms by means of a dynamometer. The object in these tests was to determine the maximum strength that could be exerted in each case. Each of the three devices was tested in the several positions shown in Figs. 191, 192, and 193.

The vertical scale represents the force in kilograms exerted by the subject, and the horizontal scale shows the particular position of the device being tested. The three sets of curves on each chart represent the three different heights at which the devices were tested, namely 580 mm. (22.8) inches), 780 mm. (30.7 inches), and 1080 mm. (42.5 inches). For example, in Fig. 191 the lever was most effective at

[11] W. P. Kühne, "Studien zur optimalen Kraftreaktion an Maschinenbedienungs elementen" (Studies on the Optimum Force Exerted on Machine Controls), *Industrielle Psychotechnik*, Vol. 3, No. 6, pp. 167–172.

the medium height, 780 mm. above the floor, and position II on the horizontal scale, which represents the position where the lever was horizontal and the operator pulled up on it.

The hand wheel was most effective when placed at the 1080-mm. height and in the vertical plane, position IV in Fig. 193. The operator, standing to one side of the wheel, pushed with his right hand and pulled with his left.

CHAPTER 20

Motion Study, Mechanization, and Automation

Over the centuries, man has been finding easier and better ways to produce the goods and services he needs and wants. Work originally consisted of activities performed with the bare hands; then simple tools were devised, and later power-driven machines. Now fully automatic equipment makes it possible to remove much of the burden of manual work from the man and transfer it to the machine. We have long recognized that an increase in productivity per man-hour was a major factor in bringing about improvement in our standard of living. Few people advocate using human labor to do work that can be done better and cheaper by machines. The day is far distant, however, when manual work in industry will disappear. Some activities are too complex to be mechanized and must therefore be performed manually. Some tasks occur so seldom that it is not economical to use machines. Moreover, other factors may affect the extent to which mechanization should be used in a given situation—factors such as quality, yield, material utilization, safety, availability of qualified workers, availability of capital, and the probable life of the product being made.

Since the application of principles of motion economy ordinarily calls for a very small capital investment and a minimum of design cost, it is suggested that the best manual method or the best combination of manual and machine methods be developed and used as a basis for evaluating a proposed mechanized or automated process. For example, one company makes it a practice to determine and evaluate alternative methods. If a large-volume fairly complex operation is to be considered, a comparison would be made of the estimated cost to do each element or each suboperation manually and also automatically. For example, in a punch-press operation the parts could be manually or automatically fed, the press could be actuated manually or automatically, and the finished parts could be removed by

307

hand or ejected automatically. The cost to install and maintain each of the automatic components would have to be evaluated against the costs of the manual method. However, an efficient manual method should be used as the basis for comparison.

In the hypothetical case shown in Fig. 194 the operation has three elements, any one of which can be done manually or automatically. There are eight possible combinations, and the cost can be estimated for each. The lowest cost combination is the one that would be recommended, others things being equal.

Combination	Elements of the Operation		
	1	2	3
A	Manual	Manual	Manual
B	Manual	Manual	Automatic
C	Manual	Automatic	Manual
D	Manual	Automatic	Automatic
E	Automatic	Manual	Manual
F	Automatic	Manual	Automatic
G	Automatic	Automatic	Manual
H	Automatic	Automatic	Automatic

Fig. 194. Eight possible combinations of manual and automatic performance of a three-element operation.

The first two cases to be described in this chapter show how output per man-hour was affected by changes in method from purely manual operation, to partial mechanization, to completely automatic production. In the third case, that of warehouse handling of the finished product, the cost of the pallet was an additional factor that affected the total warehouse handling cost.

GRADING AND PACKAGING EGGS FOR DISTRIBUTION

A careful study of the method of candling, sizing, and packaging eggs for distribution in one large plant [1] in California has resulted in a substantial increase in the output and a reduction in labor costs. Basically the process consists of three parts: (1) candling, determining quality; (2) sizing, determining weight; and (3) packaging eggs into cartons. Originally the method was as follows. The operator, called a

[1] Michel Brothers, Santa Monica, Calif.

candler, picked up two eggs in each hand and held the eggs, one at a time, directly in front of an intense light. She examined the eggs for exterior appearance (color, texture, and thickness of shell) and interior quality (size and mobility of the yoke and position and size of the air cell, and other characteristics which indicate the quality of the egg). She then placed the egg in the proper carton on her work place. In this particular plant eggs were classified into eight different quality grades and seven different sizes.

Three major changes have been made in methods.

First Improved Method. The first improvement was made by applying principles of motion economy to the job. A short belt conveyor, actuated by a motor drive controlled by a foot-operated switch, moved the eggs to be inspected into a convenient position for the operator. Semicircular shelves at several different levels were located above the conveyor to facilitate the packing of the eggs of various grades and sizes into the cartons. An air-cooled candling lamp of special design further added to the effectiveness of the work place. As a result of these improvements output increased approximately 20%.

Second Improved Method—Mechanized Sizing and Packaging. In order to further increase productivity a machine [2] was installed which performed automatically two of the three parts of the job. The operator and the machine then worked together in the following manner. The operator, working at the head of the machine, candled the eggs much as in the first improved method. However, she took the eggs directly from the cases in which they were received from the egg producer. She merely inspected the eggs in front of the light as in the old method and graded them for quality, placing the candled egg in one of eight short conveyors located directly in front of her, a conveyor for each quality grade. From this point on, the machine took over and the balance of the process was automatic. Each egg was automatically weighed and code-marked with invisible ink, and by means of a memory device and electronic controls the egg was conveyed to one of 22 carton-filling stations on the machine, where it was deposited in the proper carton. After the carton containing one dozen eggs was filled, the lid was closed and the carton was sealed, code-dated, and conveyed to the operator who placed the carton in a case. The cases were then moved by conveyor to storage. Six candlers, two machine attendants, and the foreman were able to process approximately 1620 dozen eggs per hour. This is as many eggs as were formerly processed by twelve girls.

[2] Designed and built by The FMC Corporation, Riverside, Calif.

Third Improved Method—Mechanized Candling, Sizing, and Packaging. Recently additional changes were made in the egg-processing plant which further increased the efficiency and improved the quality and uniformity of the eggs. First a plan was devised that would pay a bonus to the egg producer in the form of a higher price per dozen if the eggs would meet a specified quality standard. To

Fig. 195. New design of egg case. Fiberboard case holds 12 filler flats or trays, each containing 30 eggs. The cut-out design of the case facilitates air circulation and permits the eggs to be cooled quickly in the refrigerator at the ranch, as well as facilitating the removal of the flats at the egg processing station.

meet this quality standard, the rancher is required to remove from his flock those chickens which produce eggs of inferior quality—usually the older hens. Random samples of each rancher's eggs are inspected over the week, and the "quality level" is based on the inspection records. This quality level is in effect for the following week.

The process is as follows. Eggs are gathered by the rancher and placed directly in special filler flats holding 30 eggs and then in cases holding a total of 30 dozen (Fig. 195). The cases are placed in a cooler until they are picked up by the truck for delivery to the processing plant. When the eggs are received at the processing plant, they are placed in a cooler, each rancher's eggs being kept separate. When the eggs are to be processed, the egg cases are moved to the

Fig. 196. Mechanized egg-processing equipment. Filler flats containing 30 eggs are transferred from egg cases to conveyor by the "loader." The "scanner" removes cracked and soiled eggs as they pass over a bank of spotlights. The eggs are dropped into cartons which are automatically closed, sealed, and code dated.

front end of the processing machine[3] (Fig. 196) and a man called a loader places the flats on a conveyor which automatically distributes them evenly onto a rubber roll conveyor. The eggs are washed and dried automatically. They then pass over a bank of six 1000-watt lamps, and an operator called a scanner examines the eggs as they pass by and removes any cracked eggs or eggs containing soiled spots which were not removed by the washing process.

Next, each egg is automatically weighed and is classified into one of six different sizes as follows: jumbo, extra-large, large, medium, small, and peewee. The eggs then pass onto a blood detector—all eggs containing even the smallest bloodspot are discarded. The eggs of acceptable quality are dropped automatically from the conveyor into the proper carton. When a carton is full, it is automatically

[3] *Ibid.*

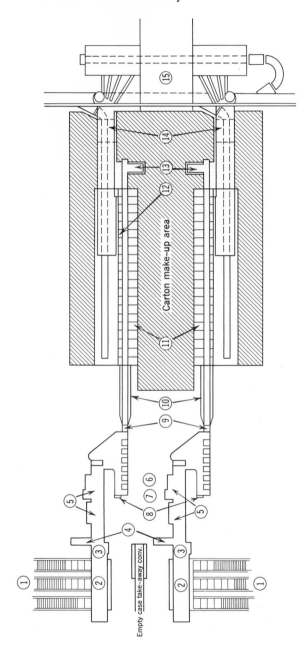

Fig. 197. Layout of mechanized egg-processing equipment. (*1*) Incoming producer case set on conveyor. (*2*) Case unloading conveyor. (*3*) Automatic egg transfer. (*4*) Automatic empty filler flat stacker. (*5*) Automatic egg cleaner—dryer (optional). (*6*) Semiautomatic mass candling unit. (*7*) Six automatic weighing stations. (*8*) Automatic printing and recording unit. (*9*) Automatic blood spot detector (optional). (*10*) Automatic processing unit (optional). (*11*) Carton and filler flat magazines. (*12*) Automatic packaging stations. (*13*) Memory control unit. (*14*) Conveying equipment through carton closing and code dating machines. (*15*) Outgoing case packing area.

Capital Invested in Building and Equipment—Four Different Methods of Grading and Packaging Eggs for Distribution

Method	Total Size of Work Force (Operators)	Total Number of Eggs Processed in Dozens		Capital Invested in Dollars (Equipment and Building)		
		Per Hour	Per Operator per Hour	Total	Per Operator	Per Dozen Eggs Processed per Hour
1. Original method of hand candling	10	1120	112	80,000	8,000	70
2. Improved method of hand candling	12	1620	135	80,000	6,667	41
3. Hand candling, mechanized sizing and packaging	9	1620	180	140,000	15,555	86
4. Mechanized washing, sizing and packaging	7	1800	257	140,000	20,000	78

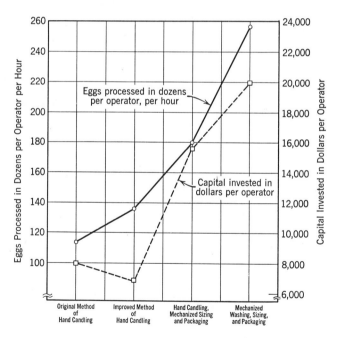

Fig. 198. Curves showing the number of eggs processed in dozens per operator per hour and the capital invested in equipment in dollars per operator for each of the four different methods of processing and packaging eggs.

	I	II
	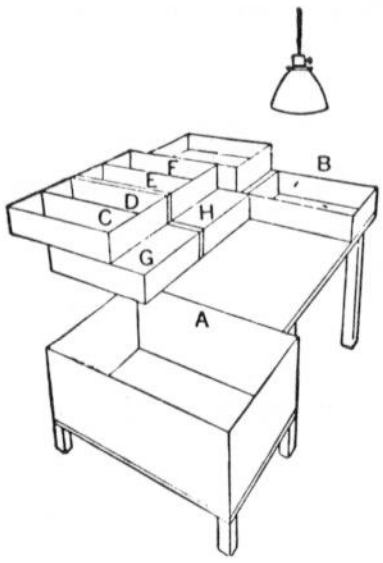	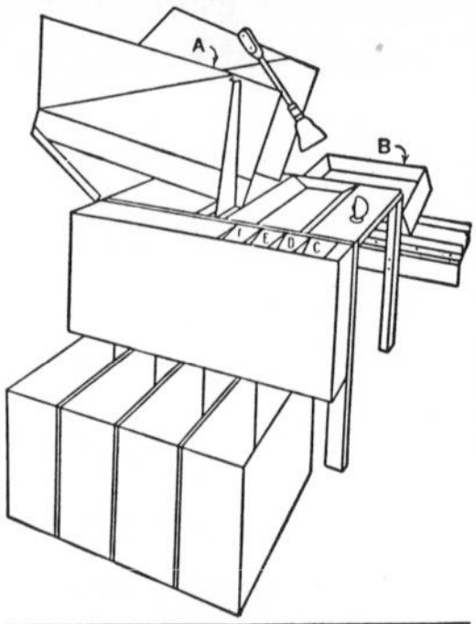
Brief statement of method used	Original method. Inspection of spools from pile on top of worktable	Improved work-place layout. Use of two light sources make it unnecessary to turn spools end-for-end to inspect them
Equipment used	Ordinary worktable	Special worktable
Method of supplying spools to be inspected	Spools to be inspected are in tote box on floor beside worktable	Spools to be inspected are fed onto work place from hopper on back of bench
Location of light source for inspection of spools	Ordinary light source above worktable	Two light sources mounted at special locations on work place
Inspection procedure	Spools inspected from one end and then turned through 180 degrees and inspected from other end	Two light sources make it unnecessary to turn spools end–for–end for inspection
Disposal of good spools	Good spools are placed in order in tray to right of operator	Good spools are placed in order in tray to right of operator
Disposal of defective spools	Rejects disposed by operator into any one of six trays on back of table for six different classes of rejects	Rejects disposed by operator into any one of four openings on left side of work place for four different classes of rejects
Increase in output of operator per hour in comparison to previous method	Original method	100% of method I

Fig. 199. Summary of five methods

314

METHOD

III	IV	V
Spool was redesigned so both ends were alike. Use of two track gauges eliminated visual inspection of spools	Mechanical vibrating bowl feeder supplies spools to four track gauges, eliminates manual feeding and visual inspection	It seems likely that rejects can be removed from the track gauges by mechanical means, thus making the operation completely automatic
Special worktable. Funnel-shaped openings in table top lead to track gauges under table	Belt conveyor, mechanical vibrating bowl feeder, four track gauges	Completely automatic equipment
Spools to be inspected are fed onto work place from hopper on back of bench (Hopper not shown in above sketch)	Spools to be inspected are brought to vibrating bowl feeders by belt conveyor from preceeding operation	Spools to be inspected are brought to vibrating bowl feeders by belt conveyor from preceding operation
General lighting	General lighting	General lighting
No visual inspection. Good spools roll through track gauges; defective spools will not pass through gauges and are removed by operator	No visual inspection. Good spools roll through track gauges; defective spools will not pass through gauges and are removed by operator	No visual inspection. Good spools roll through track gauges; defective spools will not pass through gauges and are removed by mechanical means
Good spools roll from track gauge into tray under table. Spools are the same on both ends, need not be placed in order	Good spools roll from track gauge into a chute and are carried to tray	Good spools go by conveyor directly to next operation
Rejects disposed by operator into opening on either side of work place.	Rejects disposed by operator into opening in center of working area	Rejects disposed into chute by mechanical device
108% of method II	125% of method III	Equipment automatic, no operator required. One person will attend several units. One maintenance man required for battery of machines

of inspection of metal spools.   315

Hand Truck

METHOD→

MEN REQUIRED
  Manual   35.3
  Truck operators   0
                                                 ‾‾‾‾‾
       Total   35.3

EQUIPMENT→

INVESTMENT
  Hand trucks   (36) $7,200
  Lift trucks   (7)   . .
  Pallets   (11,000)   . .
                                                 ‾‾‾‾‾‾
       Total $7,200

COST	Year	Box
Wages	$143,465	0.0197
Wage administration at 17%	24,389	0.0036
Wood pallet expense	. .	. .
Paper pallet expense	. .	. .
Operating expense	. .	. .
Maintenance	342	. .
Depreciation at 12½%	900	.0001
Insurance and taxes at 1%	72	. .
Total	$169,168	.0234
SAVINGS	Base	Base

* Paper pallet expense item.

Fig. 200. Comparison of finished-product warehouse handling methods for a
30,000-case-per-day factory. Assume: (1) 15,000 cases per day direct to shipment,
15,000 cases per day to storage; (2) time in storage, 4 weeks; (3) product
manually handled from slide to pallet or truck, and from pallet or truck into
car.

Fork Lift	Pul Pac	Load Clamp

16.5	16.5	16.5
6.2	6.8	6.3
———	———	———
22.7	23.3	22.8

. .	. .	. .
$28,000	$38,500	$38,500
49,500	*	. .
———	———	———
$77,500	$38,500	$38,500

Year	Box	Year	Box	Year	Box
$ 93,871	0.0129	$ 96,648	0.0133	$ 94,361	0.0130
15,958	0.0022	16,430	0.0023	16,041	0.0022
9,251	0.0013	. .	. .	. .	. .
. .	. .	8,833	0.0012	. .	. .
2,572	0.0004	2,838	0.0004	2,619	0.0004
3,453	0.0005	4,862	0.0007	4,486	0.0006
9,688	0.0013	4,813	0.0006	4,813	0.0006
775	0.0001	385	. .	385	. .
———	———	———	———	———	———
$135,568	0.0187	$134,809	0.0185	$122,705	0.0168
$ 33,600	0.0046	$ 34,359	0.0047	$ 46,463	0.0064

closed, sealed, code-dated, and conveyed to the operator, who places it into a rectangular wire basket holding 15 dozen eggs. The baskets are moved by conveyor to the cooler. The baskets of eggs are later delivered to the grocer and placed directly into a refrigerated self-service display case, thus eliminating a handling operation. A crew consisting of a loader, a scanner, and three packers, together with an inspector and a foreman, can process approximately 1800 dozen eggs per hour.

The curves in Fig. 198 show that the "number of eggs processed in dozens per operator per hour" increased from 112 by the original method to 257 by the present method. However, the "total capital invested in equipment and building per operator" increased from $6667 for the first improved method to $20,000 for the present method. One is nearly always confronted with the problem of evaluating both unit labor cost and initial capital invested in determining the preferred method. In this case each advance in mechanization reduced labor cost and also reduced the total cost of processing eggs. Moreover the present method results in eggs of more uniform size and quality and also enables the processor to pay the rancher a higher price per dozen for the high-quality eggs which he now knows how to produce.

New Design of Egg Case. The fiberboard egg case (Fig. 195) consists of a pallet section or bottom part with handholds in each end to facilitate lifting the case; the two sides are cut out to make it easy to load and unload the pallet. A fiberboard top telescopes over the pallet and also has handholds cut out in the ends to match those in the bottom. The case holds 12 filler flats, or trays of 30 eggs each, or a total of 30 dozen. Eggs are gathered and placed directly into the flats and then into the pallet at the ranch, thus eliminating one handling. The cutout design of the pallet facilitates air circulation and permits the eggs to be cooled quickly in the refrigerator at the ranch. The new egg case was developed by the University of California, Department of Agriculture, and has been widely adopted by ranchers and egg processors.

INSPECTION OF METAL SPOOLS

Over a period of some 25 years, four major improvements have been made in the inspection of metal spools (Fig. 199). The first improvement, Method II, resulted from a better work-place layout and the use of two light sources which made it unnecessary to turn the spools end-for-end to inspect them. For a complete description of this method see page 277. Although output was doubled, this method still required 100% inspection on the part of the operator. Some years

Fig. 201. Load clamp attachment for power-lift truck makes it unnecessary to use pallets.

later the spools were redesigned so that both ends were alike. This design change permitted the use of two track gauges which eliminated visual inspection of the spoils. All acceptable spools rolled down the two tracks into trays. Any rejects were caught in the track gauge itself, and these spools were removed by the operator. This method, Method III, resulted in an improvement of 108% in output over the previous method. The method in use today, Method IV, makes use of a mechanical vibrating bowl feeder which feeds the spools onto the track gauge instead of the operator's doing this by hand. This mechanical feeder, supplying spools to four track gauges, has resulted in an increase in output of 125% over Method III.

A mechanical device is now being designed to remove rejects from the track gauges, thus making the operation completely automatic. Spools to be inspected will be brought to the vibrating bowl feeders by a belt conveyor from the preceding operation; they will be inspected automatically, and then the good spools will be moved by conveyor to the next operation.

COMPARISON OF FINISHED-PRODUCT WAREHOUSE HANDLING METHODS

There may be situations where the controlling factor is something other than labor. Figure 200 shows four different finished-product warehouse handling methods which have been used at one time or another by the Procter & Gamble Company. Originally a flat-top hand truck was used, resulting in a total handling cost of $2.34 per 100 boxes. The substitution of a fork-lift truck and wood pallets brought about a reduction in the handling costs to $1.87 per 100 boxes. A Pul Pac pallet was designed, which reduced the pallet costs and also saved vertical space in the warehouse. This method gave a cost of $1.85 per 100 boxes. The Pul Pac method used a heavy, tough paperboard pallet instead of the heavy wood pallet previously used. The Pul Pac pallet cost approximately 50 cents in comparison to $4.50 for the wood pallet. Later a load clamp attachment for the lift truck was developed, which eliminated the pallet entirely (Fig. 201). This method of handling boxes reduced the cost to $1.68 per 100 boxes. A detailed breakdown of the various costs of each of the four methods is given at the bottom of Fig. 200.

CHAPTER 21

Standardization—Written Standard Practice

After finding the most economical way of performing an operation, it is essential that a permanent record be made of it. This record is frequently called a "standard practice." In addition to serving as a permanent record of the operation, the standard practice is often used as an instruction sheet for the operator or as an aid for the foreman or instructor in training the operator.

The Standard Practice as a Permanent Record. Once the improved method is standardized and put into effect, constant vigilance on the part of management is necessary in order to maintain this standard. Often tools and equipment get out of adjustment, belts become loose, and materials vary from specifications. When such conditions exist, standard performance cannot be expected from the operator. Only by rigid maintenance of standard conditions can there be reasonable assurance of standard performance in output and quality.

Very often time standards are used as the basis for wage incentives, and most incentive plans either imply or specifically state that time standards, or rates, will not be changed [1] unless there is a change in the method of performing the work. It is therefore essential that an accurate and complete record be made of the method at the time it is put into effect or at the time the rate is set for the operation. If no such record is kept, it will be almost impossible in the future to tell whether the method then being used is the same as that in effect at the time the standard was originally established.

One company uses the forms shown in Figs. 202 and 203 as a permanent record of each operation. This standard practice is ordinarily prepared by the person making the motion and time study, or by the person in charge of the investigation if several men are engaged in the

[1] See Ralph M. Barnes, *Industrial Engineering Survey,* University of California, **1967.**

STANDARD JOB CONDITIONS

BASE RATE NO. __27112__ CODE NO. __—__

DATE _____ STUDY NOS. __32906-32909__ SYM. NO. __—__ BINDER NO. __27__

BLDG. __148A__ DEPT. __No. 17__ DIVISION __Eastern__ OBSERVER __Davis, W.T.__

OPERATION __Label 4-oz. Bottles   Hardening Solution__

SKETCH OF WORK PLACE

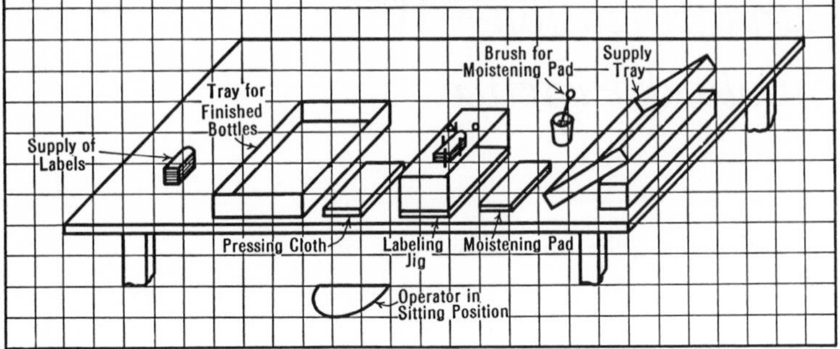

SPECIAL TOOLS, JIGS OR FIXTURES   Labeling Jig

JOB ELEMENTS

1. Moisten pad with brush.

2. Insert labels in jig.

3. Procure bottle from supply tray, moisten bottle on
   moistening pad, label, using jig, press smooth on
   pressing cloth.

AUXILIARY

Set up and clean up by
handler or operator. No allowance in standard.

4. Dispose bottle to wooden tray.

5. Upon completion of tray, make out ticket and place
   in tray as check against quality of labeling. Foreman can determine responsibility if labels are not
   up to standard.

Handler supplies bottles
and disposes of finished tray.

AUDIT   Production can be checked by order number. Foreman checks time turned in.

B118

Fig. 202. Standard job conditions form, size 8½ × 11 inches.

GENERAL JOB CONDITIONS

DATE OF ISSUE_____ BASE RATE NO.____27112_____ CODE NO._____

BLDG.____148A____ DEPT.____No. 17____ DIVISION ____Eastern____ OBSERVER_____Davis, W.T._____

TYPE OF OPERATION_____Fill and Pack Bottles of Liquid

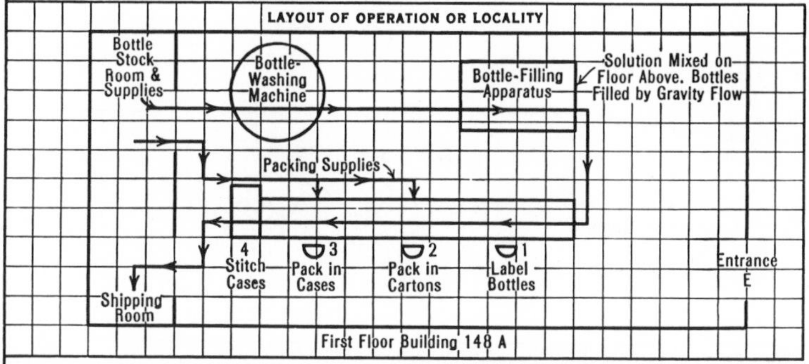

LAYOUT OF OPERATION OR LOCALITY

Bottle Stock Room & Supplies

Bottle-Washing Machine

Bottle-Filling Apparatus

Solution Mixed on Floor Above. Bottles Filled by Gravity Flow

Packing Supplies

4 Stitch Cases — 3 Pack in Cases — 2 Pack in Cartons — 1 Label Bottles

Entrance

Shipping Room

First Floor Building 148 A

RANGE OF APPLICATION Unit designed for handling bottles of liquid product from 4-oz. to 32-oz. size.

DESCRIPTION OF STANDARD EQUIPMENT Balanced production line from supply room through to finished product in shipping room. Equipment consists of: bottle-washing machine No. 3712-A, bottle-filling apparatus No. 2192-O, battery of work places on long bench for labeling, packaging, and packing, and stitching machine No. 3127-C. Bottles handled in wooden trays to prevent accidents due to broken glass.

DESCRIPTION OF WORKING CONDITIONS Regular working hours 8-12, 1-5. Jobs performed in large airy room under daylight conditions. Artificial light available if necessary. Bottle washer wears rubber apron and gloves. Filling operator wears goggles, rubber apron and gloves, and cloth sleeves.

FLOW OF MATERIAL OR SUPPLIES Bottles supplied to washing machine from stock room. Washed bottles then moved to filling apparatus. Moved by truck from filling apparatus to labeling work place. Labeled bottles are then packed in cartons, cartons are packed in cases. Finished case is stitched on stitching machine, and then flows to shipping room. Packing supplies and labels are sent from supply room to position on work place.

B117

Fig. 203. General job conditions form, size 8½ × 11 inches.

The Warner & Swasey Co.

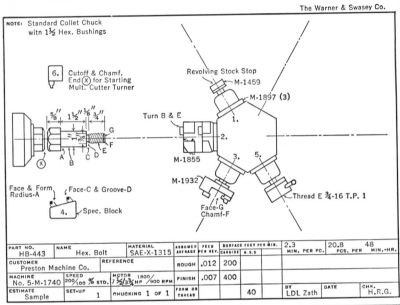

NOTE: Standard Collet Chuck witn 1½ Hex. Bushings

PART NO. HB-443	NAME Hex. Bolt	MATERIAL SAE-X-1315	ASSUMED AVERAGE	FEED PER REV.	SURFACE FEET PER MIN.		2.3 MIN. PER PC.	20.8 PCS. PER	48 MIN.-HR.
					CARBIDE	H.S.S			
CUSTOMER Preston Machine Co.		REFERENCE	ROUGH	.012	200				
MACHINE No. 5-M-1740	SPEED 200/100 % STD.	MOTOR 1800/900 RPM 1½/3¾HP	FINISH	.007	400				
ESTIMATE Sample	SET-UP 1	CHUCKING 1 OF 1	FORM OR THREAD			40	BY LDL Zath	DATE	CHK. H.R.G.

OPER. NO.	DIAM'R	FEET PER MIN.	REV. PER MIN.	FEED PER REV.	MIN. PER INCH	LENGTH OF CUT	TIME MACH	TIME HAND	TOOL NAME OR NUMBER	RT'D MACH EQUIPMENT	STANDARD TOOLS	PURCHASED EQUIP'T DURABLE	PERISH'LE	SPECIAL TOOLS DURABLE	PERISH'LE
1		LOAD & STOP					—	.20	Revolving Stock Stop						
2	1½"	215	548	.012	.15	2⅜"	.35	.12	Multiple Cutter Turner with 2 Carbide Cutters						
3	¾"	52	274	H/	—	⅛"	.15	.12	End Face & Form Tool with 1 High Speed Steel Cutter						
4	1½"	215	548	.007	.26	¼"	.10	.12	Special Block & 2 Carbide Cutters						
5	¾"	40	198	16 P	—	¾"	.05	.12	Die Head with High Speed Steel Chasers to Cut ¾-16 Threads						
6	1½"	75	198	.0045	1.10	¾"	.85	.12	High Speed Steel Cutoff Cutter						
		UNLOAD													
				TOTALS			1.50	.80							
	FLAT TIME MINUTES PER PIECE						2.30		SUB-TOTALS						

PRODUCTION — MACHINE EQUIPMENT AND TOOLING PRICE DETAIL

NOTE:

REMARKS:

(√) CHECKED ITEMS PRICED ON PREVIOUS SET-UP.

Fig. 204. Job setup and standard practice for operation on Warner and Swasey turret lathe. Size of sheet 11 × 16½ inches.

work. These two forms are prepared after the correct method has been established and put into effect. The "Standard Job Conditions" and the "General Job Conditions" forms used by this company are printed on bond paper, the first in yellow and the second in salmon color. An original and one carbon copy of each of these forms are made out in pencil. The original is placed in the folder with the original time studies of the operation and filed in the Wage Standards Department office. The carbon copy is filed in a loose-leaf binder in the office of the foreman of the department in which the operation is performed. This is used by the operator, foreman, and timekeeper.

The "Standard Job Conditions" form contains complete details of the specific operation; the "General Job Conditions" form, as the name indicates, contains more general information about the operation and the location of the work place relative to the rest of the department or building, information about the flow of material to and from the work place, working conditions, and similar matters.

Some classes of work are relatively simple, and written standard practices can be quickly prepared. On machine-tool work, for example, the speed and feed, shape and size of tools, coolant used, and method of chucking the piece are the important factors. The form in Fig. 324, developed by one company primarily as instructions for the operator, also serves as the basis for their permanent record of the operation.

In some plants where many operations are similar, methods are frequently developed for a whole class of work, and time standards are determined from tables of standard data or formulas. In such cases similar operations can be grouped into classes for which one master standard practice can be prepared. For example (Fig. 324), all sizes of gear blanks turned on turret lathe JL58 (Operation 5TR, Case D) follow the same sequence of motions of the operator and machine although the speeds, feeds, and sizes of the tools vary with the size of the blank.

Combination Computation Sheet, Instruction Sheet, and Written Standard Practice. The Jones & Lamson Machine Company uses the form shown in Fig. 205 as a combination computation sheet, instruction sheet, and standard practice. The operation referred to is that of machining a sliding gear (Fig. 206) on a No. 5 J & L universal turret lathe (Fig. 207). This part is made in the Jones & Lamson plant.

This is the way the time standard is determined and the way the instruction sheet (Fig. 205) is used. That portion of the sheet which appears above the dotted line *A–B* goes to the operator. It shows him the standard setup time and standard operation time per piece. This

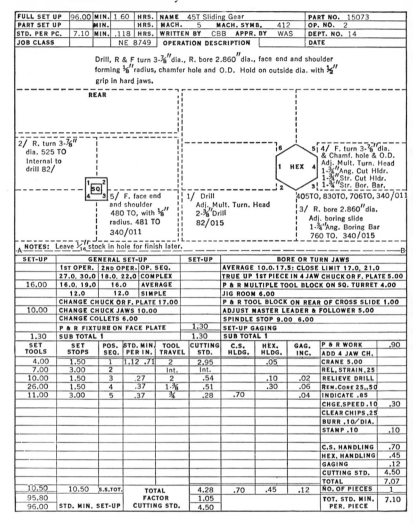

FULL SET UP	96.00	MIN.	1 60	HRS.	NAME	45T Sliding Gear		PART NO.	15073
PART SET UP		MIN.		HRS.	MACH.	5	MACH. SYMB. 412	OP. NO.	2
STD. PER PC.	7.10	MIN.	.118	HRS.	WRITTEN BY	CBB	APPR. BY WAS	DEPT. NO.	14
JOB CLASS			NE 8749		OPERATION DESCRIPTION			DATE	

Drill, R & F turn 3-⅞″dia., R. bore 2.860″dia., face end and shoulder forming ⅛″radius, chamfer hole and O.D. Hold on outside dia. with ½″ grip in hard jaws.

REAR

2/ R. turn 3-⅞″ dia. 525 TO Internal to drill 82/

5/ F. face end and shoulder 480 TO, with ⅛″ radius. 481 TO 340/011

1/ Drill Adj. Mult. Turn. Head 2-⅜″Drill 82/015

4/ F. turn 3-⅞″dia. & Chamf. hole & O.D. Adj. Mult. Turn. Head 1-¾″Ang. Cut Hldr. 1-¾″Str. Cut Hldr. 1-¾″Str. Bor. Bar. 405TO, 830TO, 706TO, 340/011

3/ R. bore 2.860″dia. Adj. boring slide 1-¾″Ang. Boring Bar 760 TO. 340/015

NOTES: Leave 1/16″ stock in hole for finish later.

SET-UP	GENERAL SET-UP			SET-UP	BORE OR TURN JAWS				
	1ST OPER.	2ND OPER.	OP. SEQ.		AVERAGE 10.0.17.5: CLOSE LIMIT 17.0, 21.0				
	27.0, 30.0	18.0, 22.0	COMPLEX		TRUE UP 1ST PIECE IN 4 JAW CHUCK OR F. PLATE 5.00				
16.00	16.0, 19.0	16.0	AVERAGE		P & R MULTIPLE TOOL BLOCK ON SQ. TURRET 4.00				
	12.0	12.0	SIMPLE		JIG ROOM 6.00				
	CHANGE CHUCK OR F. PLATE 17.00				P & R TOOL BLOCK ON REAR OF CROSS SLIDE 1.00				
10.00	CHANGE CHUCK JAWS 10.00				ADJUST MASTER LEADER & FOLLOWER 5.00				
	CHANGE COLLETS 6.00				SPINDLE STOP 9.00 6.00				
	P & R FIXTURE ON FACE PLATE			1.30	SET-UP GAGING				
1.30	SUB TOTAL 1			1.30	SUB TOTAL 1				

SET TOOLS	SET STOPS	POS. SEQ.	STD. MIN. PER IN.	TOOL TRAVEL	CUTTING STD.	C.S. HLDG.	HEX. HLDG.	GAG. INC.	P & R WORK	.90
									ADD 4 JAW CH.	
4.00	1.50	1	1.12 .71	2	2.95		.05		CRANE 5.00	
7.00	3.00	2		Int.	Int.				REL, STRAIN .25	
10.00	1.50	3	.27	2	.54		.10	.02	RELIEVE DRILL	
26.00	1.50	4	.37	1-⅜	.51		.30	.06	REM.CORE 25 .50	
11.00	3.00	5	.37	¾	.28	.70		.04	INDICATE .85	
									CHGE.SPEED .10	.30
									CLEAR CHIPS .25	
									BURR .10/DIA.	
									STAMP .10	.10
									C.S. HANDLING	.70
									HEX. HANDLING	.45
									GAGING	.12
									CUTTING STD.	4.50
									TOTAL	7.07
10.50	10.50	S.S.TOT.	TOTAL		4.28	.70	.45	.12	NO. OF PIECES	1
95.80			FACTOR		1.05				TOT. STD. MIN.	7.10
96.00	STD. MIN. SET-UP		CUTTING STD.		4.50				PER. PIECE	

Fig. 205. Combination computation sheet, instruction sheet, and written standard practice. Size of sheet 8½ × 11 inches.

information appears at the top left of the sheet. For Operation 2 (first operation on the turret lathe) the operator is allowed 96 minutes or 1.6 hours to set up his machine, and 7.10 minutes or 0.118 hour to machine each piece. Next follow instructions for machining the piece. Underneath these instructions are instructions in diagram form numbered for sequence of operation. For instance, 1/ is the first operation

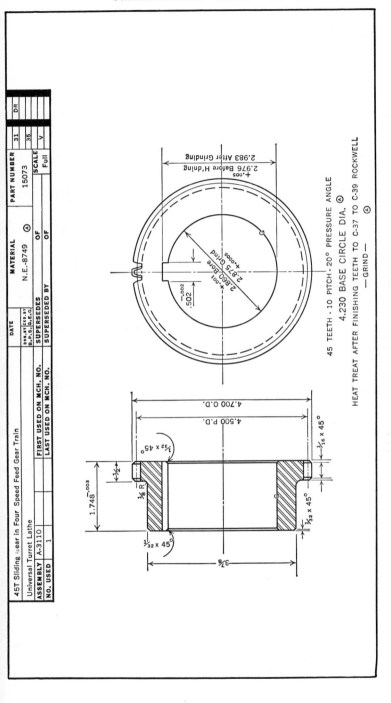

Fig. 206. Detail drawing of sliding gear, part number 15073.

on the turret lathe, 2/ is the second, etc. These instructions also give
the operator the part number of the tool that he is to use. For example
525 TO is the number of the tool to be used in the first position of the
square turret. The speeds and feeds to be used are also given. For
example, in the third position of the square turret, the operator is to
use a spindle speed of 340 r.p.m. and a feed of 0.011 inch per r.p.m.

The lower half of the sheet shows exactly how the standard time

Fig. 207. Jones and Lamson universal turret lathe.

for the operation is computed. This portion of the sheet, which is
kept in the Time Study Department, is available for reference in case
of possible complaint or error. In the left-hand corner of this section
the setup time for each operation is shown. The general setup time is
16 minutes, this being an average setup. Underneath this figure the
time is given for various operations incidental to the complete setup.
In the column adjacent to this, time is given for setting stops. The
fourth, fifth, and sixth columns show the computation of the actual
cutting time. This computation includes the actual cutting time plus
an incentive allowance. The next three columns give time allowed
for miscellaneous handling and gauging. All handling and gauging
time contains 15% allowance for rest and personal time. The last
two columns summarize the cutting and handling times, which total

7.10 minutes. The 1.05 factor by which the cutting time in column six is multiplied to get the final time of 4.05 minutes is an allowance for resharpening dull tools.

Motion Picture Records. Some complicated manual operations can be recorded best by motion pictures. In fact, it may be more economical in certain cases to make the record in this manner than to rely entirely on a written description of the job. On important operations, "before" and "after" motion pictures are frequently made for other purposes and may also serve, of course, as a supplement to the written standard practice. However, few companies as yet have seen fit to use motion pictures for standard-practice records in a general way.

CHAPTER 22

The Relation of Motion and Time Study to Wage Incentives

For many years the emphasis in the field of motion and time study was on the establishment of time standards for use with wage incentives. Although time study and wage incentives are still widely used, it has been found that motion study is also a powerful tool for cost reduction. In fact, some people believe that it is potentially more valuable than time study and wage incentives. Certainly management is encouraging and is profiting from the broader application of motion and time study. Also, employees are more likely to react favorably to this program, particularly since motion study has as its primary object finding the easiest and most satisfactory way of doing work, which usually raises output without requiring the employee to increase his effort.

Necessity for Measuring Labor Accomplishment. Labor is an important factor in the cost of producing manufactured goods, and management must consider labor costs like all other costs in the operation of a business. It is management's job to see that its employees do not do useless and unnecessary work. All operations should be subject to careful design, and the easiest and best method for the individual operation should be found. Whenever possible, the work should be measured and the employee told what a standard day's work is for his job. In all these activities, management must never forget that each employee is a person and that he should be treated as such. If management hopes to gain and keep the interest and cooperation of its employees, it must make certain that every action of the company will benefit the individual worker. There is plenty of evidence to show that the worker's mental attitude, his morale, "will to work," and enthusiasm for the job and for the company are of real value to

330

management. Wages alone, however large they may be, will not necessarily produce these desirable attributes in a working force.

Most things of value are purchased by measure; that is, a price is paid for a number of units of a given commodity of a specified quality. For example, sugar is bought by the pound, cloth by the yard, and energy by the kilowatt-hour. When a single factor is to be measured, the unit of measurement deals only with that factor. Thus distance may be measured by units of length, and contents by units of volume. When two factors are involved, however, as in electrical energy, both time and power must be included in the unit of measurement.

All work is largely a combination of mental and manual effort expended in a given period of time. Most factory work and much office work are largely manual, and it is this type of labor which is being considered in this book.

The results of work determine its value rather than the effort exerted. This is true whether a person works for himself or for someone else. It is the operator's productivity, his accomplishment, that largely measures his worth to his employer. Since accomplishment results from the application of effort and is influenced by both the duration and the intensity of effort, the unit of measurement of work done must include both quantity and time. Accomplishment can usually be measured most effectively in terms of quantity of work done per unit of time, that is, pieces per hour or tons per day. Ordinarily a standard of quality is specified, and only those units that meet the quality standard are considered as finished units.

Although some criticism has been directed at the principle of payment of labor in proportion to its productivity, there is much in favor of such a plan if properly administered. The greatest difficulty in the application of wage incentives is in the determination of the standard task. The answer to the question "What constitutes a standard day's work?" is very important indeed.

Motion and time study is the most accurate system known for measuring labor accomplishment. Although motion and time study is not a perfect tool, it will, if applied by well-qualified and properly trained persons, give results that are satisfactory both to the employee and to the employer.

In past years, and unfortunately to some extent today, time standards have been established on the basis of (1) past performance of operators, (2) estimate by the supervisor or by an "estimator," and (3) over-all time for a trial lot. These methods of "measuring" what should constitute a day's work are seldom satisfactory and should not be used as the basis for a wage incentive plan for direct factory labor. A time study carefully made by a competent analyst should be the basis for setting time standards. Of course the use of adequate elemental data, motion-time data, or work sampling is also quite acceptable.

Effects of Motion and Time Study and Wage Incentive Applications on the Worker. The two phases of motion and time study that concern the worker most are (1) improving the method of doing work, and (2) setting a time standard as the basis for a wage incentive. These two functions affect the operator in distinctly different ways. Both tend to reduce unit labor cost to the employer, mainly by decreasing the man-hours required; consequently both tend to displace labor on a given operation. That is, if windows in the factory and office buildings can be washed in half the time formerly taken, through the use of a carefully planned method and a wage incentive, then only half as many window washers will be required on the payroll of the company as would otherwise be employed. In this respect motion and time study falls into the category with tools and machinery which reduce labor costs because of their greater efficiency.

We all know that our high standard of living in this country has been attained because of our high labor productivity. We have steadily reduced the number of man-hours per acre needed to plant, cultivate, and harvest our crops. The number of man-hours of labor needed to mine a ton of coal has been cut in half during the last forty years, and the manufacturing industry has a continuous record of producing more with fewer man-hours. It has been fully demonstrated that in the long run every one benefits from increased productivity.[1] Each organization should see to it that such general benefits are obtained without asking anyone to work inordinately hard or without creating unemployment, even temporarily. Some companies are guaranteeing to their employees that no one will be laid off as the result of the introduction of new machines, processes, or methods, or because of the installation of a wage incentive system.

By the improvement of methods alone, the work is often made sufficiently easy so that with the same expenditure of energy the operator

[1] The following statement is from "Agreement Between General Motors Corporation and the UAW " January 1, 1968, p. **71**: "The improvement factor provided herein recognizes that a continuing improvement in the standard of living of employees depends upon technological progress, better tools, methods, processes and equipment, and a cooperative attitude on the part of all parties in such progress. It further recognizes the principle that to produce more with the same amount of human effort is a sound economic and social objective."

is able to produce more units per day. Thus, through the use of duplicate bins and the simple fixture for the assembling of the bolt and washers described in Chapter 17, the operator was able to do a half more work in the same time. This phase of motion and time study enables the operator to do more work without asking that he use more energy.

In contrast, the second phase of motion and time study, that of setting a time standard to be used with a wage incentive, reduces man-hours by offering to pay the operator more wages if he will do more work in a given period of time. To earn this extra reward the operator produces more, mainly through the elimination of idle time, through greater concentration on the job, and through greater expenditure of energy.

Perhaps an example is the best way to show the whole picture. This case will indicate not only how these two phases of motion and time study affect the employee through increase in earnings, but also how they affect the employer through a decrease in the direct-labor cost of the product. It will be assumed that no increase in wages is given to the operator when improvements in methods alone are made.

The operation is assembling a work-rest bracket for a bench grinder. The data in Fig. 208 show a 40% saving in time from an improvement in the method of assembling the bracket. The operator worked without incentive in both Case I and Case II; that is, she was paid a flat hourly rate irrespective of the output. She exerted approximately the same physical effort and gave approximately the same mental attention in both cases. However, in Case I she made 720 assemblies per day, whereas in Case II she made 1200 assemblies per day. This increase in output resulted not from working faster, but from a better arrangement of the work place and from the use of a special fixture which enabled the operator to use her hands to better advantage. She could do more work in the same time and with the same expenditure of energy because she could assemble a bracket with fewer motions, with no tiring holding of parts, and with an easy rhythm which was not possible in Case I.

In Case III a time standard was set by means of a stop-watch time study and a piece rate was established for the operation. The operator now had the opportunity of earning more than her guaranteed base wage of $21.60 per day. In fact, she was easily able to do 25% more work than the standard, and in return for this extra performance she earned $27.00 per day.

The increased output resulting from the application of the piece rate

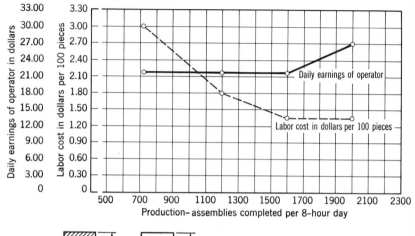

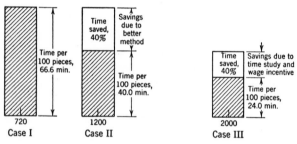

CASE I

Method. Assemble one piece at a time. Left hand holds bracket while right hand assembles parts. Method poor.

Supervision. Poor.

Operator Performance. Poor.
Method of Wage Payment. Day work. Hourly rate = $2.70. Daily earnings of operator = $21.60.

Average Production, taken from past records = 720 pieces for an 8-hour day. Average time per 100 pieces = 66.6 minutes.

Average Labor Cost per 100 pieces = $3.00.

CASE II

Method. Assemble two pieces at one time, using special fixture. Method good.

Supervision. Poor.

Operator Performance. Poor.
Method of Wage Payment. Day work. Hourly rate = $2.70. Daily earnings of operator = $21.60.

Average production, taken from past records = 1200 per 8-hour day. Average time per 100 pieces = 40.0 minutes.

Average Labor Cost per 100 pieces = $1.80.

CASE III

Method. Assemble two pieces at one time, using special fixture. Method good.

Supervision. Good.

Operator Performance. Good. Operator now on incentive.

Method of Wage Payment. Straight piece rate with guaranteed minimum rate of $2.70 per hour. Standard time per 100 pieces set by time study = 30.0 minutes. Piece rate per 100 pieces = $1.35. Standard output per day = 1600 pieces. Average number of pieces actually produced per day by this operator = 2000. Average daily earnings of this operator = $27.00

Average Labor Cost per 100 pieces = $1.35.

OPERATION: Assemble work-rest bracket for bench grinder
OPERATOR: Helen G. Meyers,
BASE WAGE: $2.70 per hour—8-hour day—40-hour week
WAGE-INCENTIVE PLAN: Straight piece rate

Fig. 208. The relation of motion and time study to wage incentives.

came because the operator "worked harder" than she did in Case II. That is, she worked more consistently through the day, eliminated idle time, took less personal time, and perhaps visited less frequently with her neighbors. She started work on time and worked until quitting time, concentrating on the work she was doing during the entire day. Although it is likely that the operator used approximately the same motions in completing a cycle in Case III as in Case II, it is certain that she used more effort in Case III than in Case II. The incentive of greater pay for greater output was responsible for this. It is also likely that the operator was more fatigued at the end of the day in Case III than in Case II.

It is therefore apparent that greater output through improved methods ordinarily causes the operator no increase in fatigue. In fact, the improved method is usually easier, more satisfying, and less fatiguing than the original one. On the other hand, the application of a wage incentive usually causes the operator to work harder. The extent of the operator's exertion will depend upon his own inclination and his fitness for the job. With straight piece rate or with a 100% premium plan of wage payment, the reward is in direct proportion to the output.

For the employer the application of motion study reduced the direct labor cost 40%, and the setting of the rate and the wage incentive reduced it another 25%. The direct-labor cost per 100 pieces in Case I was $3.00, in Case II $1.80, and in Case III $1.35. This is shown graphically by the curve at the top of Fig. 208.

Ways in Which Motion and Time Study and Wage Incentives Increase Output. The question is frequently raised as to why there is often such a great difference between the output of a person paid on a day-work basis without production standards established for his job and the output of the same person after time standards have been set and an incentive system of wage payment is used.

There are three main reasons why motion and time study and a wage incentive installation may bring greater daily output among direct labor.

1. Improved work methods enable the operator to produce more with the same effort. In some organizations it is the practice to improve methods before beginning the time study work. Even if this procedure is not followed, it is still possible that some improvement in methods will result from preliminary work incident to time study.

In some plants, particularly those with poor supervision, we may find work done in a hit-or-miss manner, inadequate planning, lack of standardization, and little or no idea of what a day's work should be. In such plants, materials that vary from standard may force operators to work at a slow pace or to perform extra operations which may result in low output per hour. Delays may be caused by machines and equipment not being kept in good repair. Lack of work, delay in sharpening tools, and inadequate supervision may cause idleness on the part of the operator. Time study would reveal such inefficiencies, and a wage incentive system would require that they be corrected. Standardization of materials, methods, tools, equipment, and working conditions *must always precede* the installation of a wage incentive system. This is management's responsibility.

2. If each employee knows what a standard day's work is and if he is paid a bonus for work produced above the standard, he will in most cases on his own accord eliminate waste time within his control, such as late starting, early quitting, and unnecessary idleness during the day. Moreover, he will put pressure on management to eliminate causes of idle time beyond his control, such as shortage of material, machine breakdowns, and delays in sharpening tools.

For example, Fig. 209 shows graphically the results of an all-day production study of a short-cycle hand-milling machine operation. The operator worked in a remote part of the plant, had little supervision on this particular operation, and on the day during which this study was made lost 38.9 minutes because of late starting and early quitting and 32.9 minutes for other personal reasons. Although an operator on such work might possibly be allowed 10% or 48 minutes per day for personal time and fatigue allowance, this operator actually took a total of 71.8 minutes or the 48 minutes allowed him and 23.8 minutes in addition. In most plants a financial incentive and a little encouragement are all that would be needed to persuade this operator to work during the 23.8 minutes that he lost.

A time standard would be set so that a qualified person working at a normal pace during 432 minutes (480 − 48 = 432) of the day would produce 480 standard minutes of work. As the data in Fig. 209 show, during the time this man was actually working he produced at a pace averaging 102%. However, he turned out only 799 pieces during the 8-hour day. Had he worked at this same pace of 102% efficiency during the 432 minutes, he would have turned out 881 pieces. Of course, the lost production that resulted from the 16.5-minute machine breakdown is beyond the control of the operator.

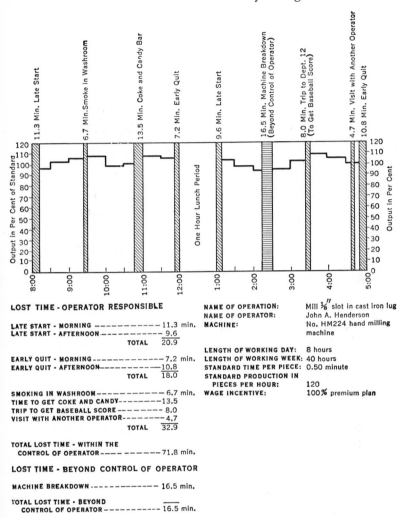

LOST TIME - OPERATOR RESPONSIBLE

LATE START - MORNING ――――――――― 11.3 min.
LATE START - AFTERNOON――――――――― 9.6
                              TOTAL   20.9

EARLY QUIT - MORNING――――――――――― 7.2 min.
EARLY QUIT - AFTERNOON――――――――― 10.8
                              TOTAL   18.0

SMOKING IN WASHROOM――――――――――― 6.7 min.
TIME TO GET COKE AND CANDY――――――― 13.5
TRIP TO GET BASEBALL SCORE ―― ― ― ― ― 8.0
VISIT WITH ANOTHER OPERATOR――――――― 4.7
                              TOTAL   32.9

TOTAL LOST TIME - WITHIN THE
    CONTROL OF OPERATOR――― ―――――― 71.8 min.

LOST TIME - BEYOND CONTROL OF OPERATOR

MACHINE BREAKDOWN―――――――――――― 16.5 min.

TOTAL LOST TIME - BEYOND
    CONTROL OF OPERATOR――――――――― 16.5 min.

NAME OF OPERATION: Mill ⅛″ slot in cast iron lug
NAME OF OPERATOR: John A. Henderson
MACHINE: No. HM224 hand milling
                         machine

LENGTH OF WORKING DAY: 8 hours
LENGTH OF WORKING WEEK: 40 hours
STANDARD TIME PER PIECE: 0.50 minute
STANDARD PRODUCTION IN
    PIECES PER HOUR: 120
WAGE INCENTIVE: 100% premium plan

Fig. 209. Production curve for one day. Output is expressed in per cent, with 100% equal to standard performance.

Some people prefer to work during some or all of the time allowed for personal needs and fatigue. When there are no fixed rest periods, the worker is paid for the output which he produces during this time. It should be pointed out that fatigue allowances are intended to permit the operator to relax and recuperate during the working day, and it is expected that most workers will take time out for this purpose. However, some workers do not seem to need such time for rest and prefer to work straight through the day with only the noon hour off.

3. Since the work standard is set so that qualified operators can easily exceed it and thus earn additional compensation, the wage incentive serves to encourage workers to increase their speed and thus turn out more work per hour than they would normally. Referring again to Fig. 209, if this operator's efficiency was 102% during his working time when paid on a day-work basis, it is to be expected that he will work at a faster pace if he receives extra pay for all work produced beyond the standard for his job.

Nearly every person finds that he can exceed the hourly output defined as "normal performance," and the average output of a group of qualified operators working on incentives usually exceeds normal by 15 to 35%. A study of 72 companies showed that the average output was 28% above normal.[2]

Work measurement establishes the correct time standard, and the wage incentive system serves to pay the worker for the extra output he produces beyond the standard. The effort that an employee chooses to exert at a given time or on any particular day is entirely a personal matter with him. Each person is guaranteed his hourly rate of pay irrespective of his output.

A Motion and Time Study and Wage Incentive Application. The demand for a new product often exceeds that predicted for it, and in order to satisfy the demand an extra shift is used or additional machines are purchased, with little or no change being made in the production methods. Eventually, however, a more careful analysis of each operation will be made, and some organizations make such an analysis at the time they put the work on incentive. The following case gives a day-by-day record of steps that were taken and the results that were obtained on one job. The operation was a rather complicated assembly, involving some gauging and adjustment. The job had been running for one week on day work, and the average output of the operator was around 22 pieces per hour (Fig. 210).

A new method was suggested by the foreman, and a special fixture was made and installed at the end of the working day on April 16. This new fixture and the new work-place layout enabled the operator

[2] Ralph M. Barnes, *Industrial Engineering Survey*, University of California, 1967.

to make the assemblies faster than formerly, but the socket wrench gave trouble. The operator liked the new layout but complained that the socket on the power wrench was too small for the nuts. Production dropped to 16 pieces per hour, mainly because of trouble with the wrench. After work on April 17, the foreman tried to have the socket

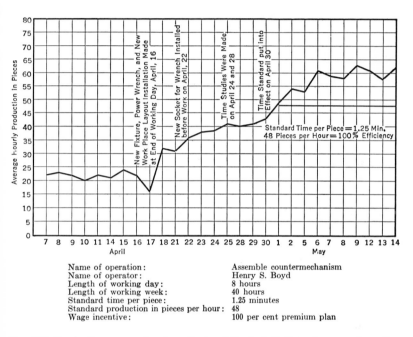

Name of operation: Assemble countermechanism
Name of operator: Henry S. Boyd
Length of working day: 8 hours
Length of working week: 40 hours
Standard time per piece: 1.25 minutes
Standard production in pieces per hour: 48
Wage incentive: 100 per cent premium plan

Fig. 210. Production curve showing effects of the installation of a wage incentive. Output is expressed in pieces per hour.

made larger, but the next day the operator still had some trouble. A new socket was ordered. Production, however, jumped to 32 pieces per hour on April 18. On the morning of April 22 a new socket was installed on the power wrench which made it easier to use, and this increased output to 36 pieces per hour. The output gradually increased to around 40 pieces per hour. The operator was somewhat amazed at the amount of work he was turning out each day. A time study was made of the job on April 24 and April 28. The standard was set at 48 pieces per hour, and the new standard was put into effect on April 30. The next day the average hourly production was 49 pieces per hour, and the following day it went up to 54 pieces. Production for

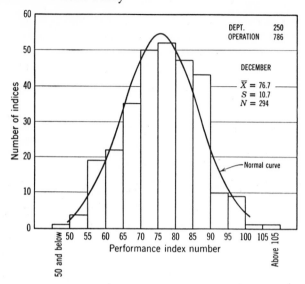

Fig. 211. Distribution curve of daily performance index for the workers in the final assembly department for the month of December, the period just before the wage incentive plan was put into effect.

this operation stabilized at around 58 to 60 pieces per hour, which represents an efficiency of 120 to 125%. At this level of production the operator earned a bonus of 20 to 25%, or 20 to 25% more money than it was possible for him to earn before the installation of the wage incentive.

Distribution of Operator Performance Index before and after Wage Incentive Application. Some interesting data on operator performance were obtained from a lock manufacturing company at the time it was installing a wage incentive plan.[3] After the methods and conditions had been standardized and the time standards were established, it became necessary to delay the actual application of the incentive plan for several months. During this period, production records were kept and the daily performance index for each operator was calculated just as though a wage incentive were being paid, but the workers were paid their regular day rate only. Figure 211 shows the distribution curve for the 294 workers in the final assembly department during the month

[3] Donald C. Demangate, "Statistical Evaluation of Worker Productivity," *Proceedings Sixth Industrial Engineering Institute,* University of California, Los Angeles-Berkeley, pp. 89–91.

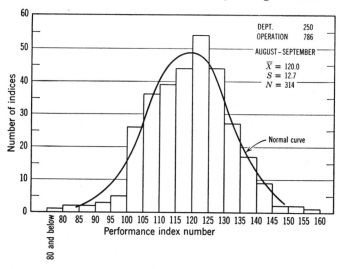

Fig. 212. Distribution curve of daily performance index for the workers in the final assembly department after the wage incentive plan had been in effect for approximately 8 months.

of December, the period just before the wage incentive plan was put into effect. The average performance index for this group was 76.7%. Figure 212 shows the distribution curve for the same department after the wage incentive plan had been in effect for approximately 8 months. Now the average performance index has increased to 120%, but the general shape of the distribution curve is not greatly different from that shown in Fig. 211.

CHAPTER 23

Time Study: Time Study Equipment; Making the Time Study

Stop-watch time study is the most commonly used method of measuring work in industry today. As will be explained later, however, there is a definite place for time standards established by elemental data, motion-time data, and work sampling. Figure 213 shows in condensed form the several methods and devices used for measuring work.

Each of the different methods of determining the standard time required to do a given task will be presented in this and the following nine chapters. This chapter describes the equipment used and explains how a time study is made. Chapters 24 and 25 show how the rating factor, allowances, and time standard are determined. Chapter 26 describes recent developments in mechanized time study and the use of electronic data-processing systems for measuring work. Chapters 27, 28, and 29 deal with elemental time data and formulas for setting time standards synthetically, and Chapters 30 and 31 show how motion-time data may also serve for obtaining the standard time for an operation without the use of a stop watch. Chapter 32 presents a rather complete description of work sampling, a relatively new but extremely useful tool for measuring work.

Definition of Time Study. Time study is used to determine the time required by a qualified and well-trained person working at a normal pace to do a specified task. This is the third part of the definition of motion and time study which appears on page 4. It should be noted that, while motion study is largely design, time study involves measurement. Time study is used to measure work. The result of time study is the time in minutes that a person suited to the job and fully trained in the specified method will need to perform the job if he works at a normal or standard tempo. This time is called the *standard time* for the operation.

342

Not Measurement	ESTIMATE *	Usually by an experienced estimator
	PAST PERFORMANCE *	From company records

Methods and Devices for Measuring Work	TIME STUDY	Data obtained by means of
		(a) Stop watch
		   1. Decimal – minute stop watch
		   2. Decimal – hour stop watch
		(b) Motion picture camera
		   1. Slow speed — 50 or 100 pictures or frames per minute. Time – lapse attachment for camera – electric – motor driven or solenoid operated.
		   2. Normal speed—960 frames per minute. Spring–driven or electric motor – driven camera.
		   3. Modified normal speed—1000 frames per minute. Electric motor – driven camera.
		   4. Sound speed — 1440 frames per minute. Spring–driven or electric motor–driven camera.
		   5. High speed — 64 to 128 frames per second. Spring–driven or electric motor – driven camera.
		   6. Very high speed—1000 to 3000 frames per second. Electric motor – driven camera.
		(c) Machine using moving tape or disk
		   1. Time study machine such as Marsto – chron.
		   2. Kymograph — such as Esterline – Angus Operation Recorder.
		   3. Servis recorder.
		(d) Electronic timer — information and time data punched on tape. Machine such as that developed by IBM and R.R. Donnelley and Sons Co.

	ELEMENTAL DATA	Information obtained from time study or motion – time data
	MOTION –TIME DATA	Some commonly used systems (listed chronologically by date system was first used or published) (a) Motion –Time Analysis (b) Work Factor (c) Motion–time data for assembly operations (Get and Place Data) (d) Methods –Time Measurement (e) Basic Motion Timestudy (f) Dimensional Motion Times

	WORK SAMPLING	Measurement by sampling methods
		(a) Observer obtains and records data
		   1. Record and analyze data manually
		   2. Record data on IBM mark – sensing cards — IBM analysis of data
		(b) Motion picture camera records information
		   1. Record and analyze data from film manually
		   2. Record data from film on IBM cards — IBM analysis of data

* Often used for cost estimating and budget purposes but not recommended for establishing time standards for direct labor for wage incentive purposes.

Fig. 213. Methods and devices for measuring work.

Uses for Time Study. Although time study has had greatest application for determining time standards in connection with wage incentive plans, it is used also for a number of other purposes. Time study may be used for:

1. Determining schedules and planning work.

2. Determining standard costs and as an aid in preparing budgets.

3. Estimating the cost of a product before manufacturing it. Such information is of value in preparing bids and in determining selling price.

4. Determining machine effectiveness, the number of machines which one person can operate, the number of men needed on a gang, and as an aid in balancing assembly lines and work done on a conveyor.

5. Determining time standards to be used as a basis for the payment of a wage incentive to direct labor.

6. Determining time standards to be used as a basis for the payment of indirect labor, such as handlers and setup men.

7. Determining time standards to be used as a basis for labor cost control.

TIME STUDY EQUIPMENT

The equipment needed for time study work consists of timing devices and auxiliary equipment. The devices for measuring time are (1) stop watch, (2) motion picture camera (with constant-speed motor drive or with a microchronometer in the picture to indicate time), and (3) time-recording machine. The auxiliary equipment consists of observation board, tachometer, and slide rule.

Decimal Stop Watches. The stop watch is the most widely used timing device for time study. The decimal-minute watch and the decimal-hour watch are the only two types of stop watches that are used for this work; the first is more widely utilized than the second. However, the camera and the time-recording machine are finding increasing application in this field.

The decimal-minute stop watch (Fig. 214) has the dial divided into 100 equal spaces, each of which represents 0.01 minute, the hand making one complete revolution per minute. A smaller dial on the watch is divided into 30 spaces, each of which represents 1 minute, the hand making one complete revolution in 30 minutes. The hands of the watch are controlled by the slide A and the winding stem B in Fig. 214. The starting and stopping of the watch are controlled by the slide. It is possible to stop the hand at any point and then start it again from that position. Pressure on the top of the stem *B* returns the hand to zero, but it starts off immediately upon releasing the stem. The hand may be held at zero either by holding the stem down or by pushing the slide *A* away from the stem.

The decimal-hour stop watch is like the decimal-minute watch in design and operation, but it has the dial divided into 100 spaces,

Fig. 214. Decimal-minute stop watch.

each of which represents 0.0001 hour, the hand making 100 revolutions per hour. The small dial on the watch is divided into 30 spaces, each of which represents 0.01 hour, the hand making $3\frac{1}{3}$ revolutions per hour. The principal advantage of this watch is that the readings are made directly in fractions of an hour, which is the common unit of time measurement in industry. The chief disadvantage of the decimal-hour watch is that it is more difficult to handle four decimal places than two decimal places. This is particularly true in recording stop-watch data on the observation sheet.

The split-second stop watch is not recommended and is seldom used for this work.

The Motion Picture Camera. The time for the elements of an operation can be obtained from motion pictures of the operation made with a synchronous motor-driven motion picture camera (Fig. 215) of known speed, or by placing a microchronometer in the picture when the operation is filmed. The method of making such pictures has been explained in Chapter 13.

The camera speed most frequently used is 1000 frames per minute, which permits the measurement of time in thousandths of a minute. A

Fig. 215. Motion picture camera with synchronous motor drive gives a constant speed of 1000 frames per minute.

motion picture of an operation forms a permanent record of the method used as well as the time taken for each element of the operation. Moreover, the film may be projected at the exact speed at which the picture was made, and a check may be made of the operator's performance. In other words, the operator's speed or tempo may be rated —that is, related to standard performance. Camera speeds greater than 1000 frames per minute may of course be used; also, a time-lapse drive may be used to take pictures at 50 or 100 frames per minute.

Time-Recording Machines. For some years machines have been used to a limited extent in this country and abroad for recording time. The time-recording machine consists of a small box through which a paper tape is drawn by an electric motor at a uniform velocity of 10 inches per minute. The tape has a printed scale in tenths of an inch, and therefore one division on the tape equals 0.01 minute. The time study machine has two keys which, when depressed, print marks on the tape. The beginning of an element is ordinarily recorded by pressing both keys, and the end of the element is recorded by pressing one key. It is necessary to have access to electric power circuits which have the correct voltage to operate the recording-machine motor.

This time-recording machine may be used instead of a stop watch, and it enables the analyst to measure shorter elements than he could with a stop watch. The machine seems to be most useful where short cycles are to be timed and where the operator follows a given routine without the introduction of many foreign elements.

The Servis recorder is a spring-driven instrument which records time on a wax-coated paper disk by means of a stylus attached to a small pendulum within the instrument. The recorder is fastened to a machine or piece of equipment, and the vibration of the machine causes the stylus to record "working time" on the disk. When the machine stops, the pendulum stops vibrating and the instrument records "idle time" on the disk. The disk, divided into hours and minutes, indicates the length of working time or idle time and also the time of day when each occurs. The instrument is most valuable for recording delays and lost time. It is not used for making time studies.

An automatic electronic timer, recorder, and computer for time study purposes has been developed and is described in Chapter **26**.

Observation Board. A lightweight board, slightly larger than the observation sheet, is used to hold the paper and the stop watch. There are many different arrangements, but it seems best to have the watch mounted rigidly somewhere near the upper right-hand corner of the board and the observation sheets held in place by some form of clamp at the side or top of the board. The observation board shown in Fig. **216** is a form commonly used. Since the analyst, in most cases, must record the data while standing, it is desirable to have his watch and paper arranged as conveniently as possible.

While taking a time study the observer should hold the board against his body and the upper left arm in such a way that the watch can be operated by the thumb and index finger of the left hand. The observer holds the board with his left hand and arm, leaving his right hand free to record the data.

By standing in the proper position relative to the work being observed, and by holding the board so that the dial of the watch falls in the line of vision, the observer can concentrate more easily on the three things demanding his attention, namely, the operator, the watch, and the observation sheet.

OBSERVATION SHEET

OPERATION		Study No.	
PART NAME		Op. No.	
Machine Name	Machine Number	Part No.	
Operator Name & No.		Dept.	
Experience on Job	Material	Date	
Begin	Finish	Elapsed	Units Finished

● ELEMENTS	Speed	Feed	1	2	3	4	5	6	7	8	9	10	Selected Time
1													
2													
3													
4													
5													
6													
7													
8													
9													
10													

Selected Time		Tools, Jigs, Gauges:-
Rating		
Normal Time		
Personal Allowance		
Fatigue Allowance		
Total Allowance		Timed By
Standard Time		

Fig. 216. Observation board with observation sheet for recording data taken by the repetitive method.

The observation sheet is a printed form with spaces provided for recording information about the operation being studied. This information usually includes a detailed description of the operation, the name of the operator, the name of the time study observer, and the date and place of study. The form also provides spaces for recording stop-watch readings for each element of the operation, performance ratings of the operator, and computations. Space may be provided for a sketch of the work place, a drawing of the part, and specifications of the material, jigs, gauges, and tools.

Observation sheets differ widely as to size and arrangement, but a sheet 8½ inches by 11 inches is widely used, mainly because it is easy to file or bind. The observation sheets shown in Figs. 217, 237, and 241 have proved to be satisfactory in industries manufacturing a diversified line of products. Some organizations find it convenient to supplement the observation sheet with a separate computation sheet (Fig. 243) and a sheet containing a more complete description of each element (Fig. 246).

Other Equipment. A speed indicator, or a tachometer, is needed where machine-tool operations are studied. It is a very good rule for the analyst to check speeds and feeds in making a time study, even though the machine has a table attached which gives this information for each setting of the speed and feed-control levers.

The ordinary slide rule is recommended as a valuable aid and time saver to every motion and time study analyst. Special slide rules may be purchased or constructed and used to advantage in connection with certain kinds of work.

MAKING THE TIME STUDY

The exact procedure used in making time studies may vary somewhat, depending upon the type of operation being studied and the application that is to be made of the data obtained. These eight steps, however, are usually required:

1. Secure and record information about the operation and operator being studied.
2. Divide the operation into elements and record a complete description of the method.
3. Observe and record the time taken by the operator.
4. Determine the number of cycles to be timed.
5. Rate the operator's performance.
6. Check to make certain that a sufficient number of cycles have been timed.
7. Determine the allowances.
8. Determine the time standard for the operation.

Request for a Time Study. A time study is not made unless an authorized person requests it. Usually it is the foreman who requests that a study be made, but the plant manager, chief engineer, produc-

OBSERVATION SHEET

SHEET 1 OF 1 SHEETS   **DATE**

OPERATION   Drill ¼" Hole   **OP. NO.**   D-20

PART NAME   Motor Shaft   **PART NO.**   MS-267

MACHINE NAME   Avey   **MACH. NO.**   2174

MALE ☑   **FEMALE** ☐

OPERATOR'S NAME & NO.   S.K. Adams 1347

EXPERIENCE ON JOB   18 Mo. on Sens. Drill   **MATERIAL**   S.A.E. 2315

FOREMAN   H. Miller   **DEPT. NO.**   DL 21

BEGIN 10:15   **FINISH** 10:38   **ELAPSED** 23   **UNITS FINISHED** 20   **ACTUAL TIME PER 100** 115   **NO. MACHINES OPERATED** 1

ELEMENTS	SPEED	FEED		1	2	3	4	5	6	7	8	9	10	SELECTED TIME
1. Pick Up Piece and Place in Jig			T	.12	.11	.12	.13	.12	.10	.12	.12	.14	.12	
			R	.12	.29	.39	.54	.66	.77	.92	8.01	14	.32	
2. Tighten Set Screw			T	.13	.12	.12	.14	.11	.12	.12	.13	.12	.11	
			R	.25	.41	.51	.68	.77	.89	7.04	.14	.26	.43	
3. Advance Drill to Work			T	.05	.04	.04	.04	.05	.04	.04	.04	.03	.04	
			R	.30	.45	.55	.72	.82	.93	.08	.18	.29	.47	
4. DRILL ¼" HOLE	980	H	T	.57	.54	.56	.51	.54	.58	.52	.53	.59	.56	
			R	.87	.99	3.11	4.23	5.36	6.51	.60	.71	.88	11.03	
5. Raise Drill from Hole			T	.04	.03	.03	.03	.03	.03	.03	.03	.04	.03	
			R	.91	2.02	.14	.26	.39	.54	.63	.74	.92	.06	
6. Loosen Set Screw			T	.06	.06	.07	.06	.06	.06	.06	.06	.07	.08	
			R	.97	.08	.21	.32	.45	.60	.69	.80	.99	.14	
7. Remove Piece from Jig			T	.08	.09	.08	.08	.09	.08	.07	.08	.09	.07	
			R	1.05	.17	.29	.40	.54	.68	.76	.88	10.08	.21	
8. Blow Out Chips			T	.13	.10	.12	.14	.13	.12	.13	.12	.12	.11	
			R	.18	.27	.41	.54	.67	.80	.89	9.00	.20	.32	
9.			T											
			R											
10. (1)			T	.12	.11	.13	.14	.12	.12	.11	.13	.12	.12	.12
			R	11.44	.56	.69	.82	.87	17.01	18.09	.21	.31	.42	
11. (2)			T	.12	.14	.12	.11	.12	.10	.13	.15	.12	.11	.12
			R	.36	.70	.81	.93	.99	.11	.22	.36	.43	.53	
12. (3)			T	.04	.04	.04	.03	.04	.04	.04	.04	.04	.04	.04
			R	.60	.74	.85	.96	16.03	.15	.26	.40	.47	.57	
13. (4)			T	.54	.53	.55	.52	.57	.54	.50	.53	.55	.54	.54
			R	12.14	13.27	14.40	15.48	.60	.69	.76	.93	21.02	22.11	
14. (5)			T	.03	.03	.03	.03	.03	.03	.03	.03	.03	.03	.03
			R	.17	.30	.43	.51	.63	.72	.79	.96	.05	.14	
15. (6)			T	.06	.06	.06	.07	.06	.05	.06	.06	.05	.06	.06
			R	.23	.36	.49	.58	.69	.77	.85	20.02	.10	.20	
16. (7)			T	.08	.08	.09	.08	.08	.07	.08	.06	.08	.08	.08
			R	.31	.44	.58	.66	.77	.84	.93	.08	.18	.28	
17. (8)			T	.14	.12	.10	.09	.12	.14	.15	.11	.12	12	.12
			R	.45	.56	.68	.75	.89	.98	19.08	.19	.30	22.40	
18.			T											1.11
			R											

SELECTED TIME 1.11   **RATING** 100%   **NORMAL TIME** 1.11   **TOTAL ALLOWANCES** 5%   **STANDARD TIME** 1.17

Overall Length 12"   Drill ¼" Hole

TOOLS, JIGS, GAUGES: Jig No. D-12-33
Use H.S. Drill ¼" Diam.
Hand Feed
Use Oil - S4

TIMED BY J.B.M.

Fig. 217. Stop-watch time study of a drilling operation made by the continuous method.

tion control supervisor, cost accountant, or other member of the organization may make such a request.

If a time standard is to be established on a new job for wage incentive purposes, in most plants it is the foreman's responsibility to make certain that the operation is running satisfactorily before requesting the study. He should also see that the operators have thoroughly learned the job and that they are following the prescribed method. The foreman should inform the operators in advance that a time study is to be made, stating the purpose of the study.

Time studies should be made only by members of the time study department. Unauthorized persons should not be permitted to make time studies even though they are not for wage incentive purposes.

Is the Job Ready for Time Study? After a request for a time study has been received by the time study department and an analyst has been assigned to make the study, he should go over the job with the foreman of the department. As they discuss each element of the operation, the analyst asks himself the question, "Is this operation ready for a time study?"

The time standard established for a job will not be correct if the method of doing the job has changed, if the materials do not meet specifications, if the machine speed has changed, or if other conditions of work are different from those that were present when the time study was originally made. The time study analyst therefore examines the operation with the purpose of suggesting any changes that he thinks should be effected before the time study is made.

Although the foreman may have set up the job originally or may have checked the method with the process engineer who set it up, the time study analyst should question each phase of the work, asking such questions as:

1. Can the speed or feed of the machine be increased without affecting optimum tool life or without adversely affecting the quality of the product?

2. Can changes in tooling be made to reduce the cycle time?

3. Can materials be moved closer to the work area to reduce handling time?

4. Is the equipment operating correctly, and is a quality product being produced?

5. Is the operation being performed safely?

It is expected that the time study analyst will be trained in motion study and that he will bring all his knowledge in this field to bear on the operation he is about to time. Any suggested changes that the foreman wishes to adopt should be made before the study is started. The foreman of course makes the decision as to the way the job is to be done, but the analyst and the foreman should discuss each element of the operation and should agree that the operation is ready for a time study. Standardization of the work has been discussed at length in the preceding chapters, and it has been emphasized that all standardization should precede the actual setting of the time standard. The possible extent of this work has been indicated in Table 1; it may range from the very elaborate Type A investigation, requiring much time and considerable expense, to Types D and E, requiring only a cursory analysis and a general check for methods.

If a major change in the operation is to be made and if considerable time will be required to put the new method into effect, it might be wise to make a time study of the present method and then, after the improvements are installed, restudy the job and set a new time standard. If only minor changes are contemplated, it is usually advisable to complete such changes before making a time study of the job.

Making the Time Study. Those phases of time study that can be carried out at the time and place of the performance of the operation will be described in this and the following chapter. They are obtaining and recording necessary information, dividing the operation into subdivisions or elements, listing these elements in proper sequence, timing them with the stop watch and recording the readings, determining the number of cycles to be timed, noting and recording the operator's tempo or performance level, and making a sketch of the part and of the work place.

Recording Information. All information asked for in the heading of the observation sheet should be carefully recorded. This is important because time studies hastily and incompletely made are of little value. The first place to practice thoroughness is in filling in all necessary information for identification. Unless this is done, a study may be practically worthless as a record or as a source of information for standard data and formula construction a few months after it has been made, because the person who made the study has forgotten the circumstances surrounding it. Ordinarily, the necessary information concerning the operation, part, material, customer, order number, lot size, etc., can be obtained from the route sheet, bill of material, or drawing of the part.

A sketch of the part should be drawn at the bottom or on the back of the sheet if a special place is not provided. A sketch of the work place should also be included, showing the working position of the operator and the location of the tools, fixtures, and materials. Specifications of the materials being worked upon should be given, and a description of the equipment being used should be recorded. Ordinarily the trade name, class, type, and size of the machine are sufficient description. If the machine has an identification number assigned to it, the number should be included. Accurate record should be made of the number, size, and description of tools, fixtures, gauges, and templets. The name and number of the operator should be recorded, and the time study should be signed by the time study analyst.

Dividing the Operation into Elements and Recording a Description of the Method. The standard time for an operation applies only to that particular operation; therefore a complete and detailed description of the method must be recorded on the observation sheet or on auxiliary sheets to be attached to the observation sheet. The importance of this description cannot be overemphasized. At any time after the standard has been set for a job, the time study department may be asked to determine whether the operator is performing the job in the same way it was being performed when the time study was originally made. The information contained on the observation sheet is the most complete description of the method that the time study department has available for such a check.

Reasons for Element Breakdown. Timing an entire operation as one element is seldom satisfactory, and an over-all study is no substitute for a time study. Breaking the operation down into short elements and timing each of them separately are essential parts of time study, for the following reasons:

1. One of the best ways to describe an operation is to break it down into definite and measurable elements and describe each of these separately. Those elements of the operation that occur regularly are usually listed first, and then all other elements that are a necessary part of the job are described. It is sometimes desirable to prepare a detailed description of the elements of an operation on a separate sheet and attach it to the observation sheet. The beginning and end points for each element may be specifically indicated (see page 403, also Fig. 245). Very often the elements taken from the time study can serve as the "standard practice" for the operation (Fig. 324). Such a list of elements also may be used for training new operators on the job.

2. Standard time values may be determined for the elements of the job. Such element time standards make it possible to determine syn-

thetically the total standard time for an operation (see Chapter **27**).

3. A time study may show that excessive time is being taken to perform certain elements of the job or that too little time is being spent on other elements. This latter condition sometimes occurs on inspection elements. Also the analysis of an operation by elements may show slight variations in method that could not be detected so easily from an over-all study.

4. An operator may not work at the same tempo throughout the cycle. A time study permits separate performance ratings to be applied to each element of the job.

When time studies are to be made of a new product or a new type of work, a careful analysis should be made of all variables of the work that are likely to occur. It is desirable to establish element time standards as soon as possible, and such standards can be obtained more quickly if the general framework of the standards is prepared before any time studies are made. It is especially important to prepare a standard definition of elements so that these same elements may be used in all time studies.

Rules for Dividing an Operation into Elements. All manual work may be divided into fundamental hand motions or therbligs, as has already been explained. These minute subdivisions are too short in duration to be timed with a stop watch. A number of them, therefore, must be grouped together into elements of sufficient length to be conveniently timed. Three rules should be borne in mind in dividing an operation into elements:

1. The elements should be as short in duration as can be accurately timed.

2. Handling time should be separated from machine time.

3. Constant elements should be separated from variable elements.

To be of value a time study must be a study of the elements of the operation, not merely a record of the total time required per cycle to do work. If elements are too short, however, it is impossible to time them accurately.

In machine work it is desirable to separate the machine time, that is, the time that the machine is doing work, from the time during which the operator is working. There are several reasons for this. Where power feeds and speeds are used on the machine, it is possible to calculate the time required for the "cut" and thus check the actual stopwatch data when the machine time is kept separately. Also, the beginning and the end of a cut are excellent beginning and ending points for an element. Where elemental time standards and formulas are to be developed, it is essential that machine time be separated from handling time. The reasons for this separation will be explained in Chapter 27.

The elements of a cycle that are constant should be separated from those that are variable. The term *constant elements* refers to those elements that are independent of the size, weight, length, and shape of the piece. For example, in soldering seams of tin cans made by hand, the time to touch the iron to the bar of solder is a constant, whereas the time to solder the side seam on the can is a variable, varying directly with the length of the seam.

The analyst trained in micromotion study technique will find it relatively easy to decide upon the elements of the operation, because they are merely combinations of fundamental motions. The analyst without such training should see that the elements begin and end at well-defined points in the cycle. These points will have to be memorized so that the analyst will always read his watch at exactly the same place in the cycle; otherwise the time for the elements will be incorrect.

Each element should be concisely recorded in the space provided on the sheet. It is advisable to use symbols to represent elements that are often repeated. In some industries a standard code of symbols is used by all time study observers. When symbols are used, their meaning should appear on each observation sheet.

Taking and Recording the Data. The three most common methods of reading the stop watch are (1) continuous timing, (2) repetitive timing, and (3) accumulative timing. The first two methods have much wider use than the last.

Continuous Timing. In the continuous method of timing the observer starts the watch at the beginning of the first element and permits it to run continuously during the period of the study (Fig. 217). The observer notes the reading of the watch at the end of each element and records this reading on the observation sheet, opposite its name or symbol. Figure 237 illustrates the continuous method of timing. The operation "Make Core for Crank Frame" was divided into four elements. The observer started his watch at the beginning of the first element, read it at the end of the first element, and recorded the reading in vertical column 1 on the lower line. In a similar manner the watch was read at the end of each element, and the readings for the first cycle were recorded in column 1. The second cycle was then timed and the data recorded in the second vertical column, and so on.

The time for each element was later determined by subtraction (Fig. 218). Thus, for the first element, .09 (.09 − 0 = .09) minute was placed in the upper line opposite element 1. In a similar way for the second element, .06 (.15 − .09 = .06) minute was placed in the first vertical column opposite the second element.

Repetitive Timing. In the repetitive or snap-back method the hands of the watch are snapped back to zero at the end of each element. At the beginning of the first element the observer snaps the hand back

STUDY NO. 8765

ELEMENTS	SPEED	FEED	1
1. FIll core box with 3 handfuls of sand. Press sand down each time.			*.09*
			0.9
2. Press sand down with one trowel stroke. Strike off with one trowel stroke.			*.06*
			.15
3. Get and place plate on core box, turn over, rap, and remove box.			*.13*
			.28
4. Carry plate with core 4 feet. Dispose on oven truck.			*.04*
			.32

Fig. 218. Part of observation sheet for operation "Make Core for Crank Frame." The watch readings and the subtracted times for the first cycle are shown. See Fig. 237 for complete study.

to zero by pressing the stem of the watch. The hand moving forward instantly begins to measure the time for the first element. At the end of the first element the observer reads the watch, snaps the hand back to zero, and then records this reading. In a like manner he times the rest of the elements. This method of timing gives the direct time without subtractions, and the data are recorded on the observation sheet as read from the watch (Fig. 241).

Some people think that there is a tendency for the observer to neglect to time and record delays, foreign elements, or false motions of the operator by simply holding down the stem of the watch. This is not a valid criticism of repetitive timing, as the observer should be taught to time and record *all elements* that occur during the study. The main advantage of the repetitive method over the continuous method is that the time for each element is visible on the observation sheet and the time study analyst can see the variations in time values as he makes the study.

Accumulative Timing. The accumulative method of timing permits the direct reading of the time for each element by the use of two stop vatches. These watches are mounted close together on the observation joard (Fig. 219) and are connected by a lever mechanism in such a way that when the first watch is started the second watch is automatically stopped, and when the second watch is started the first is stopped. The watch may be snapped back to zero immediately after it is read, hus making subtractions unnecessary. The watch is read with greater

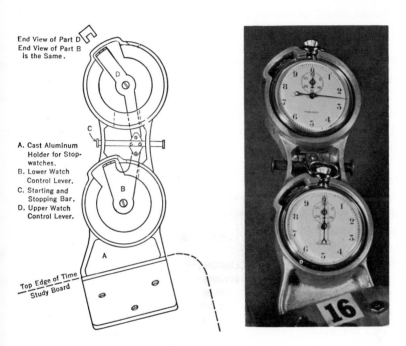

End View of Part D
End View of Part B
is the Same.

A. Cast Aluminum
Holder for Stop-
watches.
B. Lower Watch
Control Lever.
C. Starting and
Stopping Bar.
D. Upper Watch
Control Lever.

Top Edge of Time
Study Board

Fig. 219. Two stop watches connected by suitable linkage for accumulative method of timing.

ease and accuracy because its hands are not in motion at the time it is read.

Recording the Stop-Watch Readings. To the uninitiated it may seem difficult for the observer to do the several things required of him in such quick succession, namely, observe the operator, read the watch, and record the data on the observation sheet; but it is easily possible after a little experience.

A distinctive sound frequently accompanies the beginning and the ending of the element. In the study of the drilling operation (Fig. 217), as the shaft is dropped into place in the jig, there is a metallic click which denotes the end of the first element. Such sounds aid the observer in taking his readings, and he soon learns to make use of them.

The general policy of carefully timing every part of the operation should be insisted upon. If, for example, every fifth or every tenth piece is gauged, such information should be included on the observation sheet and a sufficient number of readings of this element should be made to include it in the time for the operation. The time for the element would, of course, be divided by 5 or 10, as the case might be, in order to prorate the gauging time.

Such elements as "change tools," "blow chips out of jig," "move finished parts," "replace empty tote box," "lubricate die," and the like should be considered specific parts of the operation and should be timed as such. In timing elements that occur infrequently, it is necessary to get a sufficient number of watch readings and also to obtain data as to the frequency of occurrence of such elements so that the time can be prorated.

When foreign elements occur, they should be timed and recorded on the observation sheet. These elements may or may not be included in the time standard, depending upon their nature. By foreign elements is meant elements that do not occur regularly in the cycle, such as accidentally dropping a wrench or piece of material on the floor, tightening the belt on the machine, replacing a broken tool, or placing oil on a tight screw in a jig.

Number of Cycles to Be Timed. The time required to perform the elements of an operation may be expected to vary slightly from cycle to cycle. Even if the operator worked at a uniform pace, he would not always perform each element of consecutive cycles in exactly the same time. Variations in time may result from such things as a difference in the exact position of the parts and tools used by the operator, from variations in reading the stop watch, and from possible differences in determining the exact end point at which the watch reading is made. With highly standardized raw materials, good tools and equipment, good working conditions, and a qualified and well-trained operator, the variation in readings for an element would not be great, but there would still be some variation.

Time study is a sampling process; consequently the greater the number of cycles timed, the more nearly the results will be representative of the activity being measured. Consistency of watch readings is of major interest to the analyst. For example, 20 cycles of the operation shown in Fig. 217 were studied, and the time for element 1 of the study varied from 0.10 to 0.14 minute. Had all of the 20 readings been 0.10 minute, then the consistency would have been perfect, and 0.10 would obviously have been selected as the time value for this element. The greater the variability of the readings for an element, the larger the number of observations will have to be for a desired accuracy.

Formula for Determining Number of Observations. Formulas 1 and 2 of this section provide a simple means of evaluating the error in the average time value of an element for a given number of readings.[1] It is assumed that variations in the time from observation to observation are due to chance, and this seems to be a reasonable assumption.

The standard error of the average for each element (standard error of the mean) is expressed by the formula:[2]

$$\sigma\bar{x} = \frac{\sigma'}{\sqrt{N}} \tag{1}$$

where $\sigma\bar{x}$ = standard deviation of the distribution of averages
  σ' = standard deviation of the universe for a given element
  N = actual number of observations of the element

Standard deviation is denoted by σ (sigma). By definition, it is the root-mean-square deviation of the observed readings from their average.[3] That is,

$$\sigma = \sqrt{\frac{(X_1 - \bar{X})^2 + (X_2 - \bar{X})^2 + \cdots + (X_n - \bar{X})^2}{N}} \tag{2}$$

$$= \sqrt{\frac{\Sigma(X - \bar{X})^2}{N}} = \sqrt{\frac{\Sigma X^2}{N} - \bar{X}^2}$$

where X = each stop-watch reading or individual observation
  $\bar{X}$ = (read "X-bar") average or mean of all readings of an element
  Σ = (read "sigma") sum of individual readings

[1] One of the first mathematical procedures for determining number of observations was suggested by E. B. Royer, "How Many Observations Are Necessary in Setting Wage-Incentive Standards?" *Personnel,* Vol. 13, No. 4, pp. 137–39, May, 1937.
[2] See E. L. Grant, *Statistical Quality Control,* 2nd ed., McGraw-Hill Book Co., New York, 1952, p. 87 (or any book on statistical quality control).
[3] *Ibid.,* pp. 52–53.

Since $\bar{X} = \dfrac{\Sigma X}{N}$,

$$\sigma = \sqrt{\dfrac{\Sigma X^2}{N} - \left(\dfrac{\Sigma X}{N}\right)^2} = \dfrac{1}{N}\sqrt{N\Sigma X^2 - (\Sigma X)^2} \qquad (3)$$

Combining formulas 1 and 3,

$$\sigma\bar{x} = \dfrac{\dfrac{1}{N}\sqrt{N\Sigma X^2 - (\Sigma X)^2}}{\sqrt{N'}} \qquad (4)$$

A decision must be made as to the confidence level and the desired accuracy that are to be used in determining the number of observations to make. A 95% confidence level and ±5% precision are commonly used in time study. This means that the chances are at least 95 out of 100 that the sample mean or the average value for the element will not be in error more than ±5% of the true element time. Then

$$0.05\bar{X} = 2\sigma\bar{x} \qquad \text{or} \qquad 0.05\dfrac{\Sigma X}{N} = 2\sigma\bar{x}$$

$$0.05\dfrac{\Sigma X}{N} = 2\dfrac{\dfrac{1}{N}\sqrt{N\Sigma X^2 - (\Sigma X)^2}}{\sqrt{N'}}$$

$$N' = \left(\dfrac{40\sqrt{N\Sigma X^2 - (\Sigma X)^2}}{\Sigma X}\right)^2 \qquad (5)$$

where N' is the required number of observations to predict the true time within ±5% precision and 95% confidence level.

If a 95% confidence level and a precision of ±10% are used as the criteria, then the formula will be

$$N' = \left(\dfrac{20\sqrt{N\Sigma X^2 - (\Sigma X)^2}}{\Sigma X}\right)^2 \qquad (6)$$

EXAMPLE. Assume that 30 observations have been made of an element, as shown in the first column of Fig. 220, and the observer wants to know whether he has taken a sufficient number of observations for a 95% confidence level and a precision of ±5%. Formula 5 is the one to use for this.

Operation:	Make core for Crank Frame No. 7253.
Element No. 2:	Press sand down with one trowel stroke. Strike off with one trowel stroke.

Individual Watch Readings in 0.01 Min. X	Individual Watch Readings Squared X^2
6	36
5	25
8	64
6	36
5	25
5	25
6	36
5	25
5	25
6	36
6	36
5	25
5	25
6	36
6	36
5	25
5	25
5	25
5	25
6	36
6	36
6	36
6	36
5	25
6	36
6	36
7	49
6	36
5	25
5	25
$\Sigma X = 169$	$\Sigma X^2 = 967$

Fig. 220. Values for X and X^2 for element 2 of the time study shown in Fig. 237.

Figure 220 shows the sum of the 30 observations and the sum of the squares of the 30 observations. Substituting these data in formula 5, the computations would be as follows:

$$N' = \left(\frac{40\sqrt{30 \times 967 - 169^2}}{169} \right)^2 = \left(\frac{40\sqrt{29010 - 28561}}{169} \right)^2$$

$$= \left(\frac{40 \times 21.2}{169} \right)^2 = \left(\frac{848}{169} \right)^2 = 25 \text{ observations}$$

Another formula for determining the number of cycles to be timed is

$$N' = \left[\frac{40N}{\Sigma X} \sqrt{\frac{\Sigma X^2 - (\Sigma X)^2/N}{N-1}} \right]^2 \tag{7}$$

This formula results from using the following formula instead of formula 3:

$$\sigma = \sqrt{\frac{\Sigma X^2 - (\Sigma X)^2/N}{N-1}}$$

Formula 7 tends to be more accurate as the number of cycles timed decreases.

Estimating the Number of Observations to Make. The Maytag Company uses the following procedure for estimating the number of observations to make.

1. Take readings: (*a*) ten good readings for cycles of 2 minutes or less; (*b*) five good readings for cycles of more than 2 minutes.

2. Determine the range R. This is the high time study value H minus the low time study value L ($H - L = R$).

3. Determine the average $\overline{X}$. This is the sum of the readings divided by the number of readings (either 5 or 10). This average may be approximated by high value plus low value divided by 2. That is, $(H + L)/2$.

4. Determine $R/\overline{X}$. This is the range divided by the average.

5. Determine the number of readings necessary [4] from Table 14. Read down the first column until the value of $R/\overline{X}$ is found; then read

---

[4] The values shown in Table 14 were determined in the following way. It can be shown (see Grant, *op. cit.*, p. 83) that:

$$\sigma' = \frac{\overline{R}}{d_2}$$

where $\overline{R}$ = average range, the average of the difference between the high and low observed value of subgroups of same number of readings

d_2 = factor based on the number of readings in the subgroup (see Grant, *op. cit.*, p. 512)

From formula 1,

$$\sigma\bar{x} = \frac{\sigma'}{\sqrt{N}} \quad \text{or} \quad \sigma' = \sigma\bar{x}\sqrt{N}$$

Then

$$\sigma\bar{x} = \frac{\overline{R}}{d_2\sqrt{N}} = \quad \text{or} \quad 2\sigma\bar{x} = 2\frac{\overline{R}}{d_2\sqrt{N}}$$

Table 14. Number of Time Study Readings N' Required for ±5% Precision and 95% Confidence Level

$\dfrac{R}{X}$	Data from Sample of		$\dfrac{R}{X}$	Data from Sample of		$\dfrac{R}{X}$	Data from Sample of	
	5	10		5	10		5	10
.10	3	2	.42	52	30	.74	162	93
.12	4	2	.44	57	33	.76	171	98
.14	6	3	.46	63	36	.78	180	103
.16	8	4	.48	68	39	.80	190	108
.18	10	6	.50	74	42	.82	199	113
.20	12	7	.52	80	46	.84	209	119
.22	14	8	.54	86	49	.86	218	125
.24	17	10	.56	93	53	.88	229	131
.26	20	11	.58	100	57	.90	239	138
.28	23	13	.60	107	61	.92	250	143
.30	27	15	.62	114	65	.94	261	149
.32	30	17	.64	121	69	.96	273	156
.34	34	20	.66	129	74	.98	284	162
.36	38	22	.68	137	78	1.00	296	169
.38	43	24	.70	145	83			
.40	47	27	.72	153	88			

R = range of time for sample, which is equal to high time study elemental value minus low time study elemental value.

X = average time value of element for sample.   (For ±10% precisiou and 95% confidence level, divide answer by 4.)

across to column for sample size taken (5 or 10) ; and then to the total number of readings required. (For 95% confidence level and ±10% precision divide the required number found by 4.)

6. Continue to take readings until the total of the indicated number required is obtained.

For a 95% confidence level and a precision of ±5%, then $0.05\overline{X} = 2\sigma\bar{x}$.

$$0.05\overline{X} = \frac{2\overline{R}}{d_2\sqrt{N}}$$

$$0.025d_2\sqrt{N} = \frac{\overline{R}}{\overline{X}}$$

When the number of observations in the subgroup is 5, then $d_2 = 2.326$. When the number is 10, then $d_2 = 3.078$. (See Grant, *op. cit.*, 512.)

A copy of Table 14 is attached to the time study observation board so that the observer may determine on the job the approximate number of readings necessary.

EXAMPLE. Figure 221 is a time study of ten consecutive cycles of an operation consisting of three elements. The following is the procedure used to determine the number of readings needed for a 95% confidence level and a precision of ±5%.

Element 1	.07	.09	.06	.07	.08	.08	.07	.08	.09	.07
Element 2	.12	.13	.12	.12	.11	.13	.12	.11	.13	.12
Element 3	.56	.57	.55	.56	.57	.56	.54	.56	.56	.55

Fig. 221. Time study of ten cycles of an operation. The average of the ten observations for element 1 is 0.076 minute.

1. Take readings: Ten good readings for each element are shown in Fig. 221. Element 1 will be used for this example.
2. Determine the range R for element 1.

$$R = H - L = 0.09 - 0.06 = 0.03 \text{ minute}$$

3. Determine the average $\overline{X}$.

$$\overline{X} = \frac{0.76}{10} = 0.076 \text{ minute}$$

4. Determine the value of $R/\overline{X}$.

$$\frac{R}{\overline{X}} = \frac{0.03}{0.076} = 0.395$$

5. Determine the number of readings necessary from Table 14. Since 0.395 is closer to 0.40 than 0.38, the number of readings corresponding to 40 is 27.
6. Continue the study until a total of 27 readings is obtained.

Final Check as to Number of Observations. Table 14 was used at the beginning of the study to determine the approximate number of observations required. After the study was completed (see Fig. 222), the time study observer at Maytag made a check to determine whether enough readings had been made, using the following procedure.[5]

---

[5] For an excellent explanation of a similar procedure, see John M. Allderige, "Statistical Procedures in Stop Watch Work Measurement," *Journal of Industrial Engineering,* Vol. VII, No. 4, pp. 154–163, July–August, 1956.

1. Divide the readings for the element into subgroups of 4 (Fig. 222).

2. Determine the range R for each subgroup. This is the high time study value minus the low time study value.

3. Determine the average range $\bar{R}$ of the subgroups. This is found by averaging all the ranges of the subgroups.

4. Determine the average $\bar{X}$. This is the unrated average value that is normally found in working up the time study.

5. Determine the number of readings necessary from Fig. 223. Read across on the vertical scale using average range $\bar{R}$, and up from the bottom scale using average value of element $\bar{X}$; where intersection occurs, note location with respect to diagonal lines running across graph.

6. Determine precision actually obtained from Fig. 224. Read across on the vertical scale, using the number of readings that would be necessary for $\pm 5\%$ precision, until the line intersects the number of readings actually taken; a vertical line running straight down from this intersection will fall on the actual precision obtained.

EXAMPLE. From the study shown in Fig. 222, determine (a) the number of readings necessary for a 95% confidence level and $\pm 5\%$ precision, and (b) the precision obtained from the actual number of readings observed if less than the number called for in (a).

Element 1	.07 .09 .06 .07	.08 .08 .07 .08	.09 .07 .08 .08	.07 .09 .08 .08
	.03	.01	.02	.02
Element 1	.06 .07 .08 .08	.08 .09 .09 .06	.07 .08 .08 .09	.10 .10 .07 .08
	.02	.03	.02	.03

Fig. 222. Time study data for element 1, showing subgroups of 4 observations. Average of the 32 observations is 0.0787 minute. $\bar{X} = 0.0787$.

Solution for a (number of readings necessary for $\pm 5\%$ precision):

1. Divide the readings for this element into subgroups of 4 (Fig. 222).

2. Determine the range R for each subgroup.

$$.03 + .01 + .02 + .02 + .02 + .03 + .02 + .03 = .18$$

3. Determine the average range $\bar{R}$ of the subgroups.

$$\bar{R} = \frac{0.18}{8} = 0.0225 \text{ minute}$$

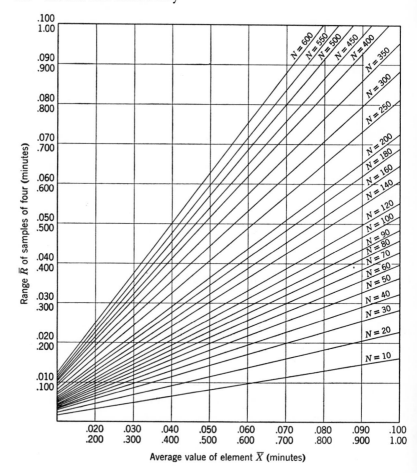

Fig. 223. Curves showing relationship between average range of samples of four observations and average value of the element being timed. All readings are in minutes.

4. Determine the average $\overline{X}$. The unrated average value for element 1, as shown on the study, is 0.0787 minute.

5. Determine the number of readings necessary for a 95% confidence level and ±5% precision from Fig. 223. (Solution for $\overline{R}$ = 0.0225 and $\overline{X}$ = 0.0787 is N = approximately 30 observations.)

Solution for b (precision obtained from the actual number of readings observed): To illustrate the use of Fig. 224, let us assume that the time study observer makes only 20 observations, as indicated in Fig. 225, and that the

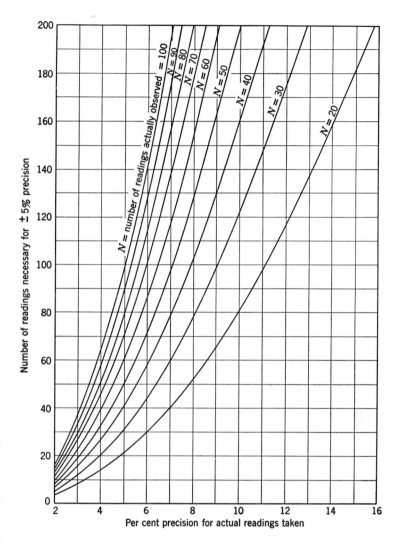

Fig. 224. Curves showing the relationship between number of readings necessary for ±5% precision, number of readings actually observed, and precision in per cent for readings actually taken.

number of observations required is 30. The precision of the 20 observations can be determined from the curves in Fig. 224 as follows. Locate 30 on the vertical scale and read across to the point where this line intersects the curve $N = 20$. Then drop straight down to the bottom scale and read precision. In this case it is ±6%. This means that with 20 observations, the precision is only ±6%, whereas it would be ±5% with 30 readings.

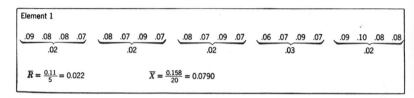

Fig. 225. Time study data for element 1, showing subgroups of 4 observations. Average of the 20 observations is 0.079 minute. $X = 0.079$.

Use of Alignment Chart for Determining Number of Observations. One large organization uses the form shown in Fig. 226 and the alignment chart in Fig. 227 for determining the number of observations. The time study analyst makes the number of observations he thinks necessary, and while still on the job he determines for each element the range for each group of four observations and the average of all observations. Then, referring to the alignment chart which he has attached to his time study observation board, he checks the required number of observations for ±5% precision and a 95% confidence level. He also determines the precision for the number of observations he has already taken.

For example, 24 observations were made of the element "Drill $\frac{1}{16}$ Hole in Control Lever," as shown in Fig. 226. The average range $\overline{R} = 0.30$ minute, and the average time $\overline{X} = 1.05$ minutes. As indicated in Fig. 227, the number of observations required is 36.

Figure 227 also shows that the 24 observations actually made gave a precision of ±5.6%.

Control Chart Analysis of Time Study Data. The control chart is an excellent device to test the consistency of the time study data. The average values $\overline{X}$ (average of group of 4) of the time study readings are plotted in sequence on the control chart in Fig. 228. The upper and lower control limits are determined as follows. Using the alignment chart in Fig. 227, find the value of $\overline{R}$ on the scale and read the corresponding value of $3\sigma\bar{x}$. When $\overline{R} = 0.30$, then $3\sigma\bar{x} = ±0.22$. Since

$\overline{x} = 1.05$, then the upper control limit $= 1.05 + 0.22 = 1.27$. The lower control limit is $1.05 - 0.22 = 0.83$.

As Fig. 228 shows, all points are within the control limits; the data are consistent, and the time value of 1.05 minutes for this element is acceptable.

Rating. As the time study analyst records his data, he is also evaluating the operator's speed in relation to his opinion of normal speed for such an operation. The observer wants enough readings for each element to give him a representative sample against which to apply his speed rating. Later the rating factor will be applied to this "representative time" to obtain the normal time for the element.

There are a number of different "systems" or methods of arriving at this rating factor, all of which depend upon the judgment of the

SHEET 1 OF 1 SHEETS

ELEMENT	EXT.	SUM (S)	AVG. (x̄)	RANGE (R)	ELEMENT	EXT.	SUM (S)	AVG. (x̄)	RANGE (R)
OPERATION Drill 1/16" hole in control lever							OP. NO. DR 12		
PART NAME Control Lever							PART NO. CL 28		
MACHINE NAME & NO. Special Drill #249					OBSERVER: T.S. Wilson				
OPERATOR NAME & NO. John Williams #16432					MALE ☑ FEMALE ☐ DATE:				
Element No. 4: Drill 1/16" hole in control lever. Hand feed. Drilling time only	1.02					1.17			
	1.32					.92			
	1.08					.90			
	.99	4.41	1.10	.33		1.05	4.04	1.01	.27
	1.04								
	.96								
	1.18								
	.90	4.08	1.02	.28					
	1.01								
	.94								
	.96								
	1.24	4.15	1.04	.30					
	1.28								
	1.13								
	.98				TOTALS		25.27		1.79
	.96	4.35	1.09	.32	NUMBER OF READINGS TAKEN 24				
	1.08				TOTAL (S) ÷ NUM- } AVERAGE TIME (x̄) BER OF READINGS } 25.27 ÷ 24			1.05	
	1.22				NUMBER OF GROUPS OF 4				6
	1.01				TOTAL (R) ÷ NUM- } AVERAGE RANGE (R̄) BER OF GROUPS } 1.79 ÷ 6				.30
	.93	4.24	1.06	.29	FROM PRECISION CHART } READINGS REQUIRED FOR ±5% PRECISION			36	
					PRECISION FROM READINGS TAKEN			±5.5%	
					RATING			105%	

Fig. 226. Recap sheet for time study data.

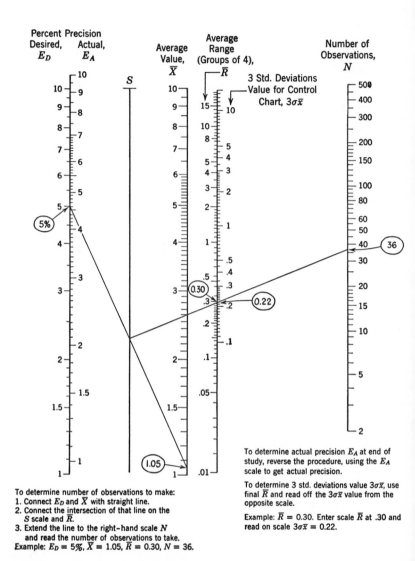

Fig. 227. Alignment chart for determining the number of observations required for 95% confidence level and ±5% precision, and also for determining the control limits for control chart.

time study analyst. One of the most common methods is for the analyst to determine a rating factor for the operation as a whole. At the beginning and at the end, and perhaps at intervals throughout the study, the observer concentrates on making ratings of the speed of the operator. It is his object to determine the average level of performance at which the operator was working while the study was being made. Such a rating is recorded on the observation sheet in the form of a rating factor (Fig. 217).

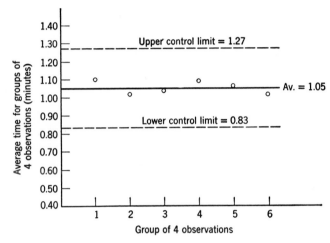

Fig. 228. Control chart for time study observations.

Another method is for the analyst to determine a rating factor for each element of the operation.[6] This is the plan most widely used today.[7] A still more refined rating plan requires the analyst to rate each element when it is timed, recording the rating for the element on the observation sheet when the stop-watch reading is recorded. Using this method, there would be a rating recorded for each stop-watch reading. Although this method is sound, it is very difficult for the time study analyst to rate each stop-watch reading unless the elements are fairly long.

The use of the rating factor will be explained more fully in Chapter 25.

Selecting the Operator to Be Timed. If more than one person is performing the same operation, the time study analyst may time one

[6] A survey of 7444 time study men showed that 34% rated the over-all study, 53% rated each element, and 13% rated each stop-watch reading.

[7] R. B. Andrews and Ralph M. Barnes, "The Influence of the Duration of Observation Times on Performance Rating," *The Journal of Industrial Engineering*, Vol. 18, No. 4, pp. 243–247, April, 1967.

or more of the operators. If all the operators are using exactly the same method, that is, the one prescribed for the job, and if there is a difference in the tempo at which the operators work, it is customary to time the operator working at nearest to normal pace. Since a rating factor is used to evaluate the operator's speed, it theoretically makes no difference whether the slowest or fastest operator is timed. However, it is admittedly more difficult to rate correctly the performance of a very slow operator. It is not desirable to time a beginner, because his method is seldom the same as it will be when he has attained greater proficiency through experience on the job (see Chapter **36**).

The fact that it is important to maintain the good will and cooperation of the employee in motion and time study work should not be overlooked. For psychological reasons it is often better to time an average operator rather than the fastest one. Workers, not fully understanding the process of rating operator performance, are likely to feel that time standards will be set directly on the output of the person timed. If that person is the best one on the job, they may think that the standard time will be so low that it will be very difficult or impossible for the average operator to meet it.

Since full micromotion studies are not generally used in industry for improving methods, the improvement of the job will in most cases tie in with the time study, so that the same operator will serve as the subject for the entire procedure. Frequently there is only one person on the job; consequently, there is no choice of operator.

Steps in Making Time Study Observations

1. Discuss the operation to be timed with the foreman of the department.

2. Make certain that the operator has been informed that a time study is to be made.

3. Secure the cooperation of the operator. Explain to him what you are going to do.

4. Make certain that the operation is ready for time study.

5. Obtain all necessary information and record it on the observation sheet.

6. Make a sketch of the piece and of the work place.

7. Divide the operation into its elements and list these on the observation sheet. If necessary describe the method more fully on a separate sheet, listing the beginning and ending points for each element.

8. Estimate the number of cycles to be timed.

9. Record the time of day as the study is begun.

10. Start the decimal stop watch at the beginning of the first element of the cycle. Read and record the time for each element of the cycle.

11. When the study is completed and when the stop watch is read at the end of the last element, read and record the time of day on the sheet.

12. Rate the operator and record these ratings on the observation sheet.

13. Sign and date the study.

14. Check to make certain that a sufficient number of cycles have been timed.

15. Plot data on control chart.

CHAPTER 24

Time Study: Determining the Rating Factor

After the time study has been taken the next step is to subtract successive watch readings in order to get the time for each element. It is advisable to record these subtracted times in ink to make them stand out from the rest of the data and also to ensure permanence.

Selecting Time Values. As the study in Fig. 237 shows, there are forty-five time values for each of the four elements. It now becomes necessary to select from these data a time value for each of these four elements that will be representative.

Occasionally there may be an abnormally high or low time value, owing to an error in reading the stop watch; such readings should not be considered in selecting the time value for the elements. However, the fact that there is considerable variation in successive times for certain elements does not mean that all high and low elements should be thrown out. In many cases there are good reasons for such data. An occasional hard casting may require longer drilling time, or a piece with a fin or a burr may take longer to place in the jig. If such time values are typical or representative of what may be expected on the job, they should not be eliminated from the study even though they happen to be abnormal. It is good policy not to eliminate any readings unless there is a definite reason for doing so.

Many organizations use the arithmetical *average* of the stop-watch readings in determining the representative time for the element. Since this is the most common method of handling data and since it is easy to explain to the worker, it is gaining in favor among time study analysts.

The *modal method,* also widely used, consists in taking the time that recurs most frequently for the element. High and low time values will have less effect upon the selected time by this method than by the average method. The selected time values in the time study shown in Fig. 217 were determined by the modal method.

It should be remembered that the observer will apply his rating factor to the selected time for the element. For this reason, as careful consideration should be given to determining the selected time as is given to the matter of determining the rating factor.

After the time value for each element is selected, the next step in establishing the time standard is to determine and apply the rating factor.

DETERMINING THE RATING FACTOR

Perhaps the most important and the most difficult part of time study is to evaluate the speed [1] or the tempo at which the person is working while the study is being made. The time study analyst must judge the operator's speed while he is making the time study. This is called rating.

Definition of Rating. Rating is that process during which the time study analyst compares the performance (speed or tempo) of the operator under observation with the observer's own concept of normal performance.[2] Later this rating factor will be applied to the time value to obtain the normal time for the job.

Rating is a matter of judgment on the part of the time study analyst, and unfortunately there is no way to establish a time standard for an operation without having the judgment of the analyst enter into the process.

We all know that there is a wide difference in the speed at which different people naturally work. For example, some people walk at a slow pace, others at a fast pace. The time required for a person to walk a given distance will of course vary directly with his walking speed. Speed or tempo can be rated. If walking at a speed of 3 miles per hour is considered normal performance or 100%, we have a definite standard or base to be used in rating the task of walking. Then walking at 2 miles per hour would equal 66⅔% of normal, and walking at 4 miles per hour would equal 133⅓%. We are measuring accomplishment. If the task is to go from one place to another, and if walking 3 miles per hour is normal or 100%, then without question

---

[1] The terms *speed, effort, tempo,* and *pace* all refer to the rate of speed of the operator's motions. *Speed* and *effort* are terms commonly used by time study analysts, and the term *performance* is gaining in favor. In this volume these terms will be used synonymously, and they will *all have but a single meaning*—speed of movement.

[2] SAM Committee on Rating of Time Studies, *Advanced Management,* Vol. 6, No. 3, p. 110.

we can use a percentage rating factor with precision in measuring performance.

Systems of Rating. Although performance rating, the system of rating described in detail in this chapter, is most widely used today, five other systems should be mentioned. However, several of these systems are little used.

1. Skill and Effort Rating. Around 1916 Charles E. Bedaux introduced the Bedaux system of wage payment and labor control in this country. His plan was based on time study, and his time standards were expressed in points or "B's." A point or B was simply another name for what we now call a standard minute. His time study procedure included the rating of the operator's skill and effort and the use of a standard table of fatigue allowances. Bedaux used 60 points equal to standard performance. In other words, an operator working at a normal pace was expected to produce 60 B's per hour, and it was expected that the average incentive pace would be around 70 to 80 points per hour.

Before Bedaux, performance rating had been done mainly by selecting stop-watch readings from the time study data. Thus, if the operator was judged to be working at a fast tempo, a watch reading considerably above average would be selected as the representative time for the element; if the operator was judged to be working at a slow tempo, then a watch reading below average would be selected. The Bedaux system was a definite improvement over this informal method of rating operator performance.

2. Westinghouse System of Rating. A four-factor system [3] for rating operator performance was developed at Westinghouse and was originally published in 1927. These four factors are (1) skill, (2) effort, (3) conditions, and (4) consistency. A scale of numerical values for each factor was supplied in tabular form (Fig. 229), and the selected time obtained from time study was normalized or leveled by applying the sum of the ratings of the four factors.

For example, if the selected time for an operation was 0.50 minute and if the ratings were as follows:

Excellent skill, B2	+0.08
Good effort, C2	+0.02
Good condition, C	+0.02
Good consistency, C	+0.01
Total	+0.13

[3] Described in *Time and Motion Study*, 3rd ed., by S. M. Lowry, H. B. Maynard, and G. J. Stegemerten, McGraw-Hill Book Co., New York, 1940, p. 233.

Skill			Effort		
+0.15	A1	Superskill	+0.13	A1	Excessive
+0.13	A2		+0.12	A2	
+0.11	B1	Excellent	+0.10	B1	Excellent
+0.08	B2		+0.08	B2	
+0.06	C1	Good	+0.05	C1	Good
+0.03	C2		+0.02	C2	
0.00	D	Average	0.00	D	Average
−0.05	E1	Fair	−0.04	E1	Fair
−0.10	E2		−0.08	E2	
−0.16	F1	Poor	−0.12	F1	Poor
−0.22	F2		−0.17	F2	
Conditions			Consistency		
+0.06	A	Ideal	+0.04	A	Perfect
+0.04	B	Excellent	+0.03	B	Excellent
+0.02	C	Good	+0.01	C	Good
0.00	D	Average	0.00	D	Average
−0.03	E	Fair	−0.02	E	Fair
−0.07	F	Poor	−0.04	F	Poor

Fig. 229. Performance rating table.

then the normal time for this operation would be 0.565 minute $(0.50 \times 1.13 = 0.565)$.

3. Synthetic Rating. Synthetic rating[4] is the name given to a method of evaluating an operator's speed from predetermined motion-time values. The procedure is to make a time study in the usual manner, and then compare the actual time for as many elements as possible with motion-time values for the same elements. A ratio can be established between the motion-time value for the element and the actual time value for that element. This ratio is the performance index or rating factor for the operator insofar as that one element is concerned. The formula for computing the performance rating factor is

$$R = \frac{P}{A}$$

where R = performance rating factor

P = predetermined motion-time standard for the element, expressed in minutes

A = average actual time value (selected time) for the same element P expressed in minutes

[4] R. L. Morrow, *Motion Economy and Work Measurement,* Ronald Press Co., New York, 1957, p. 443.

Table 15. Comparison of Average Actual Times and Times Determined from Motion-Time Data

Time Study Element	Average Actual Time (Selected Time) in Minutes	Time as Determined from Motion-Time Data in Minutes	Calculated Performance Rating Factor, $R = \dfrac{P}{A}$	Average Performance Rating Factor
1	0.12	0.13	108	110
2	0.09			*110*
3	0.17	0.19	112	110
4	0.26			*110*
5	0.32			*110*
6	0.07			*110*

Table 15 illustrates the method of making the calculation. The selected times for elements 1 and 3 were 0.12 and 0.17 minute respectively. The time values for these two elements as determined from a table of predetermined motion-time values were 0.13 and 0.19 minute respectively. In the first case the rating factor was 108% (0.13 ÷ 0.12 × 100 = 108%), and in the second case it was 112% (0.19 ÷ 0.17 × 100 = 112%). The average rating factor was the average of 108 and 112, or 110%. This average rating factor was then applied to all elements in this study. The rating factor, of course, is applied only to manually controlled elements.

4. *"Objective Rating."* Another method of rating performance has been given the name objective rating.[5] First, the operator's speed is rated against a single standard pace which is independent of job difficulty. The observer merely rates speed of movement or rate of activity, paying no attention to the job itself. After the pace rating is made, an allowance or a secondary adjustment is added to the pace rating to take care of the job difficulty. Job difficulty is divided into six classes, and a table of percentages is provided for each of these factors. The six factors or categories are (1) amount of body used, (2) foot pedals, (3) bimanualness, (4) eye-hand coordination, (5) handling requirements, and (6) weight. The following example illus-

[5] M. E. Mundel, *Motion and Time Study*, 3rd ed., Prentice-Hall, Englewood Cliffs, N. J., 1960, p. 406.

trates how the normal time for an element is determined using this system of rating.

EXAMPLE. If the selected time for an element is 0.26 minute, the pace rating is 95%, and if the sum of all secondary adjustments amounts to 20%, then the normal time will be 0.297 minute (0.26 × 0.95 × 1.20).

5. Physiological Evaluation of Performance Level. Many studies have been made which show the relationship between physical work and the amount of oxygen consumed by the subject (see pages 555 to 585). More recently it has been found that the change in heart rate is also a reliable measure of muscular activity, and moreover it is much simpler to measure heart rate than oxygen consumption.[6] An ordinary stethoscope and stop watch can be used for measuring heart rate, or a telemetering device can serve to make a continuous record of heart rate without interfering with the activities of the subject.

The procedure is to have the person work at his job for a specified period and then measure his heart rate at the end of this period, and at the end of 1, 2, and 3 minutes after stopping work, while the subject sits still in a chair. It seems entirely possible that a normal or basic heart rate can be determined, and then new jobs can be measured against this bench mark. For example, if an operator using a prescribed method worked for a 10-minute period and turned out five pieces, the change in his heart rate (from resting state) would be an index of the effort required to do this particular job. Because of individual differences it would be necessary to have this operator perform one or more "bench mark" tasks in order to relate his heart rate to the standard or norm for the plant or industry.

The fact that an increasing number of people in various parts of

[6] James H. Green, W. H. M. Morris, and J. E. Wiebers, "A Method for Measuring Physiological Cost of Work," *Journal of Industrial Engineering,* Vol. 10, No. 3, pp. 180–184, May–June, 1959; C. J. Anson, "The Physiological Measurement of Effort," *Time and Motion Study,* London, Vol. 3, No. 2, pp. 26–31, February, 1954; Lucien Brouha, "Physiological Approach to Problems of Work Measurement," *Proceedings Ninth Annual Industrial Engineering Institute,* University of California, Los Angeles-Berkeley, February, 1957; J. A. C. Williams, "Physiological Measurements in Work Study," *Time and Motion Study,* London, Vol. 3, No. 11, pp. 18–21, November, 1954; H. H. Young, "The Relationship between Heart Rate and the Intensity of Work for Selected Tasks," *Journal of Industrial Engineering,* Vol. 7, No. 6, pp. 300–303, November–December, 1956; W. E. Splinter and C. W. Suggs, "Instrument Records Heart Rate for Energy Studies," *Agricultural Engineering,* Vol. 37, No. 9, pp. 618–619, September, 1956. Also see Chapter 34, "Measuring Work by Physiological Methods."

the world are working on this problem would suggest that measurement of heart rate may eventually take its place along with the other methods of measuring work. For very light work and for physical activities that will not effect a change in heart rate, the force platform may prove to be a reliable measuring device.[7] See Chapter 35.

6. Performance Rating. By far the most widely used system of rating in this country is that of rating a single factor—operator speed, pace, or tempo. This system is called "performance rating." The rating factor may be expressed in percentage, in points per hour, or in other units.[8] Here we shall use the percentage system, with normal performance equal to 100%; for example, walking 3 miles per hour equals 100%. Since this system is so commonly used, it will be discussed more fully in the remaining pages of this chapter.

The Range of Human Capacities. From our own observations and experience we know that there are wide differences in capacities and abilities of individuals in every activity of life. We have seen champion athletes run the mile in 3 minutes and 51.3 seconds, and the 10,000-meter race in 27 minutes and 39.4 seconds, and we have heard of such physical feats as one man's lifting 6000 pounds unaided. These, however, are rare exceptions. Wechsler shows that the range of most physical and mental activities varies as 2 to 1, if the rare exceptions are not considered.[9] That is, the best has roughly twice the capacity of the poorest.

In the factory this means that if a large group of people did exactly the same manual task using the same method, the fastest operator would produce approximately twice as much in a given time as the slowest operator. Table 16 shows the average performance for one day of 121 girls operating semiautomatic lathes, the work being iden-

[7] Lucien Brouha, "Physiological Techniques in Work Measurement," *Proceedings Eleventh Annual Management Engineering Conference*, SAM-ASME, New York, pp. 129–158, April, 1956; Lucien Lauru, "Introduction de la Mesure dans l'Étude et la Simplification des Mouvements," *Travail et Méthodes*, No. 73, pp. 27–35, December, 1953, and pp. 27–37, January, 1954; Lucien Lauru and Lucien Brouha, "Physiological Study of Motions," *Advanced Management*, Vol. 22, No. 3, pp. 17–24, March, 1957.

[8] An investigation of time study practices among 72 companies showed that 90% used the 100% system, 12% used the point system, 7% used the Westinghouse system, and 1% used other systems. Ralph M. Barnes, *Industrial Engineering Survey*, University of California, 1967.

[9] David Wechsler, *The Range of Human Capacities*, 2nd ed., Williams & Wilkins, Baltimore, 1952, p. 94.

Table 16. Difference in the Performance of Operators Working on Semiautomatic Lathes

(Average performance of operators for one day)

Number of Operators	Average Output in Pieces Per Hour for the Day	Distribution		
		No.	Per Cent	Range
1	104	1	1	100 to 109
2	98			
1	91	5	4	90 to 99
2	90			
2	89			
2	87			
1	86			
2	85			
6	84	25	21	80 to 89
2	83			
1	82			
1	81			
8	80			
4	79			
1	78			
3	77			
4	76			
4	75			
1	74	40	33	70 to 79
2	73			
4	72			
3	71			
14	70			
3	69			
2	68			
3	67			
4	66			
3	64	45	37	60 to 69
6	63			
4	62			
6	61			
14	60			
1	58			
1	55			
1	54	5	4	50 to 59
1	52			
1	51			
Total 121	Average 72	121	100	

Table 17. Performance of People Placing Wood Blocks in a Hole

(32 blocks ⅜ inch by ⅜ inch by 2 inches, pre-positioned in 4 rows on worktable; hole, 2 inches by 4 inches, 4½ inches from edge of table)

Number of People	Time for Cycle in Minutes	Distribution	
		Number in Interval	Per Cent in Interval
1	0.28		
5	0.30	30	6
4	0.31		
20	0.32		
20	0.33		
27	0.34		
33	0.35	150	30
33	0.36		
17	0.365		
20	0.37		
16	0.375		
31	0.38		
5	0.385		
35	0.39	199	40
47	0.40		
24	0.41		
25	0.42		
16	0.425		
13	0.43		
11	0.435		
18	0.44		
27	0.45	109	22
10	0.46		
15	0.47		
11	0.48		
4	0.49		
1	0.495		
2	0.50		
2	0.52	12	2
6	0.54		
1	0.60		
Total 500	Average 0.395	500	100

tical for all operators.[10] All were experienced operators. They worked under a wage incentive plan and were paid a premium for all work produced above 60 pieces per hour, this being the normal performance level established by time study. As the table shows, the poorest operator produced 51 pieces per hour and the best 104 pieces per hour, or a ratio of 1 to 2.04.

T. R. Turnball conducted an experiment in his plant by having 500 employees toss 32 blocks ⅜ inch by ⅜ inch by 2 inches (pre-positioned in 4 rows on a work table) into a 2-inch by 4-inch hole 4½ inches from the edge of the table. The exact method of doing the task was first explained to each "operator." The operator then watched the person ahead of him perform the task, and finally he was asked to toss the 32 blocks into the hole as fast as he could. An observer recorded the time taken to perform the operation. As Table 17 shows, the slowest operator took 0.60 minute (100 cycles per hour), whereas the fastest took only 0.28 minute (214 cycles per hour), or a ratio of 1 to 2.14.

This range of 1 to 2 would be expected only if we considered a large number of working people just as they would be found in the factory. In any large group it is expected that an occasional misfit or an occasional star performer might fall outside the range.

Frequency Distribution. With the range of working speeds or operator tempo now established, we are interested in knowing what the distribution would be for a group of factory workers all doing the same job.

Figure 230 shows one way to arrange the average hourly output data for the 121 semiautomatic lathe operators. Those operators who averaged between 50 and 59 pieces per hour for this particular day

[10] The 121 operators employed on this work were part of a large, well-managed organization that had an excellent reputation over a long period of years. This particular job paid a guaranteed hourly base rate equal to that in the community, and the operators also had an opportunity to earn a bonus. Time standards were carefully and accurately set by time study, and they were guaranteed against change. At the time the production record shown in Table 16 was taken, this operation had been in existence many years, and the time standard had also been in effect for a long time. The operators working on this job had been selected and trained with considerable care, and they seemed to be convinced that there was no "top on earnings." That is, there was no evidence that the operators were pegging production because of fear that the rate on the job would be cut if an operator earned "too much." Therefore, the output of these 121 operators presents a cross section of performance that might be expected of a group of people employed on a job for which they were reasonably well fitted.

were tallied in the top horizontal line. There were 5 people in this group. Those who averaged between 60 and 69 pieces per hour were tallied in the second line. There were 45 people in this group. In a similar manner the operators in each of the six ranges from the slowest to the fastest were tallied, the total being 121 people. Such a graphical tally is called a frequency distribution.

Range Average Hourly Output In Pieces		Number
50 to 59	卌	5
60 to 69	卌 卌 卌 卌 卌 卌 卌 卌 卌	45
70 to 79	卌 卌 卌 卌 卌 卌 卌 卌	40
80 to 89	卌 卌 卌 卌 卌	25
90 to 99	卌	5
100 to 109	I	1

Fig. 230. Tally showing number of operators in each range. Average output in pieces per hour of 121 girls working on semiautomatic lathes.

Figure 231 presents a more convenient way to show the frequency distribution. The total range (on the X-axis) is the interval between the lowest hourly production (51) and the highest hourly production (104). The production records were kept to the nearest whole unit completed each day. The range on the X-axis is divided into six convenient intervals. The smooth curve shown in Fig. 231 is called a frequency distribution curve.

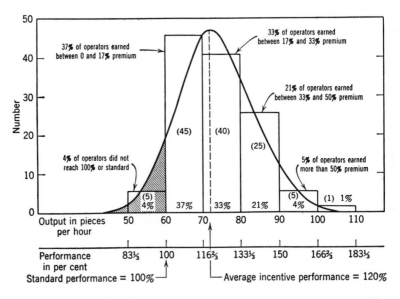

Fig. 231. Normal curve fitted to the distribution of output in pieces per hour of 121 operators working on semiautomatic lathes. Data taken from Table 16.

Figure 232 shows a normal curve fitted to the time taken by 500 people who performed the block-tossing operation.

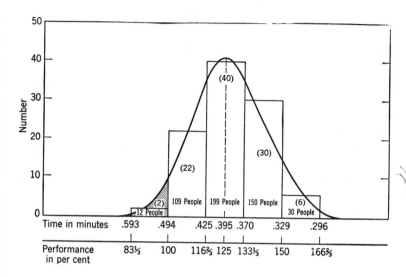

Fig. 232. Normal curve fitted to the distribution of time taken by 500 people who performed block-tossing operation. Data taken from Table 17.

Establishing a Standard as the Basis for Rating. The data obtained by a time study show the actual time taken by the operator to perform a series of consecutive elements of work. They tell nothing of the pace at which the operator worked while the study was being made. The operator might have been working at a level similar to that of the operator at the top of the column in Table 16, or at a level similar to that of the operator at the bottom of the column. It is necessary to consider the operator's speed in order to determine a standard that will permit an operator working at a normal pace to do the task in the time set for the job.

The need for rating has been pointed out, and the way the rating factor is used has been indicated. It is obvious, however, that some bench mark or some standard of comparison is required if rating is to be used as a measuring device. We must define our normal or standard. To say that normal speed is that speed expected of a qualified person working without incentive, or at a day-work pace, using a standardized method, does not define the term adequately. In fact, there seems to be no written definition that is entirely satisfactory. However, normal speed or normal rate of movement can be demonstrated; motion pictures can be made of typical factory jobs with the operator working at a normal tempo or at a known level above or below normal. Almost any person with average intelligence can be taught to rate operator tempo in terms of the established standard.

Walking on the level at 3 miles per hour [11] is frequently used to represent normal tempo. A person dealing a deck of cards into 4 equal piles in 0.50 minute is often considered to be exhibiting normal pace.[12] Filling a standard pinboard with 30 pins, using the two-handed method, in 0.41 minute is also said to represent normal speed.[13]

The General Motors Corporation has made a set of eleven films, each containing a different sequence of body and arm motions commonly found in factory work. The operator in each of these films is shown working at ten different speeds from 75% to 150%, with 100% as the normal speed. This set of films is used in connection with training programs for time study analysts and as a means of acquainting supervisors and foremen with time study technique.[14]

The Society for Advancement of Management (SAM) has made a set of "Rating of Time Study Films" consisting of 24 factory and clerical operations which have been rated by experienced time study analysts.

Some companies have made community time study surveys so that each participating company will know the position of its performance standard with relation to the average for the community.[15]

[11] Ralph Presgrave, *Dynamics of Time Study,* 2nd ed., McGraw-Hill Book Co., New York, 1945, p. 154.

[12] A standard deck of 52 cards is dealt in the following way by a person seated at a table. The deck is held in the left hand, and the top card is positioned with the thumb and index finger of the left hand. The right hand grasps the positioned card, carries and tosses it onto the table. The four piles of cards are arranged on the four corners of a 1-foot square. The only requirements are that the cards shall all be face down and that each of the four piles shall be separate from the others.

[13] For specifications of the pinboard and for a description of the method, see Chapter 11.

[14] The author has 11 silent films (Unit I Work Measurement Films—5 reels, and Unit II Work Measurement Films—6 reels) and 3 sound films showing 25 different operators working at a total of 80 different speeds. The silent films have been rated by some 5000 people from over 350 different companies in the United States and Canada, and a standard rating factor has been established for each operation shown in these films.

A study of 72 companies shows that 64% use motion picture films as a means of checking the rating ability of their time study analysts. Ralph M. Barnes, *Industrial Engineering Survey,* University of California, 1967.

[15] See *Work Measurement Manual* by Ralph M. Barnes, 4th ed., pp. 253–297.

Rating Film. Perhaps the most common form of rating film is made by having experienced operators, performing the same operation, work at a number of different speeds. Then the several sections of film are spliced together, separated from each other by a few feet of blank film, and each section is identified by a code number. Thus, the *film* of one operation might consist of 10 or 12 different sections representing 10 or 12 different working speeds.

Another common form of rating film is the *film loop*. Each section of the film just described is formed into a loop by cementing the front end to the back end. This permits the film to be placed in the projector and shown as long as the viewer wants to see it. The film loops, of course, can be shown in any order desired.

The film may take still another form. Four images, 6 images, or 12 images, for example, may be printed on one frame of the film. That is, the area of one frame may be subdivided into rectangular sections and the film for each different speed printed in one of these areas. They would be arranged in order according to speed, from slowest to fastest. It would thus be possible to have 12 operators working at 12 different speeds simultaneously on the screen. Such a film is called a *multi-image film*.

Film in any of the three forms may be used for training people in performance rating, and it may also be shown as a refresher. Either the loop film or the multi-image film can be used for comparison purposes. If a film has been made of the operation under study, it can be projected on a screen beside the multi-image film. The analyst can easily refer to the multi-image film in rating the "unknown" film.

The Relation of "Normal Pace" to "Average Incentive Pace." Since time standards are often used as the basis for some form of wage incentive plan, we are interested in the relationship between normal pace and the average pace expected of those on incentive. Most incentive earnings in this country fall between 15% and 45%, with the average around 25% to 30%.

The following explanation will show how the relationship between normal pace and average incentive pace may be established in a plant. It will also serve to emphasize the point that the performance of the great majority of workers on incentive should be fairly close to the average for the group. If the average incentive pace is 125%, it is expected that the average hourly output of two thirds of all workers would fall in a range extending from 15% below this point to 15% above this point. Only 3 or 4% of the group would be expected to exceed the 150% performance level, and only rarely would an operator exceed the 160 to 165% level.

Reference will be made to the normal distribution curve (Fig. 233) for data to support the above statements. There is considerable evidence to show that if the working speed of each member of a large

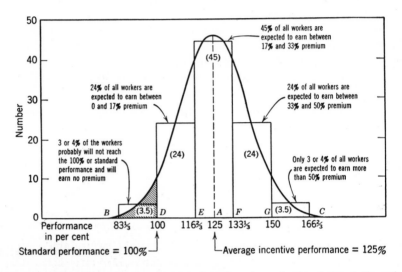

Fig. 233. Chart showing the number of workers in each class interval through entire performance range, as determined from the normal distribution curve. Ratio of slowest worker (83⅓%) to fastest worker (166⅔%) is 1 to 2. A 100% premium plan of wage payment is used. It is assumed that a premium is paid for all production above standard. That is, for each 1% increase in production above standard, the worker is paid 1% additional wage above his hourly base rate. The base rate is the guaranteed wage paid to all workers whether they reach 100% performance or not. This hourly base rate is established by job evaluation.

group of people, such as would be found in a factory, were arranged along the base line according to magnitude in per cent of normal, and if the vertical scale indicated frequency, the shape of the curve would fit fairly closely the normal bell curve.

This assumption having been made, a normal distribution curve can be drawn (Fig. 233), with five intervals covering a total speed range of 200%. The pace of the slowest operator is one half that of the fastest operator, which gives a ratio of 1 to 2. The next step is to establish the point on the curve that will represent normal speed.[16]

[16] This method of determining the relationship between normal performance and average incentive performance, suggested by Ralph Presgrave, seems to be

If we assume that this group of people is already working on incentive, and if their average incentive pace is 25% above normal, then point A in Fig. 233 can be called 125%. This point would represent the average for the group. If the length of the line B–C represents a speed range from $83\frac{1}{3}$ to $166\frac{2}{3}\%$, point D represents 100% or normal speed. In a similar manner, point B represents $83\frac{1}{3}\%$, E represents $116\frac{2}{3}\%$, F represents $133\frac{1}{3}\%$, G represents 150%, and C represents $166\frac{2}{3}\%$. Thus there is a range between $83\frac{1}{3}\%$ and $166\frac{2}{3}\%$ of 1 to 2. The number in each of the vertical bars represents the number of people per hundred who would fall in each of the five intervals.[17]

It is not expected, of course, that any group of workers would exactly fit the normal curve, although an examination of Fig. 231 will show that the output of this group of 121 operators tends to fit the normal curve. It should be noted that the standard established by time study was 60 pieces per hour; that is, 60 pieces per hour was equal to normal performance or 100%. The average output for the entire group was 72 pieces per hour, which is 20% above the standard. Thus 120% would be the average incentive performance for this group of workers for this particular day. There were five operators who worked at such a slow pace that they did not reach the 100% or normal performance level. Incidentally, these slow operators were paid their guaranteed day rate, even though they did not earn it. The very fastest operator turned out 104 pieces per hour, which was 73% above normal. In other words, this operator worked at a pace equal to 173%.

Figure 234 shows the distribution and earnings curve for 121 semi-automatic lathe operators. Figure 232 shows a normal curve fitted to the time taken by 500 people who performed the block-tossing operation.

Establishing a Company Standard. After the basic reasoning back of rating is fully understood, each company should establish a standard for its own use. Agreement should be reached as to what the normal or standard tempo, or performance level, should be in the plant. The first step would be to establish a standard for walking, card dealing, and other similar operations used generally throughout the country. The standards given on page 392 are widely used.

Then some simple operations from the plant, which can be performed by anyone, should be selected for demonstration. The method should

the most logical of any yet presented. See *Dynamics of Time Study*, 2nd ed., by Ralph Presgrave, McGraw-Hill Book Co., New York, 1945, Chapter 10.

[17] For information concerning the normal distribution curve see any book on statistics.

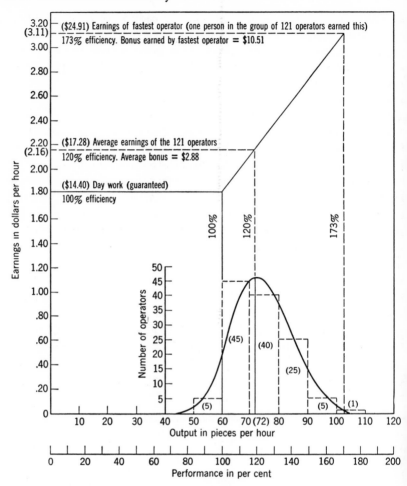

Fig. 234. Distribution and earnings curve for 121 semiautomatic lathe operators. The average output is 72 pieces per hour, and average earnings for the group are 20% above the guaranteed hourly rate.

be standardized, and the time for each job, with the operator working at normal pace, should be established. Motion pictures at 1000 frames per minute should be made of typical factory jobs, and the operator's tempo in per cent of normal should be established for each of them. Thus a library of standard rating films can be built up over a period of time for use as a bench mark for rating in the plant. Not only can time study analysts be taught to rate, but also foremen, supervisors, and the operators themselves can do this; and they are doing it in many plants today.

Rating Scales. There are several different rating scales in general use, and undoubtedly a competent and well-trained time study analyst can obtain satisfactory results with any one of them. A recent survey shows that the percentage system has greatest use and the point system comes next.

A study of the four different rating scales shown in Fig. 235 may help to show the difference between these systems. Just as we can read temperature on both Fahrenheit and centigrade thermometers although there is a difference in their scale, so we can rate operator speed whether we use percentage, points, or some other unit of measure. Since the percentage system is the plan having widest use in this country, it will be used in most of the illustrations in this volume.

Scale A—100% Equals Normal Performance. Normal performance [18] (that is, normal speed, tempo, or pace) equals 100% on rating Scale A. When this scale is used, it is expected that the average incentive pace will fall in the range of 115 to 135%, and the average for the entire group will be around 125%. This means that those operators who turn out between 15 and 35% per day more than normal will earn 15 to 35% extra pay for this extra performance. It is also expected that an occasional person, perhaps one in a thousand, would work at a pace twice as fast as normal. His performance rating would thus be 200%, and consequently he would earn twice the hourly base rate.

Scale B—60 Points Equals Normal Performance. Scale B illustrates the point system, with 60 points equal to normal performance and with the average incentive pace around 70 to 80 points. The maximum expected performance is around 100 to 120 points. This scale is similar to Scale A, 60 points being equal to 100% performance rating.

Scale C—125% Equals Incentive Performance. Some time study analysts use the "average incentive pace" as their bench mark. One company has adopted 125% as the point at which it would like to have the average output fall. Therefore, it tries to determine this point, at which to set its "incentive time standard," and then adds 25% to the hourly base rate in computing the amount of earnings that a person should receive at this point. For example, instead of stating that the time standard is 1.00 minute per piece and the base rate is, say, $2.40

---

[18] A study of 72 companies showed that 85% use "normal" operator performance as the basis for rating time studies, and 14% use "average incentive" operator performance. One per cent of the companies studied did not reply to this question. Ralph M. Barnes, *Industrial Engineering Survey,* 1967.

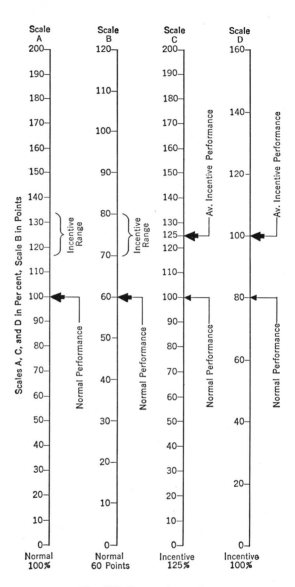

Fig. 235. Four rating scales.

per hour, giving a piece rate of 4 cents per piece, this company would state that the expected incentive output is 75 pieces per hour and that, when the operator reaches this point, he will be paid $3.00 per hour (which is, of course, at the rate of 4 cents per piece). Although this plan is perhaps as sound as any other, some people think it is not so easy to explain to the operators and that it has no advantages over a plan using Scale A.

Scale D—100% Equals Incentive Performance. A few organizations use a scale having 100% equal to "averge incentive pace," and this point is usually set 25% above normal performance. Therefore, 80% equals normal performance on this scale.

Speed and Method as They Affect Output. To summarize, there are two main factors that affect the number of units of work that a person on manual operations can produce in a given time. They are (1) speed of muscular movements, and (2) method of doing the task.

Speed or tempo, which refers to the rate of physical activity of the worker, can be measured by the rating factor, as has already been described. Method is ordinarily defined as the specified motion pattern required to perform a given operation. From the practical point of view, the method for a particular operation must be one which can be maintained in the shop day in and day out, and it must be one that the worker can be trained to follow. With this definition of method it is obvious that different individuals working on a job will, with practice, develop some refinements. Some would say that they have become more highly skilled. However, if a careful analysis were made of the operation, it generally would be found that a skilled person uses a different method from the one he used when he was less skilled on the job. Evidence to support this point is presented in Chapter 36.

Since there is neither a bench mark nor a unit for measuring differences in method, the only satisfactory way to handle this factor is to standardize the method and establish a standard time for this specified method. The time standard would not apply if a method other than the standard were used.

Although some variation in method is expected in the average factory, if care is used in developing the proper method of doing the job and if the operators are trained to do the work in the specified manner before the time study is made, the problem of variations in output due to variations in method can be minimized.

Applying the Rating Factor. The rating factor is applied to the selected time to give the normal time. Assume that in a particular operation of assembling an electric switch the operator gave a consistent performance throughout the cycle and throughout the study, and that the total selected time was 0.80 minute. With a rating factor for the study of 110%, the normal time would be as follows:

$$\text{Normal time} = \text{selected time} \times \frac{\text{rating in per cent}}{100}$$

$$= 0.80 \times \tfrac{110}{100} = 0.88 \text{ minute}$$

This value of 0.88 represents the time that a qualified and well-trained operator working at a normal pace would need to complete one cycle of the operation. This value is not the standard time for the job, since allowances must be added to the normal time to give the standard time. The determination and the application of allowances will be explained in Chapter 25.

CHAPTER 25

Time Study: Determining Allowances and Time Standard

DETERMINING ALLOWANCES

The *normal time* for an operation does not contain any allowances. It is merely the time that a qualified operator would need to perform the job if he worked at a normal tempo. However, it is not expected that a person will work all day without some interruptions. The operator may take time out for his personal needs, for rest, and for reasons beyond his control. Allowances for such interruptions to production may be classified as follows: (1) personal allowance, (2) fatigue allowance, or (3) delay allowance.

The standard time must include time for all the elements in the operation, and in addition it must contain time for all necessary allowances. Standard time is equal to the normal time plus the allowances. Allowances are not a part of the rating factor, and best results are obtained if they are applied separately.

Personal Allowance. Personal allowance will be considered first because every worker must be allowed time for his personal needs. The amount of this allowance can be determined by making all-day time studies or work sampling studies of various classes of work. For light work, where the operator works 8 hours per day without organized rest periods, 2 to 5% (10 to 24 minutes) per day is all that the average worker will use for personal time.

Although the amount of personal time required will vary with the individual more than with the kind of work, it is a fact that employees need more personal time when the work is heavy and done under unfavorable conditions, particularly in hot humid atmosphere. Under such conditions studies might possibly show that more than 5% allowance should be made for personal time.

395

Fatigue Allowance. In the modern well-managed plant in this country so many steps have been taken to eliminate fatigue that it is not of as great concern as formerly. In fact, fatigue is of such little consequence in some kinds of work that no allowance is required at all. There are many reasons for this. The length of the working day and the length of the working week have been shortened; machinery, mechanical handling equipment, tools, and fixtures have been improved so that the day's work is more easily done and the employee works in greater physical comfort than formerly. Accident hazards have also been reduced so that the fear of physical injury is less.

There are, of course, some kinds of work that still involve heavy physical exertion and are performed under adverse conditions of heat, humidity, dust, and accident hazards, and therefore require rest for the operator. Fatigue results from a large number of causes, some of which are mental as well as physical.

At the present time there is no satisfactory way of measuring fatigue. Physiological measurements promise to provide objective means of determining the time and duration of periods of work and rest during the day. However, in this country we have not yet done the research and testing required to validate this procedure.

We know from experience that a person needs time for rest when his work is arduous. The problem of determining the amount of time to be allowed for rest is very complex. Time needed for rest varies with the individual, with the length of the interval in the cycle during which the person is under load, with the conditions under which the work is done, and with many other factors. Some companies have from long experience arrived at fatigue allowances which seem to be satisfactory (Fig. 236). A few organizations having arduous physical work, such as stacking heavy boxes in warehouses or in freight cars, have tried out various combinations of periods of rest and work until they have arrived at satisfactory allowances.

Organized rest periods during which time all employees in a department are not permitted to work provide one of the best solutions to the problem. The optimum length and number of rest periods must be determined. Perhaps the most common plan is to provide one rest period during the middle of the morning and one during the middle of the afternoon. The length of these periods ordinarily varies from 5 to 15 minutes each.[1]

If no wage incentive plan is used, some companies pay for the rest

---

[1] For a more complete discussion of rest periods see Chapter 34.

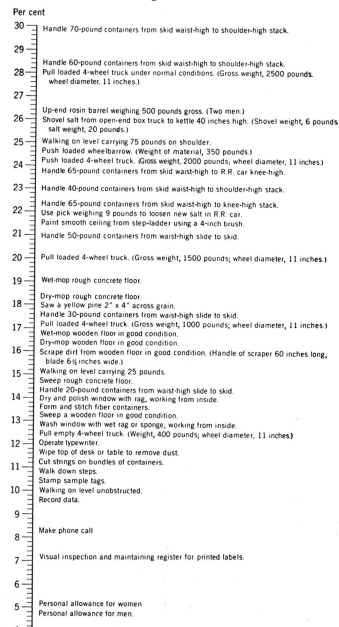

Fig. 236. Personal and fatigue allowances used by one company having mainly handling and hand-truck operations. The allowances given include personal time.

periods at the employee's regular hourly base rate. If a wage incentive plan is used and if fatigue allowances have been incorporated in the time standard, employees are not paid for the rest periods as such. The worker merely takes his fatigue allowance during the specified rest period rather than at intervals during the day at his own choosing.

It should be repeated that a fatigue allowance does not need to be made for much light factory work, and organized rest periods during the day provide sufficient rest for another group of factory operations. The amount of heavy work in factories is gradually decreasing because of the greater use of machinery and power-handling equipment; consequently the problem of fatigue allowances becomes one of decreasing importance to the time study analyst.

Delay Allowance. Delays may be avoidable or unavoidable. Those delays that the operator makes intentionally will, of course, not be considered in determining the time standard. Unavoidable delays do occur from time to time, caused by the machine, the operator, or some outside force.

It is expected that machines and equipment will be kept in good repair. When there is a breakdown or when repairs are necessary, the operator is usually taken off the job and such delays do not enter into the time standard. In such cases the operator is usually paid for waiting time at his hourly base rate. Sometimes there are minor adjustments, breakage of tools such as drills and taps, or lost time due to occasional variation in material and interruptions by supervisors, and they must be included in the standard. Each unavoidable delay should be considered as a challenge by the analyst and the foreman, and every reasonable effort made to eliminate these delays. The kind and amount of delays for a given class of work can best be determined from all-day time studies or work sampling studies made over a sufficient period of time to give reliable data.

Applying the Allowances. Personal allowance is applied as a percentage of the normal time, and affects both handling time and machine time alike. For convenience, fatigue allowance is sometimes applied in the same way, although some believe that this allowance should apply to only those elements during which the operator works, and not to the machine time during which the machine works. Delays are applied as a percentage of the normal time, or if entirely a machine-delay allowance, then on the machine elements only. If these three allowances are applied uniformly to all elements, they may be added together and applied together, necessitating but a single computation.

Although allowances have traditionally been applied as a percentage

of the normal time to be added to the normal time to obtain the standard time, there is a trend toward considering allowances in terms of
minutes allowed per working day. Thus, instead of referring to personal allowance as 5%, it would be referred to as 24 minutes per 8-hour
day ($480 \times 5\% = 24$). If this were the only allowance made, the
working time in this case would be 456 minutes per day ($480 -
24 = 456$).

If an allowance of 5% for personal time were made on the assembly
operation referred to on page 400, 5% would be added to the normal
time for this operation in the following way:

Standard time = normal time + (normal time $\times$ allowances in per cent)

$$= 0.88 + (0.88 \times 0.05) = 0.88 + 0.044 = 0.924 \text{ minute}$$

To summarize:

Selected time $= 0.80$ minute
Rating factor $= 110\%$
Personal allowance $= 5\%$
Normal time $= 0.80 \times \frac{110}{100} = 0.88$ minute
Standard time $= 0.88 + (0.88 \times 0.05) = 0.924$ minute

Another way to compute this is:

Standard time $= 0.88 \times 1.05 = 0.924$ minute

Although this method of applying the personal allowance is the
most common one in use today, it is not absolutely correct. If by a
5% allowance it is understood that 24 minutes per 8-hour day are to
be available to the worker for his personal needs, and if the normal
time for one assembly of the electric switch is 0.88 minute, then during
the 456 minutes available for work ($480 - 24 = 456$) the operator
could produce 518 pieces ($456 \div 0.88 = 518$). Since the 8-hour day
consists of 480 minutes, the standard time per piece would be 0.926
minute ($480 \div 518 = 0.926$). Another way to state this is:

$$\text{Standard time} = \text{normal time} \times \frac{100}{100 - \text{allowance in per cent}}$$

$$= 0.88 \times \frac{100}{100 - 5}$$

$$= 0.88 \times \frac{100}{95} = 0.926 \text{ minute}$$

Not only is this method of incorporating allowances into the time standard correct, but also there is considerable value in stating the total time in minutes per 8-hour day for each type of allowance. To the foreman or operator the statement that 24 minutes per day is allowed for personal time means more than merely to say that 5% has been added to the normal cycle time for personal needs.

DETERMINING TIME STANDARDS

Stop-Watch Time Study of a Core-Making Operation. The study shown in Fig. 217 is a very common type of stop-watch time study used today, although many organizations have the policy of rating each element separately instead of making an over-all rating for the study, and they also determine and apply a fatigue and personal allowance factor for each element. Figures 237 and 238 show such a study.

The operation studied was the making of a dry sand core in a wood core box. The core was 7½ inches long and 1¾₁₆ inches in diameter, taking the general shape of half a cylinder.

A full description of the operation is given in the left-hand column of Table 18, and the abbreviated description recorded on the observation sheet is given in the right-hand column. The end points used in reading the watch are also given.

Before recording stop-watch readings the back of the observation sheet (Fig. 238) was filled out, a drawing of the layout of the work place was made, and a sketch of the core was placed in the lower right-hand corner of the sheet.

The continuous method of timing and a decimal-minute stop watch were used. As the analyst made the study, he evaluated the speed of the operator for each element of the operation. In making studies such as this one the analyst may occasionally record a rating factor above the stop-watch reading as the study progresses. Then, after the study is completed, he will record the rating for each element of the study. These values are recorded on the front of the observation sheet (Fig. 237) in the vertical column headed "Rating." An over-all rating factor for the entire study is also recorded in the space provided in the lower right-hand section of the observation sheet. It may be used in connection with the elasped time and the number of pieces finished, to check the time standard after it has finally been determined.

The selected time value for each element is determined by tallying the data as shown on the bottom of the observation sheet. If there is not space on this sheet, the tally is made on a plain sheet of paper and

OBSERVATION SHEET

STUDY NO. 8765

SHEET 1 OF 1 SHEETS

ELEMENTS	SPEED	FEED	UPPER LINE: SUBTRACTED TIME / LOWER LINE: READING	MIN. TIME	AV. TIME	SELECTED TIME	OCC. PER CYCLE	RATING	NORMAL TIME
1. Fill core box with 3 handfuls of sand. Press sand down each time.			*.09* .09 / .09 .41 / *.09* .71 / *.09* 1.07 / *.08* .38 / *.08* .67 / *.10* .98 / *.08* .57 / *.09* .28 / *.08* .87 / *.07* .18 / *.08* .46 / *.08* .76 / *.09* 4.05 / *.06* .32	.06	.081	.081	1	115	.093
2. Press sand down with one trowel stroke. Strike off with one trowel stroke.			*.06* 15 / *.05* .46 / *.08* .79 / *.06* .13 / *.05* .43 / *.05* .72 / *.06* 2.04 / *.05* .62 / *.05* .33 / *.06* .93 / *.06* .24 / *.05* .51 / *.05* .81 / *.06* .11 / *.06* .38	.05	.059	.059	1	125	.074
3. Get and place plate on core box, turn over, rap, and remove box.			*.13* .28 / *.13* .59 / *.15* .94 / *.14* .27 / *.13* .56 / *.13* .85 / *.14* .18 / *.14* .76 / *.13* .46 / *.13* 3.06 / *.12* .36 / *.14* .65 / *.12* .93 / *.13* .24 / *.13* .51	.10	.126	.126	1	135	.170
4. Carry plate with core 4 feet. Dispose on oven truck.			*.04* .32 / *.03* .62 / *.04* .98 / *.03* .30 / *.03* .59 / *.03* .88 / *.03* .21 / *.03* .49 / *.03* .79 / *.03* .09 / *.03* .39 / *.03* .68 / *.03* .96 / *.02* .26 / *.03* .54	.02	.032	.032	1	125	.040
(1)			*.07* .61 / *.10* .95 / *.08* .25 / *.08* .53 / *.08* .83 / *.08* .12 / *.07* .41 / *.08* .71 / *.08* 7.01 / *.08* .28 / *.07* .55 / *.07* .84 / *.08* .16 / *.09* .48 / *.09* .77						
(2)			*.05* .66 / *.05* 5.00 / *.05* .30 / *.05* .58 / *.06* .89 / *.06* .18 / *.06* .47 / *.06* .77 / *.05* .06 / *.06* .34 / *.06* .61 / *.07* .91 / *.06* .22 / *.05* .53 / *.05* .82						
(3)			*.14* .80 / *.13* .13 / *.12* .42 / *.13* .71 / *.12* 6.01 / *.13* .31 / *.13* .60 / *.12* .89 / *.11* .17 / *.12* .46 / *.13* .74 / *.13* 8.04 / *.14* .36 / *.13* .66 / *.13* .95						
(4)			*.05* .85 / *.04* .17 / *.03* .45 / *.04* .75 / *.03* .04 / *.03* .34 / *.03* .63 / *.04* .93 / *.03* .20 / *.02* .48 / *.03* .77 / *.03* .08 / *.03* .39 / *.02* .68 / *.02* .98						
(1)			*.07* 9.05 / *.07* .34 / *.08* .64 / *.08* .93 / *.07* .21 / *.08* .50 / *.07* .78 / *.08* 11.07 / *.09* .39 / *.08* .69 / *.09* .99 / *.09* .29 / *.08* .59 / *.08* .89 / *.09* .19						
(2)			*.05* .10 / *.06* .40 / *.05* .69 / *.06* .99 / *.06* .27 / *.07* .57 / *.06* .84 / *.07* .14 / *.08* .47 / *.07* .76 / *.06* 12.05 / *.06* .35 / *.07* .66 / *.06* .95 / *.08* .27						
(3)			*.14* .24 / *.13* .53 / *.13* .82 / *.11* 10.10 / *.12* .39 / *.11* .68 / *.11* .95 / *.12* .26 / *.10* .57 / *.12* .88 / *.12* .17 / *.13* .48 / *.12* .78 / *.12* 13.07 / *.11* .38						
(4)			*.03* .27 / *.03* .56 / *.04* .85 / *.04* .14 / *.03* .42 / *.03* .71 / *.04* .99 / *.04* .30 / *.04* .61 / *.03* .91 / *.03* .20 / *.03* .51 / *.03* .81 / *.03* .10 / *.03* .41						

FOREIGN ELEMENTS:

Tally-by elements

No. 1 — .06-*I* / .07-*III*
No. 2 — .05-*IIIII* / .06-*IIIIIIII*
No. 3 — .10-*I* / .11-*III*
No. 4 — .02-*II* / .03-*IIIIIII*

.08-*IIIIIIII*/√ .07-*III*
.09-*IIII*/ .08-*III*
.12-*IIII* / .13-*IIIIIIII*
.04-*III* / .05-*I*
.10-*II*
.14-*II* / .15-*I*

TOOLS, JIGS, GAUGES, PATTERNS, ETC.:
Core box No. C-ID-7253, Size 1⅝ x 3½ x8½, Wt. 1 lb., 5" Molder's trowel
Plates 4 x 9"; weight with core 3½ lb. Core sand No. A16

OVERALL RATING: 125

	BEGIN	END	ELAPSED
	9:18	9:32	14:00

UNITS FINISHED	ACTUAL TIME PER PIECE
45	0.31 Min.

Fig. 237. Front of observation sheet—core-making operation. Size of form 8½ × 11 inches.

OPERATION:	Make core for crank frame No. 7253
	OP. NO.: C-10-A
PART NAME:	Core for crank frame No. 7253 — PART NO.: ——
MACH. NAME:	Bench No. 62 — MACH. NO.: ——
OPERATOR'S NAME & NO.:	S.R. Martin — MALE ☑ FEMALE ☐
EXPERIENCE ON JOB:	Six months — FOREMAN: M.L. Ray
NO. MACHINES OPER'D: —	MACH. SPEED —— DEPT. NO.: 17
MATERIAL:	Dry core sand-specification No. A16

SKETCH OF WORK PLACE   SCALE: One square = 4 inches

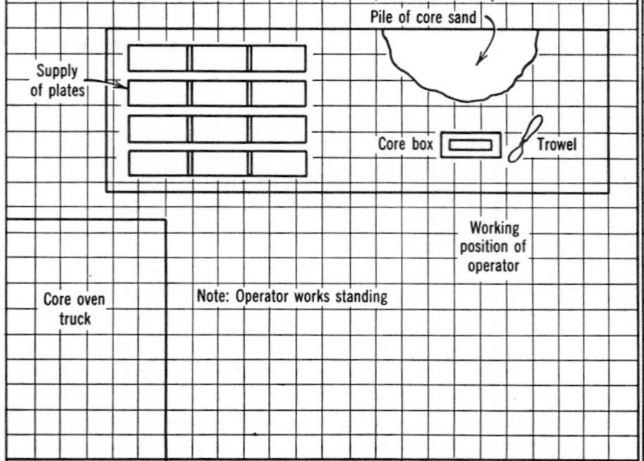

Pile of core sand

Core box   Trowel

Working position of operator

Supply of plates

Note: Operator works standing

Core oven truck

DATE OF STUDY	OBSERVER C.A. Clark	APPROVED J.S.R.

SUMMARY

NO.	ELEMENTS	NORMAL TIME	FAT'G. & PER'L ALLOW.	OTHER ALLOW.	TOTAL ALLOW.	STD. TIME
1.	Fill core box with 3 handfuls of sand. Press sand down each time.	.093	12	—	12	.106
2.	Press sand down with one trowel stroke. Strike off with one trowel stroke.	.074	15	—	15	.087
3.	Get and place plate on core box, turn over, rap, and remove box.	.170	15	—	15	.200
4.	Carry plate with core 4 feet. Dispose on oven truck.	.040	12	—	12	.046
						.439

TOTAL STD. TIME PER CYCLE:	.439
NO. PIECES PER CYCLE: 1	STD. TIME PER PIECE: Use .44

DRAWING OF PART:

Core:

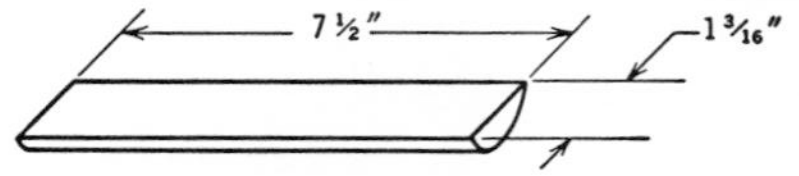

7 ½"

1 ³⁄₁₆"

One half of cylinder 1 ³⁄₁₆" x 7 ½"

Wt. of core before baking = ¼ lb.

Fig. 238. Back of observation sheet—core-making operation.

Table 18. Elements of Core-Making Operation

Detailed Description	Condensed Description Recorded on Observation Sheet, with End Points for Reading Stop Watch
1. Walk 4 feet from core-oven truck to bench, pick up core box with both hands, push loose sand on front edge of bench back against pile with edge of core box. Hold core box with left hand and fill core box with three handfuls of sand, pressing sand down in core box each time.	1. Fill core box with 3 handfuls of sand. Press sand down each time. *End of element* as right hand begins to grasp trowel.
2. Pick up trowel with right hand, press sand down with one stroke of trowel across top of box, strike off (draw edge of trowel across top of box), removing excess sand with edge of trowel. Dispose of trowel on bench at right of core box.	2. Press sand down with one trowel stroke. Strike off with one trowel stroke. *End of element* as trowel is dropped on bench (hits bench).
3. Get plate and carry from pile on bench 3 feet to left of core box, turn plate upside down, and place on top of core box. Turn plate and core box over. Pick up trowel with right hand and rap core box twice with handle of trowel, and dispose of trowel on bench to right of core box. Using both hands, carefully lift core box upward from plate, allowing core to remain on plate. Place core box on bench to right of plate.	3. Get and place plate on core box, turn over, rap, and remove box. *End of element* as core box is placed on bench (hits bench).
4. Walk, carrying plate and core, 4 feet to left, and place on shelf of core-oven truck.	4. Carry plate with core 4 feet. Dispose on oven truck. *End of element* as plate is placed on (touches) shelf of core-oven truck.

attached to the observation sheet. The tally shows that for element 1 the minimum time value is 0.06, the time value that occurs most frequently is 0.08, and the average is 0.081. In a similar manner values are determined for the other three elements of the study. The number of times that the element occurs in the study is recorded in the column marked "Occurrence per Cycle."

The normal times are then calculated in the following way:

$$\text{Normal time} = \text{selected time} \times \frac{\text{rating in per cent}}{100}$$

For element 1 (Fig. 239) the normal time is determined in the following manner:

Selected time = 0.081 minute
Rating factor = 115%
Normal time = 0.081 × $\frac{115}{100}$ = 0.093 minute

OBSERVATION SHEET

ELEMENTS	UPPER LINE: SUBTRACTED TIME														LOWER LINE: READING
	1	2	3	4	5	6	7	8	9	10	11	12	13	14	15
1. Fill core box with 3 handfuls of sand. Press sand down each time.	.09	.09	.09	.09	.08	.08	.10	.07	.08	.08	.09	.07	.08	.09	.06

															MIN. TIME	AV. TIME	SELECTED TIME	OCC. PER CYCLE	RATING	NORMAL TIME
.07	.10	.08	.08	.08	.08	.07	.08	.08	.08	.07	.07	.08	.09	.09	.06	.081	.081	1	115	.093

Fig. 239. Part of observation sheet for operation "Make Core for Crank Frame." The minimum time, average time, selected time, rating factor, and normal time for one element are shown. See Fig. 237 for complete study.

Time Study Summary. After the normal time values are calculated for each element, a summary is made on the back of the observation sheet in the space provided. A fatigue and personal allowance is determined for each element and recorded in the appropriate column. Twelve per cent is allowed for the first and fourth elements, and 15% for the second and third elements. These allowances are obtained from a table similar to the one shown in Fig. 236. No other allowances are made.

The standard time is determined for each element in the following manner:

$$\text{Standard time} = \text{normal time} \times \frac{100}{100 - \text{allowance in per cent}}$$

For element 1 this is:

$$\text{Standard time} = 0.093 \times \frac{100}{100 - 12} = 0.093 \times \frac{100}{88} = 0.106 \text{ minute}$$

In a like manner the standard time is determined for each of the four elements (Fig. 240). Then these are added together to give the standard time for the cycle. Since one piece is produced per cycle, the standard time per piece is the same as the standard time per cycle.

A helper supplies the core maker with core sand and plates and provides empty core-oven trucks; therefore no time for this work is included in the standard. Had this work been part of the core maker's job, it would have been timed and included as additional elements in the operation.

	SUMMARY					
NO.	ELEMENTS	NORMAL TIME	FAT'G & PERS'L ALLOW.	OTHER ALLOW.	TOTAL ALLOW.	STD. TIME
1.	Fill core box with 3 handfuls of sand. Press sand down each time.	.093	12	—	12	.106
2.	Press sand down with one towel stroke. Strike off with one towel stroke.	.074	15	—	15	.087
3.	Get and place plate on core box, turn over, rap, and remove box.	.170	15	—	15	.200
4.	Carry plate with core 4 feet. Dispose on oven truck.	.040	12	—	12	.046
						.439

Fig. 240. Part of observation sheet for "Make Core for Crank Frame." The normal time, allowances, and the standard time for each element are shown. See Fig. 237 for complete study.

Stop-Watch Time Study of an Assembling and Cementing Operation. Figures 241 and 242 show the observation sheet for an assembling and cementing operation in a rubber-footwear plant. Figure 243 shows the Computation Sheet, and Fig. 244 the Piece-Work Rate Sheet, which is the authorization to put the piece rate into effect. These forms are a part of the time study manual shown in Appendix A.

Left- and Right-Hand Operation Description. Many time study problems are created because the method used in performing the operation has not been recorded on the observation sheet in sufficient detail. One aid in solving this problem is to make a left- and right-hand description of each element of the operation. It may take the form of a record of each motion of each hand (Figs. 84 and 100), or it may consist of a record of get, place, use, hold, dispose, and wait for each hand. Figure 246 contains such a description of the elements of a cast-iron plate assembly (Fig. 245). This left- and right-hand description would be attached to the observation sheet, and the time study would be made in the usual way.

OBSERVATION SHEET

DEPARTMENT Shoe Room
FOREMAN W.M. Wilson
OPERATION Assemble and cement heel plugs on swing boot insoles
OPERATOR Betty Walker
DATE
OBSERVER R.J. Parson

NO.	ELEMENTS	UNITS PER ELEMENT	1	2	3	4	5	6	7	8	9	10	11	12	TOTAL TIME	NO. OF OBS'NS	AVERAGE TIME	MOST FREQUENTLY READING OCCUR.	MINIMUM TIME	REPRESENTATIVE TIME	RATING	NORMAL TIME
1	Get supply of heel plugs	20 Pr.	.06/.10	.07/.08	.10/.09	.07/.08	.08/.09	.08/.08	.07	.08	.09	.10	.08	.09	1.49	18	.083	.08	.06	.083	100	.083
2	Get supply of insoles	20 Pr.	.12/.15	.15/.13	.10/.12	.14	.13	.14	.13	.14	.13	.15	.13	.13	1.99	15	.133	.13	.10	.133	100	.133
3	Get, loosen, and lay out insoles in 15 piles	7½ Pr.	.41/.43	.43/.44	.42	.40	.44	.43	.42	.43	.45	.44	.45	.43	6.02	14	.430	.43	.40	.430	110	.473
4	Get, pick, and spot heel plugs on insole	½ Pr.	.06/.05/.07	.05/.08	.05/.05	.06/.05	.06/.05	.05/.06	.07/.05	.05/.05	.05/.06	.05/.05	.06/.05	.05/.06	2.15	39	.056	.05	.04	.056	100	.056
5	Get brush of cement, cement, and aside brush	7½ Pr.	.24/.24	.24/.23	.23/.25	.23/.24	.22/.23	.22/.23	.24/.24	.23/.22	.22/.24	.26/.23	.23/.24	.23/.23	4.62	20	.231	.23	.18	.231	95	.219
6	Stack completed work	30	.26/.24	.24/.25	.25/.26	.24/.24	.25	.23	.23	.25	.24				3.88	16	.242	.24	.18	.242	100	.242
7	Mark size on stack	30	.04/.04	.04/.04	.04/.05	.04/.04	.04/.06	.04/.05	.04/.06	.05/.05	.04	.04	.05	.04	1.02	23	.044	.04	.02	.044	100	.044
8	Aside completed work	30	.08/.10	.07/.08	.08/.09	.10/.08	.09	.08	.08	.08	.08	.08	.09	.08	1.33	16	.083	.08	.06	.083	100	.083
9	Get cement supply	2000	1.14/1.20	1.20	1.27	1.16	1.23								6.00	5	1.20	1.14	1.14	1.20	100	1.200
10	Empty and clean cement pan	2000	1.23	1.11	1.18	1.20									4.72	4	1.18	1.02	1.02	1.18	100	1.180
11	Clean up work place and cover work	2000	1.90	1.95	2.11	1.83									7.79	4	1.948	1.83	1.83	1.948	100	1.948
12	Record production	120	(From standard data)																			.070

INFREQUENT ELEMENTS	OCCURANCE	TIME	RATING	NORM. TIME

REMARKS

STOP 10:24
START 11:06
ELAPSED
NO. UNITS
UNITS PER HOUR

TS 103

Fig. 241. Front of observation sheet—stop-watch time study of assembling and cementing operation, made by the repetitive method. Size of form 8½ × 11 inches.

EQUIPMENT USED

Work Bench
Pan of Cement and Brush

PART DRAWING

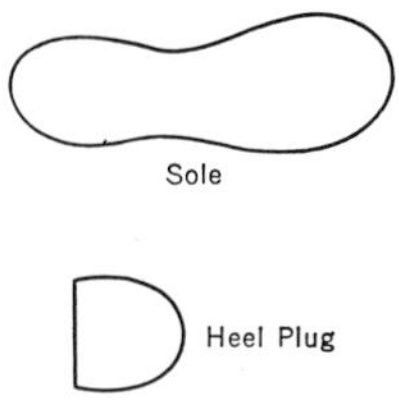

Sole

Heel Plug

DISPOSITION OF COMPLETED WORK: Onto Truck with Shelves (T431)

MATERIALS AND SUPPLIES

DESCRIPTION	PAIRS PER UNIT	SOURCE	SUPPLIED BY	HOW SUPPLIED
Cement # CT 1031		Tank	Operator	

SKETCH OF WORKPLACE   **SCALE:** $\frac{1}{2}'' = 1'$

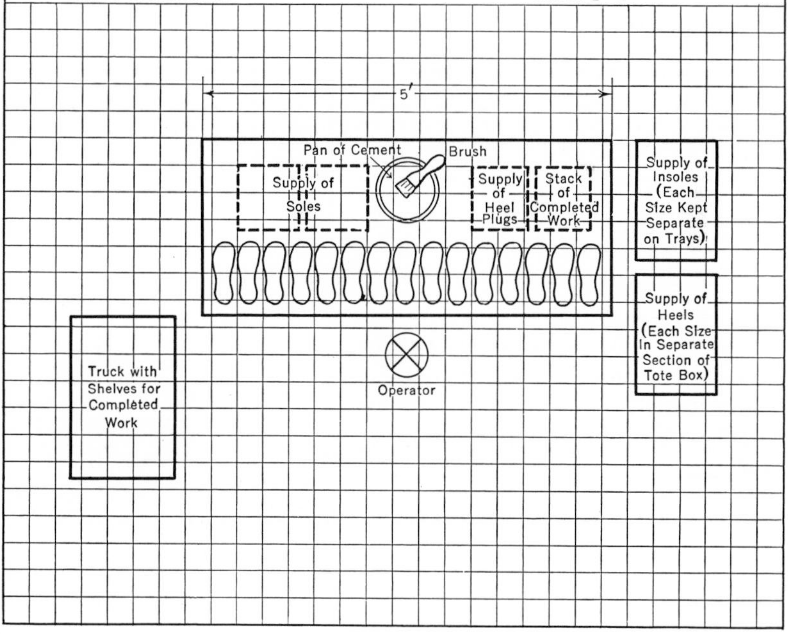

←————— 5' —————→

Pan of Cement   Brush

Supply of Soles

Supply of Heel Plugs

Stack of Completed Work

Supply of Insoles (Each Size Kept Separate on Trays)

Supply of Heels (Each Size In Separate Section of Tote Box)

Truck with Shelves for Completed Work

Operator

Fig. 242. Back of observation sheet—assembling and cementing operation.

COMPUTATION SHEET

OPERATION __Assemble and Cement Heel Plugs on Swing Boot Insoles__

DEPARTMENT __Shoe Room_____ DATE____

NO.	ELEMENTS	NORMAL TIME PER ELE- MENT	UNITS PER ELE- MEMT (Pr.)	OCCUR. OF ELE- MEMT PER 100 Pr.	NORMAL TIME PER 100 Pr.
1	Get Supply of Heel Plugs	.083	20	5	.415
2	Get Supply of Insoles	.133	20	5	.665
3	Get, Loosen, Layout Insoles in 15 Piles	.473	7½	13⅓	6.292
4	Get, Pick, and Spot Heel Plugs on Insoles	.056	½	200	11.200
5	Get Brush of Cement, Cement, and Aside Brush	.219	7½	13⅓	2.919
6	Stack Completed Work	.242	30	3⅓	.807
7	Mark Size on Stack	.044	30	3⅓	.147
8	Aside Completed Work	.083	30	3⅓	.276
9	Get Cement Supply	1.200	2000	.05	.060
10	Empty and Clean Cement Pan	1.180	2000	.05	.059
11	Clean Up Work Place and Cover Work	1.948	2000	.05	.098
12	Record Production	.07	120	.83	.058

(A) TOTAL NORMAL TIME IN MINUTES_____	22.996
(B) ALLOWANCES (10%) IN MINUTES_____	2.299
(C) TOTAL STANDARD TIME PER 100 Pr. (A + B = C)_____	25.3
(D) DAY WORK HOURLY PRODUCTION_____	237 Pr.

TS104

Fig. 243. Computation sheet for assembling and cementing operation.

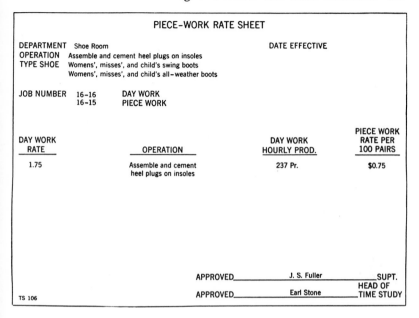

PIECE-WORK RATE SHEET

DEPARTMENT   Shoe Room   DATE EFFECTIVE
OPERATION   Assemble and cement heel plugs on insoles
TYPE SHOE   Womens', misses', and child's swing boots
             Womens', misses', and child's all−weather boots

JOB NUMBER   16–16   DAY WORK
             16–15   PIECE WORK

DAY WORK RATE	OPERATION	DAY WORK HOURLY PROD.	PIECE WORK RATE PER 100 PAIRS
1.75	Assemble and cement heel plugs on insoles	237 Pr.	$0.75

APPROVED_____ J. S. Fuller _____SUPT.
                                                 HEAD OF
APPROVED_____ Earl Stone _____TIME STUDY

TS 106

Fig. 244. Piece-work rate sheet for assembling and cementing operation.

As more time study analysts are trained in micromotion study and in the use of motion-time data, there is certain to be greater use of the left- and right-hand operation description. This more complete description of the operation will make for a better time study.

Production Studies. Although a motion and time study may have been made with care and the instruction sheet may have been prepared and given to the operator, there is sometimes a complaint that the operator is unable to perform the task in the time called for on the instruction sheet. If, after a preliminary check, it appears that the inability to do the task in the time set is not the fault of the operator, it is essential that a new study be made to check the original time study. This new study, sometimes called a production study, covers a longer period of time than the original study—sometimes as long as a day or two. Figure 247 shows a production study summary.

The inability of the operator to perform the task in the time specified may be due to any one or a combination of these causes: conditions of material, tools, or equipment are different from those existing at the time the original study was made; there has been a change in method, layout, or working conditions; operator has not had sufficient experience on the job or is unsuited to the work; or the time study itself contained errors. The production study should be made in such detail as to permit the checking of elemental times.

Although every effort should be made to prevent errors in setting

Fig. 245. Layout of work place for assembly of cast-iron plates. Power wrench is suspended above the fixture.

the original time standard, it is essential that the management be willing at all times to rectify errors or to demonstrate the correctness of the time standard. The workers must have confidence in the standards and in the men who set them.

Work Sampling Studies. Work sampling is being used to an increasing extent to supplement or replace the production study. As will be explained more fully in Chapter 32, work sampling is often a better and more economical method of getting facts than all-day time study.

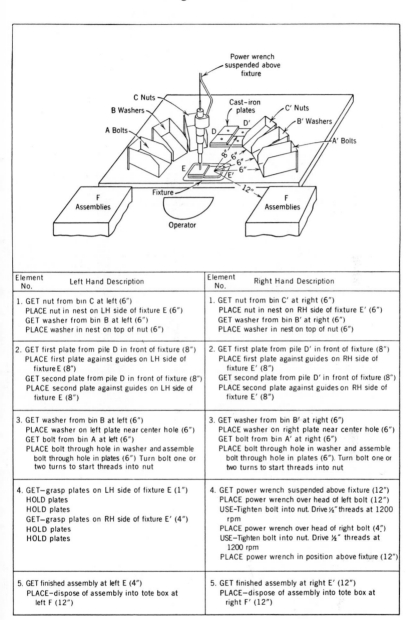

Element No.	Left Hand Description	Element No.	Right Hand Description
1.	GET nut from bin C at left (6″) PLACE nut in nest on LH side of fixture E (6″) GET washer from bin B at left (6″) PLACE washer in nest on top of nut (6″)	1.	GET nut from bin C′ at right (6″) PLACE nut in nest on RH side of fixture E′ (6″) GET washer from bin B′ at right (6″) PLACE washer in nest on top of nut (6″)
2.	GET first plate from pile D in front of fixture (8″) PLACE first plate against guides on LH side of fixture E (8″) GET second plate from pile D in front of fixture (8″) PLACE second plate against guides on LH side of fixture E (8″)	2.	GET first plate from pile D′ in front of fixture (8″) PLACE first plate against guides on RH side of fixture E′ (8″) GET second plate from pile D′ in front of fixture (8″) PLACE second plate against guides on RH side of fixture E′ (8″)
3.	GET washer from bin B at left (6″) PLACE washer on left plate near center hole (6″) GET bolt from bin A at left (6″) PLACE bolt through hole in washer and assemble bolt through hole in plates (6″) Turn bolt one or two turns to start threads into nut	3.	GET washer from bin B′ at right (6″) PLACE washer on right plate near center hole (6″) GET bolt from bin A′ at right (6″) PLACE bolt through hole in washer and assemble bolt through hole in plates (6″). Turn bolt one or two turns to start threads into nut
4.	GET—grasp plates on LH side of fixture E (1″) HOLD plates HOLD plates GET—grasp plates on RH side of fixture E′ (4″) HOLD plates HOLD plates	4.	GET power wrench suspended above fixture (12″) PLACE power wrench over head of left bolt (12″) USE-Tighten bolt into nut. Drive ½″ threads at 1200 rpm PLACE power wrench over head of right bolt (4″) USE-Tighten bolt into nut. Drive ½″ threads at 1200 rpm PLACE power wrench in position above fixture (12″)
5.	GET finished assembly at left E (4″) PLACE—dispose of assembly into tote box at left F (12″)	5.	GET finished assembly at right E′ (12″) PLACE—dispose of assembly into tote box at right F′ (12″)

Fig. 246. Left- and right-hand operation description for the assembly of cast-iron plates.

PRODUCTION STUDY SUMMARY

		KEY	ACTIVITY	MIN.	%
OPERATION	Mold Steering Gear Housing		Productive Time	219.1	81.1
			Personal Time	25.1	9.3
	OP. NO. M-27		Get Ready and Clean Up	23.8	8.8
			Unnecessary Idle Time	1.8	.7
PART NAME Steering Gear Housing	PART NO. VT-179A		Idle Time Beyond Control of Op.	.2	.1
MACH. NAME ——	MACH. NO. ——				
LOCATION Foundry	DATE				
TIME STARTED 7:00 A.M.	CHARTED				
TIME STOPPED 11:30 A.M.	BY L.M.K.		Total	270.0	100

Fig. 247. Production study summary of a foundry operation covering a 4½-hour period. Size of sheet 8½ × 11 inches.

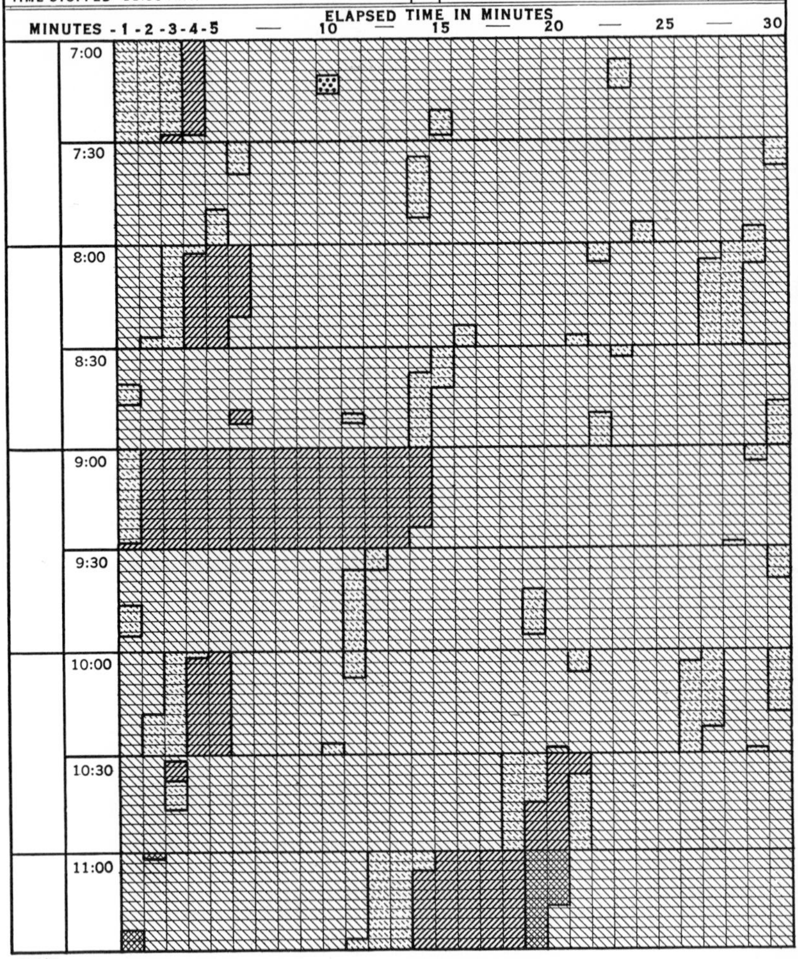

Recording and Filing. When a time study is to be made of a series of similar operations, it is desirable to define carefully each of the elements in order that standard time data for each element may finally be determined. For example, in the operation "solder side seam of rectangular can" the elements are defined (page 450), and irrespective of the person making time studies of soldering work this uniform division of the operation into elements will be made. A master form is prepared, and the essential data from each time study of soldering side seams are recorded on this sheet. After sufficient data have been accumulated, they will be used for setting up formulas for synthetically determining time standards on soldering operations, as illustrated in Chapter **28.**

Time studies, together with other data and information concerning the operation, should be filed in such a way that they may be readily located when needed. Cross-indexing is often worthwhile.

Guaranteed Time Standard. The time standard should be guaranteed against change unless there has been a change in method, materials, tools, equipment, layout, or working conditions, or when there has been a clerical error or a mistake in the calculation of the standard.

When a time standard is set for a job, it is understood that the operator must perform the operation exactly as specified in the standard practice or on the instruction sheet. If the operation is not performed in this manner, the time standard is not in effect. However, so long as the operator does the job in the prescribed manner, the company guarantees that the time standard will not be changed. The company should adhere to this policy so completely that every operator will feel free to work at whatever pace he chooses. The operator should have no fear that the time standard for the job will be reduced if he "earns too much" under a wage incentive system.

Methods Change. When there is a change in method, materials, tooling, or other factors affecting the time of the operation, the job should be restudied and a new time standard should be established. If the operator suggests a change which reduces the operation time, improves the quality, or makes the job safer, he should be compensated immediately for his suggestion. If the company has a suggestion system, he should be rewarded through the regular channels. When the new standard for the improved job has been established. the operator should find it just as easy to earn his accustomed incentive premium or bonus as he did before the method was improved. A change in method should not be used as an excuse to reduce a time standard. If management expects to get and maintain the cooperation of its employees, it must make certain that the employee gains and does not lose as a result of his suggestions.

Auditing of Methods, Time Standards, and Wage Incentive Plans. Auditing is a procedure to determine how well standard policies and techniques are being followed. Ordinarily the audit report would include a statement of variations from standard practice, an evaluation of these variations, and recommendations for changes and corrections. The design of a work measurement and wage incentive plan should provide for a periodic audit. A wage incentive program is based upon carefully established policies and procedures, and unless these procedures are followed precisely, the plan may soon become unfair to management or to the employees, depending upon the nature of the error. In most cases the error favors the employee, and management consequently stands to gain most by thorough auditing procedures. If a wage incentive plan is not carefully maintained, the earnings of some employees may get out of line, that is, may become too high as compared with those of other employees. This inequity is undesirable both to management and to the employees. Management wants to be certain that year in and year out its standards will be maintained, and the employees who are on incentive want to be assured that the plan will continue on a fair and sound basis. A good auditing program can contribute much to achieving these ends.

One well-known company has managed its work measurement and wage incentive program so well that the average performance index of some 9000 employees who are on incentive has not varied more than plus or minus 3% in the past 25 years. The average current performance index for all employees on incentive is 132%. Some 25 years ago this company started the plan of computing average performance index every 3 months, by operators, by departments, by divisions, and company-wide. In connection with this analysis, the company plots a distribution curve (Fig. 248) showing the performance index of workers on incentive by departments, and superimposes on this the average performance index for all incentive employees for the past 10 years. This company places great emphasis on auditing, and uses the most experienced and most competent industrial engineers for conducting their audits.

The following schedule for auditing methods, time standards, and wage incentives is used by the Procter and Gamble Company.[2]

[2] Richard A. Forberg, "Effective Control of the Industrial Engineering Function," *Proceedings Twelfth Management Engineering Conference*, SAM-ASME, New York, April, 1957, p. 214.

Hours per Year	Frequency of Audit
0–10	3 years
10–50	2 years
50–600	1 year
Over 600	Twice per year

This means that an operation which is not performed more than a total of 10 hours per year would be audited at least once every 3 years, whereas an operation performed more than 600 hours per year would

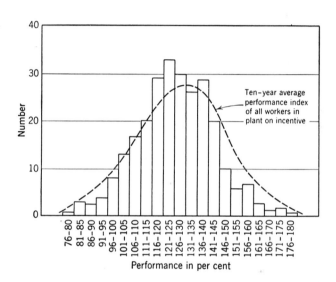

Fig. 248. Distribution of performance index of 237 workers on incentive in Department 24 during 3-month period ending March 31. Superimposed is a distribution curve of the 10-year performance index of all workers in the plant on incentive.

be audited at least twice each year. Procter and Gamble adheres strictly to this schedule, for the company considers it more important to maintain present standards than to extend standards to new operations.

Time Study as a Staff Activity. The time study department is a staff department and not a line or operating agency (Fig. 249). It is important that every industrial engineer keep this fact fixed firmly in his mind. Since a staff department must work through the foreman and supervisory groups, it is important that the line personnel be thoroughly acquainted with principles, techniques, and methods of the time study department. Foremen should be so well acquainted with time study that they can explain to an operator in the department how a time study is made, what elements are included in the operation, and exactly how the time standard for an operation is determined. The foreman should be able to do this without having to call on the time study department for help. In unusual situations, of course, the foreman may have to ask the time study department for additional information, or in some cases it might be wise for the time study analyst to supplement the information given to the operators by the foreman.

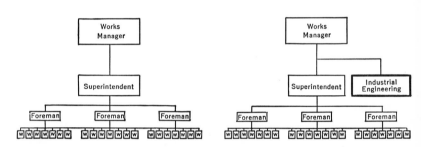

Fig. 249. Organization charts: *A*, typical line organization; *B*, line and staff organization, showing typical staff department.

Sometimes it is helpful to draw a parallel between two unrelated activities in order to clarify a situation. Guy J. Bates of General Motors has used the following analogy. When the number of rejects in a manufacturing department suddenly increases and the operator on the machine states that the cause is an inspection gauge that is out of adjustment, the foreman will ordinarily examine the gauge to see if anything is obviously wrong, and will measure some parts himself to make certain that the operator is using the gauge properly. If no cause for the difficulty can be found, the foreman will call the head of the inspection department and have him check the gauge. The foreman may be present while the check is being made, but the foreman will expect the chief inspector to make any measurements that seem necessary to determine whether the excessive number of rejects is due to a faulty gauge.

So with time study work—just as the inspector and the tool department build, service, and check all gauges and inspection devices, so the time study department sets all time standards and maintains them. If an operator complains that the time standard is too low and that he cannot earn a premium, the foreman will be expected to check the operation against the instruction sheet to see whether the operation is being performed according to the prescribed method. This involves checking the materials, speeds, and feeds of the machine, and other job conditions. If, after checking these things, the foreman is unable to find a cause for the apparent error in the standard, he will request that a time study analyst be sent to the department to check the standard. The foreman may or may not remain with the time study analyst while the check study is being made, but he will certainly follow in detail the checking procedure and will know the cause of the difficulty and the means that are finally used to correct the situation.

As for all staff functions in the plant, it is necessary to have a careful balance between the duties of the industrial engineering department and line departments. Although the time study department is definitely responsible for establishing and maintaining time standards in a plant, the industrial engineer works through the foreman and does not replace the foreman. An indifferent, lazy, or antagonistic foreman may not give wholehearted cooperation. In such situations, this attitude is a challenge to the time study department and to top management to show the foreman why it is to his advantage, to the advantage of the company, and to the advantage of the operators in his department to understand time study and to carry out the procedures established by the organization with regard to the administration of time study and wage incentives.

CHAPTER 26

Mechanized Time Study and Electronic Data Processing

Stop-watch time study is often too costly and too time consuming for measuring long-cycle operations, complex activities of groups of men and machines, and highly mechanized equipment. All too often in the past no attempt has been made to measure such work, because time study was not satisfactory for the job. Today, however, work sampling and the time-lapse camera are being used to an increasing extent for studying such operations. Also, a new electronic time recorder promises to make these studies easily possible, and may point the way to mechanized time study itself.

Work Element Timer and Recorder for Automatic Computing. R. R. Donnelley and Sons Co., working with the International Business Machines Co., developed special equipment (Fig. 250) which they call WETARFAC (Work Element Timer and Recorder for Automatic Computing), especially for studying long-cycle operations in their factories.[1] This equipment is a console unit containing standardized IBM components, designed to record on a five-channel punched paper tape the time study data fed to it. An electric typewriter mounted on top of the machine makes a copy of all data for visual examination while the time study is in progress.

This machine contains a manual data panel consisting of 27 ten-position rotary dial switches. Each rotary switch is capable of entering a single digit from 0 to 9. By utilizing a number code system all the constant data, such as date, year, shift, operation number, study number, machine number, and code group, may be read into the

---

[1] Charles W. Lake, Jr., "Automeasurement—Mechanized Time Study," *Proceedings Ninth Industrial Engineering Institute,* University of California, Los Angeles-Berkeley, February, 1957; also, Gordon R. Ewing, "Automeasurement," *Proceedings Seventh Annual Conference of the American Institute of Industrial Engineers,* Washington, D. C., pp. 12-1 to 12-10, May, 1956.

recorder. The keyboard contains six rows, with nine buttons in each row. The first three columns are used to enter the element code, the next two columns to enter frequency or subcodes, and the last column to enter the rating factor. At the end of each element the observer presses the element button.

The recorder will produce on the printer and in the tape the element

Fig. 250. WETARFAC being used to make a time study.

code, the item frequency or subcode, the rating factor, and the time at which the element is completed. Time is measured to the nearest hundredth of a minute. If an element is observed and for some reason the analyst desires this entry flagged or disregarded, the button marked X may be depressed. This enters the data in the printer and in the perforated tape, but flags the entry for special automatic handling.

After the time study is completed, the five-channel tape is removed and sent to the tabulating department. There a tape-to-card converter is used to produce one punched card for each element of the study. Each card contains the constant data information, the element code, the time for the element, and the rating factor. With the time study data on punched cards, it is easily possible for the IBM computer to determine the element mean, the standard deviation, the coefficient of

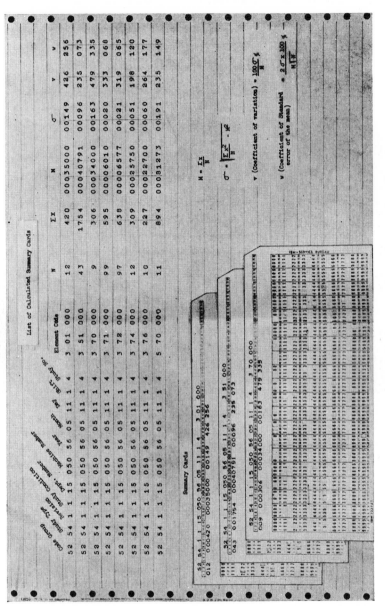

Fig. 251. Time study summary sheet as it comes from the IBM Tabulating Department.

standard error of the mean, and the coefficient of variation. In other words all the desired information about each element can be presented in statistical terms, as shown in Fig. 251, and this information can be summarized on a typed sheet of paper by an IBM automatic printer.

Not only can the time study machine increase the effectiveness of

Fig. 252. Frequency and time recorder which automatically gathers information about operating time and down time of large automatic stitchers.

the time study observer and transfer the task of making tedious calculations to the IBM computer; in some cases it can be connected directly to production machines and process equipment, eliminating the time study observer entirely.

Automatic Frequency and Time Recorder. The frequency and time recorder shown in Fig. 252 was designed by Donnelley engineers to reduce the number of man-hours required to gather time study data. This recorder is tied into one of the large automatic stitchers. It automatically compiles the following data: total running time, total down time, nonbonus waiting time, frequency of nonbonus waiting time, total delay time, frequency of delays, total time for each of 27 types of delays, frequency for each of 27 types of delays, running speed, and time of speed changes.

Automeasurement. The people at Donnelley believe that time study and production recording must be automated. They have coined the term *automeasurement*, which they define as a system for the automatic collection and processing of data for the purpose of measuring productivity in a statistically scientific manner.

It seems apparent that equipment such as the WETARFAC will find many uses in industry. With machine speeds, down time, yield and quality of output becoming so important in the operation of large and expensive machines and equipment, it is not difficult to visualize automatic time recorders connected directly to machines or production lines in many different industries. Such machines could provide a continuous record of activities, with information about interruptions and facts concerning the steps that were used to correct them. This would indeed provide management with control information the like of which it does not have at the present time.

Electronic Data Processing for Determining Time Standards from Standard Data. Some companies have available in their files standard data to cover many different activities performed in the factory and office. When a new job is to be put into production, a time standard can be established for the job by reference to the file of standard data, but the task of applying the standard elemental data is tedious and time-consuming. The Pratt and Whitney Aircraft Division, United Aircraft Corporation, is now using an electronic data-processing machine (IBM 702) to compute time standards for inspection operations and machining operations.[2]

Starting with the inspection time study standard data manual of tabular time values, a program containing approximately 3000 individual instructions, together with the tabular time values, occupies approximately 25,000 memory positions in the computer. Figure 253 shows the organization of the data processing and the general procedure for establishing inspection time standards. The use of the computer makes it possible to provide information that was considered

---

[2] Joseph Motycka, "New Advances in Time Study—Electronic Data Processing," *Proceedings National Conference American Institute of Industrial Engineers,* Detroit, pp. 159–163, May, 1961; Joseph Motycka and Travers Auburn "Electronic Data Processing Comes to Time Study," *Journal of Industrial Engineering,* Vol. 8, No. 1, pp. 11–18, January–February, 1957.

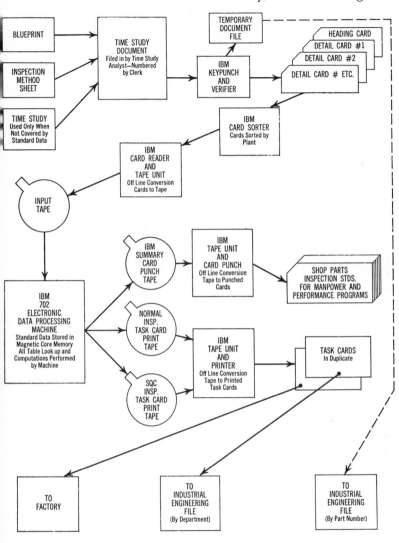

Fig. 253. Organization of inspection standards for electronic data processing.

impractical to get by the conventional method. For example, the setup or preparation for the job is performed by the inspector, and this is a part of the regular operation. Using the computer, the setup time can be incorporated into the standard, irrespective of the lot size, without requiring extra computer time.

The automatic data processing has reduced the time to calculate time standards from standard data by two thirds. In addition to applying the data, the computer can be used to calculate the machine time. Given the dimensions of the piece and the type of cut, the computer, making use of stored data such as the material and the feeds and speeds, can determine the standard time for the operation for any type of machine tool. Thus, the tedious routine tasks formerly performed by the analyst will now be taken over by the machine, releasing the person for more productive assignments.

An Advanced EDP System for Determining Methods and Establishing Time Standards. The Service Bureau Corporation, a subsidiary of the International Business Machine Corporation, has recently developed a data-processing system designed to assist the industrial engineer in establishing methods and setting time standards for a wide variety of manufacturing operations.[3] The SBC has named this system AUTORATE.[4]

The AUTORATE system consists of computer programs incorporating two master files in which information for future processing is stored. The first file, designated as the code file, consists of codes representing elementary manual motions, that is, a descriptive phrase and associated time value for each code. Any system of predetermined motion-time data can be used in compiling the code file. Thus the industrial engineer would merely submit a sequential list of coded data elements and their frequencies. The computer would make a complete analysis of the operation, containing descriptive phrases for each element and the corresponding standard time. This information is printed directly on a special form designed for this purpose.

The second master file of the AUTORATE system is designated as the rate file. It consists of a summary record of each method and time analysis established for each required operation; they are automatically stored in the rate file as they are generated. The existence of the rate file permits the regeneration of any previously established methods and time analysis on command. The rate file's principal function is to enable the industrial engineer to make changes in previously established standards with a minimum of effort. The engineer merely

---

[3] Reproduced with permission from "A Report on the Autorate System," The Service Bureau Corporation, New York, 1962. Also see "Now You Can Set Standards with a Computer," *Mill & Factory,* Vol. 71, No. 5, pp. 68–69, November, 1962.

[4] A service mark of The Service Bureau Corporation.

T	PART NO.	OPER.	SUB 0	REV.	SUB 0 REV.	SEQ.	REV. CAT.	PFD. % U	MACH. GROUP	E.C. NO.	DEPT.	DEPT. RESP.	ANAL.	APP.	D R	DATE	SET UP TIME	CARD COUNT	TEAR DN. TIME	TRAN. TIME	NEW OPERATION IDENTIFICATION			DEV. TIME
																					OPER.	SUB. 0 PLT.		
																					PART NO.	SERIAL NO.		
MN	987654	0015	00 P	00		bbbA	01	15 H	1974	69696	333	110	9	13		05 17 62								:7

DUPLICATE ON ALL CARDS

	MACH. GROUP	ALT. MACH. GROUP 1	ALT. MACH. GROUP 2	ALT. MACH. GROUP 3	N	PROD. TIME
bbbB	R	1974	1972			

OPERATION DESCRIPTION

bbbC	MILL .250 DIM. TO PRINT
bbbD	
bbbE	

REMARKS

bbbF	INSP. CHARACTERISTICS		CK	FEED	PL	INSP. GAGE
bbbG	CK 13/16 DIM.		1	1/10	WH	SCALE
bbbH	CK .250 DIM.		2	1/10	WH	MICS
bbbZ						

M/E CD	CODE	FREQUENCY	M	P	S	C	DESCRIPTION
1.0	MGLC01	1.0					(F/TL. NO. 24242A)
2.0	MFC500	1.0					
3.0	MFNW05	1.0					
4.0	MALM01	1.0	5				START MACHINE
5.0	MGUC01	1.0					
6.0	MCA501	1.0				C	OF FIXTURE
7.0	I;IMIGS01	.1				C	13/16
8.0	I;IMIGM01	.2				C	.250 DIM.
9.0	FORM0001	1974.					A/1.450, B/1.250, C/1.350, D/135, E/5,
9.0	FORM0001		1				F/.003, 6/26, H/8, I.1.17

NOTES

T—TYPE CODE

MN—NEW OPERATION
MR—REVISION TO EXISTING
   OPERATION

O—OPERATION CODE

P—PRODUCTION OPERATION

S—SET UP OPERATION
T—TEAR DOWN OPERATION

U—UNIT CODE

U—RATE GIVEN IN HR.
H— HR./100
T— HR./1,000
D— HR./10,000

D—CONDENSE CODE

B—PRINT NORMAL MTA AND
   CONDENSED MA
N—PRINT ONLY CONDENSED MA
OTHER—PRINT ONLY NORMAL MTA

R—REQUEST FIELD

P—DO NOT PUNCH STANDARD OR
   MIP
E—DO NOT PUNCH MIP
OTHER—PUNCH BOTH MIP AND
   STANDARD

E—ESTIMATE CODE

F—REGULAR
P—PROVISIONAL
R—TEMPORARY
C—CHECK LIST

Fig. 254. Input for a new time standard for a typical machining operation.

PRIME MFCT. CORP.   DATE 05-17-01
                                                       PAGE 9

TEMPORARY METHOD ANALYSIS

PART NO. 987654		E.C. NO. 69696
OPER. NO. 0015-00-P		ANALYST 9
DEPT. NO. 333		APPROVAL 13
MACH. GRP. 1974 ALT. MACH. GRP. 1972		RESP. DEPT. 110
REV. NO. 00		

MILL .250 DIM. TO PRINT

MACHINE SETTINGS AND DATA

CUTTER DIAMETER	5
NO. TEETH IN CUTTER	26
TRAVEL	3.050
AVAIL. R.P.M.	95
AVAIL. FEED IN./MIN.	7.0

FIXTURES

TL. NO. 24242A

REMARKS

INSP. CHARACTERISTICS	CK.	FEED	PL	INSP. GAUGE
CK. $13\!\!/\!_{16}$ DIM.	1	$\frac{1}{10}$	WH	SCALE
CK. .250 DIM.	2	$\frac{1}{10}$	WH	MICS

SEQUENCE OF OPERATIONS

SEQ.	DESCRIPTION	CLAIM
1.0	PLACE PART IN JIG T/TL. NO. 24242A	
2.0	FASTEN AND UNFASTEN SWING TYPE CLAMP	
3.0	FASTEN NUT WITH OPEN END OR BOX WRENCH	
4.0	START MACHINE	
5.0	UNLOAD JIG	
6.0	BLOW OFF SURFACE OF FIXTURE	
7.0	CHECK DIM. WITH SCALE $13\!\!/\!_{16}$ DIM.	
8.0	CHECK DIM. WITH MICROMETER .250 DIM	
9.0	A/APP, B/CLR, C/LOC, D/SFM, E/CD, F/CLT, G/NT, H/ADV, I/TF, A/1.450, B/.250, C/1.350, D/135, E/5, F/.003, G/26, H/8, I/.17	

PRIME MFCT. CORP.   DATE 05-17-01
                                                       PAGE 9

TEMPORARY TIME ANALYSIS

PART NO. 987654		E.C. NO. 69696
OPER. NO. 0015-00-P		ANALYST 9
DEPT. NO. 333		APPROVAL 13
MACH. GRP. 1974		RESP. DEPT. 110

SEQUENCE OF OPERATIONS

SEQ.	CODE	TMU	FREQ.	HR./100	CUM. TOTAL
1.0	MGLC01	64	1.0000	.0640	.0640
2.0	MFCS00	30	1.0000	.0300	.0940
3.0	MFNW05	210	1.0000	.2100	.3040
4.0	MALM01	26	1.0000	.0260	.3300
5.0	MGUC01	42	1.0000	.0420	.3720
6.0	MCAS01	71	1.0000	.0710	.4430
7.0	11MIGS01	187	.1000	.0187	.4430
8.0	11MIGM01	173	.1000	.0173	.4430
9.0	FORM0001 MACHINE TIME			.1848	2.2278
	TOOL ALLOWANCE			.1235	2.3513
	PFD — 15 PCT			.3527	2.7040
	***** STD. TIME—HR./100				2.704

Fig. 255. The methods and time analysis report prepared from the input.

submits changes or revisions in coded form, and the computer will produce a complete, updated version ready for immediate use. Each revision and change of an existing standard is of course immediately recorded in the rate file.

Figure 254 shows the input for a new time standard for a typical machining operation, and Fig. 255 is the analysis report prepared from the input.

CHAPTER 27

Determining Time Standards from Elemental Time Data and Formulas

Many time studies are made of a single operation, with little or no thought that the data taken will be of value on any other operation. Some kinds of work, however, have certain elements which are alike. For example, in a given class of machine-tool work all elements may be virtually alike except for the machine time or the cutting time. The same jig used for drilling the ¼-inch hole in the end of the shaft (Fig. 217) might also be used for drilling many other sizes of shafts. If the length and the diameter of the shafts fall within a limited range, handling time for drilling all shafts would be practically constant, and the only variable in the operation would be the time required to drill the hole, which would vary with its diameter and depth. Other operations using jigs similar to this one would have certain elements in common, such as "tighten set screw" or "lower drill to work."

Where motion and time studies are to be made of many different operations of a similar class of work, such as that on sensitive drill presses, lathes, and gear hobbers, it is best to consider the entire class of work as a unit, working out such improvements in methods as seem advisable, and standardizing all factors for the entire class of work. When time studies are begun on this work, the elements should be selected in a way that will make it possible eventually to construct tables of standard time data that may be applied to all elements likely to appear continually in that particular class of work.

Use of Time Values for Constant Elements. The data shown in Tables 19 to 21 were obtained from a sufficient number of time studies of representative kinds of work to guarantee their being reliable. With such data available in the time study department, it is possible to set

428

time standards for the handling elements of any job on a sensitive drill falling within the classes listed in Tables 20 and 21. These data do not give the time required to drill the hole in the piece; consequently this information must be obtained by means of a time study of this element.

Assuming that Tables 19, 20, and 21 were available and that it was necessary to determine the standard time to drill the ¼-inch hole in the end of the shaft (Fig. 217), the procedure would be as follows:

Chuck and remove piece (from Table 20)	0.50
(Class B, work held by set screw)	
Machine manipulation (from Table 21)	0.07
(Class A, drilling, one drill and no bushing)	
DRILL ¼-INCH HOLE	
(stop-watch data obtained as in Fig. 217)	0.54
Total normal time per piece	1.11
5% allowance	0.06
Total standard time per piece	1.17 minutes

Setup time (from Table 19) = 15.00 minutes

The value of standard time data such as those illustrated is evident. They reduce the number of time studies needed, shorten the time

Table 19. Time-Setting Data for Sensitive Drills
Setup Time

Description of Work	Time, Minutes
1. Small work held in jig which can be handled very easily by hand	15.00
2. Small work held in vise	15.00
3. Small work held to table by one or two straps	15.00
4. Small work held in jig having a number of drilled, tapped, and reamed holes	30.00
5. Small work held in jig and jig held in vise	30.00
6. Work of medium size held by one or two straps	30.00
7. Work of medium size prevented from turning on table by a stop in T-slot	15.00
8. Work of circular type such as washers, collars, bushings, and sleeves held to table by a draw bolt through center	15.00

Table 20. Elemental Time Data for Sensitive Drills
Chucking and Removing Time

Classes, work held in jig:
  - A. Held by thumb screw
  - B. Held by set screw
  - C. Held by thumb and set screw
  - D. Held by cover strap and thumb screw
  - E. Held by cover strap and set screw
  - F. Held by cover strap, thumb screw, and set screw

Elements	Time, Hundredths of a Minute					
	A	B	C	D	E	F
1. Pick up piece and place in jig	12	12	12	12	12	12
2. Swing cover strap and tighten lock screw	..	..	..	10	10	10
3. Tighten thumb screw	08	..	08	08	..	08
4. Tighten set screw	..	12	12	..	12	12
5. Loosen set screw	..	06	06	..	06	06
6. Loosen thumb screw	05	..	05	05	..	05
7. Swing cover strap back and loosen lock screw	..	..	..	08	08	08
8. Remove piece from jig	08	08	08	08	08	08
9. Blow out chips	12	12	12	12	12	12
Total	45	50	63	63	68	81

Note. Add 0.32 minute when jig is strapped to table. Add 0.07 minute for each additional thumb screw. Add 0.08 minute for each additional set screw.

required to set the standard, and tend to bring greater accuracy and uniformity in time standards for a given class of work.

The next step in this direction is the preparation of formulas which will make it possible to calculate quickly time values for the machine elements. Thus, with standard time data for the handling elements and calculated time values for the machine elements, it is possible to determine the time standard for a given operation without the necessity of making a time study. By using this procedure the time standard can readily be determined in advance of the actual production of the part. In this case a detailed drawing of the part to be made and the operation sheet or the route sheet should be supplied to the time study department in advance.

Determining Time Standards for Variables. On all kinds of machine-tool work the time for manipulating the machine and for chucking and removing the piece is likely to remain constant for each element, provided the size and shape of the piece are within reasonably close limits. The time for making the cut is the variable. This machine time can often be calculated, particularly when positive power feeds are used. For example, in milling-machine work with power feed, if the feed of the table in inches per revolution of the cutter is known, and if the speed of the cutter in revolutions per minute is known, it is a simple arithmetical problem to find the time required to mill a piece of a given length. Allowances must be added to the length of the piece for the approach and for the overtravel of the cutter; however, they can also be calculated easily. In a similar manner, if a shaft of a given

Table 21. Elemental Time Data for Sensitive Drills
Machine Manipulation Time

Classes:

A. Drilling, one drill and no bushing
B. Drilling, placing and removing bushing
C. Drilling, placing and removing drill
D. Drilling, placing and removing drill and bushing

Elements	Time, Hundredths of a Minute			
	A	B	C	D
1. Place bushing in jig	..	06	..	06
2. Place drill in chuck	..	..	04	04
3. Advance drill to work	04	04	04	04
4. Raise drill from hole	03	03	03	03
5. Remove bushing from jig	..	05	..	05
6. Remove drill from chuck	..	..	03	03
Total	07	18	14	25

Note. Add 0.15 minute when quick-change chuck is not used (cases *B* and *C*). Add 0.06 minute for advancing work to next spindle. Add 0.05 minute when reamer is oiled before entering hole.

length is chucked in a lathe and if the speed and the feed are known, it is a simple matter to calculate the length of time required to make a cut across the piece. Therefore, on machine tool work the handling time (a constant) plus the machine time (a variable) plus allowances will equal the standard time for the performance of a given operation.

In the operation of soldering the side seam in making rectangular cans (see Chapter 29) the principal variable is the element "solder the full length of the seam." The time for this element varies directly as the length of the seam.

SETTING TIME STANDARDS FOR MILLING SQUARE OR HEXAGON ON BOLTS, SCREWS, OR SHAFTS

With the aid of the following four tables it is possible to determine the time standard for setting up a milling machine and for milling a square or hexagon on the end of bolts, screws, or shafts. The data in Tables 22 and 23 were determined from stop-watch time studies. A sufficient number of representative jobs were studied to give reliable data. Tables 24 and 25 were compiled to facilitate the determination of the time standard for a given operation. Typical examples at the bottom of these two tables show how they are used.

There are two methods of milling squares and hexagons: (1) using a single mill, which requires a separate cut for each side (use Table 24), and (2) using a gang mill, which cuts two sides at a time (use Table 25).

Computation of Data for Table 24—Milling by Use of 6-Lip Mill. This milling operation is performed with a single milling cutter; hence four cuts are required to mill a square and six cuts to mill a hexagon. The dimension B (see sketches in Tables 24 and 25) is given as the turned diameter of the shaft rather than the width of the face to be cut, for the reason that detailed drawings (Fig. 256) are dimensioned in that manner. Since the side of a square is equal to 0.7071 times the diameter of the circumscribed circle, and since the side of a hexagon is equal to the radius of the circumscribed circle, it is easy to make the conversion.

When a single mill is used, the cut is made across the face, and the time for the cut varies as B and is independent of the dimension A, provided it falls within the scope of the data, that is, within $\frac{5}{8}$ inch to $1\frac{3}{4}$ inch. The total handling time (HT) plus the total cutting time (M) plus the allowances equal the total time for the operation.

The handling time is composed of machine manipulation time and

Table 22. Time-Setting Data for Milling Machines
Setup Time—Machine Class 36

1. Base Setup Times—Minutes

Type		Work Sizes		
		Small	Medium	Large
A.	Strapped to table or angle plate (4 straps)	25	25	25
B.	Held in vise..........................	25	25	25
C.	Held in 2 vises.......................	..	30	30
D.	Vise with false jaws...................	35	35	35
E.	Held in fixture.......................	35	45	60
F.	Held in dividing head chuck............	35	35	..
G.	Held in dividing head and tail stock......	45	45	..

2. Additional Parts Used—Time to Be Added to Base Setup Time

Part		Part Sizes		
		Small	Medium	Large
H.	Each additional strap.................	5	5	5
J.	Angle plate...........................	10	15	15
K.	Gang mills			
	(1) fractional limits..................	10	10	10
	(2) decimal limits....................	15	15	15
L.	False table...........................	10	15	20
M.	Round table—hand feed...............	10	20	20
N.	Round table—power feed..............	20	30	30
P.	High-speed head......................	..	40	..
Q.	Universal head.......................	..	60	..

Table 23. Elemental Time Data for Milling Square or Hexagon on Bolts, Screws, or Shafts

Machine Class 36—Milling Table 1A

1. *Setup time*—Complete—See setup table
        Change size — End mill   — 10 min.
                 Gang mills — 20 min.

2. *Specifications*—
  A. Method of chucking—1—3-jaw chuck (small screws and bolts)
                        2—Held on centers (shafts)
                        3—Thread arbor in dividing head (small pieces with thread on or in end)
  B. Cutters—1—6-lip mill
            2—Gang mills (6-in. stagger tooth—side milling cutters)
  C. Length of flat   — ⅝ in.–1¾ in.
  D. Size of sq. or hex. — ½ in.–1⅝ in.
  E. Number of cuts   — 1 per side
  F. Material   — SAE2315
  G Indexing   — Use rapid index plate wherever possible

3. *Operation Time.*—Machine manipulation and chucking time

Time in Minutes

		Method of Chucking									
		1				2		3			
Elements	Speed r.p.m.	Square		Hex.		Sq.	Hex.	Square		Hex	
		214	58	214	58	214	214	214	58	214	58
	Type of Mill	6-Lip Mill	Gang Mill	6-Lip Mill	Gang Mill	6-Lip Mill	6-Lip Mill	6-Lip Mill	Gang Mill	6-Lip Mill	Gang Mill
1. Stop machine..		0.04	0.04	0.04	0.04	0.04	0.04	0.04	0.04	0.04	0.04
2. Loosen dog in holder......		....	....	....	....	0.08	0.08	....	....	....	....
3. Loosen center..		....	....	....	....	0.08	0.08	....	....	....	....
4. Loosen work..		0.04	0.04	0.04	0.04	....	....	0.08	0.08	0.08	0.08
5. Remove work..		0.06	0.06	0.06	0.06	0.10	0.10	0.08	0.08	0.08	0.08
6. Remove dog...		....	....	....	....	0.08	0.08	....	....	....	....
7. Place dog.....		....	....	....	....	0.08	0.08	....	....	....	....
8. Clear chips....		....	....	....	....	0.05	0.05	0.08	0.08	0.08	0.08
9. Place piece....		0.08	0.08	0.08	0.08	0.08	0.08	0.12	0.12	0.12	0.12
10. Tighten.......		0.12	0.12	0.12	0.12	0.18	0.18	0.10	0.10	0.10	0.10
11. Start machine..		0.02	0.02	0.02	0.02	0.02	0.02	0.02	0.02	0.02	0.02
12. Advance to cut		0.04	0.06	0.04	0.06	0.04	0.04	0.04	0.06	0.04	0.06
13. Change depth.		*	*	*	*	*	*	*	*	*	*
14. Mill..........	2⅞† or 3⅝	M	M	M	M	M	M	M	M	M	M
15. Index‡.......		0.15	0.05	0.25	0.10	0.15	0.25	0.15	0.05	0.25	0.10
16. Return table...		0.05	0.07	0.05	0.07	0.05	0.05	0.05	0.07	0.05	0.07
TOTALS....		0.60	0.54	0.70	0.59	1.03	1.13	0.76	0.70	0.86	0.75

* Allow 0.08 when necessary.
† Feed in inches per minute—depending upon finish required.
‡ Above time is for rapid indexing. (Double indexing time when using precision crank.)

$$M = \text{Cutting time} = \frac{(L + OT) \times \text{No. cuts}}{\text{Feed}} \qquad \text{Base time} = H.T. + M$$
$$\text{Standard time} = \text{Base time} + \text{Allowances}$$

Table 24. Time-Setting Table for Milling Square and Hexagon on Bolts, Screws, and Shafts

Machine Class 36—Milling Table 1B

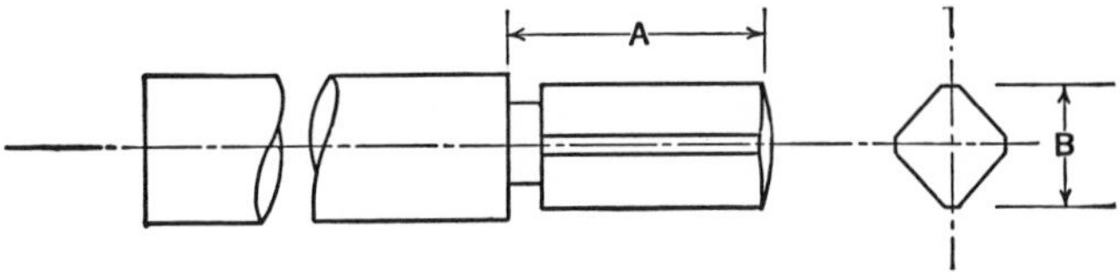

Case 1—Using a 6-Lip Mill (See Table 1*C* for Gang Mills)

Time per Piece in Minutes

Symbol	*A* (See Sketch Above)	*B* (See Sketch Above)	3-Jaw Chuck				On Centers				On Thread Arbor in Div. Head			
			Square		Hex.		Square		Hex.		Square		Hex.	
			Feed 2⅞	Feed 3⅝	Feed 2⅞	Feed 3⅝	Feed 2⅞	Feed 3⅝	Feed 2⅞	Feed 3⅝	Feed 2⅞	Feed 3⅝	Feed 2⅞	Feed 3⅝
			C	*D*	*E*	*F*	*G*	*H*	*J*	*K*	*L*	*M*	*N*	*O*
1	⅝ in. to 1¾ in.	½	2.4	2.1	3.2	2.7	2.9	2.5	3.7	3.2	2.6	2.3	3.4	2.9
2		9/16	2.5	2.1	3.3	2.8	3.0	2.6	3.7	3.2	2.7	2.3	3.5	2.9
3		⅝	2.6	2.1	3.3	2.8	3.0	2.6	3.8	3.3	2.8	2.4	3.5	3.0
4		11/16	2.6	2.2	3.5	2.9	3.1	2.7	3.9	3.4	2.8	2.4	3.6	3.1
5		¾	2.7	2.3	3.5	3.0	3.2	2.8	4.0	3.4	2.9	2.5	3.7	3.1
6		13/16	2.8	2.3	3.6	3.0	3.3	2.8	4.1	3.5	3.0	2.5	3.8	3.2
7		⅞	2.8	2.4	3.6	3.1	3.3	2.8	4.1	3.6	3.0	2.6	3.9	3.3
8		1	3.0	2.5	3.8	3.2	3.4	3.0	4.2	3.6	3.1	2.7	4.1	3.3
9		1⅛	3.1	2.6	4.0	3.3	3.5	3.0	4.4	3.7	3.3	2.8	4.2	3.4
10		1¼	3.2	2.7	4.1	3.4	3.7	3.2	4.5	3.9	3.4	2.9	4.3	3.6
11		1⅜	3.3	2.8	4.2	3.5	3.8	3.3	4.7	4.0	3.5	3.0	4.4	3.7
12		1½	3.5	2.9	4.4	3.6	4.0	3.4	4.8	4.1	3.7	3.1	4.5	3.8
13		1⅝	3.6	3.0	4.5	3.7	4.1	3.5	5.0	4.2	3.8	3.2	4.7	3.9

1. Values in this table vary as *B*, hence see that *A* falls within limits ⅝ in. to 1¾ in. before using data.
2. These time standards are based on:
   (*a*) Length of travel = *B* + overtravel
   (*b*) Handling time from Table 1*A*
   (*c*) Allowance of 5 per cent
3. Examples for reading above table:
   (*a*) Let *B* = ⅝ in., *A* = 1 in., Square head shaft, held on centers, 6-lip mill, 3⅝ in. feed. Since *A* lies between limits given this table applies. Read from table under 3-*H*, standard time = 2.6 min. per piece.
   (*b*) Let *B* = 1¼ in., *A* = 1½ in., Hexagon head bolt, held on thread arbor in div. head, 6-lip mill, 2⅞ in. feed.
   Since *A* lies between limits given this table applies. Read from table under 10-*N*, standard time = 4.3 min. per piece.

Table 25. Time-Setting Table for Milling Square and Hexagon on Bolts, Screws, and Shafts

Machine Class 36—Milling Table 1C

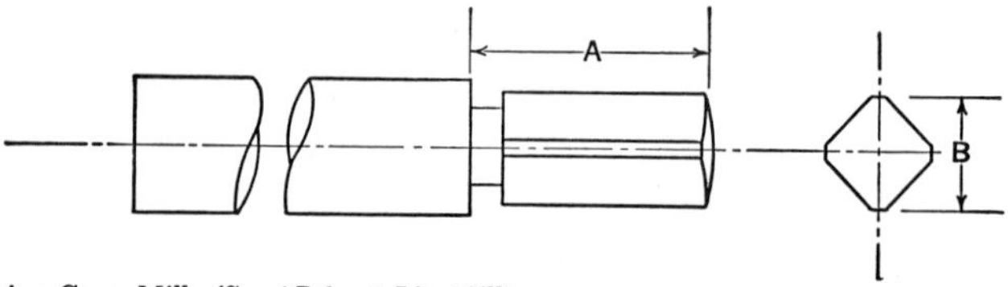

Case 2—Using Gang Mills (See 1*B* for 6-Lip Mill)

Time per Piece in Minutes

Symbol	*A* (See Sketch Above)	*B* (See Sketch Above)	3-Jaw Chuck				On Thread Arbor in Div. Head			
			Square		Hex.		Square		Hex.	
			Feed 2⅞	Feed 3⅝	Feed 2⅞	Feed 3⅝	Feed 2⅞	Feed 3⅝	Feed 2⅞	Feed 3⅝
			C	*D*	*E*	*F*	*G*	*H*	*J*	*K*
1	⅝	½ in. to 1⅝ in.	1.05	0.95	1.35	1.2	1.15	1.10	1.5	1.4
2	11/16		1.10	1.0	1.40	1.25	1.25	1.15	1.6	1.4
3	¾		1.15	1.05	1.5	1.25	1.35	1.20	1.7	1.5
4	13/16		1.20	1.05	1.6	1.30	1.35	1.25	1.7	1.5
5	⅞		1.20	1.10	1.6	1.35	1.40	1.25	1.8	1.6
6	15/16		1.25	1.15	1.7	1.45	1.45	1.30	1.8	1.7
7	1		1.35	1.20	1.8	1.50	1.50	1.35	1.9	1.7
8	1 1/16		1.35	1.20	1.9	1.5	1.6	1.35	2.0	1.8
9	1⅛		1.40	1.25	1.9	1.6	1.6	1.40	2.1	1.8
10	1 3/16		1.45	1.30	2.0	1.7	1.7	1.45	2.2	1.9
11	1¼		1.50	1.30	2.1	1.8	1.7	1.5	2.2	2.0
12	1⅜		1.60	1.40	2.2	1.9	1.8	1.6	2.4	2.1
13	1½		1.7	1.45	2.3	2.0	1.9	1.6	2.5	2.2
14	1⅝		1.8	1.6	2.5	2.1	2.0	1.7	2.6	2.3
15	1¾		1.9	1.6	2.6	2.2	2.1	1.8	2.7	2.4

1. Values in this table vary as *A*, hence see that *B* falls within limits ½ in. to 1⅝ in. before using data.
2. These time standards are based on:
   (*a*) Length of travel = *A*
   (*b*) Handling time from Table 1*A*
   (*c*) Allowance of 5 per cent
3. Examples for reading above table:
   (*a*) Let *A* = 1 in., *B* = 1 3/16 in., Square head bolt, held in chuck, gang mill, 3⅝ in. feed. Since *B* lies between limits given this table applies. Read from table under 7–*D* standard time = 1.20 min. per piece.
   (*b*) Let *A* = 11/16 in., *B* = 0.578 in., Hexagon head adjusting screw, held in chuck, gang mill, 3⅝ in. feed. Since *B* lies between limits given this table applies. Read from table under 2–*F*, standard time = 1.25 min. per piece.

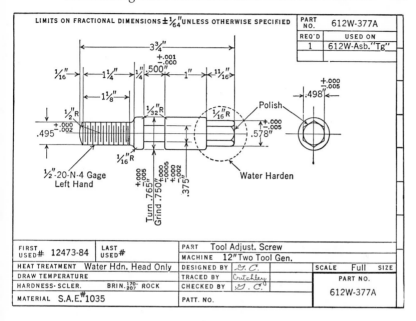

Fig. 256. Detail drawing of a tool adjusting screw, part 612W-377A.

chucking and removing time as shown in Table 23. The cutting time can be calculated from the formula

$$M = \frac{(L + OT)N}{F}$$

where M = cutting time in minutes

L = length of cut in inches

    (a) $L = 0.707 \times B$ for square

    (b) $L = 0.5 \times B$ for hexagon

OT = overtravel = $\frac{1}{2}$ diameter of mill in inches

N = number of cuts per piece

    (a) $N = 4$ for a square

    (b) $N = 6$ for a hexagon

F = table feed in inches per minute

    (a) for fine finish use $2\frac{7}{8}$ inches

    (b) for ordinary finish use $3\frac{5}{8}$ inches

EXAMPLE. Assume that the shaft at the top of Table 24 has these dimensions: $A = 1\frac{3}{4}$ inches; $B = 1$ inch. Ordinary finish (feed = $3\frac{5}{8}$ inches per minute); mill square with $1\frac{3}{4}$-inch diameter, 6-lip mill; piece held in a 3-jaw chuck.

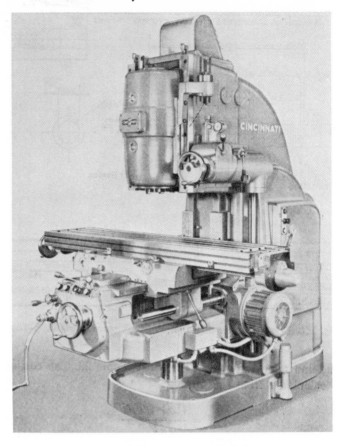

Fig. 257. Cincinnati vertical milling machine. (Courtesy of Cincinnati Milling Machine Co.)

The handling time (HT) is obtained from Table 23, under method of chucking 1, square, 6-lip mill, and is 0.60 minute.

The cutting time is calculated from the formula.

$$M = \frac{(0.707 + 0.875)4}{3.625} = \frac{6.328}{3.625} = 1.748$$

$$
\begin{aligned}
L &= 0.707 \times 1 = 0.707 \\
OT &= \tfrac{1}{2} \text{ of } 1\tfrac{3}{4} = \tfrac{7}{8} \\
N &= 4 \\
F &= 3.625
\end{aligned}
$$

$$
\begin{aligned}
HT &= 0.60 \\
M &= 1.748 \\
\hline
\text{Total normal time} &= 2.348 \\
5\% \text{ allowance} &= 0.117 \\
\hline
\text{Total standard time} &= 2.465, \text{ use } 2.5 \text{ minutes}
\end{aligned}
$$

Now with reference to Table 24, since a single cutter is used and since A lies between the limits given, this table applies. Reading under symbol 8-D, the standard time equals 2.5 minutes, which checks with that calculated above.

Computation of Data for Table 25—Milling by Use of Gang Mill.

This milling operation is performed with a gang milling cutter; hence two sides are cut at one time, two cuts being required for a square and three cuts for a hexagon. The direction of travel of the mill in making the cut is from the end toward the shoulder; hence the time for the cut varies as A and is independent of B, provided it falls within the scope of the data, that is, provided the diameter of the shaft is between $\frac{1}{2}$ inch and $1\frac{5}{8}$ inches.

The total standard time for the operation is equal to the handling time (HT) plus the cutting time (M) plus the allowances. The handling time is taken directly from Table 23, and the cutting time can be calculated from the formula

$$M = \frac{(L + OT)N}{F}$$

where M = cutting time in minutes
$\quad\quad L$ = length of cut = A
$\quad OT$ = overtravel—since the direction of travel of the gang mill is from the end to the shoulder, no overtravel is allowed.
$\quad\quad N$ = number of cuts per piece
$\quad\quad\quad$ (a) N = 2 for a square
$\quad\quad\quad$ (b) N = 3 for a hexagon
$\quad\quad F$ = table feed in inches per minute

EXAMPLE. The operation is that of milling the hexagon on the tool adjusting screw shown in Fig. 256. The following information is taken from the drawing (Fig. 256) and from the operation sheet (not shown): $A = 1\frac{1}{16}$ inch, $B = 0.578$ inch. Ordinary finish, $3\frac{5}{8}$ inches per minute feed, 7-inch gang mill, piece held in 3-jaw chuck.

Handling time from Table 23 = 0.59 minute.

Cutting time is calculated from the formula.

$$M = \frac{(0.6875 + 0)3}{3.625} = 0.569 \text{ minute} \qquad
\begin{aligned}
L &= 1\frac{1}{16} = 0.6875 \\
OT &= 0 \\
N &= 3 \\
F &= 3\frac{5}{8} = 3.625
\end{aligned}$$

Therefore the total time for the operation is

$$HT = 0.59$$
$$M = \underline{0.569}$$
$$\text{Total normal time} = 1.159$$
$$5\% \text{ allowance} = \underline{0.058}$$
$$\text{Total standard time} = 1.217, \text{ use } 1.25 \text{ minutes}$$

Now with reference to Table 25, since a gang mill is used and since B lies between the limits given, this table applies. Reading under symbol 2-F, the standard time equals 1.25 minutes, which checks with that calculated above.

CHAPTER 28

Use of Elemental Time Data and Formulas: Gear Hobbing, Soldering Cans

TIME STANDARDS FOR GEAR HOBBING

The following example of the use of elemental time data and formulas for setting time standards for gear hobbing demonstrates how the principles already explained may be applied to rather complicated work. The data and procedure given here have been in constant use in a well-known machine-tool plant for a number of years and still serve their purpose satisfactorily.

Although the data apply to the cutting of both straight and helical spur gears, only those pertaining to cutting straight spur gears will be given here. These data are applicable to straight spur gears varying from 6 to 32 diametral pitch, of steel or cast iron, and with round or spline bore. Barber-Colman hobbers, such as that shown in Fig. 258, were used. The lot sizes of the gear blanks were small.

The following explanations concerning Tables 26, 27, and 28 may make them more easily understood.

Table 26. Handling Time—Machine Manipulation. The time for machine manipulation will depend upon the method used in cutting the gears. Three different methods are shown for spur gears. The time required to chuck and remove the gear (Table 27) is independent of the method of cutting.

Table 27. Handling Time—Chucking and Removing. The data show that the time needed to chuck and remove the gears varies with the different types of mounting.

Table 28. Approach Allowance. The approach allowance required for hobbing is determined in the same manner as for milling. It is affected by the diameter of the hob and the depth of the cut. The

Fig. 258. Barber-Colman gear hobber.

⅛-inch allowance for finish hobbing is sufficient for clearance at the beginning and at the end of the cut.

Cutting Time Formula

$$M = \frac{N \times L}{F \times S \times H}$$

where M = cutting time in minutes
N = number of teeth
L = total length of cut (length of face plus approach allowance)
F = feed in inches per revolution of work
S = speed of hob in revolutions per minute
H = lead of hob
 (a) single = 1
 (b) double = 2

Hobbing is a continuous cutting action from the start to the finish of the travel of the hob across the entire gear face. One revolution of the work advances the hob a distance equal to the feed.

Table 26. Handling Time—Straight Spur Gears,
Barber-Colman Hobber
Machine Manipulation
(Time in minutes)

Operation	Spur Gears			Helical Gears	
	Reg. Hob 1 Cut	Reg. Hob 2 Cuts	Rough and Finish Comb. Hob.	1 Cut	2 Cuts
1. Advance carriage	0.08	0.08	0.08	0.25	0.25
2. Unlock hob	....	....	0.04	....	....
3. Move to rough side of hob	....	....	0.15	....	....
4. Lock hob	....	....	0.04	....	....
5. Change gears	....	....	0.05	....	....
6. Start machine	0.02	0.02	0.02	0.02	0.02
7. Cut	T	T	T	T	T
8. Loosen over-arm nuts (2)	0.04	0.04	....	0.04	0.04
9. Loosen upright nuts (4)	0.06	0.06	....	0.06	0.06
10. Raise piece	0.03	....	....	0.03	0.03
11. Back carriage	0.07	0.07	....	0.25	0.25
12. Lower piece to depth	....	0.08	....	....	0.08
13. Tighten upright nuts (4)	0.08	0.08	....	0.08	0.08
14. Tighten over-arm nuts (2)	0.06	0.06	....	0.06	0.06
15. Advance carriage	....	0.08	....	....	....
16. Unlock hob	....	....	0.04	....	....
17. Turn to finish side of hob	....	....	0.15	....	....
18. Set hob	....	....	0.25	....	....
19. Lock hob	....	....	0.04	....	....
20. Change gears	....	....	0.05	....	....
21. Start machine	....	0.02	0.02	....	0.02
22. Cut	....	T	T	....	T
23. Loosen over-arm nuts (2)	....	0.04	....	....	0.04
24. Loosen upright nuts (4)	....	0.06	....	....	0.06
25. Raise piece	....	0.03	....	....	0.03
26. Back carriage	....	0.07	0.07	....	0.25
27. Tighten upright nuts (4)	....	0.08	....	....	0.08
28. Tighten over-arm nuts (2)	....	0.06	....	....	0.06
Standard time	0.44	0.93	1.00	0.79	1.41

T = cutting time in minutes
Note: Allow 0.55 minute to set hob in alignment with keyway when required.

Table 27. Handling Time—Straight Spur Gears, Barber-Colman Hobber

Chucking and Removing

Operations	Time, minutes	
	A	B
1. Place blanks	$0.05N$	$0.05N$
2. Place washers and arbor nut	0.23	....
3. Oil center	0.10	0.10
4. Advance tailstock	0.03	0.03
5. Lock tailstock	0.02	0.02
6. Tighten center	0.04	0.04
7. Tighten arbor nut	0.10	....
8. Place washer and lock washer	....	0.08
9. Tighten draw rod	....	0.10
10. Loosen arbor nut	0.06	....
11. Loosen draw rod	....	0.06
12. Unlock tailstock	0.02	0.02
13. Back tailstock	0.03	0.03
14. Remove lock washer and washers	....	0.06
15. Remove arbor nut and washer	0.12	....
16. Remove gears	0.09	0.09
Standard time	$0.84 + 0.05N$	$0.63 + 0.05N$

A = nut-locked arbor. Use nut-locked arbor for arbor diameters up to $1\frac{5}{16}$ inches.

B = draw arbor. Use draw arbor for arbor diameters $1\frac{5}{16}$ in. and larger.

N = number of pieces per chucking.

Note: Allow 0.20 minute to insert key when keys on blanks must be aligned with teeth.

Table 28. Approach Allowance

Diam. Pitch	Full Depth	Outside Diameter of Cutter																
		2″	2¼″	2½″	2¾″	3″	3¼″	3½″	3¾″	4″	4¼″	4½″	4¾″	5″	5¼″	5½″	5¾″	6″
20	0.108	0.46	0.48	0.51	0.54	0.56	0.59	0.61	0.63	0.65	0.67							
16	0.135	0.50	0.54	0.56	0.60	0.62	0.65	0.67	0.70	0.73	0.74	0.78						
15	0.144	0.52	0.55	0.58	0.61	0.64	0.67	0.70	0.72	0.75	0.77	0.79	0.81					
14	0.154	0.53	0.57	0.60	0.63	0.66	0.69	0.72	0.74	0.76	0.79	0.81	0.84	0.85				
13	0.166	0.55	0.59	0.62	0.66	0.69	0.72	0.75	0.77	0.80	0.82	0.85	0.87	0.90	0.92			
12	0.180	0.57	0.59	0.65	0.68	0.71	0.74	0.77	0.80	0.83	0.86	0.88	0.91	0.93	0.96	0.99		
11	0.196	0.59	0.64	0.67	0.71	0.74	0.77	0.81	0.84	0.86	0.89	0.92	0.95	0.98	1.00	1.03	1.05	
10	0.216	0.62	0.66	0.70	0.74	0.78	0.81	0.85	0.87	0.90	0.94	0.95	0.99	1.02	1.04	1.06	1.09	1.12
9	0.240	0.65	0.70	0.74	0.78	0.82	0.85	0.89	0.92	0.95	0.98	1.01	1.04	1.07	1.10	1.12	1.15	1.17
8	0.270	0.68	0.73	0.78	0.82	0.86	0.90	0.93	0.97	1.00	1.04	1.07	1.10	1.13	1.16	1.19	1.22	1.25
7	0.308	0.72	0.77	0.82	0.86	0.90	0.95	0.98	1.03	1.06	1.10	1.14	1.17	1.19	1.23	1.25	1.30	1.32
6	0.360	0.76	0.82	0.88	0.93	0.97	1.02	1.06	1.11	1.15	1.19	1.22	1.26	1.29	1.33	1.36	1.39	1.43
5	0.432	0.82	0.89	0.95	1.00	1.05	1.10	1.15	1.20	1.24	1.28	1.32	1.37	1.40	1.44	1.48	1.51	1.55
4	0.540		0.96	1.03	1.09	1.15	1.21	1.26	1.32	1.37	1.41	1.47	1.51	1.55	1.60	1.64	1.71	1.72
3	0.720			1.05	1.21	1.28	1.35	1.42	1.48	1.54	1.59	1.65	1.70	1.76	1.80	1.86	1.90	1.95
2½	0.863				1.27	1.36	1.44	1.51	1.55	1.64	1.71	1.77	1.83	1.89	1.95	2.00	2.06	2.11
2	1.079					1.44	1.53	1.62	1.70	1.78	1.85	1.93	1.99	2.06	2.12	2.18	2.25	2.31
1¾	1.232						1.58	1.68	1.77	1.86	1.94	2.01	2.08	2.16	2.23	2.30	2.36	2.43
1½	1.438							1.72	1.83	1.92	2.01	2.10	2.19	2.27	2.32	2.42	2.50	2.57
1¼	1.726								1.87	1.99	2.10	2.19	2.29	2.30	2.47	2.50	2.64	2.72
1	2.157									1.99	2.13	2.26	2.37	2.48	2.59	2.70	2.79	2.89

$$\frac{N \text{ (number of teeth)}}{H \text{ (lead of hob)}} = \text{revolutions of hob per revolution of work} \quad (1)$$

$$\frac{\left\{\begin{matrix} N/H \text{ (revolutions of hob per} \\ \text{revolution of work)} \end{matrix}\right\}}{S \text{ (speed of hob in r.p.m.)}} = \text{time in minutes per revolution of work}$$

Since $\hspace{20em}$ (2)

$$\frac{L \text{ (total length of face)}}{\left\{\begin{matrix} F \text{ (feed in inches per} \\ \text{revolution of work)} \end{matrix}\right\}} = \text{number of revolutions of work required} \quad (3)$$

then
$$M = \frac{N/H}{S} \times \frac{L}{F} = \frac{N \times L}{F \times S \times H} \quad (4)$$

EXAMPLE. In order to show how the data and the formula are applied, the time required to hob a spur gear will be determined. It will be assumed that an order has been received for 24 index change gears as shown in Fig. 259. The procedure is as follows.

1. The following data are taken from the drawing of the gear, Fig. 259: length of face, 1.005 inches; diametral pitch (D.P.), 16; number of teeth (N), 70; diameter of bore, 1.3125 inches; material, 4620; hob HBG 573, ground plain spur gear.

2. Method of cutting, from Table 29. Hobbing ground-tooth spur gears, 16 D.P.—take one cut, 2-lead hob.

3. Setup time, from Table 30. Setup time, 35.0 minutes.

4. Number of gears per chucking, from Table 31. Bore diameter 1⅝₆ to 1⅜₆ inches; usable length of arbor, 6 inches; 1-inch face; nut-locked arbor—6 gears per chucking.

5. Outside diameter of hob, from Table 32. Hob HBG 573, 2.50 inches.

6. Speed of hob, from Table 29. Hobbing ground-tooth spur gears, 16 D.P. —204 r.p.m.

7. Feed, from Table 29. Hobbing ground-tooth spur gears, 16 D.P.—0.050 inch per revolution of work.

8. Approach allowance, from Table 28. Diametral pitch, 16; outside diameter of hob, 2.50 inches—0.56 inch.

9. Calculation of cutting time, from formula:

$$M = \frac{N \times L}{F \times S \times H} \qquad \begin{aligned} N &= 70 \\ L &= (6 \times 1.005) + 0.56 \\ F &= 0.050 \\ S &= 204 \\ H &= 2 \end{aligned}$$

$$M = \frac{70 \times (6.030 + 0.56)}{0.050 \times 204 \times 2} = 22.61 \text{ minutes}$$

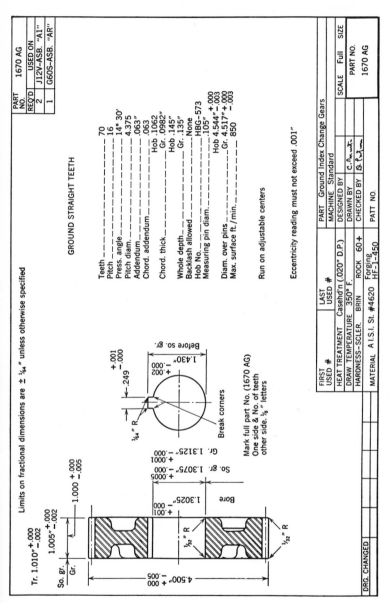

Fig. 259. Detail drawing of a ground index change gear.

Table 29. Speed and Feed Table—Method of Cutting

Diam. Pitch	Cutter Speed, feet per minute	Speed of Hob, rpm	Feed, inch per rev.	Method of Cutting
Finished hobbed gears				
6–7	90	103	0.050	2 cuts, reg. single-lead hob
8–20	100	139	0.050	2 cuts, reg. single-lead hob
20–32 incl.	115	174	0.036	1 cut, reg. single-lead hob
Hobbing ground tooth-spur gears				
16	150	204	0.050	1 cut, 2-lead hob
20	160	204	0.050	1 cut, 2-lead hob
32	170	240	0.045	1 cut, 2-lead hob
Hobbing ground tooth helicals				
8	130	139	0.040	2 cuts, 2-lead hob
10	120	150	0.040	1 cut, 2-lead hob
12	110	150	0.040	1 cut, 2-lead hob
24	100	150	0.040	1 cut, 2-lead hob

Table 30. Setup Time—Straight Spur Gears, Barber-Colman Hobber

Machines: 183, 315, 531, 906, 908

Operations	Time, minutes
1. Get drawing and ring clock	2.0
2. Remove hob	1.0
3. Remove arbor	1.0
4. Obtain hob and arbor	3.0
5. Place hob	3.0
6. Set angle	1.0
7. Place arbor	2.0
8. Indicate arbor	2.0
9. Change index gears	3.0
10. Change speed gears	1.0
11. Change feed gears	3.0
12. Check all gearing	2.0
13. Measure piece for size and set for depth	2.0
14. Check index and size	7.0
15. Set stops	1.0
Standard setup time	
Splines	35.0
Spur gears	35.0
Helical gears	40.0
Standard time—change hob	10.0

Table 31. Number of Gears per Chucking

Bore Diameter, inches	Usable Length of Arbor, inches
$1\frac{5}{16}$–$1\frac{9}{16}$	6
$\frac{3}{4}$–$1\frac{5}{16}$	3
$\frac{1}{2}$–$\frac{3}{4}$	$1\frac{1}{2}$

Table 32. Hob List

Hob Number	Diam. Pitch	Hand	Outside Diameter, inches	Press. Angle, degrees PA	Lead
HB708	5	R	$3\frac{1}{2}$	$14\frac{1}{2}$	S
HB550	5	R	4	20	S
HB710	6	R	$3\frac{1}{4}$	$14\frac{1}{2}$	S
HB734	6	L	$3\frac{1}{4}$	20	S
HB542	7	R	3	$14\frac{1}{2}$	S
HBS592	8	R	$3\frac{1}{2}$	20	2
HBS593	8	L	$3\frac{1}{2}$	20	2
HB712A	8	R	3	$14\frac{1}{2}$	S
HB737A	8	L	3	$14\frac{1}{2}$	S
HB713	9	R	3	$14\frac{1}{2}$	S
HBS586	10	R	3	20	2
HBS587	10	L	3	20	2
HBS590	10	R	$2\frac{3}{4}$	20	S
HBS591	10	L	$2\frac{3}{4}$	20	S
HB526	11	R	$2\frac{3}{4}$	$14\frac{1}{2}$	S
HB744	12	R	$2\frac{3}{4}$	20	S
HB745	12	R	3	20	2
HB605	12	R	$2\frac{3}{4}$	$14\frac{1}{2}$	2
HB606	12	L	$2\frac{3}{4}$	$14\frac{1}{2}$	2
HB571	13	L	$2\frac{3}{4}$	20	S
HB716	14	R	$2\frac{1}{2}$	$14\frac{1}{2}$	S
HBG573	16	R	$2\frac{1}{2}$	$14\frac{1}{2}$	2
HBS597	16	R	$2\frac{3}{4}$	20	2
HB742	16	R	$2\frac{1}{2}$	$14\frac{1}{2}$	S
HB580	20	R	$2\frac{1}{2}$	20	2
HB721	24	R	$2\frac{1}{2}$	$14\frac{1}{2}$	2
HB746	24	L	$2\frac{1}{2}$	$14\frac{1}{2}$	2
HBS608	30	R	$2\frac{3}{4}$	20	S
HB579	32	R	$2\frac{3}{4}$	20	2
HB701	48	R	$1\frac{7}{8}$	$14\frac{1}{2}$	S

10. Determination of total handling time:

Machine manipulation, from Table 26, column 1 (spur gear, regular hob, 1 cut)	0.44
Chucking and removing, from Table 27, nut-locked arbor A, $0.84 + (6 \times 0.05)$	1.14
Total handling time	1.58 minutes

11. Determination of total standard time:

Total handling time, 6 gears	1.58
Total cutting time, 6 gears	22.61
Total normal time, 6 gears	24.19 minutes
Allowances	
5% of handling time	0.08
5% of cutting, up to 20 minutes	1.00
Total standard time, 6 gears	25.27 minutes
Total standard time, 1 gear	4.21 minutes

TIME STANDARDS FOR SOLDERING SIDE SEAMS ON BODY OF CAN

Rectangular cans similar in shape to the one shown in Fig. 260 are made for export shipment of drawing and surgical instruments. Lot

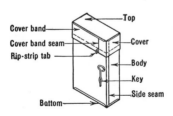

Fig. 260. Rectangular can for export shipment of instruments.

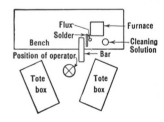

Fig. 261. Layout of work place for soldering rectangular cans.

sizes are usually small, and sixty different can sizes are made, ranging in volume from a few cubic inches to one cubic foot. The total production of cans of any one size is not large enough to justify special can-making equipment.

Operations in the Manufacture of Cans

On body:
1. Cut or slit body to length and width.
2. Make 4 breaks on the bar folder.
3. Solder the side seam.

On cover:
1. Cut or slit cover band to length and width.
2. Punch hole in rip-strip tab.
3. Mark and make cut for loose end of the rip-strip and fold tab back 90 degrees.
4. Make first break on bar folder.

On tops and bottoms:
1. Cut to length and width.
2. Miter four corners.
3. Fold four sides.

Assembly:
1. Form cover band to body and solder band seam.
2. Solder the top to the cover, solder the bottom to the body, and then solder rip-strip key to the body.
3. Inspect, wash, and dry.

Since it is the primary purpose of this case to illustrate the application of principles, only one of the operations listed above will be considered in detail here: Solder the side seam of the body.

Determination of Standard Time for Soldering the Side Seam of the Body. It was found most satisfactory to use stop-watch studies for securing all the data except those for the soldering element. Micromotion studies were used for that element.

Definition of Standard Elements for Soldering the Side Seam

1. Position piece on rod and apply the flux to the seam. The *time begins* as the last finished can is released by the hand to a tote box. The *time ends* as the hand completes the application of the flux and starts to move toward the soldering iron in the furnace.

2. Tack [1] the seam and then pick up the holder and position it on

---

[1] "Tack" refers to placing a drop of solder on the seam to hold it in the correct position while it is being soldered. A long seam requires more tacks than a short seam.

the can so as to hold the seam tightly together. The *time begins* as the hand starts toward the soldering iron from the preceding element. The *time ends* as the hand again starts toward the soldering iron in the furnace after positioning holder.

3. Solder the full length of the seam. The *time begins* as the hand starts toward the soldering iron in the furnace. The *time ends* as the hand releases the iron after disposing of it in the furnace.

4. Wipe the seam with a damp cloth and dispose. The *time begins* as the iron is released upon disposal in the furnace. The *time ends* as the hand releases the can into the tote box.

Standard Time for Elements

	Minutes per 100 Cans
1. Position piece on rod and apply the flux to seam. This element is a constant for cans of all sizes.	14.0
2. Tack the seam and then pick up the holder and position it on the can to hold the seam tightly together.	
(a) Untacked (seams under 3 inches in length)	0
(b) Two tacks per seam (seams 3.1 inches to 12 inches long)	16.0
(c) Three tacks per seam (seams 12.1 inches to 24 inches long)	23.0

If the seam is under 3 inches in length, this element is not required. If the seam is between 3.1 and 12 inches in length, it must be tacked at two places and clamped with a special holder before soldering. If between 12.1 and 24 inches long, the seam must be tacked at three points and clamped. Time standards were determined by time studies.

3. Solder the full length of the seam. The operator grasps the soldering iron, dips it in the pot of cleaning solution, touches it against the bar of solder, moves it to the seam, and draws the tip of the iron along the seam, soldering it until the supply of solder on the iron is used up. The iron is then returned to the bar of solder for a new supply, and the soldering operation is repeated. If the seam has been tacked, fewer contacts of the iron against the bar of solder will be needed.

(a) Grasp iron, dip in cleaning pot, and dispose of iron to furnace = 0.08 minute, a constant per seam or series of seams (from micromotion study).

(b) Solder seam. From a micromotion study of soldering seams of various lengths it was found that there was a straight-line relationship between the time in minutes to solder a seam and the length of the seam in inches: time for soldering = $L \times 0.014$, where L = length of seam to be soldered, in inches.

(c) Touch iron to solder and move to seam = $N \times 0.04$ minute. N = the number of dips necessary for seam (Table 33).

4. Wipe the seam with a damp cloth and dispose = 10.0 minutes per 100 cans —a constant for cans of all sizes.

Auxiliary Elements

5. Time for handling empty and full tote boxes varies with the size of the can. Handling time for the cans = $P \times 0.0009$ minute. (P = the sum of the length, width, and depth of the can in inches.) From all-day stop-watch studies.

6. Filing, forging, and retinning soldering irons require 22 minutes per 480-minute day, or 4.6% of the day. From all-day time studies.

The Sum of All Constant Elements

	Minutes per 100 Cans
1. Position piece on rod and apply flux to the seam.	14.0
2. Tack seam and then pick up the holder and position it on the can.	
(a)	0.0
(b)	16.0
(c)	23.0
3. Wipe the seam with a damp cloth and dispose.	10.0

Formula for Determining Time Standard

Standard time in minutes per 100 cans = 100 (time for constant elements + time for soldering + time for handling can) + time for maintaining iron

$$= 100 \ \{D + [0.08 + (L \times 0.014) + (N \times 0.04)] + (P \times 0.0009)\}$$
$$+4.6\{D + [0.08 + (L \times 0.014) + (N \times 0.04)] + (P \times 0.0009)\}$$
$$= 104.6\{D + [0.08 + (L \times 0.014) + (N \times 0.04)] + (P \times 0.0009)\}$$

where L = the length of the seam to be soldered, in inches
N = the number of dips necessary to complete the seam (from Table 33)
P = the sum of the length, width, and depth of the can, in inches
D = the sum of all constants (from Table 34)

Application of the Formula. Can 439 has the following dimensions: length, $8\frac{5}{8}$ inches; width, $\frac{3}{4}$ inch; depth, $10\frac{1}{8}$ inches. Values of the terms are: $L = 10.125$; $N = 2.0$; $P = 19.5$; $D = 0.40$.

Table 33. Values of N

(N = number of dips of iron against solder per seam)*

Value of N	Seam Lengths in Inches	
	Untacked	Tacked
1	0 –6.0	0–10.0
2	6.1–12.0	10.1–20.0
3	12.1–18.0	20.1–30.1
4	18.1–24.0	
5	24.1–30.0	

* From stop-watch studies and micromotion studies.

Substituting these values in the formula, we have

$$\begin{aligned}
\text{Std.} &= 104.6\{0.40 + [0.08 + (10.125 \times 0.014) + (2 \times 0.04)] \\
&\quad + (19.5 \times 0.0009)\} \\
&= 104.6(0.40 + 0.08 + 0.142 + 0.08 + 0.0176) \\
&= 104.6 \times 0.7196 = 75.27, \text{ use } 75.3 \text{ minutes}
\end{aligned}$$

Although the method of setting up a formula for determining the time standard for soldering the side seam has been presented in some detail, it is not necessary to go through this rather long procedure for each new lot or for each new can size. This formula applies to the operation "solder side seam" for all rectangular cans of any size falling within the range of the studies. In fact, it is not even necessary to

Table 34. Values of D

(D = sum of all constant elements)

Values of D in Min. per Can	Seam Lengths in Inches	No. of Tacks
0.24	under 3.0	0
0.40	3.0–12.0	2
0.47	12.1–24.0	3

use the formula, for tables have been constructed from computations made with the formula, and from these tables it is a very simple matter to determine the time standard for soldering operations on a can of any size. These tables are as easy to use as the mileage chart on a road map.

Results. Before standardizing the arrangement of the work place and the method of making the cans, all time standards were set by individual time studies. Since only a few can sizes had been studied, most of this work was not on wage incentive.

After the completion of the standardization program and the computation of the tables for setting time standards, it was possible to determine quickly the time standard for the soldering operations on a can of any size.

Decrease in labor costs resulting from improved methods and from the application of time standards and wage incentives to soldering operations in can making brought a saving of approximately 4000 man-hours of direct labor per year. A total of 510 hours was spent by the analyst in completing the motion and time study work on this project.

CHAPTER 29

Determining Time Standards
for Die and Tool Work

Toolroom work is one of the most difficult of all industrial operations to standardize and place on wage incentive. The making of tools and dies requires a high degree of accuracy, the work is nonrepetitive, and rarely is more than one tool of a given design required. Highly skilled die makers are essential for this work, and on some operations a considerable amount of hand filing, scraping, and fitting may be required along with the machine work. Nevertheless, although it may seem that no two tools are alike, all tools of a given class have similar parts, and each part requires similar operations for its manufacture. The variation in the time required for the same operation on similar parts is due to different characteristics of the particular part, such as its size and the kind of material. It is possible here, as it was in the several cases in preceding chapters, to separate each operation into those elements that remain constant and those elements that vary with the size, shape, or other characteristic of the part.

The material to be presented in this chapter is taken from an outstanding piece of work done by Floyd R. Spencer in standardizing the die and tool work in a large industrial plant. This included compiling elemental time data and constructing charts and formulas for determining time standards. Using these standards, a wage incentive system was installed which materially reduced the cost of this work. The toolmakers had confidence in this method of setting time standards and believed it to be far superior to the rather crude method of "estimating" time on such work, as practiced in most toolrooms. The incentive plan based on these time standards permits the toolmakers to earn substantially more wages than formerly.

The time required to determine the standard time by means of charts, curves, and formulas is one fourth to one half that required for estimating as formerly. For example, it takes 3 to 5 minutes to set the time standards for all operations required in making the plain blanking die shown in Figs. 262 and 263, whereas it would take 10 to 15 minutes to "estimate" the time by the old procedure. One person now determines time standards for a department employing 125 toolmakers, whereas in most shops one estimator is needed for every 30 toolmakers employed.

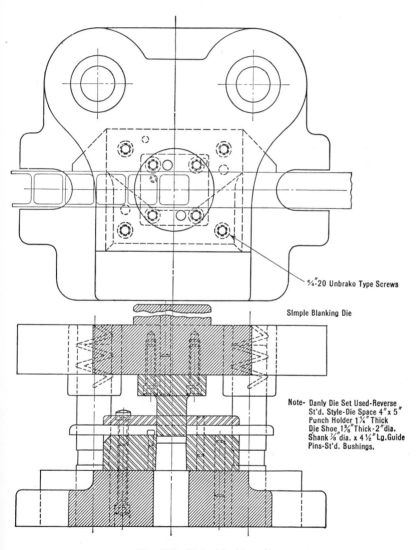

¼"-20 Unbrako Type Screws

Simple Blanking Die

Note- Danly Die Set Used-Reverse
St'd. Style-Die Space 4"x 5"
Punch Holder 1¼" Thick
Die Shoe 1⅜"Thick-2"dia.
Shank ⅞"dia. x 4½"Lg.Guide
Pins-St'd. Bushings.

Fig. 262. Plain blanking die.

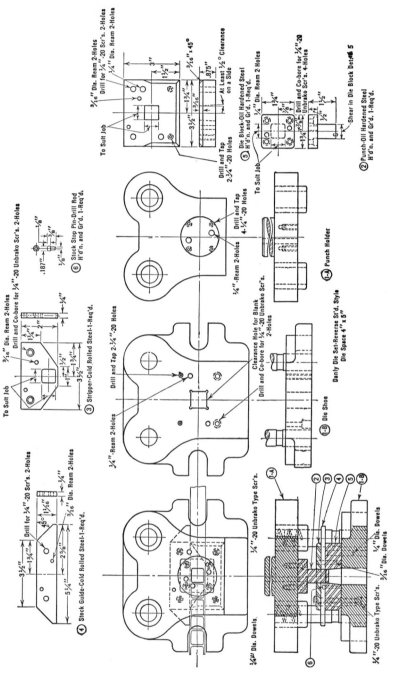

Fig. 263. Details of the parts of plain blanking die. (Courtesy of E. W. Bliss Co.)

Types of Dies. The various types of dies for which standard data are now available are (1) plain blanking, (2) compound blank and perforate, (3) single and multiple perforator, (4) shearing, (5) blank and draw, (6) forming, and (7) miscellaneous.

Plain Blanking Dies. The plain blanking die, such as the one shown in Fig. 262, will serve as an illustration because of its simplicity and general use. An explanation will be given of the method used in classifying all blanking work, as well as the way the elemental time data, charts, and formulas were established, and finally a specific case will be presented to show how the standard time is established for making a particular die.

The toolroom for which these data apply normally employs 125 toolmakers and serves a manufacturing plant employing 4000 to 5000 workers. The products manufactured are widely diversified in kind and are fairly small in size.

The first step in establishing a method for setting time standards on making blanking dies was to classify all blanks that would normally be manufactured by this type of die. A survey of all such work led to the following general classification:

1. Round blanks.
2. Square or rectangular blanks.
3. All blanks of other shapes.

Under the general classifications of pieces with round outlines and those with square or rectangular outlines, a further classification must be made of projections or indentations. It is generally advisable to use inserts for projections or indentations on the piece in question. These inserts would be used in the pad and punch (see Figs. 262 and 263) for projections, and in the die block for indentations. The inserts are used because of the low repair cost in breakage of these projections. A complete classification of blanks is given in Table 35.

It is customary to purchase the die sets from manufacturers specializing in making these parts. The die set consists of two parts, as Figs. 262 and 263 show, namely the punch holder, part 1A, and the die shoe, part 1B. The classification of die sets is given in Table 36.

Parts of a Blanking Die. The principle of a plain blanking die is very simple. The strip of sheet metal stock to be blanked is fed into the die from right to left, as shown in Fig. 262. The punch, which is the shape of the blank to be made, is moved downward by the action of the press and punches a piece from the stock, the blank being

Table 35. Classification of Work Done on Plain Blanking Dies

Classification	No. 1	No. 2	No. 3
Shape of Blank	Round blanks	Square or rectangular blanks	All blanks of other shapes
	Round..........	Plain outline......	(a) Blanks whose outline is made up of smooth curves and straight lines or a combination of the first two classifications.
	Round with indentations.........	With indentations	
	Round with projections..........	With projections..	(b) Blanks whose outline is very irregular, having no similarity with blanks of round or straight-sided outline.
Size of blank	½-in. diam. to 4 in. diam.	⅜ × 1½ in. to 5½ × 7 in.	Same range as in classifications No. 1 and 2

Table 36. Purchased Die Sets Classified by Size Range

(Size in inches)

Symbol	B1	B2	B3	B4	B5	B6	B7	B8
Maximum size round blank	1⅜	1⅞	2⅜	3⅛	3⅞	4¼	5	6
Maximum size rectangular blank	⅜×1½	½×2	1⅛×2⅜	2¼×3¼	3×4	4¼×4¼	4½×7	5½×7

forced through the die block by the punch. The stripper removes the stock from the punch as it returns to its normal position.

Figure 262 shows an assembly drawing of a plain blanking die, and Fig. 263 shows the component parts of this die. A list of all the parts of this die is as follows:

1A	Punch holder	die set (purchased)
1B	Die shoe	
2	Punch	
3	Stripper	
4	Stock guide	
5	Die block	
6	Stock stop pin (a standard part)	

Operations on Punch Holder—Part 1A. Although the die sets, composed of punch holder and die shoe, were purchased from an outside source, the following operations were performed on these parts:

Operation 1. Set up mill, mill stem to height.
Operation 2. Lay out, drill and tap for screw holes, drill and ream for dowels.

The next step was to determine the variables or the basic factors that governed the time required for each operation, and the percentage of the extent of total time that each factor controlled.

		Percentage
Operation 1	Size and weight of punch holder	100
Operation 2	Number of screw and dowel holes	85
	Size and weight of punch holder	15

Table 37. Governing Factor—Size of Die Set Required to Blank Part

(Time in hours)

Operation No.	Percentage of Time Controlled by Factor	Symbol of Purchased Die Set Sizes							
		B1	B2	B3	B4	B5	B6	B7	B8
1	100	0.70	0.75	0.78	0.80	0.88	0.96	1.08	1.22
2	15	0.10	0.15	0.21	0.27	0.32	0.39	0.44	0.50

Table 38. Governing Factor—Number of Screws and Dowels

(Time in hours)

Operation No.	Percentage of Time Controlled by Factor	3 Screws 2 Dowels	4 Screws 2 Dowels	6 Screws 2 Dowels	8 Screws 3 Dowels	12 Screws 4 Dowels	14 Screws 6 Dowels	18 Screws 8 Dowels	20 Screws 10 Dowels
2	85	0.50	0.65	0.85	1.10	1.39	1.70	2.10	2.30

In the toolroom actual studies were made of the operations required to machine the punch holder. Data were taken for many different sizes and weights, and finally, after being checked and tested for accuracy, these data were compiled in tabular form for convenient use (Tables 37 and 38).

Operations on Die Block—Part 5. The die block is perhaps the most important part of the tool, and it must be made accurately in order to produce blanks to meet production requirements. In determining time standards for making the die block the following procedure was used.

1. A list was made of all variables or governing factors that would in any way affect the time required to make the die block. They were:

a. Length of outline of blank.
b. Number of inside angles.
c. Number of sides (0.250 inch in length or over).
d. Number of radii.
e. Curves on blank whose center is outside of blank.
f. Specifications of stock to be blanked.

2. A list was made of the operations which the toolmaker must perform to make the die block, and the factors which have a bearing on the time required were noted (Table 39).

3. Time standards were determined for each operation over the entire range of the governing factors. This was a much more difficult task than determining time standards for making the punch holder, because so many more variables entered into making the die block. Approximately 100 die blocks of every conceivable shape were studied, and a considerable period of time was required for securing and classifying these data and for establishing correct relationships.

Curves for Setting Time Standard for Operation 4. Of the seven operations required to make the die block, perhaps the most interesting one for illustrative purposes is operation 4, "Work out shape through block." The five variables or factors which govern the time required for performing this operation are listed in Table 39. The curves in Fig. 264 show the relationship between the variables and the time required. These data are also given in Table 40.

EXAMPLE. To determine the time required to perform operation 4, "Work out shape through die block," for the blank shown in Fig. 265, the procedure would be as follows:

Table 39. Operations and Governing Factors for Die Block—
Part 5

Operation No.	Operation	Governing Factor
1	Cutoff	(a) Grade of material
		(b) Size of piece
2	Machine to size—Grind all over	(a) Grade of material
		(b) Size of piece
3	Layout shape on block surface	(a) Length of outline
		(b) Number of inside angles
		(c) Number of sides
		(d) Number of radii
		(e) Number of curves with centers outside blank
4	Work out shape through block	(a) Same factors as above—Operation 3
5	Drill and tap screw holes, drill and ream dowel holes	(a) Number of screw and dowel holes
		(b) Size of holes
6	Harden	(a) Grade of material
7	Grind	(a) Grade of material
		(b) Size of piece

		Curve (Fig. No.)	Standard Time, hours
(a) Outline in inches [1]	10.75	264a	9.00
(b) Inside angles	4	264b	2.50
(c) Number of sides	12	264c	5.00
(d) Number of radii	6	264d	1.75
(e) Number of radii with centers outside	0	....	....
Total standard time in hours	4	....	18.25

The standard curves shown in Fig. 264 cover all possible combinations of contour and size affecting operation 4. Space does not permit the inclusion of the curves and data for the other operations on the die block. The same procedure was followed not only for the remain-

[1] The outline of irregular dies is determined by means of a map measure—a simple instrument reading distance directly in inches.

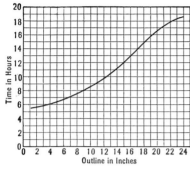

A. Length of outline of blank.

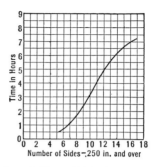

C. Number of sides, 0.250 inch or over.

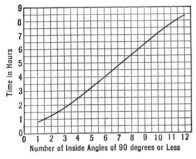

B. Number of inside angles, 90 degrees or less.

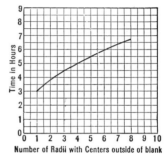

E. Number of curves with centers outside of blank.

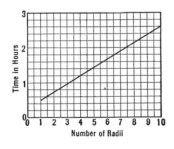

D. Number of radii.

Fig. 264. Curves for setting time standard for operation 4 (work out shape through block) on die block, Part 5—plain blanking die.

**Table 40. Standard Time for Performing Operation 4,
Work Out Shape through Die Block**

(Time in hours)

(a) Outline in inches	1	3	6	10	12	14	18	20	24
Time	5.5	5.9	6.9	8.5	9.6	11.3	15.0	16.7	18.7
(b) Inside angles, 90° or less	1	2	3	4	5	6	8	10	12
Time	0.8	1.2	1.75	2.6	3.3	4.1	5.6	7.1	8.4
(c) Number of sides (length 0.250 inch and over)	5	6	7	8	9	10	12	14	16
Time	0.5	0.7	1.1	1.7	2.3	3.3	4.9	6.2	6.9
(d) Number of radii	1	2	3	4	5	6	8	10	
Time	0.5	0.7	0.9	1.2	1.4	1.7	2.1	2.6	
(e) Number of radii, centers outside	1	2	3	4	5	6	7	8	
Time	3.0	3.8	4.5	5.0	5.5	5.9	6.3	6.7	

ing operations on the die block but also for all operations on the other parts of the blanking die.

In a similar manner elemental data, charts, curves, and formulas were developed for the seven classes of dies listed on page 465.

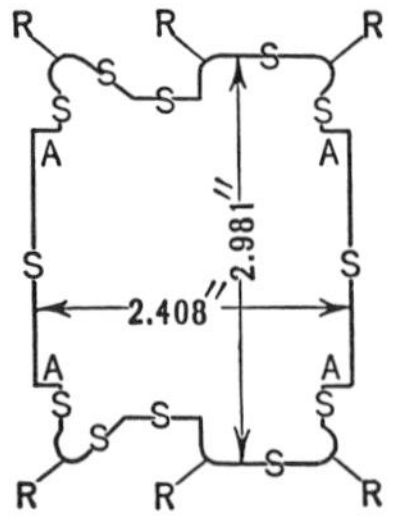

Quality Classification. It is necessary to show how the quality of the tool enters into the determination of the time standard for making the die. The quality requirements for a die depend upon the following factors:

Fig. 265. Shape of blank: A, inside angles; R, radii; S, sides.

1. Appearance of the product.
2. Total production requirements of the product.
3. Use of the product.
4. Working action of the parts produced.
5. Cost factors of the products.

Table 41. Punch and Die Parts—Base of Time-Setting Curves

Compound Dies—Blank and Perforate—Class 2
Blanks with Straight Sides or Smooth Curved Outline

Chart No.	Part of Punch and Die	Base of Chart (abscissa)
20–21–22	Pad—blank punch—stripper—punch plate	Outline in inches—Number of sides with angles 90° or less
5–7–13	Die shoe—punch holder—assembly work—hardening and miscellaneous	Size of die set
8	Perforators A. Round B. Square and rectangular C. Odd-shaped	A. Diameter of perforator B. Outline of perforator C. Outline of rough stock—Number of angles Outline of perforator—Length of slots
10	Placing round center holes through pad and punch—not center	Diameter of perforator
11	Placing square, square with indentations and projections, and odd-shaped holes through pad—punch—die shoe	Outline of perforator—Number of angles—Number of curves—Number of projections—Length of slot 0.125 wide and under
12	Bushings	Diameter of perforator—Number required
13	Minor curves	Number of curves requiring additional filing
13A	Inside radii	Number of radii requiring additional filing
14	Projections and indentations in blanking or perforating punch (with or without inserts)	Number of projections or indentations
14A	Projections and indentations (continued)	Length of outline for stock removed

Table 42. Calculation of Standard Time for Blank and Perforate Punch and Die

Compound Die Shown in Fig. 266

Chart Number	Information Required	Information Applicable to Die in Fig. 266	Reading from Curves — Standard Time, hours
20 and 21	Outline of blank	$10\frac{3}{4}$	34
22	Number of sides	18	$17\frac{3}{4}$
	Outside angles 90 degrees or less	6	$4\frac{3}{4}$
5–7–13	Size of die set	C$1\frac{1}{2}$	$27\frac{1}{4}$
	General outline	Square	
	Total number of perforations	5	$1\frac{1}{4}$
8A	Number of perforators	4	$2\frac{1}{4}$
	Diameter of perforators	0.068	
	Type—Perforator	Type B	
8B	Outline of perforator	$5\frac{1}{4}$	$5\frac{3}{4}$
	Number of perforators	1	
8C	Outline of stock from which blank is made		
	90-degree angles or less		
	Width up to 0.125 inch		
	Outline of perforator		
10	Number of perforators	4	6
	Diameter of perforators	0.068	
11A	Outline of perforator	$5\frac{1}{4}$	$9\frac{1}{4}$
	Number of projections on perforator punch	10	11
11B	Number of 90-degree angles or less		
	If form of slot 0.125 wide or less— length		
12	Number of bushings	4	$2\frac{1}{2}$
	Size of inside diameter	0.068	
13	Number of minor curves		
13A	Number of radii	6	$8\frac{1}{2}$
14	Number of projections or indentations	10	$50\frac{1}{2}$
14A	Length of outline for stock removed		

| | Standard time | | 180.75* |

* Quality classification C = 180.75 hours.
   Quality classification B = 180.75 $\times$ 112 = 202.44. Use 205 hours.

A study of these factors resulted in the establishment of the following quality classification:

Class C. Reading from curves and charts—direct.
Class B. Reading from curves and charts multiplied by 112%.
Class A. Reading from curves and charts multiplied by 130%.

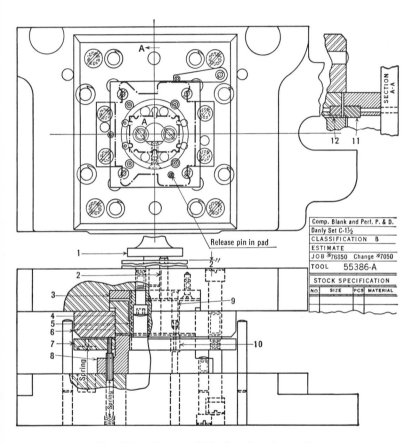

Fig. 266.   Compound blank and perforate die.

The decision as to the classification of the various punches and dies is made before they are designed, and the classification is noted on the drawings when they are made.

The Wage Incentive Application. Since punch-press parts usually require a series of operations for their manufacture, the several dies required to make a part completely are designed at one time. It is customary to design each tool in assembly, showing the construction of the tool, listing the material required, and indicating any special features, but not showing the individual parts in detail. Time standards are then established by the method already described. The order for the series of tools is given to a group leader in the toolroom, to whom are assigned as many assistants as can do the job efficiently. These toolmakers complete the entire series of tools for a given part as a group, and share in the savings through a bonus based upon the difference between the time actually used and the time standard set for the job. At the completion of this series of tools the group dissolves, and new groups are formed for other tools.

Compound Blank and Perforate Dies. Compound blank and perforate dies are more complicated than plain blanking dies in that the compound dies have more parts. The method, however, for standardizing and establishing standard data for making these dies is exactly like that already described for the plain blanking die.

Space does not permit the presentation of similar material for compound dies. Table 41, however, gives a list of the parts, and the summary in Table 42 gives the standard time required to make complete the compound die shown in Fig. 266.

CHAPTER 30

Systems of Predetermined Motion-Time Data: Motion-Time Data for Assembly Work

Although systems of motion-time data have been used to a limited extent for many years, their application has increased as the different systems available have become more numerous and as people have reached a better understanding of the advantages and limitations of such data.

Nine systems of motion-time data are listed in Fig. 267, and four of these systems are described in this and the following chapter. A number of well-designed systems now in operation are not included in Fig. 267 because information about them is not available in published form. Moreover, some companies have modified through their own research the system or systems with which they may have started. Therefore, there is no way of knowing how many different systems of predetermined motion-time data are being applied in this country. However, all the systems have much in common.

Perhaps the major advantage of motion-time data in comparison to time study is that standard data make it possible to predetermine the standard time for a job or activity if the motion pattern is known. One can determine in advance how long it will take to perform an operation in the shop, merely by examining a blueprint of the work place layout and a description of the method. Likewise, an accurate evaluation can be made of several different work methods or different tool designs. Considerable use also has been made of motion-time data for establishing elemental standard data for various classes of machines and equipment, thus expediting the actual setting of time standards for jobs to be performed on such equipment. The use of elemental data often results in greater consistency in time standards.

Name of System	First Applied Date	First Publication Describing System
Motion-Time Analysis (MTA)	1924	Data not published, but information concerning MTA published in *Motion-Time Analysis Bulletin*, a publication of A. B. Segur & Co.
Body Member Movements	1938	*Applied Time and Motion Study* by W. G. Holmes, Ronald Press Co., New York, 1938
Motion-Time Data for Assembly Work (Get and Place)	1938	*Motion and Time Study*, 2nd ed., by Ralph M. Barnes, John Wiley & Sons, New York, 1940, Chs. 22 and 23
The Work-Factor System	1938	"Motion-Time Standards" by J. H. Quick, W. J. Shea, and R. E. Koehler, *Factory Management and Maintenance*, Vol. 103, No. 5, pp. 97–108, May, 1945
Elemental Time Standard for Basic Manual Work	1942	"Establishing Time Values by Elementary Motion Analysis" by M. G. Schaefer, *Proceedings Tenth Time and Motion Study Clinic*, IMS, Chicago, pp. 21–27, November, 1946
Methods-Time Measurement (MTM)	1948	*Methods-Time Measurement* by H. B. Maynard, G. J. Stegemerten, and J. L. Schwab, McGraw-Hill Book Co., New York, 1948
Basic Motion Timestudy (BMT)	1950	Manuals by J. D. Woods & Gordon, Ltd., Toronto, Canada, 1950
Dimensional Motion Times (DMT)	1952	"New Motion Time Method Defined" by H. C. Geppinger, *Iron Age*, Vol. 171, No. 2, pp. 106–108, January 8, 1953
Predetermined Human Work Times	1952	"A System of Predetermined Human Work Times" by Irwin P. Lazarus, Ph.D. thesis, Purdue University, 1952

Fig. 267. Summary of facts concerning

Publication Containing Information about System	How Data Were Originally Obtained	System Developed by
"Motion-Time-Analysis" by A. B. Segur, in *Industrial Engineering Handbook*, H. B. Maynard, editor, McGraw-Hill Book Co., New York, pp. 4-101 to 4-118, 1956	Motion pictures, micromotion analysis, kymograph	A. B. Segur
Applied Time and Motion Study by W. G. Holmes, Ronald Press Co., New York, 1938	Not known	W. G. Holmes
Motion and Time Study: Design and Measurement of Work, 6th ed., by Ralph M. Barnes, John Wiley & Sons, New York, 1968, Ch. 30	Time study, motion pictures of factory operations, laboratory studies	Harold Engstrom and H. C. Geppinger of Bridgeport Plant of General Electric Co.
Work-Factor Time Standards, by Joseph H. Quick, James H. Duncan, and James A. Malcolm, Jr., McGraw-Hill Book Co., New York, 1962 *Ready Work-Factor Time Standards*, by J. A. Malcolm, Jr. et al., Haddonfield, N. J., 1966.	Time study, motion pictures of factory operations, study of motions with stroboscopic light unit	J. H. Quick W. J. Shea R. E. Koehler
"Establishing Time Values by Elementary Motions" by M. G. Schaefer, *Proceedings Tenth Time and Motion Study Clinic*, IMS, Chicago, November, 1946. Also "Development and Use of Time Values for Elemental Motions" by M. G. Schaefer, *Proceedings Second Time Study and Methods Conference*, SAM-ASME, New York, April, 1947	Kymograph studies, motion pictures of industrial operations, and electric time-recorder studies (time measured to 0.0001 minute)	Western Electric Co.
Methods-Time Measurement by H. B. Maynard, G. J. Stegemerten, and J. L. Schwab, McGraw-Hill Book Co., New York, 1948	Time study, motion pictures of factory operations	H. B. Maynard G. J. Stegemerten J. L. Schwab
Basic Motion Timestudy by G. B. Bailey and Ralph Presgrave, McGraw-Hill Book Co., New York, 1958	Laboratory studies	Ralph Presgrave G. B. Bailey J. A. Lowden
Dimensional Motion Times by H. C. Geppinger, John Wiley & Sons, New York, 1955	Time study, motion pictures, laboratory studies	H. C. Geppinger
"Synthesized Standards from Basic Motion Times," *Handbook of Industrial Engineering and Management*, W. G. Ireson and E. L. Grant, editors, Prentice-Hall, Englewood Cliffs, N. J., pp. 373–378, 1955	Motion pictures of factory operations	Irwin P. Lazarus

various systems of motion-time data.

The main uses of motion-time data may be divided into the following two classes:

Evaluation of Method
1. Improving existing methods.
2. Evaluating proposed methods in advance of actual production.
3. Evaluating suggested designs of tools, jigs, and equipment.
4. Aiding in the design of the product.
5. Training members of the staff to become motion-minded.

Establishing Time Standards
1. Direct use of motion-time data for establishing time standards.
2. Compilation of standard data and formulas for specific classes of work, for more rapid establishment of time standards.
3. Checking standards established by time study.
4. Auditing time standards.

It is important that motion-time data be applied only by capable and well-trained people. The statement often is made that motion-time data are superior to time study because judgment is not required on the part of the analyst in evaluating the pace of the operator being studied. This, of course, is true. There are, however, numerous points in applying systems of motion-time data at which judgment of the analyst does come into play. Thorough training in the use of a particular system of motion-time data is needed in order to minimize such judgment factors. If several people in a department are establishing time standards by motion-time data, it is especially important that they all be trained to use the data in the same way insofar as possible.

The four systems of motion-time data described in this chapter and the one following are presented in chronological order according to the time they were first used. They are (1) Motion-Time Data for Assembly Operations, (2) the Work-Factor System, (3) Methods-Time Measurement, and (4) Basic Motion Timestudy.

MOTION-TIME DATA FOR ASSEMBLY WORK

The system of motion-time data described in this chapter was developed by Harold Engstrom and his associates while he was Motion Study Supervisor at the Bridgeport Plant of the General Electric Company. These data were successfully used by this company, first for estimating labor costs on new products, and later for establishing time standards. The system was designed especially for establishing time standards on assembly operations in the electrical appliance di-vision, and was not intended for universal application.

Layout of Work Place. Where factory operations do not dictate otherwise, the assembly work place should be set up closely resembling the sketch shown in Fig. 268. Parts are to be supplied to the operator in *well-designed* bins, trays, hoppers, or other containers located in the areas indicated. In progressive assembly operations, subassemblies will be delivered to the operator within the work area outlined (24 inches from the edge of fixture or bench nearest the operator). This

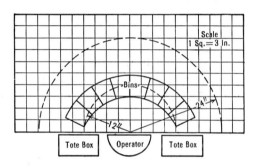

Fig. 268.   Layout of typical work place.

setup provides the operator with an approved motion study work place and is the basis on which corrected standard time values have been established.

With the work place arranged as shown above, all but a very few of the operator's transport empty or transport loaded motions will be confined to distances less than 24 inches. As the majority of assembly parts are in the small or medium class, average transport empty or transport loaded distances will be much less than 24 inches. Consequently transport empty and transport loaded corrections are calculated slightly in excess of the anticipated average, and distances to 24 inches are allowed.

Determining the Time Standard for an Assembly Operation. The data shown in Figs. 269 and 270, and the Standard Times Computation Sheet (Figs. 272 and 273), may be used for establishing time standards for assembly operations.

In an assembly operation a variety of parts are supplied to an operator who assembles them in definite positions. The operator must *get* each part and *place* it in proper position in relation to the rest of the assembly. When parts are fastened together, hand tools or machines may be *used*. Last, after an assembly cycle is completed, the device or assembly must be *placed aside* or *disposed*. All assembly operations are composed of a sequence of these elements. For purposes of practical analysis it is unnecessary to reduce these four divisions to still smaller elements if the variables which affect them are recognized and properly evaluated.

"Grasp" a Primary Variable. Throughout any assembly operation, time is consumed in obtaining, maintaining, or releasing the control of parts, tools, or machines. In the assembly operations studied, this control was largely manual.

It was determined that the type of grasp used in controlling the part being assembled was perhaps the most important variable affecting get or place times. Although special features of design or relative difficulty of assembly may affect the type of grasp used, this is largely a function of the size of the part. Hence, for practical purposes, the size of parts may be used as one base for evaluating the variation of get or place times.

Size of Part and Type of Grasp. The size of parts is a satisfactory base for evaluating variations of get and place times when size is defined according to the type of grasp employed. It must be remembered, however, that one grasp may be used to *get* a given part and another used to *place* it efficiently.

Four Types of Grasp. Analysis of many individual operations indicates that grasp may be classed readily into four divisions or types, each indicating the size of the parts which *normally* employ each type of grasp.

1. Three fingers and thumb (3*F*). This grasp is used on any object large enough to permit placing three fingers and the thumb around it (in at least two dimensions) without crowding, and not large enough to require extension of the fingers to control it. This grasp, on parts of this size, is the easiest to obtain and in most cases provides maximum control. (See Fig. 269.)

2. Extend hand (*H*). This grasp is used on any object where size requires extension of the hand, and weight, finish, or control requirements do not necessitate use of two hands. Control is good and readily obtained.

3. Two fingers and thumb (2*F*). This grasp is used where it is impossible to obtain a three-finger grasp on an object because of its small size.

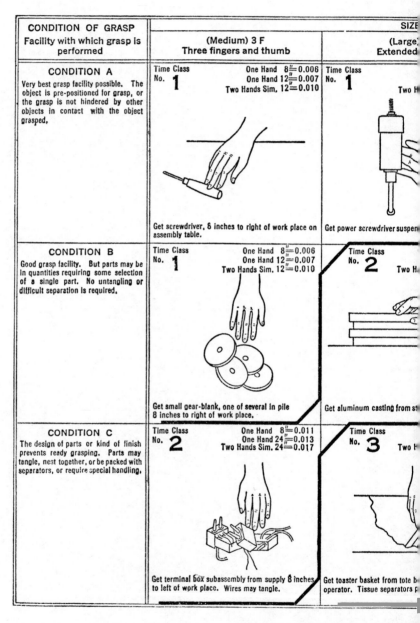

CONDITION OF GRASP	SIZE	
Facility with which grasp is performed	**(Medium) 3 F** **Three fingers and thumb**	**(Large)** **Extended**
CONDITION A Very best grasp facility possible. The object is pre-positioned for grasp, or the grasp is not hindered by other objects in contact with the object grasped.	Time Class No. **1**　　　One Hand　8″=0.006 　　　　　　　　One Hand 12″=0.007 　　　　　Two Hands Sim. 12″=0.010 Get screwdriver, 8 inches to right of work place on assembly table.	Time Class No. **1** Two H| Get power screwdriver suspen
CONDITION B Good grasp facility. But parts may be in quantities requiring some selection of a single part. No untangling or difficult separation is required.	Time Class No. **1**　　　One Hand　8″=0.006 　　　　　　　　One Hand 12″=0.007 　　　　　Two Hands Sim. 12″=0.010 Get small gear-blank, one of several in pile 8 inches to right of work place.	Time Class No. **2** Two H| Get aluminum casting from s
CONDITION C The design of parts or kind of finish prevents ready grasping. Parts may tangle, nest together, or be packed with separators, or require special handling.	Time Class No. **2**　　　One Hand　8″=0.011 　　　　　　　　One Hand 24″=0.013 　　　　　Two Hands Sim. 24″=0.017 Get terminal box subassembly from supply 8 inches to left of work place. Wires may tangle.	Time Class No. **3** Two H| Get toaster basket from tote b| operator. Tissue separators p

Fig. 269. Standard time values for

PE OF GRASP		GET TIME Class No.	STD. TIME Corrected for Transport Distance
(Small) 2 F Two fingers and thumb	(Very Large) 2 H Two hands		

SS One Hand 8″=0.006 One Hand 12″=0.007 Two Hands Sim. 12″=0.010	Time Class No. **1** Two Hands 8″=0.006 Two Hands 12″=0.007	1	0.007

I machine bolt (one of several held in left om left hand which positions bolt for grasp. used.	Get completed waffle iron by handles for disposal.

| **SS**
2 One Hand 8″=0.011
One Hand 24″=0.013
Two Hands Sim. 24″=0.017 | Time Class No. **2** Two Hands 8″=0.011
Two Hands 24″=0.013 | 2 | 0.013 |

inch brass washer from bin beside work	Get waffle-iron grid assembly from table.

| **SS**
One Hand 8″=0.019
One Hand 24″=0.021
Two Hands Sim. 24″=0.028 | | 3 | 0.021 |

| | Time Class No. **4** Two Hands 8″=0.024
Two Hands 24″=0.026 | 4 | 0.026 |

ch steel lock washer from bin beside work	Get casting from tote box. Note: This combination is seldom encountered. This condition applies when weighty parts are removed from tote box or carton. The positioning of the hands in entering a constricted area before obtaining the actual grasp is the deciding factor in classifying a 2H "Get" in Condition C.

nsport distances. Time in minutes.

CONDITION OF PLACE Amount of positioning required	S (Medium) 3 F Three fingers and thumb	(L Exter
CONDITION A Positioning is normally little more than releasing the object on the work place.	Time Class No. **1** One Hand 8″=0.006 One Hand 12″=0.007 Two Hands Sim. 12″=0.011 Place small gear blank in other hand.	Time Class No. **1** T Place toaster body in adjustment.
CONDITION B Positioning of parts on or into definite locations with ample tolerances, simple open nests or fixtures, or assemblies with one point of location.	Time Class No. **1** One Hand 8″=0.006 One Hand 12″=0.007 Two Hands Sim. 12″=0.011 Dispose of screwdriver into funnel-type holder 8 inches to right of work place.	Time Class No. **2** Tv Dispose of toss
CONDITION C Positioning of parts on or into difficult or complicated locations. Assemblies or fixtures requiring the positioning of parts with respect to two definite points, or location in two directions.	Time Class No. **2** One Hand 8″=0.011 One Hand 24″=0.013 Two Hands Sim. 24″=0.020 Position in two directions 1 2 Place screwdriver wrench over standard nut.	Time Class No. **3** T Place power screwdriv screw.
CONDITION D Positioning is much the same as Condition C but in addition may involve close tolerances, greater care of finishes, three or more points or directions of location, or application of force to assemble.	Time Class No. **3** One Hand 8″=0.019 One Hand 24″=0.921 Two Hands Sim. 24″=0.031 Position in three directions 3 2 1 Place screwdriver on screw in assembly.	This combi

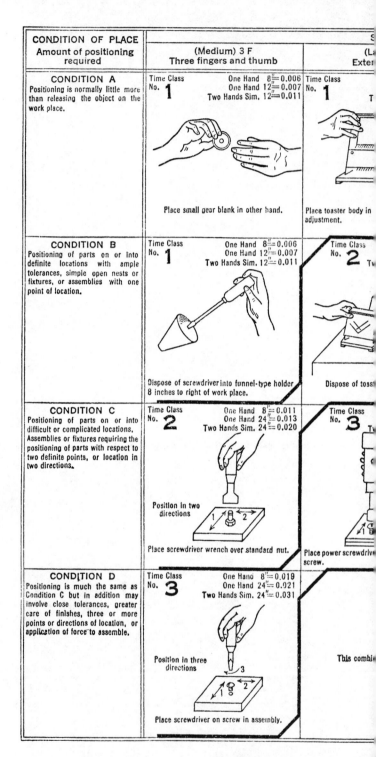

Fig. 270. Standard time values f

TYPE OF GRASP		PLACE TIME Class No.	STD. TIME Corrected for Transport Distances
(Small) 2 F Two fingers and thumb	(Very Large) 2 H Two hands		

me Class **1** One Hand 8″=0.006 One Hand 12″=0.007 Two Hands Sim. 12″=0.011		1	0.007

Time Class No. **2**   Two Hands 8″=0.011   Two Hands 12″=0.013

| | | 2 | 0.013 |

ce small machine bolt in other hand. ely used other than for placing part into er hand.

Move waffle iron partial assembly to rough position under power screwdriver.

me Class **2**   One Hand 8″=0.011   One Hand 24″=0.013   Two Hands Sim. 24″=0.020

Time Class No. **3**   Two Hands 8″=0.019   Two Hands 24″=0.021

| | | 3 | 0.021 |

e flat steel washer over stud or pin where rances are large.

Place rectangular aluminum casting in fixture.

me Class **3**   One Hand 8″=0.019   One Hand 24″=0.021   Two Hands Sim. 24″=0.031

| | | 4 | 0.026 |

Time Class No. **5**   Two Hands 8″=0.030   Two Hands 24″=0.036

Position in two directions

Place screw in tapped hole.

| | | 5 | 0.036 |

Place toaster fixture on locating pin for driving screw.

ae Class **4**   One Hand 8″=0.024   One Hand 24″=0.026   Two Hands Sim. 24″=0.039

Time Class No. **6**   Two Hands 8″=0.042   Two Hands 24″=0.048

| | | 6 | 0.048 |

ace nut on terminal in limited space where gers are cramped by design.

Place unit cover plate over unit on waffle-iron grid assembly.

or transport distances. Time in minutes.

4. Two hands (2*H*). This grasp is used where size, weight, design, or finish requires the use of two hands in moving the object, or where positioning is so difficult as to require a guiding hand.

Limitations of "Get." In establishing standards for get times it seemed reasonable to include in get only two movements—transport empty and grasp (or select and grasp). Time for grasp is affected not only by the size of part grasped, as has already been discussed, but also by the variations imposed by the physical setup of the work place or the peculiar design of parts.

Conditions of Get. Each of the four types of grasp varies with the operation conditions present. These variations are grouped into three classes, depending upon the facility with which grasp may be performed under those conditions.

Figure 269 illustrates each of the four different types of grasp (size of object) for each of the three different conditions of grasp. The standard time values in minutes are also shown.

Limitations of "Place." The establishing of place times presented a more difficult problem, since place is taken to include transport loaded, position (pre-position), and release load. Based on the amount of positioning or pre-positioning required, four classes have been determined from the study (Fig. 270).

"Dispose" Is a "Place." As condition D dispose operations are in reality a placing aside of a part, tool, or fixture, they have been evaluated on exactly the same base as place operations.

Summary. The foregoing has presented a means of classifying the great majority of the assembly operations found in the Appliance Section. It establishes a means of defining operations in a manner which largely eliminates variations in judgment on the part of those using the data.

When familiarity with the basic operation conditions has been attained, the proper classification of any operation is easily and quickly recognized. It will be noted that Figs. 269 and 270 contain only *six* different time values for the total twenty-seven separate combinations of conditions.

Use of the Computation Sheet. The Standard Times Computation Sheet may be used for estimating assembly costs on pre-production designs, or for establishing time standards on assembly operations where procedure and methods are already in effect.

EXAMPLE—ASSEMBLE PARTS OF WAFFLE IRON. The following example, taken from the assembly line of the No. 119Y197 waffle iron, is used to explain the procedure.

1. *Operation*

   Assemble unit and cover to lower waffle iron grid casting.

2. *Equipment*

   The equipment provided for this operation is:

   (1) One Millers Falls power nut-driver.

   (2) Conveyorized bench.

   (3) Chair.

   (4) Two tote boxes and floor stands.

   (5) Two cardboard boxes (4″ × 5″ × 3″ deep).

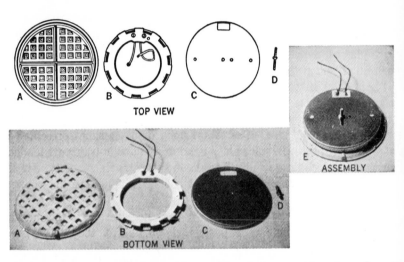

Fig. 271. Parts for waffle-iron grid assembly: *A*, lower grid casting; *B*, unit, porcelain ring and element subassembly; *C*, steel cover; *D*, stud and nut; *E*, assembly of parts *A*, *B*, *C*, and *D*.

3. *Parts*

   There are five parts (Fig. 271) used in this assembly cycle:

   (1) Lower grid-casting in tote box to operator's left.

   (2) Unit, porcelain ring and element subassembly, on conveyor belt in center of bench.

   (3) Cover, in tote box to operator's right.

   (4) Stud, ¼″−20, 1″ long, in cardboard box.

   (5) Nut, ¼″−20, in cardboard box.

4. *Fixture*

   No fixture is used.

5. *Description of Cycle*

   (1) The operator reaches into the tote box on the left and selects one grid casting.

   (2) She lifts the casting, examining the edges and face, and lays it on the bench face down, with the hinge lug toward her.

(3) She reaches with left hand to a pile of units on the conveyor, picks up the top unit, and with both hands—

(4) Places it on the grid casting, with the wire leads above the hinge lug.

(5) While the left hand straightens and raises these leads, the right hand holds the unit.

(6) The right hand then reaches into the tote box at right and obtains one cover plate, which is then—

(7) Placed over the leads onto the grid and unit.

(8) The right hand then reaches into the box containing the studs, obtains one, and—

(9) Inserts it through the cover into a tapped hole in the casting.

(10) The right hand reaches to the second box, obtains the nut, and—

(11) Starts it on the stud. During these last operations (8 through 11) the left hand has held the assembly. After starting the nut the right hand—

(12) Obtains the power driver (meanwhile the left hand has pushed the assembly to a rough position below the power driver), then—

(13) The right hand places the power driver on the stud and nut and—

(14) Operates the power driver. As the right hand—

(15) Releases the power driver, the left hand lifts the assembly from the bench. The right hand then—

(16) Grasps the assembly and—

(17) Places it on a stack of completed assemblies to the right of the work place, and returns to the work place empty.

Procedure

1. *Record All Pertinent Data for the Job*

   a. On the front or "summary" side of the computation sheet (Fig. 272) write, in the spaces indicated: the department, department number, analyst's name, drawing number of the assembly, date, and the name of the foreman in charge.

   b. Also, list under "Equipment," and note position of each item on the sketch, those pieces of equipment and all parts provided for the operation.

   c. On the back of the sheet write a brief title or description of the operation in the section headed "Operation" (section 1 of Fig. 273).

2. *Analyze the Assembly Cycle*

   a. Reduce the operation to a succession of smaller elements based on a combination of *get and place* or *get, place, use,* and *dispose* for each part and tool in the cycle.

   b. List these elements on the back of the sheet under "Elements of the Operation" (section 2 of Fig. 273).

space, tools

STANDARD TIMES COMPUTATION SHEET

DEPT.: Appliance	DEPT. NO.: 53	ANALYST: L.A. Smith
DRAWING NUMBER: 119Y197	DATE:	FOREMAN: R.T. Moore

Time to Move Tote Boxes $= \dfrac{.40 \times \text{No. Bxs.}}{\text{No. Pcs./Bx.}} = \dfrac{.40 \times 2}{20} = .040$ Min.

Fatigue Allowance $= 8\%$
Replenish Sm. Parts $= 2\%$ $\Big\} = 10\%$

Total Time in Minutes $.434 + $ Move Tote Bx. Time/Pc. $= .474 + \Big[\begin{array}{c} 10\% \\ \text{Allowance} \end{array}$ or $.047 \Big] = .521$ **STANDARD TIME IN MINUTES**

WORK PLACE   Scale: 1 unit = 3"

EQUIPMENT (Check on Work place):

1. Power Driver
2. Tote Box-Grids
3. Tote Box-Covers
4. Porcelain Units, Stacked
5. Studs
6. Nuts
7. Completed Assemblies

COMMENTS:

Studs and Nuts are Supplied in Cardboard Boxes,
   3 in. Deep x 4 in. x 5 in.

Foreman finds it necessary for operator to inspect
   all grid castings for tapped holes, grinding marks,
   and cleanliness.

GENERAL ASSEMBLY STANDARD TIMES	SIZE OF OBJECT:	M. (3F)	L. (H)	S. (2F)	V.L. (2H)	TIME CLASS
OPERATION CONDITIONS		TIME IN MINUTES				
Get Condition A	Very best grasp facility possible. The object is pre-positioned for grasp, or the grasp is not hindered by other objects in contact with the object grasped. Size of object need not be considered.	.007	.007	.007	.007	C 1
Get Condition B	Grasp is easily made but parts may be in quantities requiring some selection of a single part. No untangling or difficult separation is required.	.007	.013	.013	.013	C 2
Get Condition C	The design of parts or kind of finish prevents ready grasping. Parts may tangle, nest together, or be packed with separators, or require special handling.	.013	.021	.021	.026	C 3 / C 4
Place Condition A'	Place objects where positioning is normally little more than releasing the object or moving it slightly on the work place.	.007	.007	.007	.013	C 1 / C 2
Place Condition B	Place objects where positioning consists of some definite location, simple open nests, or fixtures with ample tolerances, or assemblies with one point of location.	.007	.013	.013	.021	C 3
Place Condition C	Place objects where positioning is in difficult or complicated locations. Assemblies or fixtures requiring the positioning of parts with respect to two definite points, or location in two directions.	.013	.021	.021	.036	C 4 / C 5
Place Condition D	Positioning is much the same as Condition C but in addition may involve close tolerances, greater care of finishes, three or more points or directions of location, or application of force to assemble.	.021	.026	.026	.048	C 6

Fig. 272. Standard times computation sheet for waffle iron assembly—front.

OPERATION: Assemble unit & cover to lower grid casting 119Y197 waffle iron.

ELEMENTS OF THE OPERATION (2)	C1 12" MAXIMUM 24" GAP .007	C1 G2 .010	C1 P2 .011	C2 24" MAXIMUM GAP .013	C2 G2 .017	C2 P2 .020	C3 24" MAXIMUM GAP .021	C3 G2 .028	C3 P2 .031	C4 24" GAP .026	C4 P2 .039	C5 24" .036	C6 24" .048	T.E.AT.LOVER 24"UP TO 36" .012	WALKING PER FOOT .007	PROCESS Power Driver	(6) TOTAL TIME IN MINUTES
1. Get & Place Grid on Table							GP							✓			.054
2. Get & Place Unit on Grid				G								P					.049
3. Get & Place Wire Leads (Adjustment Only)	GP																.014
4. Get & Place Cover							G						P	✓			.081
5. Get, Place, & Start Stud	SS						GP					P					.056
6. Get, Place, & Start Nut	SN						GP										.056
7. Get, Place, Use, & Dispose Power Driver		D G					P						P			✓	.058
8. Get & Dispose Assembly		G					D							✓			.040
9. Inspect Grid Castings (100% Inspection)																	
										.026							.434

Notes (element 9): Part of element 1 above / Examine for tapped hole, and / grinding marks on edge, and cleanliness.

G or P = "Get" or "Place" parts individually
G2 or P2 = "Get" or "Place" two parts (one with each hand) simultaneously
Walking = Transport Empty & Transport Loaded over 36 in. up to 7 ft.

(3b) PROCESS TIMES

Power Driver	.020
Tub. Riveter-Air Press	.010
Sm. Punch Press or Welder	.011
Operate Kick Press	.024
Hd. Sc. Driver/Thread	.018
Inspect/Point	.003

PROCESS

PROCESS	SMALL LIGHT	MEDIUM	LARGE HEAVY
Drive Sc.-Friction Dr.	.020	—	.036
Pitching	.015	.020	.030
Cementing	.020	.030	.040
Spinning	.020	.030	.030
Soldering (No "Place" included)	.030	.050	—

B214

Fig. 273. Standard times computation sheet for waffle iron assembly—back.

    c. Analyze each of these elements successively. Evaluate any inherent qualities or peculiarities of design or setup for each element, and determine the proper classification of each in relation to the class definitions.

    d. Check the proper class (sections 3 and 5) opposite each assembly element.

    e. Check in section 4 those elements involving transport distances greater than the allowed maximum for the class.

    f. Analyze complete assembly and note possible need for any inspection before, during, or after assembly. Indicate probable frequency of such inspection if required.

    g. Analyze the setup for allowances for replenishing both major and small parts. Make the required notations.

3 *Calculate the Assembly Operation Time*

    a. Add the check marks in each column of sections 3, 4, and 5 (Fig. 273).

    b. Multiply the standard time for each by the number of check marks in that column.

    c. Total the results and enter in column 6.

4. *Standard Time Calculation*

    a. The total time for the assembly cycle, foot of column 6 (Fig. 273), is entered on the summary page (Fig. 272), following "Total Time," in this case 0.434 minute.

    b. The allowance for replenishing parts is calculated in two parts.

      (1) Two tote boxes, 20 parts in each, supply the two large parts. Proper values are inserted in the formula

$$\frac{0.40 \text{ min.} \times \text{no. boxes}}{\text{No. pcs. per box}} = \text{time per cycle}$$

on the summary page (Fig. 272). The time per cycle is calculated and posted.

$$\frac{0.40 \times 2}{20} = 0.040 \text{ minute}$$

      (2) Replenishing small parts, nut and stud, is handled on a percentage basis. Two per cent is added to the operator's fatigue and personal allowance for this item.

    c. To the total time (*a*) is added the time per cycle for moving tote boxes. The sum of these two values is entered in the center of the fourth line (0.434 + 0.040 = 0.474 minute).

  d. The total percentage allowance given the operator for fatigue and personal time and for other reasons is then recorded to the right on the third line—in this instance, 8% fatigue and personal plus 2% for parts replenishing, or a total of 10%.
  e. This percentage (*d*) is added to the total time (*c*), and the sum in minutes is entered on the fourth line at the right, preceding "Standard Time." 0.474 + (0.10 × 0.474) = 0.474 + 0.047 = 0.521 minute.

DETAILS OF ANALYSIS

Analysis of the Cycle. In "Description of Cycle" (page 484) the operation has been outlined in step form. Breaking this outline down into elements of the operation in a form appropriate for recording on the back of the computation sheet (Fig. 273), we have:

1. Get (1) and place (2) grid on table.
2. Get (3) and place (4) unit on grid.
3. Get and place (5) wire leads. Adjustment only. (This is a simple position-adjusting operation.)
4. Get (6) and place (7) cover.
5. Get (8), place and start (9) stud.
6. Get (10), place and start (11) nut.
7. Get (12), place (13), use (14), and dispose (15) of power driver.
8. Get (16) and dispose (17) of assembly.
9. Inspect grid castings (included in 2). An inspection element is kept separate from the cycle elements, as unusual conditions may alter its importance from time to time. In this particular instance the inspection is 100%.

Analysis of Elements of the Operation. Section 1 (Operation) and section 2 (Elements of Operation) of the breakdown side of the computation sheet (Fig. 273) have now been filled in. The method of evaluating each element will now be described. In this example, the letters *G*, *P*, *D*, and *S* were used for the sake of clarity instead of check marks. They indicate the nature of the check, get, place, dispose, and start, respectively.

1. *Get and Place Grid Casting on Work Place*
   Get: The grid casting is supplied in a tote box to the operator's left. The distance from the work place is 30 inches. The grid is a *large* part leaning against others. This is a Class 3 get, as the parts are not supplied with easy grasp provided. Check *G* and *P*, Class 3, 0.021 minute.

Place: The grid is placed on the work place bottom side up, with the hinge lug toward the operator. This is a Class 3 place, as the grids are not supplied in a manner to permit this position automatically; hence, although the part does not go into a fixture, Class 3 is preferred to Class 2. Check G and P, Class 3, 0.021 minute. Since the grid casting is supplied at 30 inches, an additional transport empty and transport loaded allowance is required. Check transport empty and transport loaded over 24 inches up to 36 inches, 0.012 minute.

2. *Get and Place Unit on Grid*

Get: The porcelain units are supplied by the previous operator, who stacks them, five or six per pile, about 20 inches from the work place. No difficulty is encountered in grasping the unit. This is a Class 2 get, as the parts are in quantities, but no tangling or difficult separation is involved. Check G and P, Class 2, 0.013 minute.

Place: The unit is placed on the grid casting, lining up two recesses in the unit with two bosses on the grid. The units are always supplied right side up. This is a Class 5 place, as the part is guided by two hands into a position involving two points of location. Check P, Class 5, 0.036 minute.

3. *Get and Place Wire Leads*

This is merely an adjustment of the leads, already assembled to the unit, to facilitate the next assembly operation.

Get: The left hand grasps the two leads, which are pointing toward the operator and afford easy grasp, a simple Class 1 get. Check G and P, Class 1, 0.007 minute.

Place: The left hand raises the leads to a vertical position and straightens them as they slide through the fingers, a simple Class 1 place. Check G and P, Class 1, 0.007 minute.

4. *Get and Place Cover*

Get: The covers are supplied in a tote box to the operator's right, at a distance from the work place of about 30 inches. The cover is a large part, leaning against or piled upon others. This is a Class 3 get, as the parts are not supplied with the easy grasp provided in Class 2. Check G and P, Class 3, 0.021 minute.

Place: The cover is placed over the vertical leads from the unit (one point of location), and with two hands is placed on the assembly, lining up two holes in the cover with two bosses on the unit (second and third point of location). Large parts placed with two hands, location on 3 points of assembly, require

Class 6 place. Check *P*, Class 6, 0.048 minute. Since the cover is supplied at 30 inches, an additional transport empty and transport loaded is required. Check transport empty and transport loaded over 24 inches up to 36 inches, 0.012 minute.

5. *Get, Place, and Start Stud*

Get: Screws, nuts, and other nontangling hardware items normally are provided with Class 2 get. In this case well-designed trays have not been provided for the stud. Grasp is hindered by the size and shape of the containers, and Class 3 get is used. Check *G* and *P*, Class 3, 0.021 minute.

Place: The stud is inserted through a large hole in the cover, into a tapped hole in the grid. The stud and tapped hole are designed to facilitate assembly and would normally be Class 2, but the difficulty of placing through the hole in the cover justifies use of Class 3. Check *G* and *P*, Class 3, 0.021 minute.

Start: Place includes insertion of the stud, but rigidity requires an additional turning of the stud. Allow a simple Class 1 get and place for turning the stud an additional one or two threads. Check *twice G* and *P*, Class 1, 0.007 minute.

6. *Get, Place, and Start Nut*

Get: The nut is also supplied in an improper container. This is also a Class 3 get for the same reasons as 5 above. Check *G* and *P*, Class 3, 0.021 minute.

Place: The nut requires as much location as the stud, but in addition must be kept perpendicular to the stud in starting. This element is usually Class 3. Check *G* and *P*, Class 3, 0.021 minute.

Start: Check *twice G* and *P*, Class 1, for same reason as 5 above.

7. *Get, Place, Use, and Dispose of Power Driver*

Get: The power driver is suspended over the work place, providing nearly perfect grasp facility. This is properly a Class 1 get. However, the left hand is positioning the assembly during the get, so the get is properly a Class 1 *simultaneous* get. Check *G2*, Class 1, 0.010 minute.

Place: Placing the power driver on the stud and nut requires locating the driver in two directions (Class 3 definition). Check *G* and *P*, Class 3, 0.021 minute.

Use: Driving the stud and nut with the power driver has been determined on an average basis and a separate column incorporated for this element. Check Process Times, Power Driver, 0.020 minute.

Dispose: As the power driver is spring suspended, disposing of it is a simple motion to approximate position, and a release. This is Class 1 place. Check *G* and *P*, Class 1, 0.007 minute.

8. *Get and Dispose of Assembly*

Get: The left hand has been holding the assembly during the use of the power driver. While the right hand disposes of the driver, the left raises the assembly, positioning it for grasp. This is necessary, as the progressive assembly moves from left to right and the completed assembly is stacked at the right. Since the assembly is pre-positioned for grasp, the get is Class 1. Check *G* and *P*, Class 1, 0.007 minute.

Dispose: The assembly is placed on a stack of previously completed assemblies. This requires precise positioning, definite location. Use and place, Class 3. Check *G* and *P*, Class 3, 0.021 minute. The dispose distance is over 24 inches, so check transport empty and transport loaded over 24 inches up to 36 inches, 0.012 minute.

9. *Inspect Grid Castings, 100%*

In each cycle it is necessary to give the grid casting a quick inspection. This element is an addition to the place element (1). It has been evaluated on therblig time values.

(1) Examine casting for tapped hole:
    a. Focus eyes on hole   0.002
    b. Examine hole   0.003

(2) Examine edge for grinding marks and face for cleanliness:
    a. Turn casting over   0.003
    b. Focus eyes three times (3×0.002)   0.006
    c. Examine three points (3×0.003)   0.009
    d. Turn casting over   0.003

         Total   0.026 minute

CHAPTER 31

Systems of Predetermined Motion-Time Data: The Work-Factor System, Methods-Time Measurement, Basic Motion Timestudy

THE WORK-FACTOR SYSTEM

The Work-Factor system [1] was one of the first systems of predetermined motion-time data to have wide general use. The first actual shop application was made in 1938, and the time values were first published [2] in 1945.

This system makes it possible to determine the normal or select time for manual tasks by the use of motion-time data. First, a detailed analysis of each task is made, based on the identification of the four major variables of work and the use of Work-Factors as a unit of measure. Then the proper standard time from the table of motion-time values is applied to each motion.

A basic motion is defined as that motion which involves the least amount of difficulty or precision for any given distance and body member combination; for example, tossing a bolt into a box. Work-Factor is a unit used as the index of additional time required over and above the basic time when motions are performed involving the following variables: (1) manual control, (2) weight or resistance.

---

[1] Reproduced with permission from *Work-Factor Time Standards*, by Joseph H. Quick, James H. Duncan, and James A. Malcolm, Jr., McGraw-Hill Book Co., New York, 1962. Also see "The Work-Factor System," by Joseph H. Quick, James H. Duncan, and James A. Malcolm, Jr., *Industrial Engineering Handbook*, 2nd ed., Harold B. Maynard (editor), McGraw-Hill Book Co., New York, 1963, pp. 5-39 to 5-96.

[2] J. H. Quick, W. J. Shea, and R. E. Koehler, "Motion-Time Standards," *Factory Management and Maintenance*, Vol. 103, No. 5, pp. 97–108, May, 1945.

Four Major Variables. According to the Work-Factor system there are four major variables which affect the time to perform manual motions: (1) body member used, identified by exact definition; (2) distance moved, measured in inches; (3) manual control required, measured in Work-Factors, defined or dimensional; and (4) weight or resistance involved, measured in pounds, converted to Work-Factors.

Body Member. Work-Factor recognizes six definite body members and provides motion times for each: finger or hand, arm, forearm swivel, trunk, foot, and leg. Time values for these body members are shown in Table 43.

Distance. All distances except those with a change in direction are measured as a straight line between the starting and stopping points of the motion arc described by the body member. The point at which the distance should be measured for the various body members is shown in Table 43.

Manual Control. The following classification of. the types and degrees of control reflects the difficulty involved: Definite Stop Work-Factor, Directional Control Work-Factor (Steer), Care Work-Factor (Precaution), and Change of Direction Work-Factor.

Weight or Resistance. The effect of weight on time varies with (1) the body member used, and (2) the sex of the operator. The two variables, distance and body member, are measured in terms of inches and the member used, respectively. They are not modified or affected by Work-Factors.

To facilitate an understanding of the Work-Factor principle, the Work-Factor can be considered as merely a means of describing the motion according to the amount of control or weight (or resistance) involved in its performance.

Since the value of a Work-Factor in terms of time has been established in tabular form, it remains only for the analyst to become familiar with the specific dimensions and rules necessary to determine the number of Work-Factors involved in a given motion. Since the simplest or basic motion involves no Work-Factors, it is apparent that, as complexities are introduced to a motion, they add Work-Factors and consequently time.

The Work-Factor Motion-Time Table. The Work-Factor motion-time table (Table 43) includes all Work-Factor motion-time values in tabular form. They are so arranged that, when a motion has been identified according to the four major variables, the correct time value can be selected quickly.

Standard Elements of Work. Work-Factor recognizes the following standard elements of work:

1. Transport (Reach and Move) (TRP).
2. Grasp (GR).
3. Pre-position (PP).
4. Assemble (ASY).
5. Use (Manual, Process, or Machine Time) (US).
6. Disassemble (DSY).
7. Mental Process (MP).
8. Release (RL).

Work-Factor Notation. The symbols employed for the body members and Work-Factors are shown in Fig. 274.

Body Members	Symbol	Work-Factors (written in this sequence)	Symbol
Finger	F	Weight or Resistance	W
Hand	H	Directional Control (Steer)	S
Arm	A	Care (Precaution)	P
Forearm Swivel	FS	Change Direction	U
Trunk	T	Definite Stop	D
Foot	FT		
Leg	L		
Head Turn	HT		

Fig. 274.   Symbols for body members and Work-Factors.

Recording the Analysis. Symbols are used for recording a motion analysis. The body member is indicated first; the distance moved, second; and the Work-Factors, third. For example:

Description of Motion	Motion Analysis	Time, minutes
1. Toss small part aside 10 inches (Basic Motion)	A10	0.0042
2. Reach 20 inches to bolt in bin (Definite Stop Motion)	A20D	0.0080
3. Move 4-pound brick 30 inches from pile to place on worktable (Weight, Definite Stop Motion)	A30WD	0.0119

EXAMPLE. The analysis of motions required to get pen from holder, mark X on paper, replace pen in holder, and return hand to paper is shown in Fig. 275. The penholder is located about 12 inches from the center of the writing area.

Table 43. Work-Factor * Motion-Time Table for Detailed Analysis

(Time in ten-thousandths of a minute)

(A) ARM—Measured at Knuckles

DISTANCE MOVED	BASIC	WORK FACTORS			
		1	2	3	4
1″	18	26	34	40	46
2″	20	29	37	44	50
3″	22	32	41	50	57
4″	26	38	48	58	66
5″	29	43	55	65	75
6″	32	47	60	72	83
7″	35	51	65	78	90
8″	38	54	70	84	96
9″	40	58	74	89	102
10″	42	61	78	93	107
11″	44	63	81	98	112
12″	46	65	85	102	117
13″	47	67	88	105	121
14″	49	69	90	109	125
15″	51	71	92	113	129
16″	52	73	94	115	133
17″	54	75	96	118	137
18″	55	76	98	120	140
19″	56	78	100	122	142
20″	58	80	102	124	144
22″	61	83	106	128	148
24″	63	86	109	131	152
26″	66	90	113	135	156
28″	68	93	116	139	159
30″	70	96	119	142	163
35″	76	103	128	151	171
40″	81	109	135	159	179
Weight in Lbs. Male 2 / Fem. 1	2 / 1	7 / 3½	13 / 6½	20 / 10	UP / UP

(L) LEG—Measured at Ankle

DISTANCE MOVED	BASIC	WORK FACTORS			
		1	2	3	4
1″	21	30	39	46	53
2″	23	33	42	51	58
3″	26	37	48	57	65
4″	30	43	55	66	76
5″	34	49	63	75	86
6″	37	54	69	83	95
7″	40	59	75	90	103
8″	43	63	80	96	110
9″	46	66	85	102	117
10″	48	70	89	107	123
11″	50	72	94	112	129
12″	52	75	97	117	134
13″	54	77	101	121	139
14″	56	80	103	125	144
15″	58	82	106	130	149
16″	60	84	108	133	153
17″	62	86	111	135	158
18″	63	88	113	137	161
19″	65	90	115	140	164
20″	67	92	117	142	166
22″	70	96	121	147	171
24″	73	99	126	151	175
26″	75	103	130	155	179
28″	78	107	134	159	183
30″	81	110	137	163	187
35″	87	118	147	173	197
40″	93	126	155	182	206
Weight in Lbs. Male 8 / Fem. 4	8 / 4	42 / 21	UP / UP	—	—

(T) TRUNK—Measured at Shoulder

1"	26	38	49	58	67
2"	29	42	53	64	73
3"	32	47	60	72	82
4"	38	55	70	84	96
5"	43	62	79	95	109
6"	47	68	87	105	120
7"	51	74	95	114	130
8"	54	79	101	121	139
9"	58	84	107	128	147
10"	61	88	113	135	155
11"	63	91	118	141	162
12"	66	94	123	147	169
13"	68	97	127	153	175
14"	71	100	130	158	182
15"	73	103	133	163	188
16"	75	105	136	167	193
17"	78	108	139	170	199
18"	80	111	142	173	203
19"	82	113	145	176	206
20"	84	116	148	179	209
Weight in Lbs. Male	11	58		UP	—
Fem.	5½	29		UP	—

(F, H) FINGER-HAND—Measured at Finger Tip

1"	16	23	29	35	40
2"	17	25	32	38	44
3"	19	28	36	43	49
4"	23	33	42	50	58
Weight in Lbs. Male	⅔	2½	4	UP	—
Fem.	⅓	1¼	4	UP	—

(FT) FOOT—Measured at Toe

1"	20	29	37	44
2"	22	32	40	48
3"	24	35	45	55
4"	29	41	53	64
Weight in Lbs. Male	5	22	UP	—
Fem.	2½	11	UP	—

(FS) FOREARM SWIVEL—Measured at Knuckles

45°	17	22	28	32	37
90°	23	30	37	43	49
135°	28	36	44	52	58
180°	31	40	49	57	65
Torque Lbs. Ins. Male	3	13		UP	—
Fem.	1½	6½		UP	—

Work-Factor SYMBOLS

W — Weight or Resistance
S — Directional Control (Steer)
P — Care (Precaution)
U — Change Direction
D — Definite Stop

WALKING TIME

TYPE	30" PACES		
	1	2	OVER 2
General	Analyze from Table	260	120 + 80/Pace
Restricted		300	120 + 100/Pace

Add 100 for 120° — 180° Turn at Start or Finish

Up Steps (8" Rise — 10" Flat)	126
Down Steps	100

VISUAL INSPECTION

Focus	20
Inspect	30/Point
React	20
Head Turn	45° 40, 90° 60

1 Time Unit	= .006 Second
	= .0001 Minute
	= .00000167 Hour

Table 44. Work-Factor Grasp Table

COMPLEX GRASPS FROM RANDOM PILES

| SIZE (Major dimension or length) | | THIN FLAT OBJECTS — THICKNESS | | | | | | CYLINDERS AND REGULAR CROSS SECTIONED SOLIDS — DIAMETER | | | | | | | Add for Entangled, Nested or Slippery Objects * |
| | | SOLIDS & BRACKETS THICKNESS (over 3/64" .0469") | | (less than 1/64" 0-.0156") | | (1/64 to 3/64) .0156"-.0469" | | 0-.0625" (3/64) | .0626"-.125" (1/8) | .1251"-.1875" (3/16) | .1876"-.5000" (1/2) | | .5001" & up (over 1/2) | | |
		Blind —Simo	Visual —Simo	Blind —Simo	Visual —Simo	Blind —Simo	Visual —Simo	Blind —Simo	Blind —Simo	Blind —Simo	Blind —Simo	Visual —Simo	Blind —Simo	Visual —Simo	—Simo
.0000"-.0625"	1/16" & less	120 172	B B	—	B	131 189	B B	S S	S S	S S	S S	S S	S S	S S	17 26
.0626"-.1250"	over 1/16" to 1/8"	79 111	B B	108 154	B B	85 120	B B	85 120	S S	S S	S S	S S	S S	S S	12 18
.1251"-.1875"	over 1/8" to 3/16"	64 88	B B	102 145	B B	74 103	B B	79 111	74 103	S S	S S	S S	S S	S S	12 18
.1876"-.2500"	over 3/16" to 1/4"	48 64	B B	72 100	B B	56 76	B B	79 111	68 94	64 88	S S	S S	S S	S S	8 12
.2501"-.5000"	over 1/4" to 1/2"	40 52	B B	64 88	B B	48 64	B B	62 85	56 76	56 76	44 58	B 58	44 58	S S	8 12
.5001"-1.0000"	over 1/2" to 1"	40 52	B 32	64 88	60 82	48 64	B 58	62 85	56 76	56 76	48 64	44 58	40 52	S 32	8 12
1.0001"-4.0000"	over 1" to 4"	37 48	20 20	53 72	36 44	45 60	28 34	56 76	48 64	40 52	40 52	36 46	37 48	20 22	8 12
4.0001" & up	over 4"	46 61	20 22	70 97	44 58	62 85	36 46	56 76	48 64	40 52	40 52	36 46	37 48	20 22	9 14

B = Use Blind column since visual grasp offers no advantage.   S = Use Solid Table.

* Add the indicated allowances when objects: (a) are entangled (not requiring two hands to separate); (b) are nested together because of shape or film; (c) are slippery (as from oil or polished surface). When objects both entangle and are slippery, or both nest and are slippery, use double the value in the table.

Note: Special grasp conditions should be analyzed in detail.

Table 44 (Continued)

GRIPPING DISTANCE

Distance from Gripping Point to Alignment Point	% Addition to Alignments	Length of Upright Motion
0 – 1.99" 2 – 2.99"	Neg. 10%	1" 1"
3 – 4.99" 5 – 6.99"	20% 30%	2" 2"
7 – 9.99" 10 – 14.99"	40% 60%	3" 5"
15 – 19.99" 20" & up	80% 100%	6" 7" & up

GENERAL RULES FOR ASSEMBLY

1. When required add W and P Work Factors to all Assembly Motions according to rules for Transports.
2. Reduce number of Alignments by 50% when hand is rigidly supported.
3. Where Gripping Distance, Two Targets and Blind Targets are involved, add each percentage to Original Alignment. Don't pyramid percentages.
4. Alignments for Surface Assembly are taken from .224 column and are A1SD Motions.
5. Index is F1S, A1S or FS45°S.

DISTANCE BETWEEN TARGETS

Distance Between Targets	% Addition to Alignments	Method of Alignment
0 – .99" 1 – 1.99"	Neg. 10%	Simo Simo
2 – 2.99" 3 – 4.99"	30% 50%	Simo Simo
5 – 6.99"	70%	Simo
7 – 14.99"	Align 1st, Insert 1st, Align 2nd (1) Insert 2nd.	
15" & up	Align 1st, Insert 1st, Focus and Inspect, Align 2nd (1), Insert 2nd.	

(1) If connected, treat 2nd Assembly as open target with no upright.

BLIND TARGETS

Distance from Target to Visible Area	% Addition to Alignments	
	Permanent (Blind at all times)	Temporary (Blind during assembly)
.0 – .49" .5 – .99"	20% 30%	0% 10%
1.0 – 1.99" 2.0 – 2.99"	40% 70%	20% 30%
3.0 – 4.99" 5.0 – 6.99"	130% 250%	50% 70%
7.0 – 10.00"	380%	120%

Table 45. Work-Factor Assembly Table

TARGET DIAMETER	AVERAGE NO. OF ALIGNMENTS (A1S Motions)											
	CLOSED TARGETS						OPEN TARGETS					
	Ratio of Plug Dia. ÷ Target Dia.						Ratio of Plug Dia. ÷ Target Dia.					
	To .224	.225 to .289	.290 to .414	.415 to .899	.900 to .934	.935 to 1.000	To .224	.225 to .289	.290 to .414	.415 to .899	.900 to .934	.935 to 1.000
.875" & up	(D*) 18	(D*) 18	(D*) 18	(¼) 25	(¼**) 51	(¼***) 59	(D*) 18	(D*) 18	(D*) 18	(D*) 18	(¼**) 51	(¼***) 59
.625" to 874"	(D*) 18	(D*) 18	(SD*) 18	(¼) 25	(¼**) 51	(¼***) 59	(D*) 18	(D*) 18	(D*) 18	(SD*) 18	(¼**) 51	(¼***) 59
.375" to .624"	(SD*) 18	(SD*) 18	(¼) 25	(½) 31	(½**) 57	(½***) 65	(SD*) 18	(SD*) 18	(SD*) 18	(½) 31	(½**) 57	(½***) 65
.225" to .374"	(½) 31	(1) 44	(1) 44	(1½) 57	(1½**) 83	(1½***) 91	(¼) 25	(½) 31	(½) 31	(¾) 38	(¾**) 64	(¾***) 72
.175" to .224"	(1) 44	(1) 44	(1) 44	(1½) 57	(1½**) 83	(1½***) 91	(½) 31	(½) 31	(½) 31	(¾) 38	(¾**) 64	(¾***) 72
.125" to .174"	(1) 44	(1¼) 51	(1½) 57	(1½) 57	(1½**) 83	(1½***) 91	(¾) 38	(1) 44	(1) 44	(1) 44	(1**) 70	(1***) 78
.075" to .124"	(2½) 83	(2½) 83	(2½) 83	(2½) 83	(2½**) 109	(2½***) 117	(1¼) 51	(1¼) 51	(1¼) 51	(1¼) 51	(1¼**) 77	(1¼***) 85
.025" to .074"	(3) 96	(3) 96	(3) 96	(3) 96	(3**) 122	(3***) 130	(1½) 57	(1½) 57	(1½) 57	(1½) 57	(1½**) 83	(1½***) 91

* Letters indicate Work Factors in move preceding Assembly.

** Requires A(X)S Upright for all ratios of .900 and greater. (Table value includes A1S Upright.)

*** Requires A(Y)S Upright and A(Z)P Insert for all ratios of .935 and greater. (Table value includes A1S Upright and A1P Insert.)

Simplified and Abbreviated Work-Factor Systems. Time standards may be established by any one of the following three Work-Factor systems: Detailed, Simplified, and Abbreviated. The Detailed system has been described in the preceding paragraphs. There are some situations where one cannot justify the use of the Detailed Work-Factor system. Long-cycle operations are in this category, and studies made

Element Number	Element Description	Motion Analysis	Time, minutes
1	Reach to pen (12 inches)	A12D	0.0065
2	Grasp pen	1/2 F1	0.0008
3	Move to paper (12 inches)	A12D	0.0065
4	Position pen on paper	F1SD	0.0029
5	Make 1st stroke of X	F1D	0.0023
6	Position pen for 2nd stroke	F1D	0.0023
7	Make 2nd stroke of X	F1D	0.0023
8	Move pen to holder (12 inches)	A12SD	0.0085
9	Align pen to holder	1/4 VA1S	0.0007
10	Insert pen in holder	F1P	0.0023
11	Release pen	1/2 F1	0.0008
12	Move arm to paper (12 inches)	A12D	0.0065
		Total time	0.0424

Fig. 275.   Example of a Work-Factor analysis.

for cost estimates. The Simplified system is based on appropriate averages of the Detailed data and enables the analyst to apply the data more quickly. When the Simplified system is applied to operations to which it is suited, it is expected to provide select time values which allow 0 to 5% more time than those resulting from Detailed study.

The Abbreviated system was developed to fill the need for a very simple system of predetermined time standards. It provides a rapid measurement procedure, inasmuch as it makes use of a special time study form which contains the time data. This makes it unnecessary to refer to a separate table of time values. The Abbreviated time unit is 0.005 minute rather than the 0.0001 minute used in the other systems. When correctly applied to appropriate types of work, the accuracy of the Abbreviated system is expected to average within ±12% of the Detailed system.

METHODS-TIME MEASUREMENT

The Methods-Time Measurement (MTM) system [3] of motion-time standards was developed from motion picture studies of industrial operation, and the time standards were first published in 1948. This system is defined as a procedure which analyzes any manual operation or method into the basic motions required to perform it, and assigns to each motion a predetermined time standard which is determined by the nature of the motion and the conditions under which it is made.

Tables 46 through 55 give the motion-time data for each basic element. The unit of time used in these tables is one hundred-thousandth of an hour (0.00001 hour), and is referred to as one time-measurement unit (TMU). Thus, one TMU equals 0.0006 minute.

Reach. Reach is the basic element used when the predominant purpose is to move the hand or finger to a destination. The time for making a reach varies with the following factors: (1) condition (nature of destination), (2) length of the motion, and (3) type of reach.

Classes of Reach. There are five classes of reach (Table 46). The time to perform a reach is affected by the nature of the object towards which the reach is made.

Case A reach: to object in fixed location, or to object in other hand, or on which other hand rests.

Case B reach: to object whose general location is known. Location may vary slightly from cycle to cycle.

Case C reach: to objects jumbled with other objects in group.

Case D reach: to very small object or where accurate grasp is required.

Case E reach: to indefinite location to get hand into position for body balance, or next move, or out of the way.

The *length* of a motion is the true path, not just the straight-line distance between the two terminal points.

There are three *types* of reach to be considered: (1) hand is not moving at beginning and at end of reach, (2) hand is moving at either beginning or end of reach, and (3) hand is in motion at both beginning and end of reach.

[3] Reproduced with permission. MTM Association for Standards and Research, and "Methods-Time Measurement" by Harold B. Maynard, G. J. Stegemerten, and John L. Schwab, McGraw-Hill Book Company, New York, 1948. Also see "Methods-Time Measurement" by John L. Schwab. *Industrial Engineering Handbook,* 2nd ed., Harold B. Maynard (editor), McGraw-Hill Book Company, New York, 1963, pp. 5-13 to 5-38.

Table 46. Reach—R

(Tables courtesy MTM Association)

Distance Moved Inches	Time TMU				Hand In Motion		CASE AND DESCRIPTION
	A	B	C or D	E	A	B	
¾ or less	2.0	2.0	2.0	2.0	1.6	1.6	**A** Reach to object in fixed location, or to object in other hand or on which other hand rests.
1	2.5	2.5	3.6	2.4	2.3	2.3	
2	4.0	4.0	5.9	3.8	3.5	2.7	
3	5.3	5.3	7.3	5.3	4.5	3.6	**B** Reach to single object in location which may vary slightly from cycle to cycle.
4	6.1	6.4	8.4	6.8	4.9	4.3	
5	6.5	7.8	9.4	7.4	5.3	5.0	
6	7.0	8.6	10.1	8.0	5.7	5.7	
7	7.4	9.3	10.8	8.7	6.1	6.5	
8	7.9	10.1	11.5	9.3	6.5	7.2	**C** Reach to object jumbled with other objects in a group so that search and select occur.
9	8.3	10.8	12.2	9.9	6.9	7.9	
10	8.7	11.5	12.9	10.5	7.3	8.6	
12	9.6	12.9	14.2	11.8	8.1	10.1	
14	10.5	14.4	15.6	13.0	8.9	11.5	**D** Reach to a very small object or where accurate grasp is required.
16	11.4	15.8	17.0	14.2	9.7	12.9	
18	12.3	17.2	18.4	15.5	10.5	14.4	
20	13.1	18.6	19.8	16.7	11.3	15.8	
22	14.0	20.1	21.2	18.0	12.1	17.3	**E** Reach to indefinite location to get hand in position for body balance or next motion or out of way.
24	14.9	21.5	22.5	19.2	12.9	18.8	
26	15.8	22.9	23.9	20.4	13.7	20.2	
28	16.7	24.4	25.3	21.7	14.5	21.7	
30	17.5	25.8	26.7	22.9	15.3	23.2	

Move. Move is the basic element used when the predominant purpose is to transport an object to a destination. There are three classes of moves:

Case A move: object to other hand or against stop.
Case B move: object to approximate or indefinite location.
Case C move: object to exact location.

The time for move is affected by the following variables: (1) condition (nature of destination), (2) length of the motion, (3) type of move, and (4) weight factor, static and dynamic.

The time for move is affected by its *length,* in a manner similar to reach. The three *types* of moves are the same as those described for reach. Additional time is needed when an object is moved or a force is applied (above 2.5 pounds), as indicated in Table 47.

Turn. Turn is the motion employed to turn the hand, either empty or loaded, by a movement that rotates the hand, wrist, and forearm about the long axis of the forearm. The time for a turn depends on two variables: (1) degrees turned, and (2) weight factor, as indicated in Table 48.

Grasp. Grasp is the basic element employed when the predominant purpose is to secure sufficient control of one or more objects with the

Table 47. Move—M

Distance Moved Inches	Time TMU				Wt. Allowance			CASE AND DESCRIPTION
	A	B	C	Hand In Motion B	Wt. (lb.) Up to	Fac- tor	Con- stant TMU	
¾ or less	2.0	2.0	2.0	1.7	2.5	0	0	**A** Move object to other hand or against stop.
1	2.5	2.9	3.4	2.3				
2	3.6	4.6	5.2	2.9	7.5	1.06´	2.2	
3	4.9	5.7	6.7	3.6				
4	6.1	6.9	8.0	4.3	12.5	1.11	3.9	
5	7.3	8.0	9.2	5.0				
6	8.1	8.9	10.3	5.7	17.5	1.17	5.6	
7	8.9	9.7	11.1	6.5				**B** Move object to approximate or in- definite location.
8	9.7	10.6	11.8	7.2				
9	10.5	11.5	12.7	7.9	22.5	1.22	7.4	
10	11.3	12.2	13.5	8.6				
12	12.9	13.4	15.2	10.0	27.5	1.28	9.1	
14	14.4	14.6	16.9	11.4				
16	16.0	15.8	18.7	12.8	32.5	1.33	10.8	
18	17.6	17.0	20.4	14.2				
20	19.2	18.2	22.1	15.6	37.5	1.39	12.5	
22	20.8	19.4	23.8	17.0				**C** Move object to ex- act location.
24	22.4	20.6	25.5	18.4	42.5	1.44	14.3	
26	24.0	21.8	27.3	19.8				
28	25.5	23.1	29.0	21.2	47.5	1.50	16.0	
30	27.1	24.3	30.7	22.7				

fingers or hand to permit the performance of the next required basic element. The classes of grasps, with a description of each type and the time values for each, are given in Table 49.

Position. Position is the basic element employed to align, orient, and engage one object with another object, where the motions used are so minor that they do not justify classification as other basic elements. The time for position is affected by (1) class of fit, (2) symmetry, and (3) ease of handling, as shown in Table 50.

Release Load. Release load is the basic element to relinquish control of an object by the fingers or hand (Table 51). The two classifications of release are (1) normal release, simple opening of the fingers; and (2) contact release, the release begins and is completed at the instant the following reach begins (no time allowed).

Table 48. Turn and Apply Pressure—T and AP

Weight		Time TMU for Degrees Turned										
		30°	45°	60°	75°	90°	105°	120°	135°	150°	165°	180°
Small—	0 to 2 Pounds	2.8	3.5	4.1	4.8	5.4	6.1	6.8	7.4	8.1	8.7	9.4
Medium—	2.1 to 10 Pounds	4.4	5.5	6.5	7.5	8.5	9.6	10.6	11.6	12.7	13.7	14.8
Large—	10.1 to 35 Pounds	8.4	10.5	12.3	14.4	16.2	18.3	20.4	22.2	24.3	26.1	28.2
APPLY PRESSURE CASE 1—16.2 TMU. APPLY PRESSURE CASE 2—10.6 TMU.												

Table 49. Grasp—G

Case	Time TMU	DESCRIPTION
1A	2.0	**Pick Up Grasp**—Small, medium or large object by itself, easily grasped.
1B	3.5	Very small object or object lying close against a flat surface.
1C1	7.3	Interference with grasp on bottom and one side of nearly cylindrical object. Diameter larger than ½″.
1C2	8.7	Interference with grasp on bottom and one side of nearly cylindrical object. Diameter ¼″ to ½″.
1C3	10.8	Interference with grasp on bottom and one side of nearly cylindrical object. Diameter less than ¼″.
2	5.6	**Regrasp.**
3	5.6	**Transfer Grasp.**
4A	7.3	Object jumbled with other objects so search and select occur. Larger than 1″ x 1″ x 1″.
4B	9.1	Object jumbled with other objects so search and select occur. ¼″ x ¼″ x ⅛″ to 1″ x 1″ x 1″.
4C	12.9	Object jumbled with other objects so search and select occur. Smaller than ¼″ x ¼″ x ⅛″.
5	0	Contact, sliding or hook grasp.

Disengage. Disengage is the basic element used to break contact between one object and another. It includes an involuntary movement resulting from the sudden ending of resistance. The time for disengage is affected by the following three variables: (1) class of fit, (2) ease of handling, and (3) care of handling (Table 52).

Eye Times. In most work, time for moving and focusing the eye is not a limiting factor and consequently does not affect the time for the operation. However, when the eyes do direct the hands or body movements, eye times must be considered. There are two types of eye time, eye focus time and eye travel time.

Eye focus time is the time required to focus the eyes on an object and look at it long enough to determine certain readily distinguishable

Table 50. Position *—P

CLASS OF FIT		Symmetry	Easy To Handle	Difficult To Handle
1—Loose	No pressure required	S	5.6	11.2
		SS	9.1	14.7
		NS	10.4	16.0
2—Close	Light pressure required	S	16.2	21.8
		SS	19.7	25.3
		NS	21.0	26.6
3—Exact	Heavy pressure required.	S	43.0	48.6
		SS	46.5	52.1
		NS	47.8	53.4

*Distance moved to engage—1″ or less.

Table 51. Release—RL

Case	Time TMU	DESCRIPTION
1	2.0	Normal release performed by opening fingers as independent motion.
2	0	Contact Release.

Table 52. Disengage—D

CLASS OF FIT	Easy to Handle	Difficult to Handle
1—Loose—Very slight effort, blends with subsequent move.	4.0	5.7
2—Close — Normal effort, slight recoil.	7.5	11.8
3—Tight — Considerable effort, hand recoils markedly.	22.9	34.7

characteristics within the area which may be seen without shifting the eyes.

Eye travel time is affected by the distance between points from and to which the eye travels, and the perpendicular distance from the eye to the line of travel, as indicated in Table 53.

Body, Leg, and Foot Motions. The body, leg, and foot motions are described in Table 54, and the time values are also shown in this table.

Limiting Motions. In performing most industrial operations, it is desirable to have more than one body member in motion at a time. Usually the most effective method of performing an operation can be approached when two or more body members are in motion at the same time. If two or more motions are combined or overlapped, all can be performed in the time required to perform the one demanding the greatest amount of time, or the limiting motion. When two motions are performed at the same time by the body member, they are called *combined motions.* When the two motions are performed by different body members they are called *simultaneous motions.* Table 55 is a guide to limiting motions, although it does not apply in every case.

Conventions for Recording MTM. It has been found convenient to develop a code for referring to the various classes of motions. It would be awkward, for example, to have to refer to a "Case B reach 10 inches long with hand in motion at the end" in so many words every time a

Table 53. Eye Travel Time and Eye Focus—ET and EF

Eye Travel Time $= 15.2 \times \dfrac{T}{D}$ TMU, with a maximum value of 20 TMU.

where $T =$ the distance between points from and to which the eye travels.
$D =$ the perpendicular distance from the eye to the line of travel T.

Eye Focus Time $= 7.3$ TMU.

Table 54. Body, Leg, and Foot Motions

DESCRIPTION	SYMBOL	DISTANCE	TIME TMU
Foot Motion—Hinged at Ankle.	FM	Up to 4″	8.5
With heavy pressure.	FMP		19.1
Leg or Foreleg Motion.	LM —	Up to 6″	7.1
		Each add'l. inch	1.2
Sidestep—Case 1—Complete when lead-ing leg contacts floor.	SS-C1	Less than 12″	Use REACH or MOVE Time
		12″	17.0
		Each add'l. inch	.6
Case 2—Lagging leg must contact floor before next motion can be made.	SS-C2	12″	34.1
		Each add'l. inch	1.1
Bend, Stoop, or Kneel on One Knee.	B,S,KOK		29.0
Arise.	AB,AS,AKOK		31.9
Kneel on Floor—Both Knees.	KBK		69.4
Arise.	AKBK		76.7
Sit.	SIT		34.7
Stand from Sitting Position.	STD		43.4
Turn Body 45 to 90 degrees—			
Case 1—Complete when leading leg contacts floor.	TBC1		18.6
Case 2—Lagging leg must contact floor before next motion can be made.	TBC2		37.2
Walk.	W-FT.	Per Foot	5.3
Walk.	W-P	Per Pace	15.0

motion of that sort was encountered. Therefore, for convenience, this is coded as R10Bm. Table 56 gives the coding for all types of motions.

When these symbols are recorded, they are written down in such a way as to indicate the hand making the motions, the sequence, and the time values.

LH	TMU	RH
R12C	14.2	
G4A	7.3	
M10A	11.3	
G3	5.6	G3
	5.2	M2C
	5.6	P1SE
	2.0	RL1
Total	51.2	

This indicates that the following motions take place. The left hand makes a 12-inch Case C reach followed by a G4A to pick up an object. The left hand then moves the object back to the other hand. A trans-

Table 55. Simultaneous Motions

REACH				MOVE							GRASP						POSITION							DISENGAGE											
											G1A G2 G5		G1B G1C		G4		P1S		P1SS P2S		P1NS P2SS P2NS		D1E D1D		D2										

□ = EASY to perform simultaneously.

⊠ = Can be performed simultaneously with **PRACTICE.**

■ = DIFFICULT to perform simultaneously even after long practice. Allow both times.

MOTIONS NOT INCLUDED IN ABOVE TABLE

TURN—Normally EASY with all motions except when TURN is controlled or with DISENGAGE.

APPLY PRESSURE—May be EASY, PRACTICE, or DIFFICULT. Each case must be analyzed.

POSITION—Class 3—Always DIFFICULT.

DISENGAGE—Class 3—Normally DIFFICULT.

RELEASE—Always EASY.

DISENGAGE—Any class may be DIFFICULT if care must be exercised to avoid injury or damage to object.

CASE / MOTION column: A, E — B — C, D (REACH); A, Bm — B — C (MOVE); G1A, G2, G5 — G1B, G1C — G4 (GRASP); P1S — P1SS, P2S — P1NS, P2SS, P2NS (POSITION); D1E, D1D — D2 (DISENGAGE)

* W = Within the area of normal vision.
O = Outside the area of normal vision.
**E = EASY to handle.
D = DIFFICULT to handle.

fer grasp puts the object in the right hand, which then moves it 2 inches to an exact location, positions it, and releases it.

EXAMPLE. The analysis shown in Fig. 276 includes the motions required in order to dispose of one part and obtain the next in a given layout.

Simplified Data. A table of simplified data is available for use where ease of application is an important factor. These data include 15% allowances, and the figures have been rounded to the nearest whole number.

BASIC MOTION TIMESTUDY

Basic Motion Timestudy (BMT) is a system of predetermined motion-time standards [4] developed by Ralph Presgrave, G. B. Bailey,

[4] Reproduced with permission from *Basic Motion Timestudy*, by G. B. Bailey and Ralph Presgrave, McGraw-Hill Book Co., New York, 1958. Also see "Basic Motion Timestudy," by Ralph Presgrave and G. B. Bailey, *Industrial Engineering Handbook*, 2nd ed., H. B. Maynard (editor), McGraw-Hill Book Co., New York, 1963, pp. 5-97 to 5-106.

and other members of the staff of J. D. Woods and Gordon, Ltd., of Toronto, Canada, and first used in 1950.

A basic motion is defined as a single complete movement of a body member. A basic motion occurs every time a body member, being at rest, moves and again comes to rest.

The factors that BMT takes into consideration include (1) the distance moved, (2) the visual attention needed to complete the motion, (3) the degree of precision required in grasping or positioning, (4) the amount of force needed in handling weight, and (5) simultaneous performance of two motions.

Classification of Motions. The degree of muscular control of the fingers, hands, and arms is divided into three types or classes.

Class A motion—stopped without muscular control by impact with an object. This is the simplest in type and has the lowest time value.

EXAMPLE. A motion stopped by impact with a solid object, as in the downstroke of a hammer or slamming an open drawer shut, keeping the hand on the drawer until it strikes the stops.

Table 56. Conventions for Recording MTM

Table	Example	Significance
46	R8C	Reach, 8 inches, Case C
	R12Am	Reach, 12 inches, Case A, hand in motion at end.
47	M6A	Move, 6 inches, Case A, object weighs less than 2.5 pounds
	mM10C	Move, 10 inches, Case C, hand in motion at the beginning, object less than 2.5 pounds
	M16B15	Move, 16 inches, Case B, object weighs 15 pounds
48	T30	Turn hand 30 degrees
	T90L	Turn object weighing more than 10 pounds 90 degrees
	AP1	Apply pressure, includes regrasp
49	G1A	Grasp, Case G1A
50	P1NSD	Position, Class 1 fit, nonsymmetrical part, difficult to handle
51	RL1	Release, Case 1
52	D2E	Disengage, Class 2 fit, easy to handle
53	EF	Eye focus
	ET14/10	Eye travel between points 14″ apart where line of travel is 10″ from eyes
54	FM	Foot motion
	SS16C1	Sidestep, 16 inches, Case 1
	TBC1	Turn body, Case 1
	W4P	Walk four paces

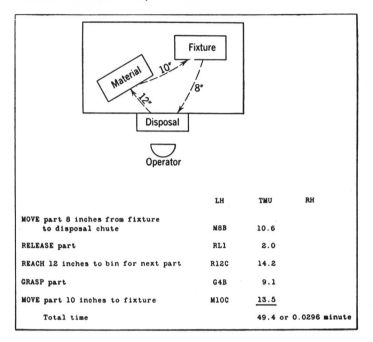

	LH	TMU	RH
MOVE part 8 inches from fixture to disposal chute	M8B	10.6	
RELEASE part	RL1	2.0	
REACH 12 inches to bin for next part	R12C	14.2	
GRASP part	G4B	9.1	
MOVE part 10 inches to fixture	M10C	13.5	
Total time		49.4 or 0.0296 minute	

Fig. 276. Example of MTM analysis—dispose of one part and obtain the next.

Class B motion—stopped entirely by the use of muscular control. Because the element of deceleration is introduced, class B motions take more time than do class A motions.

EXAMPLE. A motion stopped in mid-air by muscular control without coming in contact with any object—the upstroke in hammering or tossing objects aside.

Class C motion—stopped by the use of muscular control both to slow down the motion and to end it in a grasping or placing action. Muscular effort is used here to slow down the motion before the object is grasped or placed in position.

EXAMPLE. Reaching for a telephone or desk pad, or carrying and placing desk pad on top of desk.

Visual Direction. The time to perform a motion is affected by whether or not the eyes move as the arm moves. If the eyes move to the ending point of a motion as it is taking place, the motion time is

Table 57. Reach or Move

(Time in ten-thousandths of a minute)

Inches	½	1	2	3	4	5	6	7	8	9	10	12	14	16	18	20	22	24	26	28	30
A	27	30	36	39	42	45	47	50	52	54	56	60	64	68	72	76	80	84	88	92	96
B	32	36	42	46	49	52	55	58	60	62	64	68	72	76	80	84	88	92	96	100	104
BV	36	42	48	53	57	60	63	66	68	70	73	77	81	85	89	93	97	101	105	109	113
C	41	48	55	60	64	68	71	74	77	79	81	86	90	94	98	102	107	111	115	119	123
CV	45	54	62	67	72	76	79	82	85	87	90	95	99	104	108	112	116	120	124	128	132

greater than if there is no eye movement. When an eye movement is needed to complete the motions, they are said to be visually directed.

Class BV motion: a class B motion visually directed.
Class CV motion: a class C motion visually directed.

It should be noted that class BV and CV motions occur only when the eyes move with the hand. If the eyes can be fixed on the ending point of the motion before it starts, the basic arm motion is not delayed and no allowance is necessary for visual direction.

Reach or Move. Table 57 shows the times for reaches and moves (transport empty and transport loaded) for varying distances in inches, and for the different classes of motions. The time values in this and the following tables are expressed in ten-thousandths of a minute (0.0001). They are net; they do not include allowances for personal time, fatigue, or incidental delays.

Turn. The times for the turn motion are shown in Table 58. Turns represent a specialized phase of the move and reach motions, requiring higher time values for corresponding distances because of differences in the degree of control required. They are measured in degrees rather than inches, and are classified like other arm motions.

Table 58. Turn

(Time in ten-thousandths of a minute)

Degrees	30	45	60	75	90	120	150	180
A	26	29	32	34	37	43	49	54
B	33	36	40	43	47	54	60	67
BV	40	44	48	52	56	65	72	80
C	56	60	64	68	72	81	88	96
CV	73	77	81	85	89	98	105	113

EXAMPLE. Rotating the arm in using a screw driver, or turning a knob or door handle.

Precision Requirements in Grasping and Positioning. Precision is the term applied to the extra muscular control required when a motion ends in grasping a small object or in placing an object in an exact location.

The degree of precision needed in any motion situation can be stated in quite definite terms. In the case of motions that end with a grasp, this is done by determining the limits within which the fingertips must be located in order to make a satisfactory grasp.

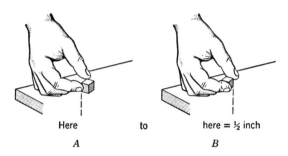

Here   to   here = ½ inch

A   *B*

Fig. 277. Grasping a single ¼-inch cube.

Illustrations *A* and *B* in Fig. 277 show the grasping of a single ¼-inch cube. In *A* the fingers are grasping the cube as close to the left side as possible, and in *B* they have moved along to grasp as far to the right side as possible. The distance between these two positions (½ inch) is a measurement of the precision.

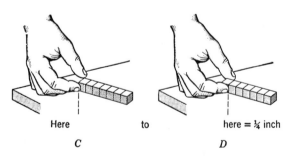

Here   to   here = ¼ inch

C   *D*

Fig. 278. Grasping the end cube from a row of ¼-inch cubes.

Illustrations *C* and *D* in Fig. 278 show the grasping of the end cube from a row of ¼-inch cubes. To grasp the end without disturbing the others, the movement of the fingers is now restricted to a distance of ¼ inch. Under these circumstances, ¼ inch is the measurement of the precision.

Precision requirements for motions that end in placing an object are found by measuring the differ-
ence in the size of the objects at the point of contact. Figure 279 shows how precision is measured in placing a ¼-inch round peg in a ½-inch round hole. In this case, precision is ¼ inch.

A precision allowance is made whenever (*a*) the object to be grasped measures less than ⅛ inch above the surface that supports it, or (*b*) the object to be grasped provides a horizontal target of not more than ½ inch in length, or (*c*) the tolerance or clearance between objects to be fitted together is not over ½ inch.

Time values for precision allowances are given in Table 59.

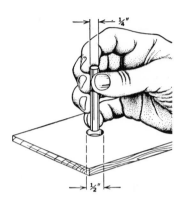

Fig. 279. Placing a ¼-inch round peg in a ½-inch round hole. The precision is ¼ inch.

EXAMPLE. The allowance for a 20-inch arm motion completed within pre-
cision limits of ¼ inch is 48. The time values obtained from this table should be added to those of Table 57, depending upon the degree of precision.

Simultaneous Motions. The time for moves and reaches may be affected when motions are performed simultaneously. Time must be added when the end points of both motions require visual direction and one hand has to wait for the eyes to direct the other to the end of the motion. The amount of time that must be added depends upon (*a*) the distance between the end points of the two motions, and (*b*) the degree

Table 59. Precision

(Time in ten-thousandths of a minute)

Inches	1	2	3	4	5	6	7	8	9	10	12	14	16	18	20	22	24	26	28	30
½″ tol.	3	4	6	7	8	9	10	11	12	13	14	16	17	18	19	20	21	22	23	24
¾″ tol.	13	16	18	21	23	25	27	29	31	32	36	39	42	45	48	51	53	55	57	59
⅜″ tol.	33	37	41	45	48	52	55	58	60	62	67	72	76	80	83	87	91	94	98	101
¹⁄₁₆″ tol.	60	65	69	73	76	80	83	87	90	93	98	103	107	112	115	119	123	127	131	135
½″ tol.	90	97	102	106	110	114	117	120	123	126	131	135	139	143	147	150	153	157	161	165

Table 60. Simultaneous Motions

(Time in ten-thousandths of a minute)

Separation Distance	0	2	4	6	8	10	12	14	16	18	20	22	24
¼" tol. and over	0	10	18	27	34	41	47	54	59	65	69	74	78
⅛" tol.	0	12	21	30	37	44	51	57	63	68	73	78	82
¹⁄₁₆" tol.	0	15	27	37	45	53	61	68	75	80	86	91	96
¹⁄₃₂" tol.	0	19	34	47	58	68	77	84	90	97	103	107	111

of precision required to end the motions. The time values for performing simultaneous motions shown in Table 60 are added to those for a single arm motion.

EXAMPLE. The allowance for simultaneous arm motions ending 12 inches apart within precision limits of ¼ inch is 47.

Force Factor. Whenever a heavy object must be handled or when friction must be overcome, added muscular effort is required. This extra effort is called force. Times for the force factor are shown in Table 61.

The force factor is introduced in three phases, which may occur singly or combined: (1) to apply pressure when grasping an object, in order to gain control of the weight; (2) after control is gained, to

Table 61. Force

(Time in ten-thousandths of a minute)

Apply Pressure, Start, or Stop			
Inches	6	12	24
2 pounds	2	3	3
4	6	6	7
6	8	9	10
8	10	11	13
10	13	14	16
15	18	20	22
20	23	26	28
30	31	35	38
40	38	43	47
50	45	50	55

overcome inertia and start the weight in motion; (3) toward the end of a motion, to apply restraining muscular effort to overcome momentum and bring the weight to a stop.

When a motion consists of picking up, moving, and placing an object of significant weight, all three phases of the force factor are present. For instance, to move a 10-pound object 24 inches, we have times of 16 each for the apply pressure, start, and the stop. This total of 48 would be added to the normal move time (Table 57) for 24 inches.

Where force is needed to overcome friction, a basic allowance is made in terms of the pounds of force required.

EXAMPLE. If a force equal to 6 pounds is needed to tighten or loosen an object, a basic allowance of 8 is added to the normal move time.

Body Motions. Basic Motion Timestudy (BMT) time values for body motions other than those of the fingers, hands, and arms are given in Table 62, with the symbol used for each.

Eye Time. An eye time allowance is used whenever (a) the start of an arm motion is delayed until the eyes are transferred from one point of focus to another, or (b) the new point of focus is different

Table 62. Body Motions

(Time in ten-thousandths of a minute)

Symbol	Units	Description
LM (1″–6″)	50	Leg motion
Add per inch	2	
FM	55	Foot motion
W	100	Walk one pace
SS₁ (1″–6″)	60	Side step
Add per inch	2	
SS₂ (1″–6″)	120	Side step
Add per inch	4	
TB₁	110	Turn body
TB₂	220	Turn body
B	180	Bend
S	180	Stoop
K₁	180	Kneel on one knee
AB etc.	200	Arise
K₂	440	Kneel on knees
AK₃	480	Arise from knees
SIT	220	Sit
STAND	270	Stand

from the ending point of the arm motion that is delayed. The time value for this allowance is 80.

Motion Description	Code	Motion Time
1. Reach with left hand to drawer (12 inches) and grasp drawer pull	R12C	86
2. Pull drawer open (8 inches)	M8B	60
3. Reach with right hand into drawer (6 inches) and grasp pencil	R6C	71
4. Remove pencil from drawer (6 inches)	M6B	55
5. Slam drawer shut (8 inches)	M8A	52
Total motion time		324

Normal time = 324 × 0.0001 = 0.0324 minute

Fig. 280. Example of a Basic Motion Timestudy analysis—open desk drawer, remove pencil, and close drawer.

Example. Figure 280 shows an example of the use of BMT data to determine the time required to perform the operation "open desk drawer, remove pencil, and close drawer."

CHAPTER 32

Work Sampling

Work sampling was first used by L. H. C. Tippett[1] in the British textile industry, and it was introduced into this country under the name of "ratio delay" in 1940. Work sampling is a fact-finding tool. In many cases, needed information about men or machines can be obtained in less time and at lower cost by this method than by other means.

Work sampling has three main uses: (1) *ratio delay*—to measure the activities and delays of men or machines—for example, to determine the percentage of the day that a man is working and the percentage that he is not working;[2] (2) *performance sampling*—to measure working time and nonworking time of a person on a manual task, and to establish a performance index or performance level for the person during his working time;[3] (3) *work measurement*—under certain circumstances, to measure a manual task, that is, to establish a time standard for an operation.

Work sampling is based upon the laws of probability. A sample taken at random from a large group tends to have the same pattern of distribution as the large group or universe. If the sample is large

[1] L. H. C. Tippett, "Statistical Methods in Textile Research. Uses of the Binominal and Poisson Distributions. A Snap-Reading Method of Making Time Studies of Machines and Operatives in Factory Surveys," *Shirley Institute Memoirs,* Vol. 13, pp. 35–93, November, 1934. Also, *Journal of the Textile Institute Transactions,* Vol. 26, pp. 51–55, 75, February, 1935.

[2] D. S. Correll and Ralph M. Barnes, "Industrial Application of the Ratio-Delay Method," *Advanced Management,* Vol. 15, No. 8 and No. 9, August and September, 1950.

[3] Ralph M. Barnes and Robert B. Andrews, "Performance Sampling in Work Measurement," *Journal of Industrial Engineering,* Vol. 6, No. 6, November–December, 1955; also Ralph M. Barnes, *Work Sampling,* 2nd ed., John Wiley & Sons, New York, 1957, pp. 194–221.

511

enough, the characteristics of the sample will differ but little from the characteristics of the group. *Sample* is the term used for this small number, and *population* or *universe* is the term used for the large group. Obtaining and analyzing only a part of the universe is known as *sampling*.

Simple Example of Work Sampling. The determination of the percentage of the working day that the operator or machine is working or idle is based on the theory that the percentage *number* of observations recording the man or machine as idle is a reliable measure of the percentage *time* that the operation is in the delay state. The accuracy of the result is a function of the number of observations taken.

Briefly, the work sampling procedure in its simplest form consists in making observations at random intervals of one or more operators or machines and noting whether they are working or idle. If the operator is working, he is given a tally mark under "working"; if he is idle, he is given a tally mark under "idle." The percentage of the day that the worker is idle is the ratio of the number of idle tally marks to the total number of idle and working tally marks.

State	Tally	Total
Working	ⱵⱵⱵⱵⱵⱵⱵ|	36
Idle	||||	4

Fig. 281. Tally of working time and idle time.

In Fig. 281 there are 36 working observations and 4 idle observations, or a total of 40 observations. In this example the percentage of idle time is $4 \div 40 \times 100 = 10\%$. Working time is $36 \div 40 \times 100\% = 90\%$. If this study covered one operator for an 8-hour day, the results would indicate that the operator was idle 10% or 48 minutes of the day $(480 \times 0.10 = 48)$, and was working 90% or 432 minutes of the day $(480 \times 0.90 = 432)$.

We have found that a demonstration showing how a problem is solved by sampling is of value in explaining this technique. Figure 282 shows a panel with 480 blocks representing the 480 minutes of an 8-hour day arranged across the bottom of the board. The white blocks represent working time, and the colored blocks idle time. By drawing numbers from a hat or by the use of a random number table, it is possible to simulate a work sampling study.

Figure 284 shows the working time and idle time of one operator for one day as obtained from a continuous time study. Random observations are shown on the bars.

Figure 285 shows the results of an all-day time study of one operator during five consecutive 8-hour working days. It is suggested that you determine the percentage of idle time for this operator for the week using the random sampling method—just follow the instructions given in Figs. 285 and 286.

Fig. 282.   Work sampling demonstration panel.

Random sampling requires that there be no bias in the sampling process. Each part comprising the universe must have as much chance of being drawn as any other. It is important that the concept of randomness be understood and carefully followed in work sampling studies.[4]

The Normal Distribution Curve. The normal distribution curve is typical of the kind of frequency distribution which is of importance in work sampling because it represents graphically the probability of the occurrence of certain chance phenomena. The normal curve is significant because of the relationship of the area under the curve between ordinates at various distances on either side of the mean ordinant. In the upper curve in Fig. 283, the shaded area represents 1 sigma, or one standard deviation on either side of the mean ordinate A. This area will always be 68.27% of the total area under the curve. The area

[4] There is some evidence that the intervals may be regular if the activity or process being observed is random. See Harold Davis, "A Mathematical Evaluation of a Work Sampling Technique," *Naval Research Logistics Quarterly*, Vol. 2, No. 1 and 2, pp. 11–117, March–June, 1955.

at 2 sigma equals 95.45%, and the area at 3 sigma equals 99.73% of the total area.

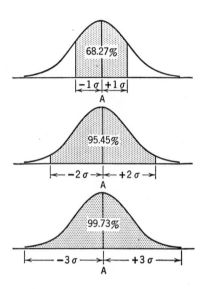

Fig. 283.   Areas under the normal curve.

Confidence Level. At the outset it is necessary to decide what level of confidence is desired in the final work sampling results. The most common confidence interval is 95%. The area under the curve at 2 sigma or two standard deviations is 95.45%, which, if rounded off, gives 95%. This means that the probability is that 95% of the time the random observations will represent the facts, and 5% of the time they will not. One sigma would give a confidence interval of 68% (68.27% rounded off to 68%). This means that the data obtained by random sampling has a 68% chance of representing the facts, and that it will be in error 32% of the time.

The formula for determining the sample size for a confidence level of 68% or 1 sigma is

$$Sp = \sqrt{\frac{p(1 - p)}{N}}$$

where S = desired relative accuracy

p = percentage expressed as a decimal

N = number of random observations (sample size)

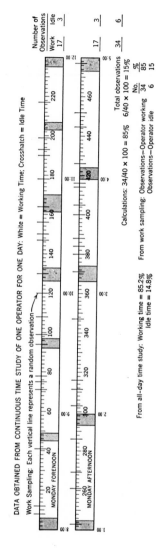

Fig. 284. Simple work sampling study. Results of random observations shown on the bars representing one working day.

DEMONSTRATION OF THE WORK SAMPLING METHOD

The bars below represent to scale the 240 minutes of the forenoon and the afternoon for five working days—Monday through Friday—a full 40-hour (2400-minute) week. The results of a continuous time study of one operator for one week are shown. White = working time; crosshatch = idle time. The total actual *working* time for the week from time study = 2035 minutes. The total actual *idle* time for the week from time study = 365 minutes.

$$\text{Percentage working time} = 2035/2400 \times 100 = 84.8\%$$
$$\text{Percentage idle time} = 365/2400 \times 100 = 15.2\%$$

Now see how you can obtain similar information by the use of random sampling. You can make your own random observations by following the instructions on Fig. 286.

RESULTS

Count the number of times your marks intersect the crosshatched portion of the bars and post this number in the box at the end of the line. Then add the number of idle observations and divide this total by 200. This gives you the percentage of the week that the operator was idle by the random sampling procedure. Now compare your answer with the actual idle percentage of 15.2%, which was originally obtained by time study.

Accuracy of Work Sampling Measurement. The accuracy of work sampling results deserves more than passing consideration, for when we determine the degree of accuracy desired, we are in effect determining the number of observations required. The number of observations, of course, affects the time and cost of making the study. The purpose of the work sampling study will suggest the degree of accuracy of the results desired, but there may be considerable latitude in specifying accuracy.

In designing the work sampling study the analyst will size up the entire situation. He will want results that will be satisfactory from an accuracy viewpoint and at the same time will not require an unreasonably large number of observations. Fortunately, in a work sampling study the analyst can determine in advance the number of observations needed for a given degree of accuracy.

One of the things the analyst will consider, consciously or unconsciously, is the inherent variability of the men, machines, or processes being measured. A department that is operating week in and week out with a steady volume of output, with raw material of uniform quality, low labor turnover, and good supervision, presents an ideal subject for work sampling, or for measurement by time study for that matter.

For many kinds of measurement an accuracy of ±5% is considered satisfactory. This is sometimes referred to as the standard error of the percentage. For the following illustrations we will assume that a confidence level of 95% and an accuracy of ±5% are satisfactory. Also assuming that the binomial distribution is used as the basis for determining the error, then the formula for determining the number of observations required is

$$Sp = 2\sqrt{\frac{p(1-p)}{N}} \tag{1}$$

where S = desired relative accuracy

  p = percentage occurrence of an activity or delay being measured, expressed as a percentage of the total number of observations or as a decimal, that is, $15\% = 0.15$

  N = total number of random observations (sample size)

Even if the desired accuracy is known, there are still two unknowns in the equation: p, the percentage occurrence, and N, the total number

Fig. 286. Demonstration of work sampling method—work sheet.

INSTRUCTIONS

Place a sheet of carbon paper over Fig. 285. Draw *at random* 20 vertical marks across each of the ten lines shown above. Do not space the marks at regular intervals—space them haphazardly along the entire length of the line. These marks represent 20 random observations made of the operator during the forenoon and the afternoon. Now turn to bottom of Fig. 285. NOTE: If you do not wish to mark the book, then reproduce the lines in Fig. 286 on tracing paper or vellum. Draw at random 20 vertical pencil marks across each of the ten lines and superimpose the sheet of drawing paper on Fig. 285

of observations. In order to find N, p is generally assumed or estimated by a preliminary study.

EXAMPLE. Suppose that we want to determine the percentage of idle time of the automatic screw machines in a department by work sampling. Further, assume that a confidence level of 95% and an accuracy of ±5% have been decided upon. We want to know how many random observations will be needed to give us the desired results. Before we can use equation 1 it is necessary to estimate the value of p. In other words, a trial study would be made of the screw machines to get a first estimate as to the percentage of idle time.

Suppose that a total of 100 observations were made, and in this preliminary study 25 observations showed the machines to be idle. The percentage of idle time would be 25% (25 ÷ 100 × 100 = 25%).

We are now ready to calculate N. Where $p = 25\% = 0.25$, and $S = \pm 5\% = \pm 0.05$,

$$Sp = 2\sqrt{\frac{p(1 - p)}{N}}$$

$$0.05p = 2\sqrt{\frac{p(1 - p)}{N}}$$

$$0.0025p^2 = 4\left[\frac{p(1 - p)}{N}\right] = \frac{4p(1 - p)}{N}$$

$$N = \frac{4p(1 - p)}{0.0025p^2} = \frac{4(1 - p)}{0.0025p} = \frac{1600(1 - p)}{p}$$

$$= \frac{1600(1 - 0.25)}{0.25} = 4800$$

Table 63 or the alignment chart in Fig. 287 could be used instead of the formula to determine the number of observations.

After the work sampling study is under way and 500 observations have been made, a new calculation might be made in order to check our original value for N. Assume that the results were as follows:

Observations of machines working	350
Observations of machines idle	150
Total observations	500

150 ÷ 500 × 100 = 30% idle time

This new information would enable us to recalculate the number of observations needed. Now $p = 30\% = 0.30$.

$$0.05(0.30) = 2\sqrt{\frac{0.30(1 - 0.30)}{N}} \quad \text{or} \quad N = \frac{0.84}{0.000225} = 3733$$

Table 63. Table for Determining the Number of Observations for a Given Degree of Accuracy and Value of p, 95% Confidence Level

Per Cent of Total Time Occupied by Activity or Delay, p	Degree of Accuracy									
	±1	±2	±3	±4	±5	±6	±7	±8	±9	±10
1	3,960,000	990,000	440,000	247,500	158,400	110,000	80,800	61,900	48,900	39,600
2	1,960,000	490,000	217,800	122,500	78,400	54,400	40,000	30,600	24,200	19,600
3	1,293,300	323,300	143,700	80,800	51,700	35,900	26,400	20,200	16,000	12,900
4	960,000	240,000	106,700	60,000	38,400	26,700	19,600	15,000	11,900	9,600
5	760,000	190,000	84,400	47,500	30,400	21,100	15,500	11,900	9,390	7,600
6	626,700	156,700	69,600	39,200	25,100	17,400	12,800	9,790	7,740	6,270
7	531,400	132,900	59,000	33,200	21,300	14,800	10,800	8,300	6,560	5,310
8	460,000	115,000	51,100	28,800	18,400	12,800	9,380	7,190	5,680	4,600
9	404,400	101,100	44,900	25,300	16,200	11,200	8,250	6,320	5,000	4,040
10	360,000	90,000	40,000	22,500	14,400	10,000	7,340	5,630	4,450	3,600
11	323,600	80,900	36,000	20,200	12,900	8,990	6,600	5,060	4,000	3,240
12	293,300	73,300	32,600	18,300	11,700	8,150	5,980	4,580	3,620	2,930
13	267,700	66,900	29,700	16,700	10,700	7,440	5,460	4,180	3,310	2,680
14	245,700	61,400	27,300	15,400	9,830	6,830	5,010	3,840	3,040	2,460
15	226,700	56,700	25,200	14,200	9,070	6,300	4,620	3,540	2,800	2,270
16	210,000	52,500	23,300	13,100	8,400	5,830	4,280	3,280	2,590	2,100
17	195,300	48,800	21,700	12,200	7,810	5,420	3,980	3,050	2,410	1,950
18	182,200	45,600	20,200	11,400	7,290	5,060	3,720	2,850	2,250	1,820
19	170,500	42,600	18,900	10,700	6,820	4,740	3,480	2,660	2,110	1,710
20	160,000	40,000	17,800	10,000	6,400	4,440	3,260	2,500	1,980	1,600
21	150,500	37,600	16,700	9,400	6,020	4,180	3,070	2,350	1,860	1,510
22	141,800	35,500	15,800	8,860	5,670	3,940	2,890	2,220	1,750	1,420
23	133,900	33,500	14,900	8,370	5,360	3,720	2,730	2,090	1,650	1,340
24	126,700	31,700	14,100	7,920	5,070	3,520	2,580	1,980	1,560	1,270
25	120,000	30,000	13,300	7,500	4,800	3,330	2,450	1,880	1,480	1,200
26	113,800	28,500	12,600	7,120	4,550	3,160	2,320	1,780	1,410	1,140
27	108,100	27,000	12,000	6,760	4,330	3,000	2,210	1,690	1,340	1,080
28	102,900	25,700	11,400	6,430	4,110	2,860	2,100	1,610	1,270	1,030
29	97,900	24,500	10,900	6,120	3,920	2,720	2,000	1,530	1,210	980
30	93,300	23,300	10,400	5,830	3,730	2,590	1,900	1,460	1,150	935
31	89,000	22,300	9,890	5,570	3,560	2,470	1,820	1,390	1,100	890
32	85,000	21,300	9,440	5,310	3,400	2,360	1,730	1,330	1,050	850
33	81,200	20,300	9,000	5,080	3,250	2,260	1,660	1,270	1,000	810
34	77,600	19,400	8,630	4,850	3,110	2,160	1,580	1,210	960	775
35	74,300	18,600	8,250	4,640	2,970	2,060	1,520	1,160	915	745
36	71,100	17,800	7,900	4,440	2,840	1,980	1,450	1,110	880	710
37	68,100	17,000	7,570	4,260	2,720	1,890	1,390	1,060	840	680
38	65,300	16,300	7,250	4,080	2,610	1,810	1,330	1,020	805	655
39	62,600	15,600	6,950	3,910	2,500	1,740	1,280	980	775	625
40	60,000	15,000	6,670	3,750	2,400	1,670	1,220	940	740	600
41	57,600	14,400	6,400	3,600	2,300	1,600	1,170	900	710	575
42	55,200	13,800	6,140	3,450	2,210	1,530	1,130	865	680	550
43	53,000	13,300	5,890	3,310	2,120	1,470	1,080	830	655	530
44	50,900	12,700	5,660	3,180	2,040	1,410	1,040	795	630	510

n										
48		555	680		1,200				10,800	43,500
49	415	515	650	850	1,160	1,660	2,600	4,630	10,400	41,600
50	400	495	625	815	1,110	1,600	2,500	4,440	10,000	40,000
51	385	475	600	785	1,070	1,540	2,400	4,270	9,610	38,430
52	370	455	575	755	1,030	1,480	2,310	4,100	9,230	36,920
53	355	435	555	725	985	1,420	2,220	3,980	8,870	35,470
54	340	420	530	695	945	1,360	2,130	3,790	8,520	34,070
55	325	405	510	670	910	1,310	2,050	3,640	8,180	32,730
56	315	390	490	640	870	1,260	1,960	3,490	7,860	31,430
57	300	375	470	615	840	1,210	1,890	3,350	7,550	30,180
58	290	360	450	590	805	1,160	1,810	3,220	7,240	28,970
59	280	345	435	565	770	1,110	1,740	3,090	6,950	27,800
60	265	330	415	545	740	1,070	1,670	2,960	6,670	26,670
61	255	315	400	520	710	1,020	1,600	2,840	6,390	25,570
62	245	305	385	500	680	980	1,530	2,720	6,130	24,520
63	235	290	365	480	650	940	1,470	2,610	5,870	23,490
64	225	275	350	460	625	900	1,410	2,500	5,630	22,500
65	215	265	335	440	600	860	1,350	2,390	5,390	21,540
66	205	255	320	420	570	825	1,290	2,290	5,150	20,610
67	195	245	305	400	545	790	1,230	2,190	4,925	19,700
68	190	230	295	385	520	750	1,180	2,090	4,705	18,820
69	180	220	280	365	500	720	1,120	2,000	4,490	17,970
70	170	210	265	350	475	685	1,070	1,900	4,285	17,140
71	165	200	255	335	455	655	1,020	1,815	4,085	16,340
72	155	190	245	315	430	620	970	1,730	3,890	15,560
73	145	180	230	300	410	590	925	1,640	3,700	14,790
74	140	175	210	285	390	560	880	1,560	3,510	14,050
75	135	165	205	270	370	535	835	1,480	3,330	13,330
76	125	155	195	255	350	505	790	1,400	3,160	12,630
77	120	145	185	245	330	480	745	1,330	2,990	11,950
78	110	140	175	230	315	450	705	1,253	2,820	11,280
79	105	130	165	215	295	425	665	1,180	2,660	10,630
80	100	125	155	205	275	400	625	1,110	2,500	10,000
81	94	115	145	190	260	375	585	1,040	2,345	9,380
82	88	110	135	180	245	350	550	975	2,195	8,780
83	82	100	130	165	225	325	510	910	2,050	8,190
84	76	94	120	155	210	305	475	845	1,905	7,620
85	71	87	110	145	195	280	440	785	1,765	7,060
86	65	80	100	130	180	260	405	725	1,630	6,510
87	60	74	93	120	165	240	375	665	1,495	5,980
88	55	67	85	110	150	220	340	605	1,360	5,450
89	49	61	77	100	135	200	310	550	1,235	4,940
90	44	55	69	90	125	175	280	495	1,110	4,440
91	40	49	62	80	110	160	250	440	990	3,960
92	35	43	54	70	96	140	220	385	870	3,480
93	30	37	47	61	83	120	190	335	750	3,010
94	26	31	40	52	71	100	160	285	640	2,550
95	21	26	33	43	59	85	130	234	525	2,110
96	17	21	26	34	46	67	105	185	420	1,670
97	12	15	19	25	34	50	78	140	310	1,240
98	8	10	13	17	23	33	51	91	205	815
99	4	5	6	8	11	16	25	45	100	405

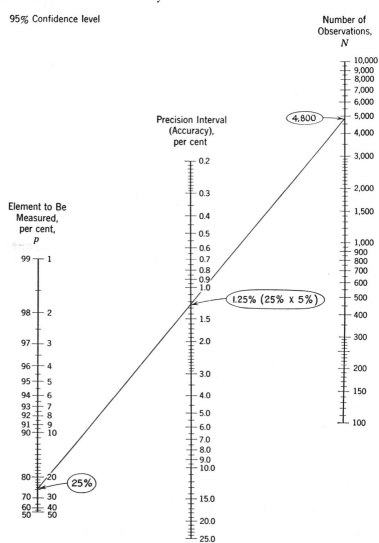

(1) Determine the average per cent p of the element to be measured, by estimate or by a trial study. (2) Decide on the desired accuracy S of the results. (3) Determine the precision interval by multiplying the average per cent of the element p by the desired accuracy S. (4) Draw a straight line on the chart from p through the precision interval, to the required number of observations line. *Example:* For a ±5% accuracy and a value of $p = 25\%$, the chart shows that 4800 observations are required.

Fig. 287. Alignment chart for determining the number of observations needed for a given degree of accuracy and value of p, 95% confidence level. (Courtesy Johns-Manville Corporation.)

As will be explained later, it is advisable to recalculate N at regular intervals, perhaps at the end of each day, in order to better evaluate the progress of the study. The control chart may also be used as explained later in this chapter.

Determination of Accuracy for a Given Number of Observations. After the study is completed, a calculation is made to determine whether the results are within the desired accuracy. This can be done by calculating S in the formula instead of N as was previously done.

Assume that the final results of the study were as follows:

Observations of machines working	2600
Observations of machines idle	1400
Total observations	4000

Then $p = 1400 \div 4000 \times 100 = 35\% = 0.35$

$$Sp = 2\sqrt{\frac{p(1-p)}{N}}$$

$$0.35S = 2\sqrt{\frac{0.35(1-0.35)}{4000}} = 2\sqrt{\frac{0.35 \times 0.65}{4000}} = 2\sqrt{\frac{0.2275}{4000}}$$

$$S = \pm\frac{0.01508}{0.35} = \pm0.043 = \pm4.3\%$$

Since $\pm4.3\%$ is below the $\pm5\%$ required accuracy, the number of observations is sufficient.

In this case the statement could be made that we are 95% confident that the automatic screw machines were idle 35% of the time. The accuracy or standard error of $\pm4.3\%$ means that the results were correct within $\pm4.3\%$ of 35% ($\pm4.3\% \times 35\% = \pm1.5\%$), or the true value was between 33.5% and 36.5%. The 95% confidence level means that the probability is that in 95 cases out of 100 the above results will represent the facts.

Table 64 could be used instead of the formula for determining the degree of accuracy.

Absolute Error or Desired Absolute Accuracy. Table 65 shows the number of observations required for the different values of p where

Table 64. Table for Determining the Degree of Accuracy for a Given Number of Observations and Value of p, 95% Confidence Level

Per Cent of Total Time Occupied by Activity or Delay, p	Number of Observations														
	500	600	700	800	900	1000	2000	3000	4000	5000	6000	7000	8000	9000	10,000
1	±89.0	±81.3	±75.2	±70.4	±66.3	±62.9	±44.5	±36.3	±31.5	±28.1	±25.7	±23.8	±22.3	±21.0	±19.9
2	62.6	57.2	52.9	49.5	46.7	44.3	31.3	25.6	22.1	19.8	18.1	16.7	15.7	14.8	14.0
3	50.8	46.5	43.0	40.2	37.9	35.9	25.4	20.7	18.0	16.1	14.7	13.6	12.7	12.0	11.4
4	43.8	40.0	37.0	34.6	32.7	31.0	21.9	17.9	15.5	13.9	12.7	11.7	11.0	10.3	9.8
5	39.0	35.6	33.0	30.8	29.1	27.6	19.5	15.9	13.8	12.3	11.3	10.4	9.8	9.2	8.7
6	35.4	32.3	29.9	28.0	26.4	25.0	17.7	14.5	12.5	11.2	10.2	9.5	8.9	8.3	7.9
7	32.6	29.8	27.6	25.8	24.3	23.1	16.3	13.3	11.5	10.3	9.4	8.7	8.2	7.7	7.3
8	30.3	27.7	25.6	24.0	22.6	21.5	15.2	12.4	10.7	9.6	8.8	8.1	7.6	7.2	6.8
9	28.4	26.0	24.0	22.5	21.2	20.1	14.2	11.6	10.1	9.0	8.2	7.6	7.1	6.7	6.4
10	26.8	24.5	22.7	21.2	20.0	19.0	13.4	11.0	9.5	8.5	7.6	7.2	6.7	6.3	6.0
11	25.4	23.2	21.5	20.1	19.0	18.0	12.7	10.4	9.0	8.1	7.3	6.8	6.4	6.0	5.7
12	24.2	22.1	20.5	19.2	18.1	17.1	12.1	9.9	8.6	7.7	7.0	6.5	6.1	5.7	5.4
13	23.1	21.1	19.6	18.3	17.3	16.4	11.6	9.5	8.2	7.3	6.7	6.2	5.8	5.5	5.2
14	22.2	20.2	18.7	17.5	16.5	15.7	11.1	9.1	7.8	7.0	6.4	5.9	5.5	5.2	5.0
15	21.3	19.4	18.0	16.8	15.9	15.1	10.6	8.7	7.5	6.7	6.2	5.7	5.3	5.0	4.8
16	20.5	18.7	17.3	16.2	15.3	14.5	10.3	8.4	7.3	6.5	5.9	5.5	5.1	4.8	4.6
17	19.8	18.0	16.7	15.6	14.7	14.0	9.9	8.1	7.0	6.3	5.7	5.3	4.9	4.7	4.4
18	19.1	17.4	16.1	15.1	14.2	13.5	9.5	7.8	6.8	6.0	5.5	5.1	4.8	4.5	4.3
19	18.5	16.9	15.6	14.6	13.8	13.1	9.2	7.5	6.5	5.8	5.3	4.9	4.6	4.4	4.1
20	17.9	16.3	15.1	14.1	13.3	12.7	8.9	7.3	6.3	5.7	5.2	4.8	4.5	4.2	4.0
21	17.4	15.8	14.6	13.7	12.9	12.3	8.7	7.1	6.1	5.5	5.0	4.6	4.3	4.1	3.9
22	16.8	15.4	14.2	13.3	12.6	11.9	8.4	6.9	6.0	5.3	4.9	4.5	4.2	4.0	3.8
23	16.4	14.9	13.8	12.9	12.2	11.6	8.2	6.7	5.8	5.2	4.7	4.4	4.1	3.9	3.7
24	15.9	14.5	13.5	12.6	11.9	11.3	8.0	6.5	5.6	5.0	4.6	4.3	4.0	3.8	3.6
25	15.5	14.1	13.1	12.3	11.6	11.0	7.8	6.3	5.5	4.9	4.5	4.1	3.9	3.7	3.5
26	15.1	13.8	12.8	11.9	11.2	10.7	7.5	6.2	5.3	4.8	4.4	4.0	3.8	3.6	3.4
27	14.7	13.4	12.4	11.6	11.0	10.4	7.4	6.0	5.2	4.7	4.2	3.9	3.5	3.5	3.3
28	14.4	13.1	12.1	11.3	10.7	10.1	7.2	5.9	5.1	4.5	4.1	3.8	3.6	3.4	3.2
29	14.0	12.8	11.6	11.1	10.4	9.9	7.0	5.7	5.0	4.4	4.0	3.7	3.5	3.3	3.1
30	13.7	12.5	11.6	10.8	10.2	9.7	6.8	5.6	4.8	4.3	3.9	3.65	3.4	3.2	3.05
31	13.4	12.2	11.3	10.6	9.9	9.4	6.7	5.5	4.7	4.2	3.85	3.60	3.3	3.1	3.00
32	13.0	11.9	11.0	10.3	9.7	9.2	6.5	5.3	4.6	4.1	3.70	3.50	3.25	3.05	2.90
33	12.7	11.6	10.8	10.1	9.5	9.0	6.4	5.2	4.5	4.0	3.60	3.40	3.20	3.00	2.85
34	12.5	11.4	10.5	9.9	9.3	8.8	6.2	5.1	4.4	3.9	3.55	3.30	3.10	2.90	2.80
35	12.2	11.1	10.3	9.6	9.1	8.6	6.1	5.0	4.3	3.85	3.50	3.25	3.05	2.85	2.70
36	11.9	10.9	10.1	9.4	8.9	8.4	6.0	4.9	4.2	3.75	3.45	3.20	3.00	2.80	2.65
37	11.7	10.7	9.9	9.2	8.7	8.3	5.8	4.8	4.1	3.7	3.35	3.10	2.90	2.75	2.60
38	11.4	10.4	9.7	9.0	8.6	8.1	5.7	4.7	4.0	3.6	3.30	3.05	2.85	2.70	2.55
39	11.2	10.2	9.5	8.8	8.3	7.9	5.6	4.6	3.95	3.55	3.25	3.00	2.80	2.65	2.50
40	11.0	10.0	9.3	8.7	8.2	7.8	5.5	4.5	3.85	3.45	3.15	2.90	2.75	2.60	2.45
41	10.7	9.8	9.1	8.5	8.0	7.6	5.4	4.4	3.80	3.40	3.10	2.85	2.70	2.55	2.40
42	10.5	9.6	8.9	8.3	7.8	7.4	5.3	4.3	3.70	3.30	3.05	2.80	2.65	2.50	2.35
43	10.3	9.4	8.7	8.1	7.7	7.3	5.2	4.2	3.65	3.25	2.95	2.75	2.60	2.45	2.30
44	10.1	9.2	8.5	8.0	7.5	7.1	5.0	4.1	3.55	3.20	2.90	2.70	2.50	2.40	2.25

48	2.10	2.20	2.30	2.50	2.70	2.95	3.30	3.80	4.65	6.6	6.9	7.4	7.9	8.5	9.3
49	2.05	2.15	2.30	2.45	2.65	2.90	3.20	3.70	4.55	6.5	6.8	7.2	7.7	8.4	9.1
50	2.00	2.10	2.25	2.40	2.60	2.85	3.15	3.65	4.45	6.3	6.7	7.1	7.6	8.2	8.9
51	1.96	2.06	2.19	2.34	2.53	2.77	3.10	3.58	4.38	6.20	6.53	6.93	7.41	8.00	8.76
52	1.92	2.02	2.15	2.29	2.48	2.71	3.04	3.51	4.29	6.07	6.40	6.79	7.26	7.84	8.59
53	1.88	1.98	2.10	2.25	2.43	2.66	2.97	3.43	4.20	5.95	6.27	6.65	7.11	7.68	8.41
54	1.84	1.94	2.06	2.20	2.38	2.60	2.91	3.36	4.11	5.82	6.13	6.51	6.95	7.51	8.23
55	1.81	1.91	2.02	2.16	2.34	2.56	2.86	3.30	4.05	5.72	6.03	6.40	6.84	7.39	8.09
56	1.77	1.87	1.98	2.12	2.29	2.50	2.80	3.23	3.96	5.60	5.90	6.26	6.69	7.23	7.92
57	1.73	1.82	1.93	2.07	2.23	2.45	2.74	3.16	3.87	5.47	5.77	6.12	6.54	7.06	7.74
58	1.70	1.79	1.90	2.03	2.19	2.40	2.69	3.10	3.80	5.38	5.67	6.01	6.43	6.94	7.60
59	1.66	1.75	1.86	1.98	2.14	2.35	2.62	3.03	3.71	5.25	5.53	5.87	6.27	6.78	7.42
60	1.63	1.72	1.82	1.95	2.10	2.30	2.58	2.98	3.64	5.15	5.43	5.76	6.16	6.65	7.29
61	1.59	1.68	1.78	1.90	2.05	2.25	2.51	2.90	3.56	5.03	5.30	5.62	6.01	6.49	7.11
62	1.57	1.65	1.76	1.88	2.01	2.22	2.48	2.87	3.51	4.96	5.23	5.55	5.93	6.41	7.02
63	1.53	1.61	1.71	1.83	1.98	2.16	2.42	2.79	3.42	4.84	5.10	5.41	5.78	6.25	6.84
64	1.50	1.58	1.68	1.79	1.94	2.12	2.37	2.74	3.35	4.74	5.00	5.30	5.67	6.12	6.71
65	1.47	1.55	1.64	1.76	1.90	2.08	2.32	2.68	3.29	4.65	4.90	5.20	5.56	6.00	6.57
66	1.44	1.52	1.61	1.72	1.86	2.04	2.28	2.63	3.22	4.55	4.80	5.09	5.44	5.88	6.44
67	1.40	1.48	1.57	1.67	1.81	1.98	2.21	2.56	3.13	4.43	4.67	4.95	5.29	5.72	6.26
68	1.37	1.44	1.53	1.64	1.77	1.94	2.17	2.50	3.06	4.33	4.57	4.84	5.18	5.59	6.13
69	1.34	1.41	1.50	1.60	1.73	1.89	2.12	2.45	3.00	4.24	4.47	4.74	5.06	5.47	5.99
70	1.31	1.38	1.46	1.57	1.69	1.85	2.07	2.39	2.93	4.14	4.37	4.63	4.95	5.35	5.86
71	1.28	1.35	1.43	1.53	1.65	1.81	2.02	2.34	2.86	4.05	4.27	4.53	4.85	5.26	5.72
72	1.24	1.31	1.39	1.48	1.60	1.75	1.96	2.26	2.77	3.92	4.13	4.38	4.69	5.06	5.55
73	1.21	1.28	1.35	1.45	1.56	1.71	1.91	2.21	2.71	3.83	4.03	4.28	4.57	4.94	5.41
74	1.18	1.24	1.32	1.41	1.52	1.67	1.87	2.15	2.64	3.73	3.93	4.17	4.46	4.82	5.28
75	1.15	1.21	1.29	1.37	1.48	1.63	1.82	2.10	2.57	3.64	3.83	4.07	4.35	4.69	5.14
76	1.12	1.18	1.25	1.34	1.45	1.58	1.77	2.04	2.50	3.54	3.73	3.96	4.23	4.57	5.01
77	1.09	1.15	1.22	1.30	1.41	1.54	1.72	1.99	2.44	3.45	3.63	3.85	4.12	4.45	4.87
78	1.06	1.12	1.19	1.27	1.37	1.50	1.68	1.94	2.37	3.35	3.53	3.75	4.01	4.33	4.74
79	1.03	1.09	1.15	1.23	1.33	1.46	1.63	1.88	2.30	3.26	3.43	3.64	3.89	4.21	4.61
80	1.00	1.05	1.12	1.20	1.29	1.41	1.58	1.83	2.24	3.16	3.33	3.54	3.78	4.08	4.47
81	0.97	1.02	1.08	1.16	1.25	1.37	1.53	1.77	2.17	3.07	3.23	3.43	3.67	3.96	4.34
82	0.94	0.99	1.05	1.12	1.21	1.33	1.49	1.72	2.10	2.97	3.13	3.32	3.55	3.84	4.20
83	0.90	0.95	1.01	1.08	1.16	1.27	1.42	1.64	2.01	2.85	3.00	3.18	3.40	3.67	4.02
84	0.87	0.92	0.97	1.04	1.12	1.23	1.38	1.59	1.95	2.75	2.90	3.08	3.29	3.55	3.89
85	0.84	0.89	0.94	1.00	1.08	1.19	1.33	1.53	1.88	2.66	2.80	2.97	3.17	3.43	3.76
86	0.81	0.85	0.91	0.97	1.05	1.15	1.28	1.48	1.81	2.56	2.70	2.86	3.06	3.31	3.62
87	0.77	0.81	0.86	0.92	0.99	1.09	1.22	1.41	1.72	2.43	2.57	2.72	2.91	3.14	3.44
88	0.74	0.78	0.83	0.88	0.96	1.05	1.17	1.35	1.65	2.34	2.47	2.62	2.80	3.02	3.31
89	0.70	0.74	0.78	0.84	0.90	0.99	1.10	1.28	1.57	2.21	2.33	2.47	2.65	2.86	3.13
90	0.67	0.71	0.75	0.80	0.86	0.95	1.06	1.22	1.50	2.12	2.23	2.37	2.53	2.74	3.00
91	0.63	0.66	0.70	0.75	0.81	0.89	1.00	1.15	1.41	1.99	2.10	2.22	2.38	2.57	2.82
92	0.59	0.62	0.66	0.71	0.76	0.83	0.93	1.08	1.32	1.87	1.97	2.09	2.23	2.41	2.64
93	0.55	0.58	0.61	0.66	0.71	0.78	0.87	1.00	1.23	1.74	1.83	1.94	2.08	2.25	2.46
94	0.50	0.53	0.56	0.60	0.65	0.71	0.79	0.91	1.12	1.58	1.67	1.77	1.89	2.04	2.24
95	0.46	0.48	0.51	0.55	0.59	0.65	0.73	0.84	1.03	1.45	1.53	1.63	1.74	1.88	2.06
96	0.41	0.43	0.46	0.49	0.53	0.58	0.65	0.75	0.92	1.30	1.37	1.45	1.55	1.67	1.83
97	0.35	0.37	0.39	0.42	0.45	0.49	0.55	0.64	0.83	1.10	1.17	1.24	1.32	1.43	1.57
98	0.28	0.30	0.31	0.33	0.36	0.40	0.44	0.51	0.63	0.88	0.93	0.99	1.06	1.14	1.25
99	0.20	0.21	0.23	0.24	0.26	0.28	0.32	0.37	0.45	0.63	0.67	0.71	0.79	0.82	0.89

Table 65. Relationship between Value of p and Number of Observations

Per cent of occurrence time, p	1	2	3	4	5	10	15	20	25	30	40	50
Number of observations, N	158,400	78,400	51,700	38,400	30,400	14,400	9,070	6,400	4 800	3,730	2,400	1,600

the confidence level is 95% and the desired degree of accuracy is ±5%. The table shows at a glance the relationship between the value of p and the number of observations required. When p is 1%, 158,400 observations are needed, whereas only 1600 are required when p is 50%. The absolute error in the first case is ±5% of 1%, or ±0.05%. In the second case the absolute error is ±5% of 50%, or ±2.5%. There seems to be little reason to require an absolute error of ±0.05% in one instance and to be satisfied with an error of ±2.5% in another. There are some people who believe that an absolute error of 2.5% or 3% or possibly 3.5% represents a compromise which is acceptable for many kinds of work sampling.[5]

A sample size of 158,000 observations, for example, for a work sampling study is simply not realistic. However, this discussion will serve to explain why the formula for standard error may be modified when values of p are small. An understanding of the meaning of absolute error will aid the analyst in designing the work sampling study to be made in a given situation.

Table 66 shows the number of observations required for a given value of p and a given *desired absolute accuracy*, at 95% confidence level. Table 67 is used for determining the absolute degree of accuracy for a given number of observations and value of p, at 95% confidence level.

Control Charts. Control charts have found extensive use in quality control practice. Inspection data obtained at random and plotted on the control chart show graphically whether or not the process is in control.

In a similar manner the control chart in work sampling enables the analyst to plot the daily or the cumulated results of the sampling

[5] A. J. Rowe, "The Work Sampling Technique," *Transactions of ASME,* pp. 331–334, February, 1954. Also, "Relative versus Absolute Errors in Delay Measurement" by A. J. Rowe, *Research Report No.* 24, University of California, **1953.**

study. If a plotted point falls outside the control limits, this is likely to indicate that some unusual or abnormal condition may have been present during that part of the study. The 3-sigma limit is ordinarily used in determining the upper and lower control limits. This means that there are only three chances in 1000 that a point will fall outside the limits owing to a chance cause. It can be safely assumed that when a point falls outside the limits there is a reason for it. For example, a minor fire in one part of a factory building might disrupt production in the adjoining departments, and the sampling data taken in these areas during the day might therefore show excessive idle time on the part of the operators studied. This would be an assignable cause for the data being out of control for that particular day. Since this is an unusual occurrence, the data taken during the day of the fire would not be used, and the results of the study would be determined from the remainder of the data. To satisfy the requirements as to the number of observations, it probably would be necessary to extend the study an extra day.

The results of a work sampling study of the idle time of the operator of a large press are shown in Fig. 288. One hundred observations were made each day for a period of 12 consecutive working days. The data indicate that the operator was idle for as little as 6% of the day on December 9, and as much as 23% on December 13.

Figure 288 shows the formula for computing the upper and lower control limits, which are +19% and +1% respectively, and the control chart for the data. The results of the study for December 13 were out of control for an "assignable cause"—this was the day of the fire in a neighboring department. The alignment chart in Fig. 289 could also be used for computing the upper and lower control limits.

The control chart is also useful in determining the length of a work sampling study. The chart in Fig. 290 shows that the percentage of "semitractor trucks available" is beginning to level off around 800 observations.[6] This indicates that a sufficient number of observations have been taken. However, to make certain that the final results are within the desired accuracy the necessary checks should be made, as explained earlier.

Use of Random Number Tables. Work sampling, to be statistically acceptable, requires that each individual moment have an equal

[6] For more information about this study see Ralph M. Barnes, *Work Sampling*, John Wiley & Sons, New York, 1957, pp. 71–79.

Table 66. Table for Determining the Number of Observations for a Given Absolute Error or Absolute Degree of Accuracy and Value of p, 95% Confidence Level

Per Cent of Total Time Occupied by Activity or Delay, p	Absolute Error					
	±1.0%	±1.5%	±2.0%	±2.5%	±3.0%	±3.5%
1	396	176	99	63	44	32
2	784	348	196	125	87	64
3	1,164	517	291	186	129	95
4	1,536	683	384	246	171	125
5	1,900	844	475	304	211	155
6	2,256	1003	564	361	251	184
7	2,604	1157	651	417	289	213
8	2,944	1308	736	471	327	240
9	3,276	1456	819	524	364	267
10	3,600	1690	900	576	400	294
11	3,916	1740	979	627	435	320
12	4,224	1877	1056	676	469	344
13	4,524	2011	1131	724	503	369
14	4,816	2140	1204	771	535	393
15	5,100	2267	1275	816	567	416
16	5,376	2389	1344	860	597	439
17	5,644	2508	1411	903	627	461
18	5,904	2624	1476	945	656	482
19	6,156	2736	1539	985	684	502
20	6,400	2844	1600	1024	711	522
21	6,636	2949	1659	1062	737	542
22	6,864	3050	1716	1098	763	560
23	7,084	3148	1771	1133	787	578
24	7,296	3243	1824	1167	811	596
25	7,500	3333	1875	1200	833	612
26	7,696	3420	1924	1231	855	628
27	7,884	3504	1971	1261	876	644
28	8,064	3584	2016	1290	896	658
29	8,236	3660	2059	1318	915	672
30	8,400	3733	2100	1344	933	686
31	8,556	3803	2139	1369	951	698
32	8,704	3868	2176	1393	967	710
33	8,844	3931	2211	1415	983	722
34	8,976	3989	2244	1436	997	733
35	9,100	4044	2275	1456	1011	743
36	9,216	4096	2304	1475	1024	753
37	9,324	4144	2331	1492	1036	761
38	9,424	4188	2356	1508	1047	769
39	9,516	4229	2379	1523	1057	777
40	9,600	4266	2400	1536	1067	784
41	9,676	4300	2419	1548	1075	790
42	9,744	4330	2436	1559	1083	795
43	9,804	4357	2451	1569	1089	800
44	9,856	4380	2464	1577	1095	804
45	9,900	4400	2475	1584	1099	808
46	9,936	4416	2484	1590	1104	811
47	9,964	4428	2491	1594	1107	813
48	9,984	4437	2496	1597	1109	815
49	9,996	4442	2499	1599	1110	816
50	10,000	4444	2500	1600	1111	816

Table 66 (Continued)

Per Cent of Total Time Occupied by Activity or Delay, p	Absolute Error					
	±1.0%	±1.5%	±2.0%	±2.5%	±3.0%	±3.5%
51	9996	4442	2499	1599	1110	816
52	9984	4437	2496	1597	1109	815
53	9964	4428	2491	1594	1107	813
54	9936	4416	2484	1590	1104	811
55	9900	4400	2475	1584	1099	808
56	9856	4380	2464	1577	1095	804
57	9804	4357	2451	1569	1089	800
58	9744	4330	2436	1559	1083	795
59	9676	4300	2419	1548	1075	790
60	9600	4266	2400	1536	1067	784
61	9516	4229	2379	1523	1057	777
62	9424	4188	2356	1508	1047	769
63	9324	4144	2331	1492	1036	761
64	9216	4096	2304	1475	1024	753
65	9100	4044	2275	1456	1011	743
66	8976	3989	2244	1436	997	733
67	8844	3931	2211	1415	983	722
68	8704	3868	2176	1393	967	710
69	8556	3803	2139	1369	951	698
70	8400	3733	2100	1344	933	686
71	8236	3660	2059	1318	915	672
72	8064	3584	2016	1290	896	658
73	7884	3504	1971	1261	876	644
74	7696	3420	1924	1231	855	628
75	7500	3333	1875	1200	833	612
76	7296	3243	1824	1167	811	596
77	7084	3148	1771	1133	787	578
78	6864	3050	1716	1098	763	560
79	6636	2949	1659	1062	737	542
80	6400	2844	1600	1024	711	522
81	6156	2736	1539	985	684	502
82	5904	2624	1476	945	656	482
83	5644	2508	1411	903	627	461
84	5376	2389	1344	860	597	439
85	5100	2267	1275	816	567	416
86	4816	2140	1204	771	535	393
87	4524	2011	1131	724	503	369
88	4224	1877	1056	676	469	344
89	3916	1740	979	627	435	320
90	3600	1600	900	576	400	294
91	3276	1456	819	524	364	267
92	2944	1308	736	471	327	240
93	2604	1157	651	417	289	213
94	2256	1003	564	361	251	184
95	1900	844	475	304	211	155
96	1536	683	384	246	171	125
97	1164	517	291	186	129	95
98	784	348	196	125	87	64
99	396	176	99	63	44	32

Table 67. Table for Determining the Absolute Error or Absolute Degree of Accuracy (%) for a Given Number of Observations and Value of p, 95% Confidence Level

Per Cent of Total Time Occupied by Activity or Delay, p	Number of Observations									
	10,000	9000	8000	7000	6000	5000	4000	3000	2000	1000
1	±0.20	±0.21	±0.22	±0.24	±0.26	±0.28	±0.30	±0.36	±0.44	±0.63
3	0.34	0.36	0.38	0.41	0.44	0.48	0.54	0.62	0.76	1.08
5	0.44	0.46	0.49	0.52	0.56	0.62	0.69	0.79	0.97	1.38
7	0.51	0.54	0.57	0.61	0.66	0.72	0.81	0.93	1.14	1.61
10	0.60	0.63	0.67	0.72	0.77	0.85	0.95	1.10	1.34	1.89
15	0.71	0.75	0.79	0.85	0.92	1.00	1.12	1.29	1.59	2.26
20	0.80	0.84	0.89	0.96	1.03	1.13	1.26	1.46	1.79	2.53
25	0.86	0.91	0.96	1.04	1.11	1.22	1.36	1.57	1.92	2.74
30	0.91	0.96	1.01	1.09	1.17	1.29	1.44	1.66	2.03	2.89
35	0.95	1.00	1.06	1.14	1.23	1.34	1.50	1.73	2.12	3.02
40	0.97	1.02	1.08	1.17	1.25	1.37	1.53	1.77	2.17	3.09
45	0.99	1.04	1.10	1.19	1.28	1.39	1.57	1.80	2.21	3.13
50	1.00	1.06	1.11	1.20	1.29	1.41	1.58	1.82	2.24	3.16

	900	800	700	600	500	400	300	200	100	
1	±0.66	±0.70	±0.75	±0.81	±0.89	±1.00	±1.15	±1.41	±1.99	
3	1.13	1.21	1.29	1.39	1.52	1.71	1.97	2.41	3.41	
5	1.45	1.54	1.65	1.78	1.95	2.18	2.52	3.08	4.36	
7	1.70	1.80	1.93	2.08	2.28	2.55	2.94	3.61	5.10	
10	1.99	2.12	2.27	2.45	2.68	3.00	3.46	4.24	6.00	
15	2.38	2.52	2.70	2.90	3.19	3.57	4.12	5.05	7.14	
20	2.67	2.83	3.02	3.27	3.58	4.00	4.62	5.66	8.00	
25	2.89	3.06	3.27	3.54	3.87	4.33	4.99	6.12	8.66	
30	3.06	3.24	3.46	3.74	4.10	4.58	5.29	6.48	9.17	
35	3.18	3.37	3.60	3.90	4.27	4.77	5.51	6.75	9.54	
40	3.26	3.46	3.70	3.99	4.38	4.90	5.65	6.92	9.80	
45	3.30	3.50	3.74	4.04	4.43	4.95	5.71	7.00	9.91	
50	3.33	3.54	3.78	4.08	4.47	5.00	5.77	7.07	10.00	

opportunity of being chosen. In other words, the observations must be random, unbiased, and independent. The use of a table of random numbers is perhaps the best method of ensuring that the sample is random. The table will serve, first of all, to determine the time of day that an observation should be made. It may also be used to indicate the order in which the operators should be observed, or the specific location in the department or plant where a reading should be taken.

In Table 68 the first number is 950622. The first digit of this number might indicate the hour, and the second and third digits the minutes. Thus, 950 would indicate 9.50, or 9:30 o'clock. The second half of this number, 622, might be read as 6.22, or approximately 6:13 o'clock. Since this plant operates only during the periods 8:00 A.M. to 12:00

Date of Study	Total Number of Observations	Number of Observations "Operator Idle"	Per Cent of Day "Operator Idle"
12-5	100	9	9
12-6	100	10	10
12-7	100	12	12
12-8	100	8	8
12-9	100	6	6
12-12	100	9	9
12-13	100	23	23
12-14	100	9	9
12-15	100	8	8
12-16	100	9	9
12-19	100	9	9
12-20	100	8	8
	1200	120	

The formula for determining the control limits for p is:

Control limits for $p = p \pm 3\sqrt{\dfrac{p(1-p)}{n}}$

N = total number of observations = 1200

n = number of daily observations = $\dfrac{\text{total number of observations}}{\text{number of days studied}} = \dfrac{1200}{12} = 100$

$p = \dfrac{\text{number of "operator idle" observations}}{\text{total number of observations}} = \dfrac{120}{1200} = 0.10$

Control limits for $p = 0.10 \pm 3\sqrt{\dfrac{0.10 \times 0.90}{100}}$

$= 0.10 \pm 3\sqrt{0.0009} = 0.10 \pm 0.09$

$= +0.19 \text{ and } +0.01 = +19\% \text{ and } +1\%$

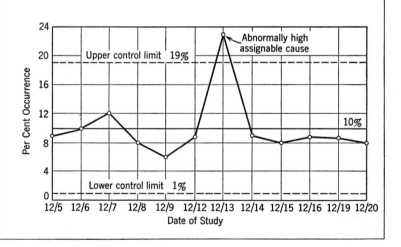

Fig. 288. Control chart.

Three standard deviations

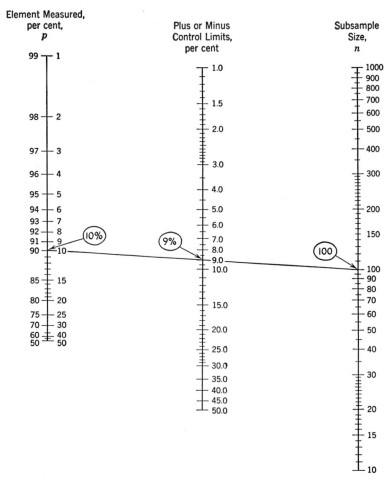

(1) Locate on left-hand line the element to be measured, in per cent. (2) Locate on right-hand line the number of observations made each day during the study. (3) Connect these two points with a straight line and read the point on the center line, "control limits." EXAMPLE: As shown in Fig. 288, the operator was idle 10% of day, or $p = 10\%$. $n =$ number of daily observations $= 100$. Therefore, the control limits are $\pm9\%$. That is, the upper control limit is $0.10 + 0.09 = 0.19$ or 19%; the lower control limit is $0.10 - 0.09 = 0.01$, or 1%.

Fig. 289. Alignment chart for determining control limits. (Courtesy Johns-Manville Corporation.)

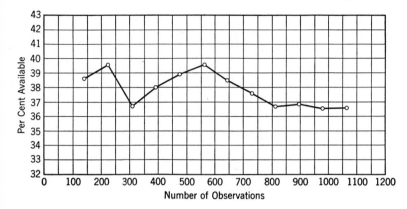

Fig. 290. Chart showing the variation in the availability of trucks during a work sampling study, by days.

and 1:00 P.M. to 5:00 P.M., this number would be discarded because it falls outside the working period. The next number, 133, would indicate that an observation should be made at 1.33, or approximately 1:20 o'clock. In a similar manner, random times would be selected as required for the particular study. If 50 observations are to be made per day, then 50 numbers would be obtained from the table of random numbers and each placed on a card. These cards, arranged in order from the start of the shift to the end of the shift, would provide the schedule to be followed by the analyst in making his observations. Observations usually are not made during the lunch period or during regularly scheduled rest periods.

Ordinarily the observer would start his trip through the department or plant at a different place each time. The location for starting could also be selected by using the random number table. The number 9506 might indicate that this trip start at 9:30 A.M., and that it begin with

Table 68. Table of Random Numbers

950622	220985	742942	783807	907093	989408	037183
133869	362686	485453	194660	687432	674192	695066
899093	785915	610163	414101	171067	096124	978142
269577	163214	211559	168942	326355	358421	268787
947189	069133	356141	679380	866478	595132	347104

Department 6. The observer also might change direction and path of travel in going from department to department, in order to achieve a greater degree of randomness.

Since it is important that a plan of random sampling be followed, it is desirable that it be as simple as possible. The table of random sampling times (Table 69) was devised from a table of random numbers. This table lists 25 chronological random sampling times for each of 14 eight-hour work days. The figures which appear in the columns are easily translated into actual clock times. They represent the hours and minutes after the start of the work shift. For example, assuming that the working period begins at 8:00 A.M., the first sampling time of the first column, 0:05, would be interpreted as 8:05 A.M. Similarly, the last sampling time of the same column, 7:25, would represent 8:00 + 7:25 = 15:25, or 3:25 P.M.

By the proper use of this table, a list of random times of any desired length can be obtained. If 25 or less sampling times are planned for a day, one column will be sufficient. After the column selected has been translated into clock times, those times falling in scheduled rest and lunch periods are eliminated. If the number of sampling times remaining is greater than planned, the numbers in parentheses to the left of certain times are used to reduce the list to the desired number. These auxiliary numbers indicate the order in which the times were originally selected from the random number table. In order to maintain the randomness of the list, numbers should be eliminated from the list in reverse order to their selection. Thus, if only 20 sampling trips were planned from column 1, times designated as (25), (24), (23), (22), and (21), which would be 3:15, 1:35, 6:20, 1:10, and 3:45 respectively, would be omitted.

Should more than 25 trips be desired in any one day, two or more columns may be combined and duplications eliminated. The same procedure as outlined above can then be applied to achieve the desired number of sampling trips. Different columns or combinations of columns should be used for planning the sampling trips for different days.[7]

PROCEDURE FOR MAKING A WORK SAMPLING STUDY

The procedure for planning and organizing a work sampling study will now be described, and in the latter part of this chapter an explana-

[7] Ralph M. Barnes and Robert B. Andrews, "Performance Sampling in Work Measurement," report of research project, presented in *Work Sampling* by Ralph M. Barnes, John Wiley & Sons, New York, 1957, Chapter 22.

Table 69. Table of Random Sampling Times

1	2	3	4	5	6	7
(19)0:05	0:20	0:10	0:15	(18)0:05	(23)0:10	0:15
0:20	(18)0:50	(16)0:35	0:25	0:25	0:25	(21)0:20
0:55	(24)1:20	0:55	(16)1:20	0:45	(21)0:30	(16)0:35
(22)1:10	(21)1:45	(24)1:00	1:40	1:05	0:40	(15)0:50
(20)1:20	1:55	1:10	1:55	(21)1:50	1:10	1:00
(24)1:35	2:00	1:45	2:00	(20)2:10	1:20	1:25
2:30	2:30	(19)2:00	2:30	2:20	1:30	(23)1:40
3:05	2:40	2:05	(15)2:50	2:30	2:25	(22)1:50
(16)3:10	3:10	(21)2:45	3:10	(19)2:35	2:35	1:55
(25)3:15	(23)3:30	2:50	(18)3:30	(17)2:50	2:40	2:45
3:25	(22)3:40	(22)3:00	3:45	(23)3:00	(24)2:55	(25)3:05
(21)3:45	3:50	3:20	3:50	(16)3:10	(19)3:05	3:50
4:00	4:05	3:30	4:30	3:40	3:15	(19)4:00
4:10	(16)4:15	(20)4:40	(20)4:40	(24)3:45	(17)3:25	4:25
(18)4:35	(17)4:20	4:45	5:10	(15)4:30	(15)3:30	(18)4:45
4:55	(19)4:25	4:55	5:20	5:00	3:40	(20)5:00
5:00	4:30	5:00	(17)5:30	5:45	(16)3:50	5:10
(15)5:05	(15)4:35	(18)5:55	(25)5:45	(22)5:50	4:00	(24)5:15
(17)5:35	5:20	(25)6:00	(19)5:50	5:55	4:15	6:20
5:55	5:35	6:05	(21)6:15	6:00	4:25	6:25
(23)6:20	6:15	(23)6:35	6:20	6:35	(18)4:35	6:50
6:45	(20)6:40	(15)6:40	(24)6:25	6:45	(22)5:40	6:55
6:50	(25)6:45	7:10	6:50	(25)7:00	(25)6:45	7:15
7:10	7:10	7:35	7:30	7:45	6:55	7:40
7:25	7:35	(17)7:50	7:55	7:55	(20)7:35	(17)7:45

8	9	10	11	12	13	14
(17)0:05	0:25	0:05	(25)0:05	(22)0:10	(25)0:10	0:10
(18)0:20	0:30	0:15	(18)0:15	0:20	0:15	(17)0:15
(15)1:05	0:40	0:40	0:20	0:30	1:10	0:20
1:25	(24)0:45	1:30	0:25	1:30	1:25	(22)0:25
1:30	1:00	1:45	0:55	(19)1:45	(21)1:30	(24)0:50
2:05	(18)1:10	(21)2:20	1:20	1:50	1:40	(18)1:25
2:25	(17)1:25	2:25	1:35	2:25	1:45	1:35
(24)2:40	1:40	(22)3:10	1:55	(25)2:35	(16)2:05	(23)2:10
(16)3:00	2:15	(20)3:40	(17)2:10	(17)3:05	2:40	(20)2:15
3:20	2:20	(15)3:50	2:30	3:10	(19)2:45	2:40
4:25	2:30	4:15	2:45	3:50	2:55	2:55
4:45	(15)2:40	(24)4:20	2:50	3:55	(22)3:40	3:35
4:50	2:45	4:30	(22)2:55	4:05	3:45	(21)3:40
(25)4:55	(21)3:05	(25)4:40	(15)3:00	4:10	(18)3:50	4:35
5:05	(16)3:30	4:55	(16)3:30	4:50	(24)4:05	(16)4:45
5:15	3:35	5:00	3:35	(21)5:10	(20)4:25	(19)5:05
5:50	4:00	5:15	(23)3:45	(16)5:25	4:55	5:10
5:55	4:15	(19)5:20	4:05	(15)5:30	5:15	5:50
(22)6:00	(23)4:50	5:25	5:00	(24)6:00	5:45	6:05
(20)6:10	(20)5:45	(23)6:05	(19)5:40	6:05	(15)6:20	6:20
(19)6:20	(22)5:50	(17)6:45	(24)5:50	6:15	6:25	7:05
6:35	6:25	(18)7:15	6:25	6:30	(17)6:30	7:10
(23)7:10	(19)6:50	7:25	7:20	(18)6:50	6:35	7:20
7:15	(25)7:05	7:35	7:40	(23)6:55	(23)7:35	(25)7:50
(21)7:30	7:30	(16)7:55	(20)7:50	(20)7:25	7:50	(15)7:55

tion is given of the procedure to be followed for measuring work by sampling.

Steps in Making the Study. The following steps are usually required in making a work sampling study.

1. Define the problem.
   A. State the main objectives or purposes of the project or problem.
   B. Describe in detail each element to be measured.
2. Obtain the approval of the supervisor of the department in which the work sampling study is to be made. Make certain that the operators to be studied and the other people in the department understand the purpose of the study—obtain their cooperation.
3. Determine the desired accuracy of the final results. This may be stated as the standard error of a percentage or desired accuracy, or as the absolute error or desired absolute accuracy. The confidence level should also be stated.
4. Make a preliminary estimate of the percentage occurrence of the activity or delay to be measured. This may be based on past experience; however, it is usually preferable to make a one-day or two-day preliminary work sampling study.
5. Design the study.
   A. Determine the number of observations to be made.
   B. Determine the number of observers needed. Select and instruct these people.
   C. Determine the number of days or shifts needed for the study.
   D. Make detailed plans for taking the observations, such as the time and the route to be followed by the observer.
   E. Design the observation form or IBM card.
6. Make the observations according to the plan. Analyze and summarize the data.
   A. Make the observations and record the data.
   B. Summarize the data at the end of each day.
   C. Determine the control limits.
   D. Plot the data on the control chart at the end of each day.
7. Check the accuracy or precision of the data at the end of the study.
8. Prepare the report and state conclusions. Make recommendations if such are called for.

Purpose of Study. Ordinarily a work sampling study would be undertaken only upon request from a line or staff department supervisor or manager. In many organizations the industrial engineering department or the methods and standards department would be asked to make the study. However, unless the study requires the rating of operator performance and the establishment of time standards, it would not be necessary to use trained analysts or industrial engineers to make it. In fact, supervisors themselves often make work sampling studies.

The objectives of the proposed study should be worked out following the initial request. A full statement of the purpose should be prepared, so that the study can be properly designed. The analyst should try to visualize in detail what the final report of the study will contain. This will aid in determining the degree of accuracy required and the length of the period over which the study should be made.

Elements to Be Measured. The purpose of the study will indicate how the activities and delays should be broken down. When over-all information is needed, a few elements may be satisfactory. In other situations a finer breakdown may be called for, and consequently each element will represent a smaller percentage of the whole. This calls for more observations, and so increases the cost of making the study. If a work sampling study is being made to aid in reducing nonworking time and increasing output per man-hour, the elements should be such that they will reveal delays within the control of the operator, such as late starting and early quitting. The elements should also reveal delays within the control of management, such as shortage of materials or machine down for repair or adjustment. If a work sampling study is being made to establish time standards, the unit of measure is considered first, as in a time study. The units produced must permit easy and positive count.

Irrespective of the purpose of the study or the nature of the break-down, each element to be measured must be carefully defined so that there can be no mistake in identifying it. A carefully prepared written definition is desirable.

Design of Observation Form. In most cases a new observation form will be designed for each work sampling study. Many of these forms, however, will follow a similar pattern. One of the purposes of the preliminary study is to determine just what information is to be obtained, and this, of course, will form the basis for the design of the observation sheet. The form should be simple and should be arranged to facilitate recording and summarizing data, and yet should contain sufficient space to record all information that may be needed to prepare the final report of the study. When the work sampling study is used for establishing time standards, the form will include essentially all the information that would appear on a time study observation sheet.

Experience shows that the time required to tabulate and compute the results of a work sampling study can be reduced from 25 to 50% by the use of IBM mark-sensing cards and IBM equipment, in comparison with doing the work manually.

The Motion Picture Camera for Work Sampling Studies. The motion picture camera can be used satisfactorily for some kinds of work sampling studies. The camera and timer shown in Fig. 291 permit pictures to be made at random intervals during the 8-hour day. The observation times can be preset by the placement of the small metal clips around the circumference of the timer dial. A random number table may be used in locating the clips. The timer can also be set to take pictures at *regular* intervals during the day if this is desired. The electric motor-driven timer operates the synchronous motor, which drives the camera at 1000 frames per minute. A separate device on the timer permits the camera "run time" to be set for intervals of 2 to 30 seconds each. The camera run time, or the length of the observation time, would be preset on the timer and maintained throughout the study. Since pictures are taken at a speed of 1000 frames per minute, when the film is projected at this same speed a performance rating of the operator can be made from the film.

A motor drive with a gear reducer or a solenoid-operated drive can be attached to the camera to take pictures at 50 frames or 100 frames per minute (*A* of Fig. 89). A single frame or a few frames can be exposed at random intervals or at regular intervals during the day. For many kinds of work such a film record permits the satisfactory analysis of activities and delays of operators or machines. With a time-lapse camera drive, all necessary data can be obtained from the film, and a relatively small footage of film is needed for an entire day's record.

Continuous Performance Sampling. Management strives to control all costs, and in most organizations labor cost control is of special importance. The usual method is to establish time standards for specific operations and then obtain a count of the number of units finished each day. Thus, the number of standard minutes earned can be compared with the number of minutes actually worked and a performance index can be determined for each worker and for the department. This plan

Fig. 291. Motion picture camera with synchronous motor drive and timer set to actuate the camera at random intervals during an 8-hour day.

of labor control is widely used and is very effective in many situations. In the example below the operator produced 600 good pieces for which the standard time was 1.0 minute per piece. He worked an 8-hour day, and his performance index was 125%, or 25% above standard.

$$\frac{\left(\substack{\text{Number of pieces} \\ \text{produced during day}}\right) \times \left(\substack{\text{standard time} \\ \text{per piece in minutes}}\right)}{(\text{Hours worked during day}) \times 60} \times 100$$
$$= \text{performance index}$$

$$\frac{(600 \text{ pieces}) \times (1.0 \text{ minute per piece})}{(8 \text{ hours}) \times 60} \times 100 = 125\%$$

However, much work does not lend itself to direct measurement. The cycles may be long and varied, methods may not be standardized, and it is often difficult to obtain a count of the units of work completed. Much indirect factory labor falls into this category. In such situations it is possible to obtain some control of labor costs by the use of work sampling. Continuous performance sampling can be carried on—that is, observations can be made of all workers in a department at random during the entire week or month and the results can be computed for this period. Then this same procedure can be repeated week after week. Management can be provided with information concerning the work force in a department such as:

1. Percentage of time working.
2. Percentage of time out of department.
3. Percentage of time idle.
4. Average performance index while working.
5. Labor effectiveness factor (Item 1 × Item 4).

Some companies present this information weekly on a graph,[8] such as that shown in Fig. 292. This may be posted in a central chart room and copies made for the use of the supervisor and for other members of management. This form of work sampling can serve a useful purpose as a means of labor cost control.

A Specific Case. The Boeing Airplane Company in Wichita, Kansas, has used performance sampling since October, 1953. At the present time this program is being used in 86 departments or shops, and 12 observers are needed to make the observations and compile the reports on a monthly basis.[9] The report (Fig. 293) shows the total number of

---

[8] D. N. Peterson, "Labor Cost Control Through Performance Sampling," *Proceedings Eleventh Industrial Engineering Institute,* University of California, Los Angeles-Berkeley, pp. 62–73, February, 1959. George H. Gustat, "Applications of Work Sampling Analysis," *Proceedings of Tenth Time Study and Methods Conference,* SAM-ASME, New York, pp. 130–146, 1955. George H. Gustat, "Incentives for Indirect Labor," *Proceedings Fifth Industrial Engineering Institute,* University of California, Los Angeles-Berkeley, pp. 80–86, February, 1953.

[9] W. R. Leighty, "Work Sampling in Engineering," *Proceedings Thirteenth Industrial Engineering Institute,* University of California, Los Angeles-Berkeley, pp. 42–48. February, 1961. Figure 293 reproduced with permission of the Boeing Airplane Company.

employees in the department, the percentage of the time of the work force on productive work, the average performance rating, and the productive efficiency, that is, the labor effectiveness index, or "tempo value" as Boeing calls it. The tempo value is the "productive elements" multiplied by the "average rating." The monthly report sheet also contains a breakdown by productive and nonproductive categories.

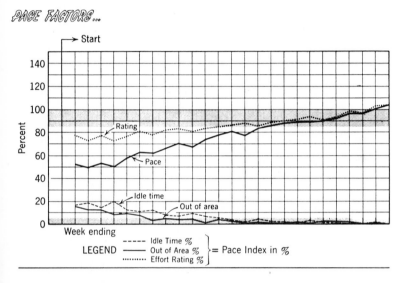

Fig. 292. Weekly labor control report—data obtained by continuous performance sampling

The data in the upper right-hand part of the report shows that, for the month ending June 30, 32 people spent 5.9% of their time operating office equipment, 4.3% filing, 19.5% preparing paper work, and so forth. Ten and three tenths per cent of the time they were out of the area, and 1.9% of the time they were idle. This information is shown graphically on the bottom of the report. The heavy line at the extreme right shows the target which the supervisor of the department has established as his goal for the balance of the year. The report shown in Fig. 293 goes to the department supervisor each month, and a composite of the reports of several departments is sent to the superintendent. A complete management report containing the summarized

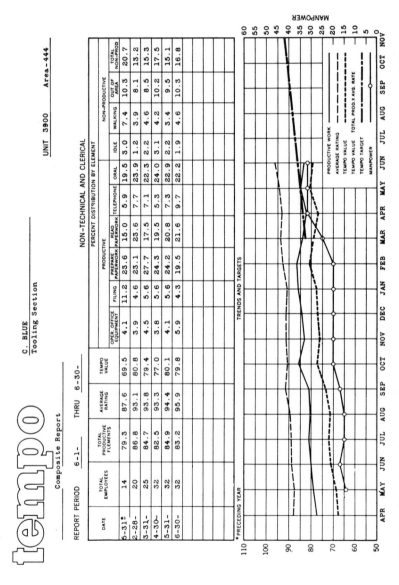

Fig. 293. Monthly labor control report—data obtained by continuous performance sampling—called "tempo program" by Boeing at Wichita.[9]

data from all 86 shops is presented each month to the manufacturing manager, together with a written and oral analysis of the data.

Work Sampling Applied to Nonmanufacturing Activities. The United California Bank has completed the installation of a work measurement and labor control program for 200 of its branch banks and is now studying by work sampling and other measurement techniques many of the operating departments in the central office.[10]

Work sampling studies have been successfully made of supervisors, draftsmen, engineers, and technical personnel.[11] One large and expanding company that constructs its own factories and warehouses regularly makes work sampling measurements of the activities of all construction personnel, such as welders, pipe fitters, and carpenters.

DETERMINING TIME STANDARDS BY WORK SAMPLING

Sampling can be used for measuring work as well as for measuring delays, idle time, and performance. On short-cycle repetitive operations, however, time study, elemental data, or motion-time data would usually be preferred for establishing time standards. Sampling can be used profitably for measuring long-cycle operations, work where people are employed in groups, and activities which do not lend themselves to time study.

It is possible to determine by work sampling the percentage of the day that a person is idle and the percentage of the day that he is working, as well as the average performance index or speed at which he worked during the working portion of the day. For example, assume that John Jones works an 8-hour day as a drill press operator. A work sampling study might show that he was idle 15% of the day or 72 minutes ($480 \times 0.15 = 72$), and that he worked the remainder of the day, or 408 minutes, at an average performance index of 110%. If the record shows that he turned out 420 pieces of acceptable quality during the day, the standard time for the operation he performed could be computed as shown in Fig. 294. The assumption is made that the allowances for this drill-press operation would be taken from the company time study manual.

[10] R. W. Lockwood, "Work Measurement in a Non-Manufacturing Organization," *Proceedings Fourteenth Industrial Engineering Institute,* University of California, Los Angeles-Berkeley, February, 1962, pp. 35–41.

[11] Hugh A. Bogle, "Work Sampling Studies of Supervisory and Technical Personnel," *Proceedings Fourteenth Industrial Engineering Institute,* University of California, Los Angeles-Berkeley, February, 1962, pp. 1–7.

In establishing a time standard, the same rigorous analyses would be required of work sampling as of time study. The method of performing the operation would have to be standardized and a detailed written description prepared. Quality standards would be required, as well as assurance from the foreman that the job was running the way it should. During the study the analyst would make an instantaneous observation as to whether the operator was working or idle,

Information	Source of Data	Data for One Day
Total time expended by operator (working time and idle time)	Time cards	480 min.
Number of parts produced	Inspection Department	420 pieces
Working time in per cent	Work sampling	85%
Idle time in per cent	Work sampling	15%
Average performance index	Work sampling	110%
Total allowances	Company time-study manual	15%

$$\begin{array}{c} \text{Standard time} \\ \text{per piece} \end{array} = \frac{\left(\begin{array}{c}\text{Total time}\\ \text{in minutes}\end{array}\right) \times \left(\begin{array}{c}\text{Working time}\\ \text{in per cent}\end{array}\right) \times \left(\begin{array}{c}\text{Performance index}\\ \text{in per cent}\end{array}\right)}{\text{Total number of pieces produced}} + \text{Allowances}$$

$$= \left(\frac{480 \times 0.85 \times 1.10}{420}\right) \times \left(\frac{100}{100 - 15}\right) = 1.26 \text{ minutes}$$

Fig. 294.   Data sheet and computation of standard time.

and when the operator was working, the analyst would rate his speed and note whether he was performing a regular part of the cycle of work. Information would be obtained as to the starting time and quitting time, and the total number of parts of acceptable quality finished during the day. The sampling study could be counted upon to measure with a preassigned degree of accuracy the percentage of the 8-hour day that the operator was working on the regular drill-press operation, and his average performance index or speed for the working portion of the day.

Establishing Time Standard for an Assembly Operation. The following case illustrates how a time standard was established on a mechanical subassembly operation by work sampling.

Ten operators regularly performed this job, and all ten were studied during a 3-day period. Each day 240 observations were made, making a total of 720 observations. Of this number the analyst found the ten operators working 711 times, and the performance index was noted and recorded for each of the 711 observations. Nine times during the 3-day period, the analyst observed the operators idle.

	DAILY SUMMARY				Computation of Average Performance Index
Performance Index	April 1	April 2	April 5	Total	
100	3	6	1	10	100 × 10 = 1,000
105	13	22	9	44	105 × 44 = 4,620
110	32	21	24	77	110 × 77 = 8,470
115	48	45	17	110	115 × 110 = 12,650
120	47	49	39	135	120 × 135 = 16,200
125	27	28	56	111	125 × 111 = 13,875
130	26	13	22	61	130 × 61 = 7,930
135	15	8	11	34	135 × 34 = 4,590
140	14	15	22	51	140 × 51 = 7,140
145	8	20	27	55	145 × 55 = 7,975
150	2	10	11	23	150 × 23 = 3,450
					711 87,900
"Working" Observations	235	237	239	711	$\frac{87,900}{711}$ = 123.6
"Idle" Observations	5	3	1	9	
Total Observations	240	240	240	720	

Fig. 295. Daily summary and computation sheet.

As the summary of the results of the study shows (Fig. 296), the total time expended by the operators, including working time and idle time, was 13,650 minutes. The working time was 98.7% (711 ÷ 720 × 100 = 98.7%). The remainder, or 1.3%, was idle time. The number of minutes of working time was 13,473 (13,650 × 98.7 = 13,473). During this time the ten operators turned out 16,314 subassemblies of acceptable quality. The average performance index of this group was 123.6%. Figure 295 shows a summary of the performance index for the 711 observations, and the way in which the average performance index was determined. The computations showing how the standard time was determined for this operation are given at the bottom of Fig. 296.

Information	Source of Data	Data for Three-Day Period
Total time expended by operator (working time and idle time)	Time cards	13,650 min.
Number of parts produced	Inspection Department	16,314 pieces
Working time in per cent	Work sampling	98.7%
Idle time in per cent	Work sampling	1.3%
Average performance index	Work sampling	123.6%
Total allowances	Company time-study manual	15%

$$\text{Standard time per piece} = \frac{\left(\begin{array}{c}\text{Total time}\\ \text{in minutes}\end{array}\right) \times \left(\begin{array}{c}\text{Working time}\\ \text{in per cent}\end{array}\right) \times \left(\begin{array}{c}\text{Performance index}\\ \text{in per cent}\end{array}\right)}{\text{Total number of pieces produced}} + \text{Allowances}$$

$$= \left(\frac{13,650 \times 0.987 \times 1.236}{16,314}\right) \times \left(\frac{100}{100 - 15}\right) = 1.20 \text{ minutes}$$

Fig. 296. Data sheet and computation of standard time.

Some Advantages and Disadvantages of Work Sampling in Comparison to Time Study

Advantages

1. Many operations or activities which are impractical or costly to measure by time study can readily be measured by work sampling.

2. A simultaneous work sampling study of several operators or machines may be made by a single observer. Ordinarily an analyst is needed for each operator or machine when continuous time studies are made.

3. It usually requires fewer man-hours and costs less to make a work sampling study than it does to make a continuous time study. The cost may be as little as 5 to 50% of the cost of a continuous time study.

4. Observations may be taken over a period of days or weeks, thus decreasing the chance of day-to-day or week-to-week variations affecting the results.

5. There is less chance of obtaining misleading results, as the operators are not under close observation for long periods of time. When a worker is observed continuously for an entire day, it is unlikely that he will follow his usual routine exactly.

6. It is not necessary to use trained time study analysts as observers for work sampling studies unless performance sampling is required. However, if a time standard or a performance index is to be established, then an experienced time study analyst must be used.

7. A work sampling study may be interrupted at any time without affecting the results.

8. Work sampling measurements may be made with a preassigned degree of reliability. Thus, the results are more meaningful to those not conversant with the methods used in collecting the information.

9. With work sampling the analyst makes an instantaneous observation of the operator at random intervals during the working day, thus making prolonged time studies unnecessary.

10. Work sampling studies are less fatiguing and less tedious to make on the part of the observer.

11. Work sampling studies are preferred to continuous time studies by the operators being studied. Some people do not like to be observed continuously for long periods of time.

12. It usually requires less time to calculate the results of a work sampling study. In fact, IBM mark-sensing cards may be used, and the results obtained from standard IBM equipment.

13. No stop watch or other timing device is needed for work sampling studies.

Disadvantages

1. Ordinarily work sampling is not economical for studying a single operator or machine, or for studying operators or machines located over wide areas. The observer spends too great a proportion of his time walking to and from the work place or walking from one work place to another. Also, time study, elemental data, or motion-time data are preferred for establishing time standards for short-cycle repetitive operations.

2. Time study permits a finer breakdown of activities and delays than is possible with work sampling. Work sampling cannot provide as much detailed information as one can get from time study.

3. The operator may change his work pattern upon sight of the observer. If this occurs, the results of such a work sampling study may be of little value.

4. A work sampling study made of a group obviously presents average results, and there is no information as to the magnitude of the individual differences.

5. Management and workers may not understand statistical work sampling as readily as they do time study.

6. In certain kinds of work sampling studies, no record is made of the method used by the operator. Therefore, an entirely new study must be made when a method change occurs in any element.

7. There is a tendency on the part of some observers to minimize the importance of following the fundamental principles of work sampling, such as the proper sample size for a given degree of accuracy, randomness in making the observations, instantaneous observation at the preassigned location, and careful definition of the elements or subdivisions of work or delay before the study is started.

CHAPTER 33

Measuring Work by Physiological Methods

Over the years many people have sought an objective method of measuring physical work. Frederick W. Taylor saw the great need for work measurement and created stop-watch time study to perform this function. In developing the time study technique, he experimented with the concept of horsepower (that is, foot-pounds of work per minute) as a measure of work but found this approach to be unsatisfactory.

Around the turn of the century, physiologists demonstrated the validity of using rate of oxygen consumption as the basis for measuring energy expenditure (indirect calorimetry). Later studies showed that change in heart rate was also a reliable measure of physical activity. During more than a half century extensive human energy expenditure studies have been carried on in various parts of the world. Studies have been made to gain new knowledge about the human machine,[1] to better understand the performance of champion athletes, and to aid the physically handicapped. Studies also have been conducted in the area of work physiology, notably in England, Holland, the Scandinavian countries, and at the Max Planck Institute in Germany. Perhaps the most extensive studies in this area in the United States have been made by Dr. Lucien Brouha.[2]

Interest in work physiology is increasing in this country, partly because a more objective method of measuring physical work is needed and also because better apparatus has become available for measuring oxygen consumption and heart rate. Sufficient evidence is now available to show that physiological measurement can supplement present work measurement techniques. Moreover, research now under way in business and industry, in colleges and universities, and in de-

---

[1] A. V. Hill, *Living Machinery*, Harcourt, Brace and Co., New York, 1927.

[2] Lucien Brouha, *Physiology in Industry*, Pergamon Press, New York, 1960.

fense and space agencies may soon bring changes in equipment and techniques that will make work physiology regularly accepted in the field of work measurement.

Physical work results in changes in oxygen consumption, heart rate, pulmonary ventilation, body temperature, lactic acid concentration in the blood, 17-ketosteroid excretion in the urine, and other factors. Although some of these factors are only slightly affected by muscular activity, there is a linear correlation between oxygen consumption, heart rate, total pulmonary ventilation, and the physical work performed by an individual.

Physical work is performed by the hands, feet, or other members of the body. For example, when a worker lifts a 50-pound carton from the floor through a vertical distance of 3 feet and places it on the workbench, 150 foot-pounds of work has been accomplished. The physiological cost to the worker to perform this operation results from the activities of the muscles of the arms, legs, back, and other parts of the body. If the carton were located on the floor under the bench, the physiological cost of dragging it out and placing it on top of the bench would be greater than if it were located directly on the floor beside the bench. Of course the task would have been much easier if the carton had been located on a pallet directly beside the workbench and at the same height as the bench. When a person is required to bend over to pick up an object from the floor, or when he must work in a cramped or unnatural position, his energy expenditure is higher than would be the case if these conditions were not present. The physiological cost of performing a task, then, is affected by the number and type of muscles involved, either to move a member of the body or to control antagonist contraction.

When a person is at rest, his heart rate and the rate of oxygen consumption are at a fairly steady level. Then when the person does muscular work, that is, when he changes from a "resting level" to a "working level," both the heart rate and the oxygen consumption increase. When work ends, recovery begins, and the heart rate and oxygen consumption return to the original resting level.

Heart Rate Measurement. The resting heart rate of the individual shown in Fig. 297 was 70 beats per minute. When he started to work, his pulse rate increased rapidly to 110 beats per minute and leveled off during the working period. When he stopped working, his heart rate dropped off and finally returned to the original resting level. The increase in heart rate during work may be used as an index of the physiological cost of the job. Also the rate of recovery immediately after work stops can be utilized in some cases in evaluating physiological cost. Incidentally, the *total* physiological cost of a task consists not only of the energy expenditure during work but also the energy expenditure above the resting rate during the recovery period, that is, until recovery is complete.

Each time the heart beats, a small electric potential is generated. By placing electrodes on either side of the chest, this potential can be picked up and transmitted by wire or by radio transmitter to a re-

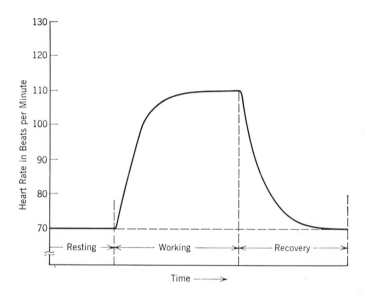

Fig. 297.   Heart rate in beats per minute before, during, and after physical work.

ceiver (Fig. 298). There the individual heart beats can be counted directly, or by means of a cardiotachometer the impulsives can be converted into heart rate, that is, heart beats per minute. These data can be recorded continuously on ruled graph paper by a milliampere recorder (Fig. 299). The graph is in effect a curve similar to the one shown in Fig. 297.

Heart beat signals can also be obtained by means of an ear lobe unit. This apparatus consists of a photo duodiode placed behind the ear and illuminated by a light source mounted on the other side of the ear. The opacity of the ear lobe changes as the blood surges through the ear with each heart beat. The impulse created by each heart beat can be transmitted by wire or by radio transmitter and recorded as described above. Most studies in this country are made by taking the heart rate from the chest by means of electrodes, although the investigators at the Max Planck Institute in Germany have had considerable success with the ear lobe device.

Information concerning rate of recovery also can be obtained by simply using a stethoscope and stop watch. Studies made at the Harvard Fatigue Laboratory showed that heart rate data obtained

Fig. 298.   Apparatus for measuring heart beat and oxygen consumption. *A*, transmitter for telemetering heart beats; *B*, respirometer for measuring volume of exhaled air; *C*, rubber football bladder for collecting random sample of exhaled air.

in this manner are reliable and easy to secure. The procedure consists of obtaining the total number of heart beats during the second half-minute after work stops. Then the number of heart beats is taken during the second half-minute of the second minute, and the second half-minute of the third minute, after work stops. Such data make it possible to compare the rate of recovery during different working conditions. For example, Fig. 308 shows such information for a person working under conditions of high temperature and humidity wearing (*A*) ordinary clothes and (*B*, *C*) a ventilated suit.

Measuring Oxygen Consumption. Change in the rate of oxygen consumption from the resting level to the working level is also a measure of the physiological cost of the work done. A person extracts oxygen from the air he breathes. In order to measure the oxygen consumed per unit of time, it is necessary to measure the volume of air exhaled and the oxygen content of this air. Oxygen consumption may be defined as the volume of oxygen expressed in liters per minute which

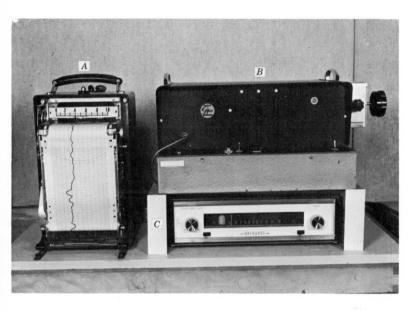

Fig. 299. Apparatus for receiving and recording heart beats. *A*, recorder; *B*, cardiotachometer; *C*, receiver.

the individual extracts from the air he inhales. The most common method of obtaining this information is by means of a portable respirometer, such as the one shown in Fig. 298. This is a lightweight (5½ pounds) gas meter which can be worn on the back. The person is equipped with a mask and a 1-inch rubber tube which carries the exhaled air from the mask to the respirometer. The respirometer indicates directly the volume of exhaled air in liters. A sample of the exhaled air is drawn off at random intervals into a rubber football bladder, and an analysis of its content is made. This permits a comparison of the oxygen content of the sample of expired air with that of the air in the room.

A curve showing the energy expenditure before, during, and after physical work would look much like the curve for heart rate in Fig. 297. In the case of a worker handling 10-pound cartons in the shipping room at the rate of 12 cartons per minute, the resting rate for one study was 1.2 calories (abbreviation for kilocalories) per minute. When he started to work, his energy expenditure increased rapidly to

Fig. 300. Oxygen analyzer, mercury manometer and vacuum pump.

5.0 calories per minute and then returned to the resting level. Thus, both heart rate and oxygen consumption can be used to measure physical work.

Although oxygen consumption has been most widely used for measuring physiological cost of muscular activity, equipment now available permits heart rate measurements to be made more easily. Moreover, certain factors causing physiological stress, such as temperature, humidity, and clothing, cannot be properly evaluated by oxygen consumption alone.

Individual Differences. There is a great difference in the ability of individuals to perform muscular work. Studies were made of a group of 2000 healthy college students, which included men of low physical efficiency as well as varsity athletes. The results of this study show that the capacity to withstand the stress of hard physical work was ten times as great in the fit as in the unfit.

Even in a more restricted and highly selected group of varsity and junior varsity trained athletes, the best men were able to perform hard physical work twice as efficiently as their less fit colleagues. These results emphasize, quantitatively, the well-known fact that men vary markedly as far as their physical capacity is concerned; and that, even in trained and selected groups, wide variations are found in the physiological price that individuals have to pay to accomplish a given task.

The physical capacity of the individual is the result of numerous factors such as the innate potential of physiological mechanisms, age, health and nutritional status, sex, specific fitness for a given job and for given environmental conditions. These factors exist in any industrial population, and similar differences have been found among workers.[3]

Practice—Fitness for Job. There is some evidence to show that the well-trained worker who is physically fit and suited to his job might be expected to expend approximately 5 calories per minute, or 2400 calories per 8-hour day, on his job. The physiological cost to this same man as a beginner on the same job would be greater if he attempted to produce the same number of units of product per day. Practice enables the worker to do his job with a lower expenditure of energy. Moreover, the better trained the worker is, the sooner his heart rate will return to the resting level after he stops work. Figure 301 shows the effect of training on heart rate for a standard task. The top curve shows the heart rate for the person pedaling a bicycle ergometer for 20 minutes with a heavy workload. The other curves show heart rate after 22 days, 56 days, and 84 days of training.[4] On the initial ride the heart rate increased to 175 beats per minute; after 84 days of practice it did not exceed 143 beats per minute. Also the recovery rate was much more rapid for the trained person.

Physiological Cost of Walking. Studies of energy expenditure in walking have been made by many different investigators. Results of these studies [5] seem to indicate that for speeds of 2 to 4 miles per hour energy expended in calories per minute is linearly proportional

[3] Lucien Brouha, "Physiological Approach to Problems of Work Measurement," *Proceedings Ninth Industrial Engineering Institute,* University of California, Los Angeles-Berkeley, p. 13, 1957.

[4] Lucien Brouha, *Physiology in Industry,* Pergamon Press, New York, p. 30, 1960.

[5] R. Passmore and J. V. G. A. Durnin, "Human Energy Expenditure," *Physiological Reviews,* Vol. 35, No. 4, p. 806, October, 1955.

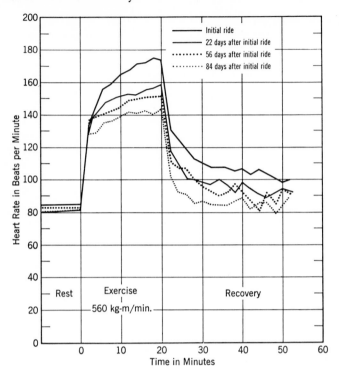

Fig. 301. Effect of training on heart rate for a standard amount of exercise: pedaling a bicycle ergometer for 20 minutes with a heavy work load. Training was achieved by riding the bicycle 4 days a week with the same work load.

to the speed of walking in miles per hour. Assuming a metabolic cost at the resting state (zero point on the abscissa of Fig. 302) to be 1.2 calories per minute, then the relationship can be expressed by the equation

$$C = 1.0V + 1.2$$

where C equals energy expenditure in calories per minute, and V equals the walking speed in miles per hour. As the curve (Fig. 302) shows, the energy expended while walking at 3 miles per hour is 4.2 calories per minute.

However, energy expenditure is also proportional to body weight. One study [5] of 50 persons walking at 3 miles per hour led to the equation

$$C = 0.047W + 1.02$$

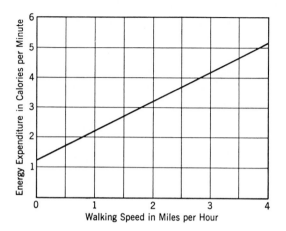

Fig. 302. Curve showing relationship between energy expenditure in calories per minute and speed of walking in miles per hour.

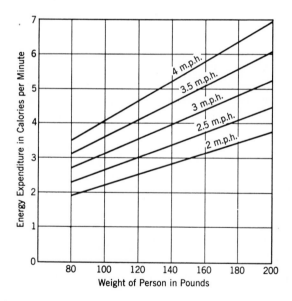

Fig. 303. Curves showing relationship between energy expenditure in calories per minute, speed of walking in miles per hour, and body weight in pounds.

where W is the weight of the person in kilograms. The curves in Fig. 303 have been drawn from the data developed from this study. The energy expenditure for a person weighing 150 pounds and walking 3 miles per hour would be 4.2 calories per minute, whereas for a 200-pound person it would be 5.3 calories per minute.

Use of Physiological Measurements in Work Methods Design. When a new plant and its production facilities are being designed or purchased, management is often confronted with the problem of whether a person can physically perform a particular operation, or how best to organize the work for each person when a group is needed to do the job, or how much rest will be required by a worker performing a specific task. The objective is to design the work method so that the operator can perform the task 8 hours per day, 5 days per week, without undue fatigue. Physiological measurements of the worker on the actual job or on a simulated operation can provide useful information pertaining to such problems.

A Specific Case—Power Shear Operation. An example is given here of the use of physiological measurements for the evaluation of the physical requirements for operating a power shear for cutting photographic paper. A power shear is commonly used in paper manufacturing and in the printing and photographic industries to cut large sheets of paper into smaller sizes. The operation consists of stacking the paper onto a pile or lift several inches high and cutting it by a power-operated knife at the back of the shear. Eastman Kodak Company was considering the purchase of a new power shear larger than any it was then using.[6] The company wanted to know whether a person could operate this new machine manually or whether a completely mechanized feeding and unloading device would be required. This special equipment would cost approximately $20,000 and would place certain constraints on the flexibility of the new power shear.

Physiological studies using changes in oxygen consumption and heart rate were conducted to make the evaluation. A mock-up of the new power shear and work place was constructed, and the operation of the new machine was simulated. The operation of the new power shear would consist of moving lifts of paper stock from a pallet located on a levelator onto the bed of the machine. Each lift was 52

[6] Harry L. Davis and Charles I. Miller, "The Use of Work Physiology in Job Design," *Proceedings Annual Conference American Institute of Industrial Engineers,* Atlantic City, N. J., pp. 281–286, May, 1962.

inches by 43 inches by 1 inch in size and weighed approximately 45 pounds. Five lifts were transferred from the pallet to the machine to form a total lift 5 inches high and weighing 225 pounds. An air table was used as the bed of the machine, thus greatly reducing the effort required to move the paper. The paper was trimmed and two cuts were made. These cuts were then transferred to another pallet located on a levelator at the other side of the machine bed. The cycle was then repeated.

The man selected for the study was an experienced power-shear operator. On the day before the actual study, the operator was asked to do some standard lifting tasks and to perform a treadmill operation. Oxygen consumption and heart rate data were obtained. The results of these tests showed the responses of this operator to be well within the range of a group of other factory workers who had been studied in the past. On the day of the study the operator performed the simulated task for 3½ hours. He was paced by signals from a tape recorder which gave a uniform cycle time of 4.65 minutes. The results of the study are shown in Table 70.

Since the average energy expenditure of 4.25 calories per minute and a heart rate of 109 beats per minute were well within the accepted range of 5.00 and 100 to 125 respectively, it was decided that the new square cutter could be purchased and that the mechanized feeding and unloading device would not be needed. This study gave the company assurance that the operation of the new machine would not place unreasonable stress on the operator. The power shear was purchased and installed (see Fig. 304), and actual experience was almost identical with that forecast by the simulation study.

Table 70. Energy Expenditure and Heart Rate for Operator While Working on Simulated Power-Shear Operation

	Average	Minimum	Maximum
Energy expenditure in calories per minute	4.25	4.00	5.00
Heart rate in beats per minute	109	92	118

Fig. 304.　New power shear for square-cutting photographic paper.

ESTABLISHING TIME STANDARDS
BY PHYSIOLOGICAL METHODS

Time standards established by time study or by predetermined motion-time data are generally set so that the average qualified, well-trained, and experienced operator working on a manual task against the time standard can produce at a level of approximately 125% day-in and day-out when employed in a plant where wage incentives are used. It is expected that approximately 96% of the working population can meet or exceed the standard (see Fig. 233). It is known that some people can attain the 100% performance level much more easily than others. As a result, some people regularly work at a level of 150 or 160%, whereas others using the same expenditure of energy may attain a level of only 110 or 115%. Time standards are set for the *task*, that is, for a specific and carefully defined job.

To illustrate this point, let us consider three men, Jones, Brown, and Smith, who performed a carton-handling operation [7] in the shipping room (see Fig. 298). This job consisted of the operator standing in front of a worktable 34 inches high, and lifting a 10-pound carton

[7] Study made by Ralph M. Barnes, Robert B. Andrews, James I. Williams, and B. J. Hamilton.

Table 71. Change in Energy Expenditure and Heart Rate for Three Operators in the Shipping Room, Working at Three Different Speeds

	Energy Expenditure in Calories per Minute			Heart Rate in Beats per Minute		
Cartons handled per minute	6	9	12	6	9	12
Operator						
Jones	3.2	3.9	5.0	87	93	99
Brown	3.5	4.4	5.5	88	92	98
Smith	2.8	3.7	4.9	89	98	105

from a conveyor on his left up 10 inches to the table, stamping the shipping address on the carton, and disposing of it onto a conveyor located 10 inches above the table. Each man worked at three different speeds, handling cartons at the rate of 6 per minute, 9 per minute, and 12 per minute. Speeds were established by the use of signals from a tape recorder. The energy expenditure in calories per minute and the change in heart rate in beats per minute are shown in Table 71 and Fig. 305. Although the heart rate of the three men is not greatly different at any given working speed, there is some difference in energy expenditure as measured by calories per minute. At the slow speed of 6 cartons per minute, Jones expended 3.2 calories per minute, Brown 3.5, and Smith 2.8. At the high speed of 12 cartons per minute, the energy expenditure was 5.0, 5.5, and 4.9 calories respectively.

Physiological measurements can be used to compare the energy cost on a job for which there is a satisfactory time standard, with a similar operation on which there is no standard, but the comparison should be made for the same person. For example, if handling 10-pound cartons at the rate of 12 cartons per minute under the conditions described above was considered normal performance,[8] and if the energy cost for Jones was 5 calories per minute, the answer to the question of what the time standard should be for handling 15-pound cartons under the same conditions might be obtained by having Jones handle 15-pound cartons

[8] This is a hypothetical case.

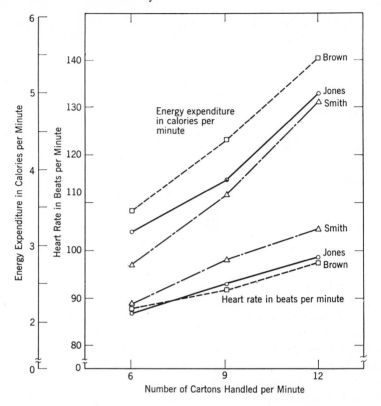

Fig. 305. Curves showing change in energy expenditure and heart rate for three operators in the shipping room, working at three different speeds.

at various speeds, and then selecting the speed that gave an energy cost of 5 calories per minute. Thus, the energy cost of the two jobs would be similar, and the time standard, that is, the number of 15-pound cartons handled per minute, could be determined.

CHAPTER 34

Fatigue

Since one of the main objectives of motion and time study is to reduce fatigue and to make the work as easy and satisfying for the individual as possible, it is desirable at this time to examine the nature of fatigue.

PHENOMENA ASSOCIATED WITH FATIGUE

The term *fatigue* has various meanings, depending upon the point of view that is taken in considering the subject. Fatigue in industry refers to three related phenomena: (1) a feeling of tiredness; (2) a physiological change in the body (the nerves and muscles fail to function as well or as fast as is normal because of chemical changes in the body resulting from work); and (3) a diminished capacity for doing work.

Feeling of Tiredness. A feeling of tiredness is commonly associated with long periods of work. It is subjective in nature, and consequently the extent of tiredness cannot be determined by an observer. Tiredness may be localized in some particular muscle, or it may be a general sensation of weariness.

This feeling of fatigue acts as a protective device in preventing exhaustion, but there is often no direct correlation with physiological fatigue that manifests itself in decreased ability to do work. A person may feel tired and yet may work as efficiently as ever, or he may feel normal and yet may be actually working at a low rate because of physiological fatigue. Therefore, the feeling of tiredness does not seem to be a valid basis for judging the effect of work on the individual.

Physiological Changes Resulting from Work. From the physiological point of view the human body may be thought of as a machine [1] which consumes fuel and gives out useful energy. The principal mech-

[1] A. V. Hill, *Living Machinery*, Harcourt, Brace and Co., New York.

563

anisms of the body that are involved are (1) the circulatory system, (2) the digestive system, (3) the muscular system, (4) the nervous system, and (5) the respiratory system. Continuous physical work affects these mechanisms both separately and collectively.

Fatigue is the result of an accumulation of waste products in the muscles and in the blood stream, which reduces the capacity of the muscles to act. Very possibly the nerve fiber terminals and the central nervous system may also be affected by work, thereby causing a person to slow down when tired. Muscular movements are accompanied by chemical reactions which require food for their activities. This food is furnished as *glycogen*, a starchlike substance which is carried in the blood stream and is readily converted into sugar. When the muscle contracts, the glycogen is changed into lactic acid, a waste product which tends to restrict the continued activity of the muscle. In the recovery phase of muscular action, oxygen is used to change most of the lactic acid back to glycogen, thus enabling the muscles to continue moving. The supply of oxygen and the temperature affect the speed of recovery. If the rate of work is not strenuous, the muscle is able to maintain a satisfactory balance. Excessive lactic acid does not accumulate and the muscle does not go into "oxygen debt," both of which diminish the capacity of the muscle to act.

An athlete running the mile might be used as an example of an individual exerting himself to the utmost. He is using the supply of fuel and oxygen at a rapid rate, and therefore will require time for recuperation, that is, time to bring his muscles back to equilibrium.

Effects of Physical Environment on the Worker. The physiological cost of doing work is affected by environmental factors, such as temperature, humidity, air movement, and atmospheric contamination. A person has certain energy requirements just to maintain his bodily functions; when he does physical work, his energy requirements increase. If the resting environment and the working environment are changed—for example, if the temperature is increased from 70 to 90° F.—then the energy cost increases at both the resting level and the working level. Change in heart rate appears to be the best means of measuring the effects of such environmental factors. Figure 306 shows the heart rate and oxygen consumption before, during, and after a standard exercise was performed under two conditions of dry-bulb temperature and relative humidity. In this study six men pedaled a bicycle ergometer at two different speeds.[2] The men worked for 30

[2] Lucien Brouha, "Physiological Approach to Problems of Work Measurement," *Proceedings Ninth Industrial Engineering Institute,* University of California, Los Angeles-Berkeley, p. 13, February, 1957.

minutes at a medium work load and then for a 4-minute period at a
heavy load. The environmental conditions were as follows:

Normal room temperature: 72° F. and 50% relative humidity
Hot wet: 90° F. and 82% relative humidity

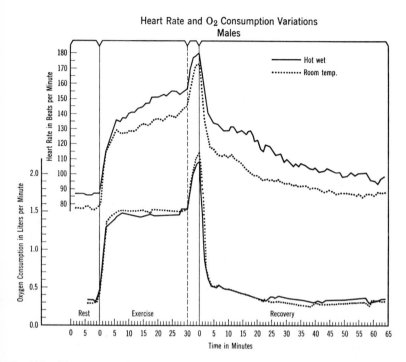

Fig. 306. Heart rate and oxygen consumption during rest, during exercise at two
different work loads, and during recovery. These experiments were performed in
two different environments.

The curves show that an immediate increase in heart rate and oxygen
consumption occurs as exercise begins. As the moderate work at room
temperature progresses there is a slight increase of the heart rate, but
the oxygen consumption remains at a steady level. As soon as the work
load becomes heavy, both heart rate and oxygen consumption increase
immediately and continue to increase until the exercise ends. During
the recovery period the oxygen consumption returns to the resting
level after 35 minutes; whereas the heart rate diminishes less rapidly
and is still well above the resting level after 65 minutes of recovery.

For any job in which the physiological expenditure is great enough to produce significant changes in heart rate, the heart rate recovery curves will

determine the physiological cost of the job and will permit evaluation of any modification that is made in attempting to reduce stress and fatigue. For example, in a job where men had to skim impurities from the surface of a liquid with a long and heavy ladle, the tanks were situated at such a height above the floor that the operation had to be performed at shoulder level. Average reactions were high, reaching 160 beats per minute for the first pulse recorded 1 minute after skimming one tank. Special platforms were built so that men could operate slightly above waist level. The average heart rate recorded 1 minute after the operation dropped to 112 beats per minute, indicating a drastic reduction in the "physiological work" necessary to perform the job.[3]

The following is a procedure for evaluating the effect of physical environment upon the operation of a drop hammer in a forge shop.[4] Heart rate and oxygen consumption data were obtained from the operator performing his regular job in the factory. The same operator then was asked to operate a bicycle ergometer which had been moved into the factory and placed beside the drop hammer. The worker operated the ergometer at a load that produced approximately the same heart rate and oxygen consumption as were present when he was working on his regular job. The ergometer was then moved into the laboratory, where the temperature was maintained at 70° F. and the relative humidity at 50%, and the operator worked at the same speed as in the factory. His heart rate and oxygen consumption in the laboratory were obtained. The difference was a measure of the effects of the shop environment, that is, high temperature, humidity, smoke, and fumes.

Effect of Protective Clothing on Pulse Rate. Although industry is striving to reduce fatigue and improve working conditions, there are still some heavy jobs that have to be performed under conditions of high temperature and humidity. Dr. Lucien Brouha's studies show that pulse rate can be used as a measure of the effectiveness of special clothing to protect the worker from heat and noxious fumes.[5] Figure 307 shows a ventilated suit which Dr. Brouha designed for use by operators working in a magnesium plant. The task consisted of removing impurities from the bottom of a magnesium cell, using an iron ladle weighing 35 to 40 pounds with a handle 10 feet long. The oper-

[3] Lucien Brouha, "Fatigue—Measuring and Reducing It," *Advanced Management,* Vol. 19, No. 1, p. 13, January, 1954.

[4] F. H. Bonjer, Netherlands Institute for Preventive Medicine, Leyden, Holland.

[5] Lucien Brouha, "Fatigue—Measuring and Reducing It," *Advanced Management,* Vol. 19, No. 1, p. 9, January, 1954.

ators were exposed to the heat from the molten magnesium, and normally wore a face shield to protect them from the heat and a respirator as protection from the chlorine fumes.

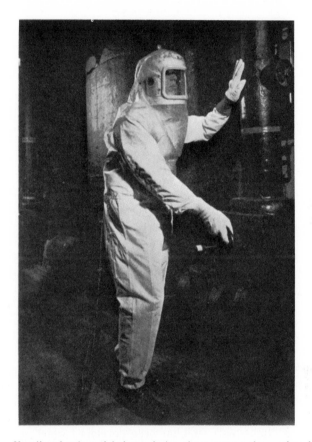

Fig. 307. Ventilated suit and helmet designed to protect the worker from heat and noxious fumes.

Figure 308 shows the effect of a ventilated suit on the heart rate. The curve at the top shows the average reaction in 45 operations without the ventilated suit. The pulse rate was 127 beats per minute at the end of the first minute, 115 at the end of the second, and 109 at the end of the third. The second curve gives the results for men working with suits inflated with air at room temperature, 90° F. The pulse rate was 111 beats per minute at the end of the first minute, 101 at the end of the second, and 96 at the end of the third. The bottom curve shows the results for men working with the suit inflated with air cooled to 70°. In this case the pulse rates were 92, 85, and 81 at the end of the first, second, and third minutes respectively.

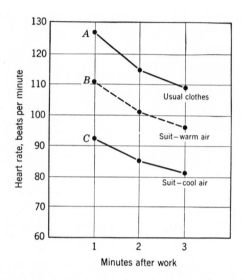

Fig. 308. Average heart rate recovery curves of workers performing the same operation in a magnesium plant: *A*, usual working clothes; *B*, suit ventilated with air at 90° F.; *C*, suit ventilated with air at 70° F.

The Force Platform as a Means of Measuring Work. Oxygen consumption and pulse rate can be used to measure work only when the physical activity is of considerable magnitude and duration. Very light hand or body movements, for example, cannot be measured with these techniques. A "force platform" has been developed which seems to be useful in measuring the physical effort involved in performing light work or activities of short duration.[6] The mechanism (Fig. 309) consists of a rigid triangular platform mounted on quartz crystals in such a manner that they react to forces that are vertical, frontal, and transverse. The quartz crystals act as sensing elements and can be used to measure loads from a fraction of an ounce to a ton or more. These measurements can be amplified and recorded. The subject stands on the platform and the system is set back to zero. Then any movement made by the subject will affect the pressure on the quartz crystals,

[6] Lucien Lauru and Lucien Brouha, "Physiological Study of Motions," *Advanced Management*, Vol. 22, No. 3, pp. 17–24, March, 1957.

Fig. 309. Force platform. *A*, triangular plate supporting worker; *B*, piezoelectric quartz crystal used for measuring force; *C*, electrometers for the vertical, the frontal, and the transverse components; *D*, balancing bridge and power supply for electrometers; *E*, Sanborn multiple-channel recorder. (Courtesy Dr. Lucien Brouha and the du Pont Company.)

and these pressure variations will be recorded as vertical, frontal, or transverse. These forces are proportional to the effort required by the subject to perform the specific task under study.

Figure 310 shows the motions of an operator loading textile bobbins onto the pegs of a special truck.[7] Figure 311 shows the forces produced in three dimensions while handling these textile bobbins at various heights. Note that the first, second, and third rows do not require as much effort as the bottom row. In the latter case the operator had to bend over and straighten up. This is especially noticeable on the vertical and frontal records. The study showed that the effort

[7] Lucien Lauru and Lucien Brouha, *op. cit.,* p. **20.**

Fig. 310. The force platform. A stroboscopic picture showing operator handling textile bobbins.

required to place the bobbins on the bottom row was greater than that required to place bobbins on the other four pins combined. Therefore, the bottom pin should not be used in this case.

Figure 312 shows the force involved in opening and closing filing cabinet drawers at various heights. The total effort (calculated from the three components) required for closing the top drawer amounted to 16 kilograms. As it happened, the same amount of effort was required in opening and closing the second drawer, 21 kilograms was required for the third, 30 kilograms for the fourth, and 42 kilograms for the bottom drawer.

Decrease in Output an Indication of Fatigue. Some people believe that the most practical and useful index of fatigue is its effect upon

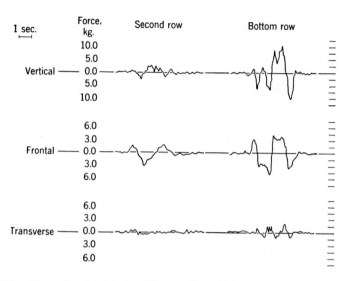

Fig. 311. Forces involved in handling textile bobbins at various heights. Note scale change for vertical force.

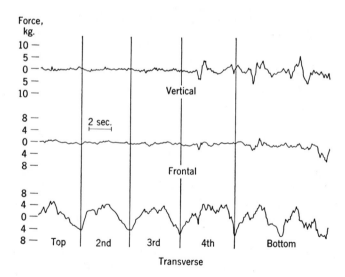

Fig. 312. Forces involved in opening and closing filing cabinet drawers at various heights.

the quantity and the quality of the individual's work; that fatigue can be measured in terms of reduced output resulting from work. However, one cannot say definitely that a reduction in output results from fatigue. That a person turns out less work during the last hour of the day may, of course, be due to the fact that he is tired. It may also be due to the fact that he has lost interest in the job, or that he is worried about some personal problem, or simply that he believes he has already done a day's work.

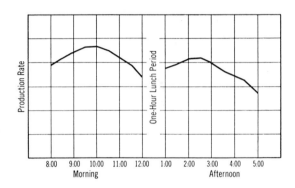

Fig. 313. Typical daily production curve for an individual engaged in very heavy muscular work.

The amount of work done per unit of time may be shown by means of a production curve, sometimes called an output curve or a work curve. It is not improbable that the production curve for *very heavy* manual work might take the shape shown in Fig. 313. Some people interpret this curve in the following way. The upward slope of the curve indicates a "warming up" period in the morning. This is followed by an increase in output until the middle of the morning, when a falling off in production occurs, possibly because of the fatigue of the worker. The curve for the afternoon is similar in shape to that for the forenoon, except that it falls off more rapidly towards the end of the day.

Much work in industry today is light and requires little physical exertion on the part of the operator. The production curve shown in Fig. 314 seems to be typical for such work, there being a fairly uniform output throughout the day. The operator has such a reserve of energy and the physical requirements of the task are so small that it is entirely possible for the operator to maintain a steady output for the entire day. In fact, it is not uncommon to find an operator actually increasing his speed during the last hour of the day when a delay has existed earlier in the day causing him to fall behind, or when a rush job has been put into production.

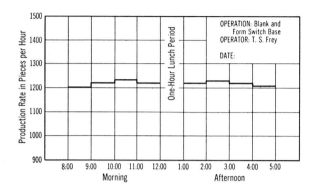

Fig. 314. Production curve for blanking and forming switch base on No. 21 Bliss press.

FACTORS AFFECTING DEGREE OF FATIGUE

Many factors affect the amount of work that an individual will do in a day and the extent of the physical fatigue that will result from this work. With a given set of working conditions and equipment, the amount of work done in a day will depend upon the ability of the worker and the speed at which he works. This latter factor depends directly upon the individual's inclination or his "will to work," which itself is affected by many things. The fatigue resulting from a given level of activity will depend upon such factors as (1) hours of work, that is, the length of the working day and the weekly working hours; (2) the number, location, and length of rest periods; (3) working conditions, such as lighting, heating, ventilation, and noise; and (4) the work itself.

Hours of Work. The findings of the Health of Munition Workers Committee organized in Great Britain in 1915 gave impetus to the movement for decreasing the length of the working day. At that time the 12- to 15-hour day was common. The reports of this committee and of many other investigations made since that time indicate the economy of shorter working hours. There is evidence to show that on most work, except for operations whose output depends mainly upon the speed of the machine, reduction of the length of the working day to 8 hours results in an increase in hourly and daily output.[8]

Since in this country the 8-hour day and the 5-day week are now in effect, it seems that little would be gained from a further reduction in the length of the working day insofar as *preventing physical fatigue* is concerned. When increased demands for output cannot be met with an 8-hour day, the possibility of two or three 8-hour shifts per day offers a practical way out. This also tends to reduce the overhead costs of operating the plant, since these expenses would be distributed over a larger number of units.

Rest Periods. When a person performs heavy physical work, it is necessary for him to stop and rest at intervals during the working day. If heavy work is done under conditions of high temperature and humidity, then the worker experiences still greater stress and consequently needs more time to recuperate. On such jobs the worker will rest a considerable part of the day whether "official" rest periods are allowed by management or not. Vernon found that on *heavy work* men rested one half to one fourth of the working time.[9] Taylor, in his classic experiment of handling pig iron, increased the output from 12½ to 47 tons per day mainly by requiring that the workmen rest 57% of the time and work but 43% of the time.[10] It should be noted that these examples are taken from *heavy work* and that much work in industry today is very light and requires little physical exertion on the part of the operator.

As previously explained, a person can work with an energy expenditure up to 5 to 7 calories per minute without going into "oxygen debt." On any task that does not tax a person beyond this level, no rest period would be required. If the work is more demanding physically, or if the temperature and humidity are high and the energy expenditure exceeds 5 calories per minute, or the pulse rate goes beyond 100 to 125 beats per minute, then rest periods should be introduced. Each person has a certain "physiological capital" upon which he can draw. For example, if a person is working on a task which requires 10 or 12 calories per minute instead of 5, he is contracting a physiological debt. Therefore, on such work a rest period should be introduced so

[8] H. M. Vernon, *Industrial Fatigue and Efficiency*, George Routledge & Sons, London, 1921; also, "Two Studies on Hours of Work," Industrial Fatigue Research Board, *Report* 47.

[9] H. M. Vernon and others, "Rest Pauses in Heavy and Moderately Heavy Industrial Work," Industrial Fatigue Research Board, *Report* 41, p. 20.

[10] F. W. Taylor, *The Principles of Scientific Management*, Harper & Bros., New York, 1911, p. 57.

Table 72. Classification of Work Loads in Terms of Physiological Reactions

Work Load	Oxygen Consumption in Liters per Minute	Energy Expenditure in Calories per Minute	Heart Rate During Work in Beats per Minute
Light	0.5–1.0	2.5– 5.0	60–100
Moderate	1.0–1.5	5.0– 7.5	100–125
Heavy	1.5–2.0	7.5–10.0	125–150
Very heavy	2.0–2.5	10.0–12.5	150–175

that the person can sit down and rest, thus allowing him to repay the physiological debt, or to bring his pulse rate and rate of oxygen consumption back to normal.[11] The difficulty of a job can be evaluated by the oxygen consumption in liters per minute, the energy expenditure in calories per minute, and the heart rate in beats per minute. Dr. Brouha has compiled a classification of work loads in terms of physiological reactions [12] as shown in Table 72. Table 73 shows energy expenditure data for a number of different tasks. This information is taken from research reports of many different investigators as summarized by Passmore and Durnin.[13]

In many kinds of work, both heavy and light, rest periods are desirable for the following reasons: (1) rest periods increase the amount of work done in a day, (2) the workers like the rest periods, (3) rest periods decrease the variability in the rate of working and tend to encourage the operator to maintain a level of performance nearer his maximum output, (4) rest periods reduce physical fatigue, and (5) rest periods reduce the amount of personal time taken during the working hours. Rest periods are particularly effective in heavy manual work, in operations that require close attention and concentration, such as fine inspection work, and in work that is highly repetitive and monotonous. Rest periods are usually placed in the middle of the morning and the middle of the afternoon, and range in length from

[11] E. A. Müller, "The Physiological Basis of Rest Pauses in Heavy Work," *Quarterly Journal of Experimental Physiology,* Vol. 38, No. 4, 1953.

[12] Lucien Brouha, *Physiology in Industry,* Pergamon Press, New York, 1960, p. 87.

[13] R. Passmore and J. V. G. A. Durnin, "Human Energy Expenditure," *Physiological Reviews,* Vol. 35, No. 4, pp. 816–834, October, 1955.

Table 73. Energy Expenditure Table

Type of Operation	Energy Cost in Calories per Minute
Sitting—idle	1.2
Watch and clock repair	1.6
Clerical work—sitting	1.65
Light assembly work	1.8
Draftsman	1.8
Clerical work—standing	1.90
Taylor hand sewing	2.0
Hand compositor	2.2
Taylor machine sewing	2.6
Sheet metal work	3.0
Punch-press operator	3.8
Taylor pressing suit	4.3
Trimming battery plates	4.4
Straightening lead contact bars	4.6
Pushing wheelbarrow at 2.8 miles per hour with 125-pound load on fairly smooth surface	5.0
Shoveling 18 pounds of sand through a distance of 3 feet with 1.5-foot lift at 12 throws per minute	5.4
Unloading battery boxes from oven	6.8
Pushing wheelbarrow at 2.8 miles per hour with 330-pound load on fairly smooth surface	7.0
Shoveling 18-pound load through a distance of 3 feet with lift up to 3 feet at 12 throws per minute	7.5
Digging ditch in clay soil	8.5
Tending furnace in steel mill	10.2

5 to 15 minutes. The proper number of rest periods, the spacing, and the proper length of each will depend upon the nature of the work and can be determined most satisfactorily by experiment. In general, several short rest periods are better than fewer long ones. When a person must work under hot, humid conditions, his rate of recovery will be much more rapid if ,he can rest in a cool, air-conditioned rest room. Where several people work as a group, the rotation of jobs at frequent intervals may serve to reduce the total physiological cost per person in that different sets of muscles may be brought into use with a change in the job.

Tests show that definite rest periods sanctioned by the management have a far greater recuperative effect than those which must be taken surreptitiously. Whether the rest is in the form of "soldiering" or whether it is enforced because of lack of materials, such hit or miss rests may have as little as one fifth the value of prescribed rests in relieving fatigue.[14]

When workers are paid by the hour, when their work is not measured, and when they are employed to perform tasks on which they can set their own pace, each worker can rest when he wishes and is free to adjust his working time and his resting time to suit his own needs. However, when an operation is measured and the worker is given an opportunity to earn a wage incentive, a time standard is established for the job, and this time standard contains an allowance for personal needs (personal allowance) and an allowance for rest or recuperation (fatigue allowance). Each company has its own tables for determining fatigue allowances, and among older well-established companies these data have been obtained by trial and error over a period of many years. There is a real need for a more systematic method of determining fatigue allowances, and it seems logical that physiological measurements can contribute to the solution of this problem.

Lighting, Heating, and Ventilation. Lighting, heating, and ventilation have a definite effect upon the physical comfort, mental attitude, output, and fatigue of the worker. Working conditions should be so adjusted as to make the shop and office a comfortable place in which to work. The requirements for proper illumination, heating, and ventilation are well understood, and equipment that will supply comfortable physical conditions for work is now available.

Of these three factors, illumination is perhaps most inadequately provided in most plants. Where the work is of such a nature that visual perception is required for its satisfactory performance, the output is invariably increased when adequate illumination is provided. Inspection operations such as those described on pages 277 to 283 are examples of work of this nature.

Noise and Vibration. Although noise is annoying to practically everyone, adaptation to noise is readily made by most people, and the psychological and physiological effects are not so serious as many believe.[15] Viteles draws the following conclusions from his study of noise:

---

[14] H. M. Vernon, *op. cit.*, p. 21.
[15] K. G. Pollock and others, "Two Studies in the Psychological Effects of Noise," Industrial Health Research Board, *Report* 65, p. 30.

(1) No experimental evidence is available to show that automatic performance is adversely affected by noise or by vibration.

(2) Nevertheless, except with certain "meaningful" noises, there is a wide agreement that both noise and vibration are "disagreeable" or "uncomfortable" accompaniments of work.

(3) A continuous noisy background often appears to have an initial stimulating effect, and this taken together with (2) appears to indicate that the noise should be regarded as an adverse condition which is met by an unwitting ncrease of effort.

(4) With constructive work involving mental effort, fairly consistent slight deterioration is observed, particularly in continued effort. Although, so far as the experiments go, the deterioration is barely or only just statistically significant, it may be "psychologically" significant. The consistency of the small deterioration seems to point to this.

(5) Discontinuous noise is more subjectively disturbing than continuous noise; "meaningful" noise may be more or less disturbing than "unmeaning" noise, according as it is interesting or familiar.[16]

Since noise and vibration are annoying, they are undesirable and should be reduced or eliminated insofar as possible. Stamping, cutting, and presswork are often segregated in one part of the factory so that the remainder of the plant may be kept relatively free of noise. Where large numbers of employees are affected and where the work requires a high degree of concentration or attention, it may be economical to reduce the noise by covering the ceilings and walls with acoustic board, as is done in many places. Some companies are completely enclosing noisy equipment such as automatic punch presses with solid walls of sound-absorbing materials. The walls are designed so that they can be opened up for servicing and adjusting the equipment.

Effect of Mental Attitude on Fatigue. Fatigue is by no means the simple, easily defined thing that many would have us believe it to be. Cathcart,[17] Dill,[18] and Mayo,[19] all of whom have written very clearly on this subject, point out the many-sided nature of fatigue.

A carefully conducted study, lasting over a period of several years, of fatigue of factory operators on regular production work at the

[16] M. S. Viteles, *Industrial Psychology*, W. W. Norton & Co., New York, 1932, p. 510.

[17] E. P. Cathcart, *The Human Fatigue in Industry*, Oxford University Press, London, 1928.

[18] D. B. Dill, "Fatigue and Work Efficiency," American Management Association, *Personnel Journal*, Vol. 9, No. 4, pp. 112–116.

[19] E. Mayo, *The Human Problems of an Industrial Civilization*, Macmillan Co., New York, 1933.

Western Electric Company showed that the mental attitude of the workers was by far the most important factor governing their efficiency.

Specific conclusions [20] relating to this point are:

(1) The amount of sleep has a slight but significant effect upon individual performance.

(2) A distinct relationship is apparent between the emotional status or home conditions of the girls and their performance.

(3) Total daily productivity is increased by rest periods, and not decreased.

(4) Outside influences tend to create either a buoyant or a depressed spirit, which is reflected in production.

(5) The mental attitude of the operator toward the supervisor and working and home conditions is probably the biggest single factor governing the employee's efficiency.

[20] G. A. Pennock, "Industrial Research at Hawthorne, an Experimental Investigation of Rest Periods, Working Conditions, and Other Influences," *Personnel Journal*, Vol. 8, No. 5, p. 311.

CHAPTER 35

Motion and Time Study Training Programs

The work of the motion and time study department in some organizations is not as successful as it should be because members of the organization do not understand how such studies are made and consequently they do not give this department the support and cooperation it should have. Often this lack of understanding extends from the president of the company to the foremen and the workmen in the plant.

One of the best ways to overcome such difficulties is to acquaint all members of the organization with motion and time study methods and procedures through well-organized and carefully conducted training programs. Some typical programs of this kind are described here.

MOTION STUDY TRAINING PROGRAMS

Before any job can be started, someone must plan it and set it up. This preliminary work includes determining the steps to be followed in doing the work, selecting the tools and equipment to be used, and training the operator.

When the production of a given article is large, staff engineers usually work out the details and aid the foremen and supervisors in putting the job into production. However, most work is not highly repetitive, and an operator may do several different jobs during the course of a day or a week. In such cases the supervisor usually decides how the job is to be done, lays out the work place, selects the tools and equipment, and instructs the operator. For this reason it is desirable that people in immediate charge of operations know the fundamentals of good work methods. Even when the production of the article is expected to be large and when industrial engineers are assigned to work out the manufacturing methods, the foremen and supervisors usually play an important part in aiding the engineers to develop the procedures to be followed. Here, too, it is desirable for the supervisor to have a working knowledge of the industrial engineer's technique as it pertains to methods improvement.

In the final analysis, however, it is the operator who does the job. It is the operator who uses the tools and equipment selected by the supervisor or the engineer, and employs the methods suggested by him. Therefore it is logical that the operator too should understand those methods and techniques which will enable him to do his job in the easiest and most efficient manner possible.

It has been demonstrated many times that both supervisors and operators can profitably do work methods design. Naturally, the supervisor should take the initiative in improving job methods. When this is done, the operator may be expected to absorb this knowledge more quickly either through instruction from the foreman or through a formal training course.

Training programs designed to present the procedures and techniques of the industrial engineer to top executives, foremen, supervisors, and operators provide an effective means of promoting better work methods in any organization. A methods design and methods improvement training program, to be most effective, should be developed to meet the needs of the particular group to receive the training.

A Preview of the Program. A methods development program, like any other important activity in an organization, must be understood and fully supported by top management if it is to be successful. In fact, every executive, manager, and supervisor must be acquainted with the philosophy, purposes, and objectives of the program and must understand the principles and approaches used in developing better work methods. For this reason it is essential that a preview of or introduction to the program be given to top management.

As has already been indicated, the program must be worked out to fit the particular needs of the organization, and the preview given at the very outset should reflect the type of program that will follow.

The Program. In many cases it has been found profitable to present the program to industrial engineers, supervisors, foremen, process engineers, tool and jig designers, mechanical engineers, group leaders, and key operators. A program 30 to 40 hours in length is quite common. Perhaps greatest success is obtained when the program is given in one continuous period of approximately 2 weeks. The forenoons may be devoted to conference room discussions and demonstrations, and the afternoons to working on projects or problems. A combination shop and laboratory (Figs. 315, 316, and 317) is needed if project work is included in the program. If it does not seem feasible to present the entire program in one continuous session, the material may be given in a series of 1-, 2-, or 3-hour sessions held once or twice a week as conditions seem to indicate.

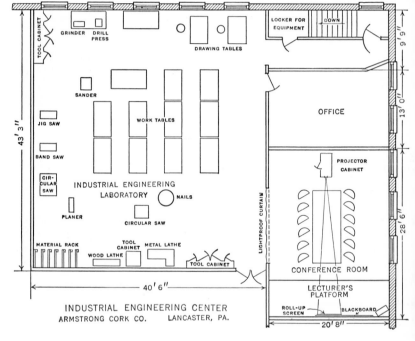

Fig. 315.   Floor plan of Industrial Engineering Center, Armstrong Cork Company.

A Specific Case. Programs for developing improved methods have been a regular part of management training at Armstrong Cork Company since 1945. This organization employs approximately **16,000** people in **20** different plants in this country. The company is well managed, and it has had a staff of well-trained industrial engineers in each of its plants for many years.

An Industrial Engineering Center was established at the main plant, and a Methods Development Program was inaugurated in 1945 in order to give greater emphasis to this phase of industrial engineering in the company, to standardize the techniques and procedures among all plants and personnel of the organization, and to lay the groundwork for a Methods Development Program for foremen and supervisors that later would be conducted in each plant.[1]

Industrial engineers, process engineers, mechanical engineers, and top management representatives from the various plants came to the Industrial Engineering Center in groups of 10 to 15 for 2 weeks' training, consisting of conferences and project work. At the completion

Fig. 316.   Conference room for Methods Development Program, Armstrong Cork Company.

of this program a shorter series of conferences for foremen and supervisors was held in each plant. This program was designed especially to fit the needs of the particular plant in which it was to be presented. An important feature of these conferences was the project work which each foreman did in his own department. The methods and techniques presented in the conferences were applied to specific problems by the foreman. The industrial engineers in the plant as well as the conference leader were available to assist the foreman when he needed help with his projects.

[1] John V. Valenteen, "A Long-Range Methods Development Program," *Modern Management,* Vol. 8, No. 5, pp. 6–9, July, 1948; also, "Stimulating and Maintaining Enthusiasm for Methods Improvement," *Proceedings Eighth Industrial Engineering Institute,* University of California, Los Angeles-Berkeley, pp. 69–74, February, 1956.

When the company established the Industrial Engineering Center in 1945, the Methods Development Program was designed as the first part of a long-range training program in the field of industrial engineering. Table 74 lists the major programs that have been conducted since the Methods Development Program was first started in the company plants in 1946.

Fig. 317. Laboratory for project work in connection with the Methods Development Program, Armstrong Cork Company.

Forty Years of Motion Study Training. The Fort Wayne works of the General Electric Company has maintained continuously for nearly 40 years a motion study training program. From the outset this company saw the importance of training all members of its supervisory personnel in motion study methods and techniques.

In 1928 representatives from the various plants of the General Electric Company were sent to Schenectady, where training in motion study was given. This training included both classroom instruction and the application of the principles in the laboratory and factory. These representatives, after thorough training, returned to their respective plants and proceeded to carry out training programs of their own.

Table 74. Participation in Major Programs, Armstrong Cork Company

Subject	Number of Plants Holding Classes	Number of Classes Held	Number of Persons Trained
Process Analysis	14	35	350
Man and Equipment Analysis	18	40	400
Material and Scrap Analysis	18	40	400
Lancaster Floor Plant Leaders	1	10	150
Production Management Methods Section for New Supervisors (at Industrial Engineering Center)	18 (Participating plants)	58	530
Totals		183	1830

L. P. Persing of the Fort Wayne works attended the first Schenectady course, and he returned to Fort Wayne with great enthusiasm and with an original and unique idea. Instead of offering training in motion study only to industrial engineers and staff people, his plan was to condense the three-month Schenectady course and offer this shorter program to all first line foremen and supervisors. As a part of the program, each supervisor would have the opportunity of applying the principles to specific jobs in his own department, and the operators working on the jobs would be consulted and encouraged to contribute their ideas in finding better work methods. Also, the supervisor would pass his knowledge of motion study on to his workers. This developed into a highly successful program, and has had a profound influence on industrial management and the practice of industrial engineering.

Mr. Persing started by acquainting management with the content of his proposed program and the way he planned to conduct it. Courses were then held for staff people in the manufacturing division. The first class of planning and time study engineers was held in January 1929. During a 3-year period the following classes were conducted:[2]

[2] L. P. Persing, "Motion Study—The Teacher," *Factory and Industrial Management,* Vol. 83, No. 9, pp. 337–340, September, 1932.

3	Classes of planning and time study engineers (beginners)	27
4	Classes of planning and time study engineers (advanced)	61
16	Classes of general foremen, foremen, assistant foremen, leading operators	268
2	Classes of special tool and machine designers	27
3	Classes of leading operators, expert workers, and personnel workers (women)	42
1	Class of plant construction engineers	22
1	Class of expert workers (assemblers)	16
30	Total number of classes Total number of persons trained	463

During the period in which the training was being carried on, new methods were devised by the application of motion study principles. In all, 96 jobs were studied and the methods revised. The new methods brought about an average reduction in time of 40%, and the tools and equipment necessary to put the improved methods into effect cost 7.4% of the total savings.

Motion study training has been given continuously since the program was started in 1929. During a recent 3-year period the following classes in motion and time study were conducted at the Fort Wayne works and its branch plants:

18	Classes in motion study and predetermined time values for methods planners, time standards men, and foreman (32 periods of $1\frac{1}{2}$ hours each, biweekly)	211
1	Class in motion study and predetermined time values for methods planners and design engineers (32 periods of $1\frac{1}{2}$ hours each, biweekly)	9
1	Class in motion study and predetermined time values for methods planners, tool planners, and foreman (32 periods of $1\frac{1}{2}$ hours each, biweekly)	10
1	Class in motion study and predetermined time values for product design engineers (36 periods of 2 hours each, biweekly)	7
1	Class in motion study and predetermined time values for methods planners, design engineers, and design draftsmen (20 periods of 2 hours each, biweekly)	11
3	Classes in motion study and predetermined time values for foremen and general foreman (20 periods of 2 hours each, biweekly)	24
2	Classes in time study for time standards men (24 periods of 2 hours each, biweekly)	16
2	Classes in motion study for apprentice toolmakers—students (20 periods of $1\frac{1}{2}$ hours each, biweekly)	24
36	Classes in motion study and predetermined time values for process and equipment planners; time standards analyst; foremen; supervisors of cost, production, purchasing; tool designers; and design engineers (30 periods of $1\frac{1}{2}$ hours each, biweekly)	432
65	Total number of classes Total number of persons trained	744

Mr. Persing, in supplying the above information, made the following statement:

Thousands of profitable projects have been worked out by members of these classes in connection with our motion study and predetermined time value training programs during this period. Moreover, this type of training has promoted a better understanding in our manufacturing organization and has made the work of all groups more effective.

Since the program was started in 1929 a total of over 624 classes have been conducted and over 6300 people have received training at the Fort Wayne plants of General Electric.

Methods Change Program. The Procter and Gamble Company is a large and successful organization with worldwide operations. It has been a pioneer in many areas, having introduced profit sharing in 1885 and guaranteed annual employment in 1923. Its industrial engineering program in the early days was mainly concerned with work measurement, wage incentives, and cost control. During the 1930's the company had placed special emphasis on improving performance in its plants and had achieved considerable success.[3] However, by the early 1940's the rate had diminished considerably, and it became apparent that if the rate of cost reduction was to be maintained at an acceptable level major emphasis on methods change would be necessary. Since it is essential to sell major changes in management practice from the top down, the methods program was begun by giving manufacturing management appreciation sessions in the concepts and techniques of work simplification. Then a special course was conducted for methods engineers—men who were college graduates with 1 to 5 years of company experience. After completion of the 1-week training course, these men returned to their respective plants to begin work as methods specialists.

1. *Specialist.* During the period 1946–1949, trained methods engineers served as specialists on methods improvement and worked more or less independently. As projects were completed and put into effect, the results were summarized in short reports. The savings during this period amounted to approximately $700 per year per member of fac-

---

[3] Richard A. Forberg, "Administration of the Industrial Engineering Activity," *Proceedings Twelfth Industrial Engineering Institute,* University of California, Los Angeles-Berkeley, pp. 22–30, 1960. Also see Richard A. Forberg, "Effective Control of the Industrial Engineering Function," *Proceedings Management Engineering Conference,* SAM-ASME, pp. 217–219, April, 1957; Arthur Spinanger, "Increasing Profits Through Deliberate Methods Change," *Proceedings Seventeenth Industrial Engineering Institute,* University of California, Los Angeles-Berkeley, pp. 33–37, 1965.

tory management. The program was considered a success, and additional men were trained and the program was extended to more plants.

2. *Coordinator.* However, it became apparent that this organizational approach had some limitations. By 1950 the position of the methods engineer had changed from specialist to coordinator. Formerly, the engineer had suggested the changes, and the foreman had not participated actively. He was inclined to take the suggested changes as criticism. Now the engineer spent approximately two thirds of his time helping the factory supervisors on their projects and the other one third of his time working on special factory projects as an individual. Each member of plant management had several selected costs to be reduced. He worked on these himself and requested help from the methods engineer as required. The methods engineer also conducted training courses at his plant for the line organization and other staff members. With the active participation of plant management, the rate of savings per member of management[4] had increased to $2300 per year by 1950. As the program grew and expanded, a bimonthly bulletin was published to promote the reapplication of successful methods changes in the other plants and to summarize achievements of the program. The company wanted all plant management, both line and staff, to participate in the plan.

3. *Methods Teams Formed.* In the beginning, any project that seemed to have savings potential was selected for study. Some of these had been small, but it was important that the men work on projects that interested them. Small successes gave them confidence to carry out savings projects regularly.

The plants experimented with various organizational approaches. One plant, for example, organized the management people into teams of four to eight men. Study of operations by groups tended to build a backlog of better projects. These projects were then ranked according to estimates of effort and potential savings, with consideration for the rate of return on any capital required. About this time the company developed "the elimination approach." This approach is applied to any cost, operation, machine, or piece of equipment by identifying the basic cause for the cost. It is an approach that asks the question, "If it were not for what basic cause, this cost could be eliminated." When this questioning shows that there is no basic cause, or that the basic cause can be eliminated, then the cost is eliminated. The process of identifying basic causes is most successful when done

---

[4] Example: If the annual savings in this plant amounted to $230,000 and if there were a total of 100 members of management (all line and staff management people in the plant), the savings per member of management would be $2300. ($230,000 ÷ 100 = $2300.$)

by a group of people. These groups were referred to as methods teams. In addition to increasing the annual savings rate to $3000 per member of management, the team approach had other values. Opportunities for recognition of good work were increased. Individuals who found it difficult to show results were aided and stimulated by the example of the more successful people on the team. There was a friendly rivalry for first place in plant standing between teams. Display boards and intraplant newsletters showed team standings.

4. *Team-Goal Approach.* Another plant experimented with a different approach. Since they were a high-cost plant they were motivated to take a total approach to cost reduction. Previously, plants had usually only concerned themselves with costs under their control. This plant began to consider all of the product costs such as materials, inbound and outbound freight, insurance and taxes, as well as the direct operating expenses. A goal of $500,000 was established for the year which represented over $5000 per member of management. With the participation of the Buying and Traffic Departments and with considerable plant effort the goal was achieved. The company-wide savings for 1954 amounted to approximately $4000 per member of management.

5. *All-Plant Team Goals.* The experience of this plant indicated that the cost-reduction program should be company-wide, and that all costs should be challenged. The central industrial engineering staff analyzed the strengths and weaknesses of the separate plant programs and recommended the most successful of these to the other plants. Thus the team-goal approach was extended to all plants. Teams committed themselves to a dollar methods change goal and then worked as a team to attain it. A periodic report showing comparison of results of the program in each plant was circulated, which further increased the desire to make a good showing.

Richard A. Forberg, Director of Industrial Engineering for Procter and Gamble, makes the following statement[5] concerning the program:

Early in the program, plants tended to think that they had "skimmed the cream." The easy projects had all been picked off. Next year would be harder, they thought; consequently a lower goal was in order. We had to sell some people on the reasonableness of a higher goal each year.

Higher goals were reasonable because of increased experience at cost reduction. The growth in business was another factor. Getting the less active team members to do more offered real potential for increased savings.

[5] Richard A. Forberg, "Administration of the Industrial Engineering Activity," *Proceedings Twelfth Industrial Engineering Institute,* University of California, Los Angeles-Berkeley, p. 27, 1960.

Although we had concluded that the goal setting process should be democratic, some guides were useful. The teams were encouraged to compare themselves with others on a dollars saved per member of management basis. Comparisons were also made in terms of goal as a per cent of operating expense and as a per cent of production value. The desire of plant management groups to show up well in all-plant comparisons was a strong factor in motivating teams to choose goals which required their best efforts for attainment.

Recognition. A successful methods change program depends upon positive recognition. The approaches to recognition generally fall into the following categories.

Methods Program Kickoff Meeting
Methods Goal Victory Celebrations
Team Status Bulletin Board
Monthly Methods Change Newsletter
Plant and Team Rotating Trophies
Small personal awards of recognition to members of a successful team
Letters of commendation

The plants have been very ingenious in devising ways to recognize successful methods change achievements. One plant, for example, conducted a methods saving contest and awarded stamps for completed projects. This contest was a takeoff on commercial trading-stamp plans. The plant printed its own stamps, stamp books, and prize catalogues. However, the savings exceeded expectations, the supply of stamps was exhausted, and vouchers had to be issued for the additional stamps earned. A victory celebration banquet was held at which stamps could be redeemed for modest prizes or exchanged for bingo cards in the hope of improving one's lot. The important point is that the recipients were recognized publicly and were made to feel important because their contribution was important.

Summary. The key points in organizing a methods change program based on Procter and Gamble's experience are the following ones.[6]

1. Form methods teams

a. Organize the program so that each manager can spend part— possibly five per cent—of his time working on deliberate changes as a member of a methods team or a profit team.

[6] Arthur Spinanger, "Increasing Profits Through Deliberate Methods Change," *Proceedings Seventeenth Industrial Engineering Institute,* University of California, Los Angeles-Berkeley, p. 37, 1965.

b. The teams should survey all of the dollars in the area. There was a time when most cost-reduction thinking was related primarily to labor costs. All costs should be challenged.

c. All management people should join in the program because cost reduction is possible wherever dollars are spent. No group should be above participating in cost reduction.

d. The responsibility for the success of the program should fall on the line organization.

2. Establish dollar goals

a. Challenging goals should be established. Procter and Gamble's experience indicates a need for building the goals democratically, starting with first-line supervision. Comparisons providing competition among plants and functional groups are a real incentive to achievement.

b. A capable person, such as a methods engineer, should be assigned to give the program continuing assistance and coordination that leads to meeting the goals. The coordinator will help with ideas, work on reports, obtain information from other technical organizations, and keep the record of achievement.

c. It is necessary to fight the delusion that cost reduction is "cream skimming"—that next year's goal will be harder to make because of success this year. The point should be emphasized that every dollar of cost is a dollar of potential savings. Past success points the way to ever greater profit increase through deliberate change.

3. Provide positive recognition

a. People like to be recognized in the presence of their associates. They want to be a member of a winning team. Procter and Gamble has found that when sincerely done it is almost impossible to overdo proper recognition.

The Industrial Engineering Division has in recent years been extending the methods change program to the technical staff groups and to the nonmanufacturing functions. For example, the Technical Packaging Division has been establishing methods change goals, and office building operating managers have added industrial engineers to their staffs and several have a full methods change program. The object is to extend the methods change program to include the entire management of the business.

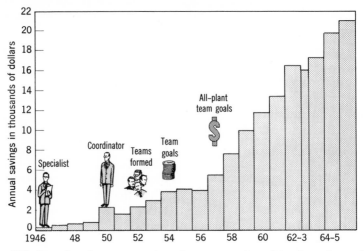

Fig. 318. Annual savings per member of management resulting from the methods change program.

Results. Figure 318 shows the entire record of savings of the methods change program.[7] The savings per member of management was $400 in 1946 and slightly over $21,000 in 1966–1967. The program provides one of the company's most attractive payouts. Since 1946 when the program started, the rate of return—using first-year savings only—has been around 1000 per cent. In other words, ten dollars of profit is returned for every one dollar spent.

Motion Study Training for Employees. Although motion study training programs for foremen and supervisors have been most common, there is increasing use of such programs for factory and office employees. A company-wide work simplification training program was inaugurated by the Maytag Company in 1949, and as a part of this program all factory workers have received 8 to 10 hours of training in this field from the Industrial Engineering Department. The training program is tied in with the company's "Employee's Idea Plan." This is a form of suggestion system whereby an employee is paid 50% of the first 6 months' net savings resulting from an idea which he has submitted. The training in work simplification is designed to aid and encourage the employee to think of and submit ideas that will

[7] The half width bar in 1963 represents a transition from accounting on a calendar year basis to a fiscal year basis.

reduce costs to the company. Last year 95% of the employees submitted one or more ideas.

Since the Employee's Idea Plan was started in 1949, a total of 34,149 ideas have been submitted and 9237 have been installed at a cost of $197,819, resulting in savings to the company of $2,586,528. The total amount of awards paid to employees for suggestions during this period was $582,807.

Incidentally, the supervisors at Maytag also submit ideas for methods improvement and cost reduction. Since the supervisors' work simplification program was started in 1949, 16,286 ideas have been submitted, and 7185 were installed at a cost of $998,818, resulting in a cost reduction to the company of $12,099,716. Since cost reduction is a regular part of the supervisor's job, no awards are made for their ideas.

Motion Study Training in Colleges and Universities. Since motion and time study occupies such an important place in American industry, many universities, colleges, and technical schools have incorporated work in this field in their curricula. Many schools now have both staff and physical facilities for presenting this subject in a very satisfactory manner. New York University and the State University of Iowa were among the first to give such training.

Motion Study Applied by Every Member of the Organization. If the foremen, supervisors, setup men, maintenance men, tool designers, cost accountants, time study analysts, production control men, and gang leaders have been trained in motion study fundamentals, they are able not only to apply them to their own particular work, but also to pass them along to the workers in the plant. And what is still more valuable, every member of the organization is available for consultation. From this large group of trained men valuable suggestions are constantly being received.

Although there are certain principles or rules of motion economy that may be used in designing preferred methods for doing work, there is no definite way of getting the most satisfactory results. Finding the best method is much like inventing or discovering something unknown. Suggestions, questions, discussions, and criticisms are all helpful. With several interested persons working on the problem, results are likely to come more quickly. Such cooperative work produces an entirely different atmosphere from the one present where the "expert" works out the method himself and puts it into effect, consulting no one and taking the full credit.

The General Electric Company was among the first organizations in this country to give motion study training to large numbers of its staff, and A. H. Mogensen was one of the first consultants in this field to advocate this practice.[5]

Cooperation. The training of all members of an organization in motion and time study principles and methods tends to bring about greater cooperation between the members of the motion and time study department and the rest of the organization. In plants where not everyone is familiar with the work of this department, and particularly where motion economy principles are not understood, there is often resentment on the part of some foremen, supervisors, tool designers, and others when new methods are put into effect. This opposition is largely due to a lack of understanding of just what is being done, and perhaps also to the failure of those installing the new methods to consult others associated with the work in order to get their suggestions.

There is no reason why every member of the organization should not be constantly on the lookout for better methods of doing work. And when a particular job is under consideration for improvement, all those connected with it in any way should be able to contribute to, or at least understand fully, what is being undertaken. The fine cooperation that usually results when all members of an organization have been trained in the use of motion study is an important and valuable by-product of such a training program.

TIME STUDY AND WORK MEASUREMENT TRAINING PROGRAMS

People with a knowledge of even the elementary principles of motion study and methods design are usually able to make valuable suggestions for improving methods, and consequently the training of foremen and supervisors in this field can be justified on this basis. However, only a qualified person, who is thoroughly trained in the fundamentals of time study and has served an "apprenticeship" under an experienced time study analyst, should be permitted to set time standards. People with only a superficial knowledge of time study should not attempt to do this work.

Time Study Training Programs for Top Executives, Foremen, and Supervisors. A time study training program designed for top executives,

---

[5] For information on work simplification see "Work Simplification—A Program of Continuous Improvement," by Allan H. Mogensen, *Industrial Engineering Handbook,* 2nd ed., Harold B. Maynard (editor), McGraw-Hill Book Co., New York, 1963, pp. 10-183 to 10-191; "Work Simplification," by Herbert F. Goodwin, *Production Handbook,* 2nd ed., Gordon B. Carson (editor), Ronald Press Co., New York, 1958, pp. 14-1 to 14-35.

foremen, and supervisors is not intended to train these men to make time studies. Rather, it is the purpose of such a training program to acquaint these people with the methods and procedures of time study so that they can assist the time study department in doing a better job. The main reasons why these groups should know how time studies are made may be listed as follows.

1. So that the planning and control department will better understand why it is important that there should be a smooth flow of parts and materials of the proper specifications to the processing departments.

2. So that the maintenance department will realize the importance of keeping all equipment in good repair, with the result that there will be a minimum of interruptions in the operation of equipment.

3. So that the inspection department will specify and maintain a definite standard of quality for each product. If quality standards change at frequent intervals, time standards cannot be used satisfactorily.

4. So that all branches of management will be on the alert to report to the time study department any changes in methods, tools, equipment, or other factors affecting the operations on incentives. When conditions beyond the control of the operator affect an operation on incentive, the job should immediately be taken off standard and put on day work, or an equitable adjustment should be made by the time study department, so as to give the operator the same earning opportunity as formerly.

5. So that these groups will understand the importance of keeping an accurate record of work done and applying the correct time standard (or piece rate) for each job completed. Each operator must be paid for the work he turns out—no more and no less.

6. So that all members of the organization will understand the procedure of rating operator speed, and will know the meaning of normal performance.

7. So that the foreman will have the operation to be timed running smoothly before requesting that a time study be made.

A Specific Case. A successful time study training program for top management, foremen, and supervisors was conducted in a midwestern plant. Since it is perhaps typical of such programs in medium-size manufacturing plants, the program is described here in some detail. This plant has a well-organized time study department, and the piece rate plan of wage incentives is used.

Size of organization: One plant with aproximately 1000 factor employees, 60% of whom are women.

Product: Complete line of rubber footwear.

Wage payment plan: (a) Hourly base rates established by job evaluation.

  (b) Time standards set by time study.

  (c) Straight piecework, with normal performance equal to 100% efficiency. At this point the hourly day-work rate is guaranteed.

Groups receiving training:

  Group 1—Top executives.[6]

  Group 2—All foremen: two groups, 7 persons in each group.

  Group 3—All supervisors: three groups, 5 persons in each group.

  Note: Junior members of the following departments also received this training: time study, production planning and control, and payroll.

The sessions were approximately $1\frac{1}{2}$ hours in length, and they were held on consecutive days. The same material was presented to all three groups. All three sessions for Group 1 were completed before the Group 2 conferences were started, and sessions for the second group were completed before the Group 3 conferences were started. All conferences were held during the working day, either in the forenoon or in the early afternoon. Meetings were held in the plant conference room, with ample space and facilities for showing motion pictures and displaying charts.

The factory manager was present through all sessions of Group 1, and introduced the program to each of the other groups. He was also present at the conclusion of each session of Groups 2 and 3, and made certain that each foreman and supervisor received complete and satisfactory answers to all his questions. In fact, the factory manager himself was familiar with every phase of time study and was fully convinced that all members of the organization should understand the detailed procedures. He encouraged the foremen and supervisors to raise questions on any points that were not perfectly clear to them.

The conferences were conducted mainly by the head of the time

---

[6] The top executive group included the factory manager, superintendent, chief chemist, chief engineer, head of production planning and control, head of cost accounting and payroll office, chief designer, head of purchasing department, and their assistants.

study department, with the assistance of the factory manager. An outline of the conference follows.

First Session. (1) Statement of the purpose of the conferences by the factory manager. Statement of benefits the company expects to receive from the conferences and the benefits the men attending the conferences should receive. General outline of the material to be covered in the conferences, by the head of the time study department.

(2) Showing of a 30-minute motion picture film presenting the complete time study procedure. This film showed step by step just how a time study is made, including the calculation of the final time standard.

(3) Discussion of the company time study manual. This manual was read section by section, and each section was carefully explained by means of specific illustrations. (See Appendix A.)

Second Session. Continuation of discussion of the time study manual. An actual job from the plant was brought into the conference room and demonstrated to the group. The time study of this job (Fig. 241) which had previously been made had been transferred to a large wall chart 4 feet by 8 feet in size, and to a computation sheet 4 feet by 8 feet in size (Fig. 243). These charts were hung in the front of the conference room. Each item on the time study sheet and on the computation sheet was explained to the group. This explanation led to the final establishment of the time standard for the job, and then to the determination of the piece rate for it. Finally, an explanation was made as to how the piece rate was put into effect.

Third Session. The first part of this session was devoted to a definition of "normal operator performance" as it pertained to this plant. Then the group was shown the introductory reel of the Unit I Work Measurement Film. After this each person rated ten different walking speeds and a film of several different factory operations. There was a general discussion of the meaning of "100% performance." The importance of the foremen and supervisors being able to evaluate operator speed accurately was emphasized.

There was a discussion of the relationship between the earnings of operators and their efficiencies above and below 100% performance. The number of people who might be expected to attain efficiencies in the range of 125 to 150% of normal was also discussed. The importance of accurately reporting work done was emphasized.

The entire time study procedure was reviewed in light of just how it affected the foremen or supervisor in the particular group. There was a question and discussion period, with the factory manager and the head of the time study department both participating.

After the conferences for the three groups had been completed, the factory manager and the head of time study held a 2-hour conference with the president of the union and the shop steward. The same material, in somewhat condensed form, was presented to these men. This session was followed by several discussion periods.

Follow-up. This series of three conferences was followed each month with a 1-hour conference for Groups 2 and 3, at which time a review of current problems pertaining to time standards and wages was made and each person was given the opportunity to make ratings of motion picture films of factory operations for which known standards were available. Actual factory operations were also rated.

This company owns its own motion picture camera, projector screen, and auxiliary equipment for making and showing 16-mm. films.

Time Study Training for Industrial Engineers from Several Plants of the Same Organization. Even though time study analysts may be doing accurate and consistent work in a given plant, if a company operates several plants it is good policy to standardize the work measurement procedure for all plants. This is especially desirable if identical operations are performed in two or more plants, if time study analysts are frequently transferred from one plant to another, if labor cost comparisons are made between plants, and if company-wide standard data are to be developed and used most effectively.

There is merit in having time study procedures so standardized and time study analysts so trained that if all time study analysts in all plants of a company were simultaneously to time the same operation independently they would establish essentially the same time standard for the job. Because the standardization of time study procedure and the training of time study analysts in this procedure are of growing importance, an actual case will be presented to show how a group of time study analysts were trained in one organization.

A Specific Case. The following case illustrates what may be accomplished through a systematic attempt to improve work measurement procedures. The company referred to here has five plants in four different midwestern states and employs a total of some 10,000 people. The company has been in business many years and has used time study in all plants for a long time. Before the inauguration of the Time Study Conferences, each plant had its own time study procedure and there was almost no exchange of information in this area between the plants. As might be expected, the time study methods differed, not all the time study forms used were alike, and there was considerable variation in time standards for identical jobs performed in the several plants.

Top management had received complaints from plant managers and from union officers about the variation in time standards, and a survey of time study procedures in all plants confirmed this point. Management decided to develop and standardize time study procedures that would best serve its needs, and a plan was inaugurated to have the time study analysts themselves, under the guidance of an able executive of the company, solve their own problems. The plan took the form of two-day Time Study Conferences, which were held at approximately 2-month intervals.

The first conference was attended by the chief industrial engineer and two or three of his assistants from each of the five plants. Work measurement procedures used in each plant were described and criticized. Everyday operating problems were presented by each person and discussed. The group agreed that they should work toward a standardized work measurement procedure, and that they should adopt the best time study practices available. A new time study form was designed, and the problem of performance rating was considered.

At succeeding conferences, the following subjects were discussed: methods of timing, the determination and application of allowances, waiting time, the development of a performance rating procedure for the company, all-day time studies, a "methods development program" for training foremen in all plants, the development of standard data for common operations in all plants, and other related subjects

The rating of walking, of dealing cards, and of films of factory operations was a part of each conference. Also, during conferences 3 to 7, several actual factory operations were timed. Each time study analyst independently made a time study of each operation. The man was given ample time to work up his data and to determine the time standard in the same way he ordinarily did this in his own plant. The completed time studies were immediately submitted to the chairman of the conference, and a summary was prepared similar to that shown in Fig. 320. The time studies were then returned to the time study analysts for use in the discussion which followed.

After all time study data were tabulated, the conference leader read off the values for each of the seven items, without mentioning from whose time study the values had been taken. Then there was a general discussion of the variations and the probable reason for them. After the discussion of a particular study was completed, the original time studies were turned in to the conference leader. Later these studies were reproduced and bound together, and a copy of the complete set of time studies was given each man for his file.

Several factors contributed to the success of the program described above. Among them the following seem to be the most important.

1. Top management was in complete sympathy with the program and was determined that the company should make use of the best time study techniques known.

2. A capable man (the assistant to the vice-president in charge of production) was selected to direct the program. It was his belief that time study analysts from all plants should assist in working out the details of the program, rather than that the program should be developed in the main office. It was felt that the training each man would receive in the process of developing the new time study procedure would be very valuable.

3. Each time study analyst was made to see the merits of having a standardized work measurement procedure for all plants, and each man contributed to the development of a workable system. At the suggestion of the time study analysts themselves, the program was expanded to include such things as determination of allowances, analysis of down time, development of a "methods improvement program for foremen and supervisors," and the determination of standard data for use in all plants.

4. The plant managers and other executives in the organization were interested in the details of the program and kept themselves informed as to the progress that was being made. The president and the vice-president of the company on several occasions attended the Time Study Conferences and took part in them.

5. A bound, stencil-duplicated volume of proceedings, prepared after each time study conference, served as a progress report. A copy went to the vice-president in charge of production and to other executives, as well as to each time study analyst. A reproduction of every time study made during the conference (but without the time study analyst's name) was included in the volume, as well as summary sheets similar to Fig. 320. These reports were successfully used by the management in discussing with the union the ability of time study analysts to determine accurate, consistent, and fair standards.

Results of a Simultaneous Time Study Made by a Group of Time Study Analysts. There has been much speculation about the variation that would be found in time standards set by a group of time study analysts if they were to study the same operation at the same time.

As a part of each Time Study Conference already referred to, the time study analyst made time studies of actual factory operations (Fig. 319). An experienced operator performed the job, and each time study analyst made his time study in his accustomed way. Figure 320 shows the summary of one of the best studies made by this group at the seventh Time Study Conference. Assembling a chain idler for a combine was the operation studied. Time study analyst B set a low time standard of 0.98 minute, and A set a high standard of 1.08

Fig. 319. Group of time-study men making a simultaneous time study in the machine shop. The results of such a study may be tabulated in a manner similar to that shown in Fig. 320.

minutes on this job. The average of the nine time study analysts was 1.03 minutes. B was 5% lower and A was 5% higher than the average of the group. All time standards fell within ±5% of the average for the group. Although it seems certain that these men will further improve their ability to set accurate and consistent time standards with more practice and experience, it might be added that their record, as shown in Fig. 320, is perhaps as good as will be found among time study analysts in general. These men worked in five different plants, and only two men had seen the operation before they studied it.

It should also be noted that the average performance rating factor for the nine men was 107%, with E using the lowest factor of 100%, and A, G, H, and I using the highest factor of 110%. E was 7% low, and A, G, H, and I were 3% high.

RESULT SHEET

Operation: Chain Idler Assembly for Combine

Information from Time Study	Industrial Engineer Who Made Study									Aver-age
	A	B	C	D	E	F	G	H	I	
1. Number of elements	9	7	9	10	9	8	11	8	13	9
2. Performance rating factor	110	105	105	105	100	105	110	110	110	107
3. Personal allowance in per cent	5	3	3	3	5	5	5	5	3	4
4. Delay allowance in per cent	2	2	3	2	2	2	2	2	2	2
5. Fatigue allowance in per cent	2	4	2	5	5	2	3	4	3	3
6. Total allowances in per cent	9	9	8	10	12	9	10	11	8	10
7. Total standard time in minutes	1.08	.98	1.04	1.02	1.00	1.04	1.06	1.06	1.03	1.03

Operation: Chain Idler Assembly for Combine

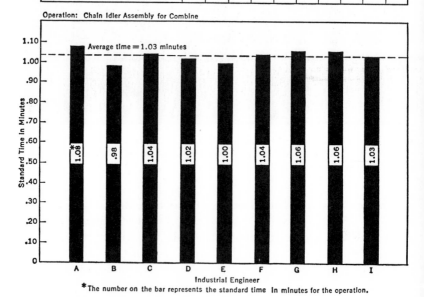

*The number on the bar represents the standard time in minutes for the operation.

Fig. 320. Table and chart showing the results of a simultaneous time study made by nine different industrial engineers from five different plants.

The total allowances varied from a low of 8% to a high of 12%, with an average of 10%.

Training in Rating Operator Performance. Practice in the rating of walking and the rating of dealing cards serves to show the importance of performance rating in time study work. Such rating studies may well be included in all time study training programs. These studies are also excellent for training beginning time study analysts and for improving the rating ability of experienced industrial engineers. The

Fig. 321. Silent motion picture proector (16-mm.) with a tachometer attached. This projector is suitable for showing rating films and time-study training films.

rating of walking and dealing cards is so widely used in industry that suggestions for making rating studies of these two activities are given on pages 759 to 765.

Rating of other simple operations, such as filling a pinboard (Figs. 86 and 87), tossing blocks, and assembling small parts, is also recommended. A motion picture film of an actual factory operation may be formed into a continuous loop and projected at constant speed for rating purposes. Considerable use is being made of such film loops for practice in rating and for time study training purposes. Stopwatch time studies can be made of operations on the screen, and standards

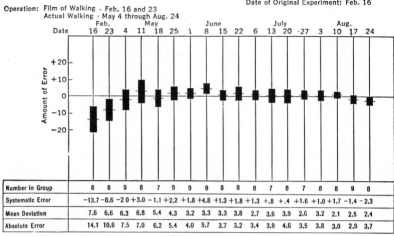

Fig. 322. Chart showing the improvement in rating over a 4-month period. This group of time-study analysts rated a film of walking on February 16 and on February 23, and then rated actual walking during the remainder of the period. Study made by Methods and Standards Department, Eli Lilly & Co.

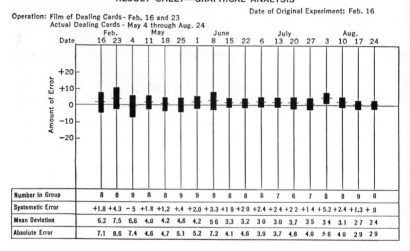

Fig. 323. Chart showing the improvement in rating over a 4-month period. This group of time-study analysts rated a film of card dealing on February 16 and on February 23, and then rated actual card dealing during the remainder of the period. Study made by Methods and Standards Department, Eli Lilly & Co.

can be established. Some companies have a library of such films to use for training purposes and to serve as standards on their important operations.

Figure 215 shows a motor-driven camera for making such motion pictures, and Fig. 321 shows a 16-mm. projector equipped with a tachometer for showing the films.

Effect of Practice on Accuracy of Rating. In order to measure the variations in ratings made by experienced time study analysts over a long period, Eli Lilly and Company repeated the walking study and the card-dealing study each week for a period of 4 months. The results shown in Figs. 322 and 323 are as follows: for walking, the systematic error was reduced from −13.7 to −2.3, and the mean deviation from 7.6 to 2.4; for card dealing, the systematic error was reduced from +1.8 to +0.8, and the mean deviation from 6.2 to 2.4.

The following general statements[7] can be made concerning training in performance rating.

1. An individual's accuracy and consistency in performance rating can be improved through proper training.

2. An individual can rate more accurately on performances that are close to normal. There is some tendency to rate too high on slow working speeds and too low on working speeds that are considerably above normal.

3. On simple work (free and unrestricted motions) a person can rate a motion picture film of an operation about as accurately as the operation itself.

4. Motion picture films serve a valuable purpose for training individuals in performance .rating.[8] They may also be used to assist in acquainting top management, foremen, supervisors, and workers with time study technique. A library of films showing operators working at normal tempo, and at known speeds above and below normal,

[7] Ralph M. Barnes, *Work Measurement Manual*, 4th ed., pp. 91–158; also, R. G. Carson, Jr., "Consistency in Speed Rating," *Journal of Industrial Engineering*, Vol. 5, No. 1, pp. 14–17.

[8] A study of 72 companies showed that 64% of them use motion pictures and 29% of them use simultaneous time studies for checking the rating ability of their time study analysts. This same study showed that these companies have conducted courses in time study, elemental standard data, or motion-time data as follows: 89% conducted such courses for foremen and supervisors, 21% for factory workers, 19% for office workers, and 42% for union officers and stewards. *Industrial Engineering Survey,* Ralph M. Barnes, University of California, 1967.

can serve as standards or bench marks for time study work in a plant.

5. Periodic time study conferences or clinics within a company, at which all phases of time study practices are discussed, are needed to keep a time study department functioning satisfactorily. Practice in making simultaneous time studies of actual factory jobs should be a part of such conferences.

CHAPTER 36

Training the Operator—Effect of Practice

It is not the purpose here to discuss the broad field of employee training, but rather to present some specific methods that have been found useful in training operators to do a particular job. Although such training is usually given by the supervisory force, the motion and time study analyst or a special instructor may handle this work.

We often have pictured for us large groups of workers performing identical routine operations over long periods, but this is not the typical situation even in large plants. Not only does the worker normally perform many different operations in the course of a month; also, with constant changes in methods, with improvements in materials, and with the rapid introduction of new models, there is a never-ending succession of new jobs which the operator must learn. It seems that the worker today, more than ever before, must be able to do a variety of work, which tends to increase the amount of training required in industry.

Training Methods on Simple Operations. The best method imaginable for doing a given task is of little value unless the operator can and will do the work in the prescribed manner. Where one or a very few persons are employed on a given job and where the work is relatively simple, the ordinary instruction sheet forms an excellent guide for training the operator. Also, on semiskilled work where the worker is familiar with the operation of the machine but needs instructions for the performance of particular operations, the simple instruction sheet is satisfactory. The example shown in Fig. 324 gives a written description of the elements required for turning the gear blank, and the drawing at the top of the sheet shows the exact location of the tools and of the parts to be machined. The time value for each element is also included, as well as the total standard time for the operation.

607

INSTRUCTION SHEET

Customer Amer. Tool Co.

Part No. 1073 A–F

Operation No. 5 TR.

Part Name Spur gear   Case D

Operation Name Drill, rough one side and ¾ of outside diameter

Dept. 11   Machine class, 58   Machine name, Jones & Lamson

Made by S. R. K.   Approved by S. M.   Date   Mat'l SAE2315

Tool layout

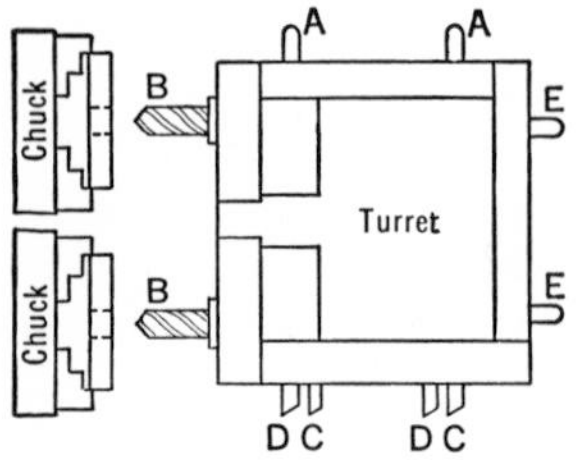

Set-up Time:
New set-up   60.00
Change of size 30.00

No.	Procedure	Tools—jigs, etc.	Speed		Feed		Base time
			Set-ting	Ft./min.	Set-ting	In./rev.	
1	Pick up and chuck 2 pieces...............	....	....	....	....	....	0.12
2	Start machine and true up (if necessary).....	....	....	....	....	....	0.10
3	Change speed.........		....	....	....	....	0.03
4	Adv. turret and throw in feed..............		....	....	....	....	0.06
5	ROUGH OUTSIDE DIAMETER (¾)...........	A. ¾ × 1¼ in. tools........	70	71	0.014	....	2.32
6	Back turret and index...		....	....	....	....	0.07
7	Advance turret, set headstock, throw in feed and change speed.		....	....	....	....	0.12
8	DRILL...............	B. 1 3/16 in. drills..........	60	71	0.014	....	0.58
9	Back turret and index...		....	....	....	....	0.07
10	Advance turret and lock		....	....	....	....	0.08
11	Advance headstock, change speed and throw in feed.......		....	....	....	....	0.08
12	ROUGH FACE 1 SIDE....	C. ¾ × 1¼ in. tools.......	70	71	0.014	....	1.65
13	ROUGH FACE HUB......	D. ¾ × 1¼ in. tools.......	30	71	0.014	....	
14	Unlock, back and index turret..............		....	....	....	....	0.07
15	Advance turret and set head stock.........		....	....	....	....	0.09
16	CHAMFER INSIDE FLANGE	E. ¾ × 1¼ in. Form tools	70	Hand	....	....	0.10
17	Advance head stock...		....	....	....	....	0.06
18	CHAMFER HUB.........	E. ¾ × 1¼ in. Form tools	30	Hand	....	....	0.10
19	Back turret and index..		....	....	....	....	0.07
20	Set head stock........		....	....	....	....	0.12
21	Stop machine.........		....	....	....	....	0.03
22	Loosen and remove 2 pieces..............		....	....	....	....	0.10
	Total handling time for two pieces..........		....	....	....	....	1.47
	Total machine time for two pieces..........		....	....	....	....	4.55
	Total base time for two pieces............		....	....	....	....	6.02
	Total base time for one piece.............		....	....	....	....	3.01
	Allowances 10 per cent..		....	....	....	....	0.30
	Standard time in minutes per piece.........		....	....	....	....	3.31

Fig. 324. Instruction sheet for turret lathe operation. Size 8½ × 11 inches.

Where the work is entirely manual, instructions prepared on the order of the operation chart shown in Fig. 70 are of value in that they indicate exactly what hand motions are required and show the layout of the work place.

Another case is taken from a chocolate factory. When a new box or a new assortment of chocolates is to be packed, the pattern is determined and the operators are required to pack by this pattern. The customary procedure was to send a sample package to the supervisor, along with the order for packing. Very often the first order was a rush one, and a number of operators were put on packing at once. Before the operators could begin work, the supervisor had to pack a sample box for each of them, often having the girls standing around waiting while this was being done. The use of an instruction sheet similar to that shown in Fig. 325, prepared in advance and reproduced by hectograph, has not only saved the operators waiting time but has also enabled them to bring their packing speed up to standard in a very short time.

Pictorial Instruction Sheets. The use of still pictures in connection with written instructions, as shown in Figs. 326 and 327, has proved very effective in supplementing the efforts of the instructor in training operators in a rubber footwear plant [1] and a glass plant. After glass bottles are made, they must be inspected for defects. The handling and inspection of bottles require considerable time to learn, and there is a knack to doing the job. The instruction sheet shown in Fig. 327 was developed by the Training Department of the Armstrong Cork Company to show the key points. It is supplemented by a motion picture showing an experienced operator inspecting bottles. Several different sizes and shapes of bottles are included, and some slow-motion shots illustrate the position of the hands in grasping and turning the bottles.

It seems that much of the skill required in doing some kinds of manual work centers around the exact way certain motions are performed, particularly grasp, hold, position, and pre-position. It appears that the transport and use motions require less attention and are more easily taught. In other words, it is more useful to show the operator how to take hold of the object before moving it, and how to position it before releasing it, than it is to show the actual transportation or movement of the object.

[1] A. Williams, "Teach It with Pictures," *Factory Management and Maintenance,* Vol. 94, No. 12, pp. 50–51.

½ Lb. BLUE RIBBON BOX (Flange) List No. 4623–12

Cups	Unit No.	Name	Cups	Unit No.	Name
Round	203	Raspberry Cup	Round	376	Caramelized Brazil
"	204	Apricot Cup	"	392	Croquante Whirl
"	221	Strawberry Creme	"	393	Vanilla Caramel
"	275	Coffee Creme	"	394	Marzipan Sandwich
"	371	Orange Marzipan	"	396	Tosca Pate

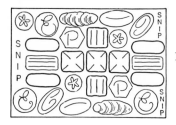

Heavy lines
= Foil Covered
Units

Make weight with Accommodation Units, one less than weight of last Chocolate.
If Light Add: 1 Croquante Whirl, 1 Apricot Cup.
If Heavy Take Out: 1 Apricot Cup.

	No.		Patt. No.	Paper No.
Linings (Center) (Emb. E. Foil)...........	1	13, ⅜ × 6, ⅞	Shaped	8795
" (Ends)...........	2	4, ⅞ × 2, ¹⁵⁄₁₆	3226	8796
Top-Pad................	1	6, ¹³⁄₁₆ × 4, ¹³⁄₁₆	....	4990
" Stock No. 04990— To be cleared first				
Cups (Round)...........	25	....	....	3569
Wrap.................	1	14, ¹³⁄₁₆ × 11, ⅛	2716	142
Wrap fastened on bottom with Gloy, ends folded and fastened on bottom with Gloy.				
Printed Identification Key.	1	8, ¾ × 6, ⅞	....	5070

Snip—Brown.
Filled on Printed Identification.
Tear-off Price Seal (Stk. No. 2878) on wrap, top-left.

Foil (Stock No. 8666) Blue and Silver E. Design—to be used when Stk. No. 08666 is cleared.

Foils

Stock No. 08666—Blue Printing on Silver.

Symbol No. F. 136.
Outer No. R. 976—Packed ¼ dozen.
Outer tied String—Single.

New Lines Office

Width of Reel 3″ for Tosca Pate, Marzipan Sandwich and Strawberry Creme.

First packing to be sent to Inspection Office.
Issued to Inspection Office from New Lines Office.

Fig. 325.   Instruction sheet for packing chocolates. Size 8½ × 11 inches.

OPERATION: Lacing
TYPES: L.T.T.—S.U.
DETAIL: 1 Spindle
MACHINE: Ensign
CODE: No. 52

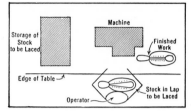

Sketch of Layout

Get next upper like this, first finger of left hand between eyelet edges.

Lift upper from stack with left hand, inserting second, third, and fourth fingers in top of shoe. Insert first finger of left hand between eyelet edges.

Line up eyelet edges by moving hands in opposite directions and closing eyelet edges together.

Get top end of eyelet edges between thumb and first finger of right hand. Move first finger of left hand out from between eyelet edges and grasp edges like this.

Position fifth eyelet over spindle.

Pull upper down onto spindle.

Press down pedal to start machine and move fingers to this position.

Hold upper while being laced. As spindle is automatically removed, upper is moved up and slightly to right while machine ties knot and cuts thread. Laced upper is finished in this position.

Move finished upper to a position over stack of finished uppers.

Place finished upper on stack.

Repeat Cycle.

Fig. 326. Pictorial instruction sheet for lacing tennis shoes.

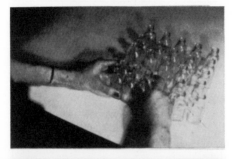

1. Pick up bottles (2 rows of 3).

   Grasp 6 bottles (2 in left hand, 4 in right hand). Hold thumbs toward you and fingers away from you.

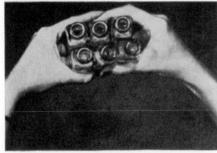

2. Inspect necks.

   Tilt necks slightly so that the light will show defects.

3. Separate bottles.

   Separate bottles so that the left hand holds 2 and the right hand 4.

4. Turn left wrist to left with palm of hand up. At the same time move the left thumb toward the left so that the top bottle falls into place to the left of the bottom bottle. This places 2 bottles in the palm ready for inspection. Use the left thumb as a stop.

Fig. 327. Pictorial instruction sheet for inspection of bottles

5. Lower upper left bottle in right hand to the fingers of left hand. To do this, tilt both hands slightly to the left, raise the right thumb, and let bottle slide to tips of fingers of left hand. (Keep hands together so that tips of fingers are touching. This prevents bottles from falling.)

6. Lower upper right bottle to palm of right hand. Slide right thumb to the right, pushing bottle into place to right of all bottles.

7. Line up bottles on tips of fingers and shift bottles toward right thumb as a guide. (Keep bottles tight together to shift more easily.)

8. To inspect sides, roll bottles one-fourth turn. Four bottles are in the right hand, the fifth bottle is on the tips of the fingers just ready to fall. With the left thumb turn the left bottle one-fourth turn to the left.
Hold the thumb as a guide.
Tilt both hands to left so that bottles roll one-fourth turn, one at a time.

9. Repeat steps 7 and 8 so that you can inspect the other side of the bottles.

10. Inspect bases and pack neck down in cartons. Keep all bottles tight together between thumbs, inspect bases, and slide between partitions. Make sure that carton is filled with bottles.

(1-ounce French squares), Armstrong Cork Company, Millville Plant.

Another example of the usefulness of still pictures is given in Figs. 141 and 142, showing how the operator grasps the bone and how she positions it at the beginning of the creasing motion in folding paper.

Training Assembly Operators. The following is a detailed description of the procedure that was followed in training a group of ten operators to perform a short-cycle assembly operation. This operation consisted of placing four small parts together, using both hands and the eyes. Figure 328 is a drawing of the finished mechanism assembly. The new operation was a combination of two old operations, which will be referred to as the superseded operations. Figure 329 shows the

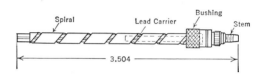

Fig. 328.   Mechanism assembly

arrangement of the work place. The only special equipment needed was a steel plate containing a small V-block to hold the lead carrier while it was placed in the stem of the assembly.

The improved method, which saved over 13,000 man-hours of direct labor per year, was developed by L. F. Youde. He describes [2] the training procedure which he used in the following way:

After the V-block was made for the new operation, the motion study analyst proceeded to run the new operation for approximately four hours in the methods laboratory. This trial run was made for the following reasons: (1) To test the equipment and eliminate any "bugs" that were present. This is important, as equipment should be fully tested before an operator is placed in training on the job. (2) To check the movements and establish a synthetic time standard for the operation from standard data. (3) To make certain that the motion study analyst who was to act as the trainer was able to perform the new operation with the correct movements.

The procedure described below was used to train the first operator on the new job.

(1) The new operation was shown and explained in general terms to the operator. Since she had been working on one of the superseded operations,

[2] L. F. Youde, "A Study of the Training Time for Two Repetitive Operations," thesis, University of Iowa, 1947.

she was told that her old job was being combined with another job so as to make a more efficient operation. Also, that the change was part of the regular methods improvement program. Other operations in her department that had been improved were pointed out to her as examples of the program. Because this operator had been on one of the superseded operations, it was not necessary to describe to her where the assembly was used.

(2) The operator was told what the time standard was on the new operation and that after a period of training she would be placed on piecework.

(3) With the operator standing slightly to his left and rear, the trainer demonstrated the new operation as follows:

(a) For 20 cycles the trainer performed the operation quite rapidly in order to give the operator an over-all picture of the new job.

(b) For 20 cycles the trainer performed quite slowly in order to give the operator a picture of the "gets" and "places" and the hand that performed them.

(c) For 10 cycles the trainer explained and performed slowly each of the "gets" and "places" in the operation. The explanation consisted of telling the operator where the eyes were used, which fingers were used in getting different parts, and how the parts were placed together. In running 10 cycles,

Fig. 329. Mechanism assembly operation—arrangement of work place.

each explanation was repeated 10 times. This repetition helped the operator retain more of the instructions than if only one cycle had been explained.

(4) The operator was seated at the work place and she was asked to perform the operation slowly the first day or two. She was told that at first the output was not important—that it was more important that she learn the correct methods, and that the speed would come naturally later.

(5) The operator was then told to go ahead and start performing the operation. The operator's performance as to the correct movements was checked as follows:

(a) The trainer watched the operator for her first ten minutes on the job to be sure she had the right idea and to correct any very bad errors.

(b) The operator was allowed to run one hour to get the feel of the operation and its parts.

(c) During the remainder of the operator's first day on the job, the trainer checked her once every hour to change any incorrect movements before they became a habit.

(d) During the second and third days, the trainer checked the operator once every two hours.

(e) For the remainder of the time until the operator reached standard time and was placed on piecework, the trainer checked twice a day. The amount of checking varied among the different operators trained on the new operation. Figure 330 shows the learning curve for the first operator on this job.

The other operators were trained in exactly the same way as the first one, except that their training was done on the production floor instead of in the methods laboratory.

Training Methods on Complex Operations. Some operations are complex in nature, and the operator may require considerable skill to perform them satisfactorily. A much longer training period is ordinarily required for this type of work than for simpler operations.

Where a sizable group of employees is engaged on such work, there is an opportunity for a more elaborate training program. Some companies find it profitable under such conditions to establish a vestibule training school or a separate training department apart from the regular production departments.

With over 100 operators on the semiautomatic lathe operation described in Chapter 4, the company established a special school for training new operators for this work. Whereas it formerly took 6 months to train these operators, it now requires but 6 to 8 weeks.

To cite another case [3] of group training on complex operations,

[3] L. P. Persing, "Motion Study—The Teacher," *Factory and Industrial Management,* Vol. 83, No. 9, pp. 337–340.

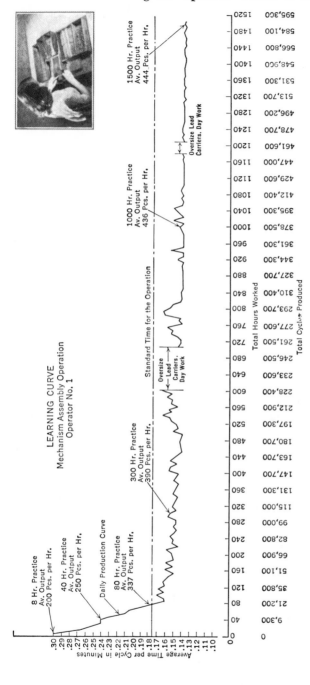

Fig. 330. Learning curve for mechanism assembly operation. The average hourly output for the first day was 200 pieces. After 590,000 cycles of practice, output increased to 444 pieces per hour.

L. P. Persing of the Fort Wayne works of the General Electric Company supervised the training of 200 new operators hired for a rush job of assembling large numbers of extremely delicate parts required in the manufacture of electric meters.

The entire assembly process was broken down into small assemblies, which were studied to determine the best way of making these subassemblies. Where special trays, fixtures, and combination tools were required, they were built, and the correct layout of the work place was arranged. One operator was trained by the instructor, and after she became proficient in the new method, motion pictures were made of the operation to use in the training of the other operators.

It was found practicable to train eighteen operators at one time. The training of this group was carried out in the following way. Eighteen duplicate sets of trays and tools were installed in exactly the same way on tables in the motion study laboratory (see Fig. 19 in Chapter 6). All the operators were seated at these tables, facing the motion picture screen at the front of the room. A general explanation of the operation was made, and instructions were given in the care that should be exercised in handling the parts so that the finish would not be marred or intricate and delicate parts damaged during the assembly operation. The motion pictures of the operation were then projected several times on the screen, both forward and backward and at reduced speed, so that the new operators could see the correct way of doing the work. With the projector running very slowly, the instructor pointed out the correct way of grasping, carrying, positioning the parts, and performing each of the other motions of the cycle. There were two instructors: one operated the projector and explained the motions, and the other, an experienced operator, gave individual instruction and inspected the work for the group.

With the ordinary method of using an experienced operator to train one or two new workers on the production floor, the experienced operator produced but 40 to 50% of her normal output. Also this method of training required an exceedingly long training period. Using the new method, two instructors trained eighteen new operators in a separate room where there was no interference with the regular manufacturing operations. At the end of one week of training, the eighteen operators were transferred to the production floor, properly trained for their task. This was but one third of the time required for training new operators by the old method.

Mr. Persing gives the following reasons why the company prefers to train the operators in a separate room: [4]

(1) We could get 100 per cent attention. There was not the confusion and noise of other activities to distract the operator's attention, such as you have on the average manufacturing floor.

(2) The operator does not get as nervous when trained in a separate room. There are not a lot of other workers watching the instructor teach the new cycle of motions, which in general are a great deal different from those they have seen before, and the layout of the trays is a curiosity.

(3) When any problem came up that was of interest to all the operators, you could get their attention at once and explain how to overcome or correct the fault.

(4) As the operators on this particular operation were to be taught the cycle of motions by watching the experienced operator assemble the register on the moving picture screen, it was necessary that the room be in semi-darkness.

The Colonial Radio Corporation was one of the first radio manufacturers to successfully operate a school for training new assembly operators. At the time of employment all girls were given 2 to 3 days' training in a separate room, under the supervision of a competent instructor.

The classroom contained assembly benches with jigs, fixtures, hand tools, and the necessary parts and bins to handle such typical factory operations as screwdriver work, assembly and benchwork with pliers, and soldering operations. Groups of 8 to 12, and never more than 15, were trained at a time. The girls were paid their regular hourly base wage during the training period. At the beginning of the training period a simple explanation of the purpose of the course was given to the group. Extracts from this explanation are given below:

As you are probably aware, the purpose of this class is to teach a better way of performing some of our more common assembly operations, which involve such familiar parts as nuts, screws, lock washers, wires, condensers, resistors, etc.

All of us realize the fact that certain ways of doing a thing are better than others. It has been established that there is a best way of performing any given act, and we have also made the discovery, which most of you have probably known all along, that the best way is almost invariably also the easiest way. Haven't you found this to be the case in your experience?

Just as you in your home attempt to find the best way of performing your

4 *Ibid.*

household duties, so in industry we attempt to find the best way of doing the things required of us.

It has been established that at least 25 per cent of the motions used by the average employee in the average factory operations are wasted motions. These wasted motions are needless motions which contribute only one thing as far as the operator or the operation is concerned, and that is fatigue.

Naturally you may ask—"What is the purpose of finding the best and easiest way of performing operations in the plant?" This can be stated briefly as follows:

"It is the desire on the part of Colonial to build a better radio set at a lower cost without, however, requiring the expenditure of any more physical effort on the part of those of us directly engaged in building them."

All of you realize the amount of work we have in our plant depends on the number of radio sets the Sales Organization of the Colonial Radio Company can sell. When you and I, and millions of other consumers, decide to buy a radio set, or any other merchandise, we always attempt to get the best product we can for the money we want to spend, and, if the best radio set we can buy for a given sum of money happens to be a Colonial radio, we will buy it. In other words, the welfare of the Colonial Radio Corporation and, coincidentally in a large measure, all of us, depends on the ability of Colonial to build at least as good a set as any other manufacturer at the same or at a lower price.

After the above explanation has been made and any questions by members of the class have been discussed, the group is given a simple assembly operation to perform. An explanation is given of what the finished job must be like, and then each person is allowed to do the task in any way that she wishes. Each girl is given a timer and pencil and paper to record the time for making ten assemblies. She continues to do this task for an hour, recording the time for each set of ten assemblies.

Then an assembly fixture and improved bins are given to the operator. The proper arrangement of the work place is made, and the girl is carefully instructed in the proper method of doing the work. An explanation is also given of the principles of motion economy employed, and why the new method is easier, faster, and safer than the old one.

After the girl understands how to do the task in the proper way, she again works for an hour or so, timing herself for groups of ten pieces and recording the time as before. The fact that the improved method saves time is obvious to her, since she has set her own pace and read and recorded her own time. She is well aware that motion study is not a "speed-up," but that it enables her to do more work with less fatigue.

After the new girl has worked on a simple assembly operation, she is given other jobs that are typical of those she will see in the factory, some of which she may work on after the training period is over.

Although the main purpose of the school is to train new operators in the principles of motion economy, Colonial has found that the school also serves another very important function. It shows the employees, in a most convincing manner, that improving methods of doing work is for their benefit as well as for the company's, and that actually the best way from a motion study angle is invariably the least fatiguing way and the most satisfactory way in every respect for the operator.

Incidentally, it requires approximately 50% less time for a girl who has been through the training school to attain standard performance than for new girls going directly onto the production floor without the training. During a 2-year period more than 700 girls were trained in the manner described above.

Audio-Visual Instruction for Operators on Complex Long-Cycle Operations. Audio-visual instructions provide one solution to the problem of training operators on long-cycle operations. The assembly of some electronic components may take as long as an hour or more, and it is not feasible to break the operations into small subassemblies. In such cases, detailed instruction in the form of 35-mm. color slides

Fig. 331. Standard work bench equipped with 35-mm color slide projector and magnetic tape player which provide the operator with visual and oral instructions. (Courtesy of Hughes Aircraft Company.)

which are shown on a translucent screen directly in front of the operator and oral instructions from a tape player are provided. The operator controls the speed of the audio-visual instructions by a foot-operated switch. It is reported that on complex long-cycle work this form of operator instruction reduces labor costs and improves the reliability and quality of the product (Fig. 331).

Training to Reduce Anxiety Among New Employees. A study made in a large manufacturing department at Texas Instruments Incorporated led to a plan for reducing causes of anxiety among new employees.[5] The unique indoctrination and training program that they developed resulted in the following gains:

1. Training time was shortened by one half.
2. Training costs were lowered to one third of their previous levels.
3. Absenteeism and tardiness dropped to one half of the previous normal.
4. Waste and rejects were reduced to one fifth of their previous levels.
5. Costs were cut as much as 15 per cent to 30 per cent.

Earlier studies and the analysis of 135 individual interviews with 405 operators made by the department manager indicated that new employees had many anxieties associated with lack of job competence during the early days of employment. The following facts were revealed: "Anxiety interfered with the training process. "New employee initiation" practices by fellow workers intensified anxiety. Turnover of newly hired employees was caused primarily by anxiety. The new operators were reluctant to discuss problems with their supervisors."

It was well known that anxiety dropped as competence was achieved. The question was then asked, "Is it possible to accelerate achievement to the competence level by reducing anxiety at a faster rate?" A study was designed to obtain factual information bearing on this question. The setting for the study was a rapidly growing department employing over 1400 people spread throughout three shifts. The department manufactured integrated circuits (microminiature circuitry units). The subjects of the study were women operators who collectively performed approximately 1850 different operations. Approximately 57 per cent of the operators worked with microscopes, and all jobs placed a premium on visual acuity, eye-hand coordination, and mechanical aptitude. Training was a continuous activity in this department—training new people hired for expansion and replacement purposes and retraining transferees and the technologically

[5] E. R. Gomersall and M. Scott Myers, "Breakthrough in On-The-Job Training," *Harvard Business Review*, Vol. 44, No. 4, pp. 62–72, July–August, 1966.

displaced. Figure 332 shows the learning curve for the ball bonders and is fairly typical of production operations in the department.

The ball bonders required approximately three months to reach what was termed the "competence" level. (The competence level is the stage at which assemblers can independently manufacture the product, but have not yet achieved the speed and accuracy ultimately expected of them to reach the labor standards set by industrial engineering. The competence level is about 85 per cent of labor standards; a position about 115 per cent of standard is termed the "mastery" level.

A group of 10 girls hired for bonding work on the second shift was chosen as the first experimental group. A control group was selected from the first and third shifts.

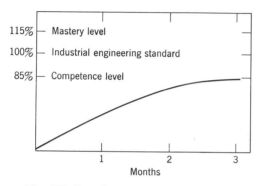

Fig. 332. Learning curve for ball bonders.

Conventional Indoctrination. The control group was given the usual first-day orientation, which consisted of a two-hour briefing on hours of work, parking, insurance, work rules, and employee services. This session included warnings of the consequences of failure to conform to organization expectations, and tended to raise rather than reduce anxieties. However, this was not intended as a threat.

Following this orientation, it was customary for a bonder to be introduced to her supervisor, who gave her further orientation and job instruction. Unfortunately, the supervisor's familiarity with the operations had desensitized him to the technological gap between them, and the following might be typical of what the operator heard him say:

Alice, I would like you to take the sixth yellow chair on this assembly line, which is in front of bonding machine #14. On the left side of your machine you will find a wiring diagram indicating where you should bond your units. On the right-hand side of your machine you will find a carrying tray full of 14-lead packages. Pick up the headers, one at a time, using your 3-C tweezers and place them on the hot substrate below the capillary head. Grasp the cam actuator on the right-hand side of the machine and lower the hot capillary over the first bonding pad indicated by the diagram. Ball bond to the pad and, by moving the hot substrate, loop the wire to the pin indicated by the diagram. Stitch bond to this lead, raise the capillary, and check for pigtails. When you have completed all leads, put. the unit back in the carrying tray.

Your training operator will be around to help you with other details. Do you have any questions?

Overwhelmed by these instructions and not wishing to offend this polite and friendly supervisor or look stupid by telling him she did not understand anything he said, the operator would go to her work station and try to learn by watching the operators on either side of her. But they, in pursuit of operating goals, had little time to assist her. Needless to say, her anxieties were increased, and her learning ability was impaired. And the longer she remained unproductive, the more reluctant she was to disclose her frustration to her supervisor, and the more difficult her job became.

Experimental Approach. The experimental group participated in a special one-day program designed to overcome anxieties not eliminated by the usual process of job orientation. Following the two-hour orientation by members of the Personnel Department, they were taken directly to a conference room. They were told that there would be no work the first day; that they should relax, sit back, and use the time to get acquainted with the organization and each other; and that they should ask questions. Throughout this one-day anxiety-reduction session, questions were encouraged and answered. This orientation emphasized four points:

1. *"Your opportunity to succeed is very good."* Company records disclosed that 99.6% of all persons hired or transferred into this job were eventually successful in terms of their ability to learn the necessary skills. Trainees were shown learning curves illustrating the gradual buildup of competence over the learning period. They were told five or six times during the day that all members of this group could expect to be successful on the job.

2. *"Disregard 'hall talk.'"* Trainees were told of the hazing game that old employees played—scaring newcomers with exaggerated allegations about work rules, standards, disciplinary actions, and other job factors—to make the job as frightening to the newcomers as it had been for them. To prevent these distortions by peers, the trainees were given facts about both the

good and the bad aspects of the job and exactly what was expected of them.

The basis for "hall talk" rumors was explained. For example, rumor stated that more than one half of the people who terminated had been fired for poor performance. The interviews mentioned earlier disclosed the fact that supervisors themselves unintentionally caused this rumor by intimating to operators that voluntary terminations (marriage, pregnancy, leaving town) were really performance terminations. Many supervisors felt this was a good negative incentive to pull up the low performers.

3. *"Take the initiative in communication."* The new operators were told of the natural reluctance of many supervisors to be talkative and that it was easier for the supervisor to do his job if they asked him questions. They were told that supervisors realized that trainees needed continuous instruction at first, that they would not understand technical terminology for a while, that they were expected to ask questions, and that supervisors would not consider them dumb for asking questions.

4. *"Get to know your supervisor."* The personality of the supervisor was described in detail. The absolute truth was the rule. A description might reveal that—the supervisor is strict, but friendly; his hobby is fishing and ham radio operation; he tends to be shy sometimes, but he really likes to talk to you if you want to; he would like you to check with him before you go on a personal break, just so he knows where you are.

Following this special day-long orientation session, members of the experimental group were introduced to their supervisor and their training operators in accordance with standard practice. Training commenced as usual, and eventually all operators went on production.

Results. The attitude and learning rate of the two groups was different from the beginning. By the end of four weeks the experimental group was significantly outperforming the control group—excelling in production and job attendance as well as in learning time. The one-month performance levels of the experimental and control groups were as follows:

	Experimental Group	Control Group
Units per hour	93	27
Absentee rate	0.5%	2.5%
Times tardy	2	8
Training hours required	225	381

Figure 333 shows the comparison of the learning curves for the two groups for the first eight weeks. The people at Texas Instruments believe that the area between the experimental curve and the control curve represents the learning time lag caused by anxiety.

When the experimental study began to show significant results, the anxiety-reduction process was used on additional groups. More than 200 people were employed in experimental and control groups for assembling, welding, and inspection operations. The performance curves for these groups are shown in Fig. 334. It should be noted that the third week's methods change in the inspection department

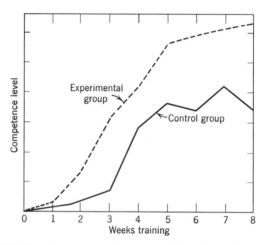

Fig. 333. Learning curves of experimental and control groups.

depressed the performance of the experimental group more than that of the control group, but the experimental group recovered more rapidly.

The motivated assemblers in the integrated circuits group without methods improvement exceeded the labor standards by approximately 15% to achieve the "mastery level" in two to three months (see Fig. 335), whereas the control groups took about five months. The area between the experimental group and the control group curves in Fig. 335 represents an improvement in the performance of approximately 50%. This was equal to a net first-year savings of at least $50,000 for 100 new employees. Considering reduced turnover, absenteeism, and training time, it was estimated that an additional savings of $35,000 have resulted.

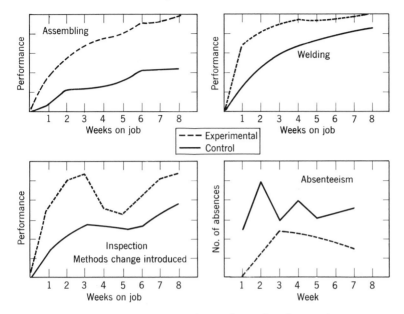

Fig. 334. Further comparisons of experimental and control groups.

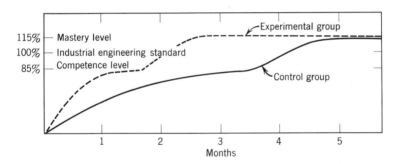

Fig. 335. Mastery attainment by experimental and control groups.

Another benefit was noted. The new trainees with less anxiety gradually influenced the performance of the work groups they joined. The
older employees were inspired by the greater confidence of the trainees.
Also, the higher performance of the new members established a new
reference point for stimulating the natural competitiveness that existed among members of the work group. There was also a definite

improvement in quality, and inspection labor costs were lowered by 30%.

EFFECT OF PRACTICE

If a person can perform a manual task at all, with practice he can reduce the time required per cycle. The shape of the learning curve will be affected by the nature of the work and by the traits, abilities, and attitude of the individual performing the task.

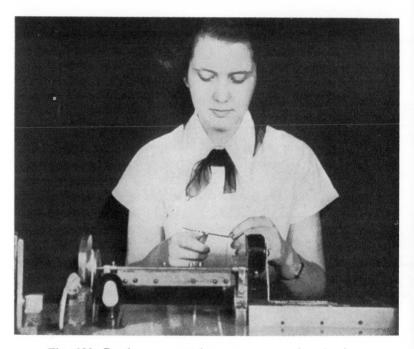

Fig. 336. Punch-press operation—arrangement of work place.

Figure 330 shows the learning curve for an operator assembling a mechanical pencil mechanism.[6] The operator on her first day of work averaged 200 assemblies per hour. After 11 days of practice she turned out 343 assemblies per hour, which was standard performance. Since the piecework plan of wage payment was used in this plant, this operator earned a bonus for all work produced above the standard of 343 pieces per hour. Although the operator was producing 444 pieces per hour after 1500 hours of practice, it is apparent that

[6] L. F. Youde, *op. cit.*

the greatest increase in production took place during the first few weeks of work on the job. At the end of the first week the operator had increased her output 25% over that of the first day, and at the end of 2 weeks she had increased it 68% over that of the first day. After 38 weeks of practice, the total increase amounted to 122%.

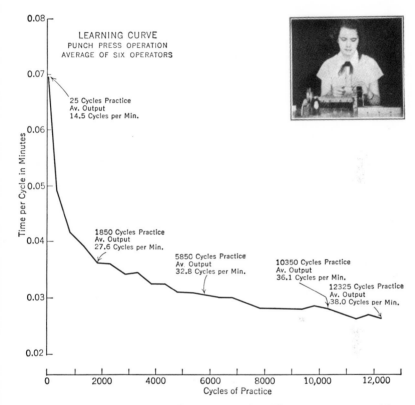

Fig. 337. Learning curve for punch-press operation. The average output without practice was 14.5 pieces per minute. This increased to 38 pieces per minute after 12,325 cycles of practice.

Of this increase, over half took place during the first 2 weeks of practice.

Figure 337 shows the average learning curve for six operators performing a fairly complicated punch-press job (Fig. 336) involving the use of both hands and one foot. The output increased 75% after 1350 cycles of practice, and doubled at the end of 3350 cycles of practice.

Figure 338 shows the learning curve for a very simple operation, filling a pinboard with 30 pins, using the two-handed method (Fig.

87). Here the output increased very rapidly because the simplicity of this operation.

A number of studies[7] have been made of typical factory operations to determine the effect of practice on the fundamental hand motions. On one job, for example, while there was a reduction in time of 40% after 3000 cycles of practice for the operation as a whole, the reduction

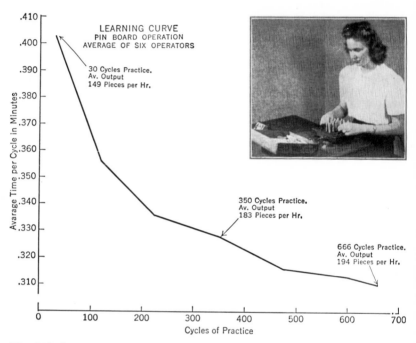

Fig. 338. Learning curve for pinboard operation. The average number of pieces per hour without practice was 149. This increased to 194 pieces per hour after 666 cycles of practice.

in time for the transport loaded motion was 15%, wereas the reduction in time for the position was 55%.

In another investigation[8] an attempt was made to determine through micromotion analysis the difference between the way the operator

[7] See studies by Harold T. Amrine, "The Effect of Practice on Various Elements Used in Screwdriver Work," *Journal of Applied Psychology,* Vol. 26, No. 2, pp. 197–209, and J. V. Balch, "A Study of Symmetrical and Asymmetrical Simultaneous Hand Motions in Three Planes," *Motion and Time Study Applications,* Section 14, pp. 70–72.

[8] Ralph M. Barnes, James S. Perkins, and J. M. Juran, "A Study of the Effect of Practice on the Elements of a Factory Operation," *University of Iowa Studies in Engineering, Bulletin* 22, p. 67.

performed the job without experience and the way she performed it after she had become proficient. Motion pictures were made of the operator as a beginner and then at intervals during the learning period until she reached a high level of proficiency. Figure 339 shows the results of this study. The upper line *A* is the actual learning curve, whereas the lower line *B* is the learning curve with the fumbles, delays, and hesitations removed. This study shows that in this case

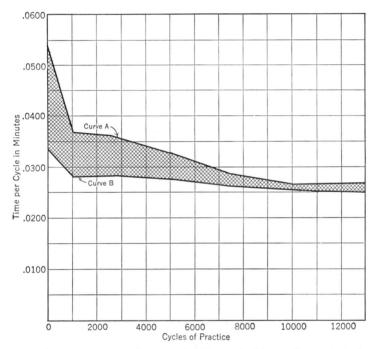

Fig. 339. Curves showing effects of fumbles and delays. Curve A is based on data including all fumbles and delays. Curve B is based on the same studies but excluding fumbles and delays.

two thirds of the increase in output during the learning period can be attributed to the elimination of fumbles, delays, and hesitations on the part of the operator, and one third perhaps to faster hand motions.

The following is an analysis of the differences of the two learning curves:

Cycle time at the outset	0.052 minute
Cycle time at the finish	0.027 minute
Improvement	0.025 minute

The improved performance can be traced to the following overlapping causes:

Reduction in fumbles and delays	0.017 minute
Faster performance	0.008 minute
Total	0.025 minute

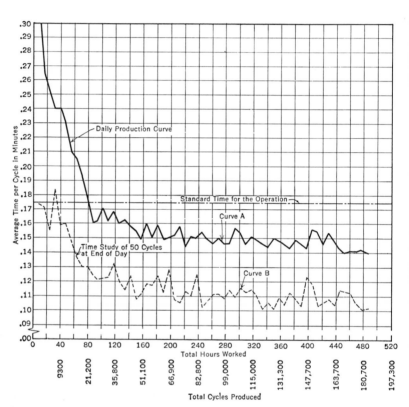

Fig. 340. Learning curves for mechanism assembly operation. Curve A is a record of the total daily output. Curve B is the actual time per piece as determined by a time study made of 50 cycles at the end of the day.

There is evidence to show that the beginner does not use the same method that he will use after he becomes proficient in performing the job. This difference in method is the biggest single factor affecting the cycle time for the job during the learning period.

Lost Time. The studies just referred to were made in the laboratory, and the conditions there are not always identical with those in the factory. Other studies seem to show that there is considerable difference between the average time per piece as determined by time study and the average time per piece as determined by dividing the number of minutes worked during the day by the number of pieces completed during the day. Figure 340 shows such information.[9] The upper curve *A* is a record of the total daily output, whereas the lower curve *B* is the actual time per piece as determined by a time study made of 50 cycles at the end of the day. This shows that as a beginner the operator "loses more time" per day than she does after she has had some practice.

If the operator is given proper instruction, the learning period can

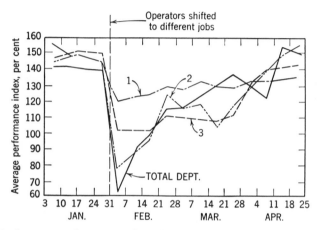

Fig. 341. Average performance index of three operators in a factory. A change in the design of the product eliminated their job on January 31. They immediately started on new work.

be reduced, thus lowering unit labor cost to the employer and giving the operator greater satisfaction on his job. In time study work, emphasis is always placed on standardizing the method before setting a time standard for the job. The discussion and the learning curves in this chapter would indicate that time standards should not be established from time studies made of inexperienced operators. Practical time study men know this only too well.

Time for Experienced Operators to Learn Another Job. Figure 341 shows the average weekly performance index of three operators in a washing machine factory for the period January 3 to April 25. During the month of January these three operators were working at a very high level, averaging 143, 150, and 154% efficiency. At

[9] L. F. Youde, *op. cit.*

the end of January the operations they were working on were discontinued, and they were transferred to other work. It became necessary for them to learn a new job, and almost 3 weeks was required to bring their performance index up to the standard of 100%, and a much longer period to reach the same level of efficiency at which they had been working before the change in jobs.

At the time these men were transferred to the new job, they stated that in their opinion the time standard on the new operation was too low—that they could not make the same performance index as they had in the past. The effect of practice was carefully explained to these operators, and they were persuaded to apply themselves to the new job as they had done in the past. Figure 341 shows the progress made by these three operators in regaining their high performance level.[10]

Learner's Progress Record. When a vacancy occurs or when a new job is established, it is management's desire to select a person for the work who has the traits and qualities that will enable him to succeed on the job and to get personal satisfaction from it. It is also management's responsibility to train the worker so that he reaches standard performance in as short a period of time as possible. In order to have definite knowledge of the progress that a learner should make on a given job, some companies have made extensive studies of learning curves for various types of work, and have prepared "normal learning curves" for their operations. The following procedure illustrates how such curves may be used.

A Specific Case. The purpose of the "learner's progress record" is to compare the learner's progress with the average performance of normal learners. This record serves also as a guide to the foreman and the instructor as to the necessary items to be included in the training of each learner. The normal learning curve is shown[11] in Fig. 342 and the "learner's progress record card" in Figs. 343 and 344. It should be emphasized that the normal learning curve developed by this company is only a rough indication of the expected output at intervals during the learning period. This curve is not sufficiently accurate for use as the basis for a wage incentive for learners.

---

[10] J. F. Biggane, "Time Study Training for Supervision and Union," *Proceedings Fourth Industrial Engineering Institute,* University of California, Los Angeles-Berkeley, p. 12, February, 1952.

[11] The shape of the curve in Fig. 343 is different from the learning curves on the preceding pages because the vertical scale for Fig. 343 is "Efficiency in Per Cent," whereas that of the preceding curves is "Time per Cycle in Minutes."

The new employee or the employee who is assigned to a new job is turned over to an instructor, who is a skilled operator. This instructor, together with the foreman of the department, is responsible for

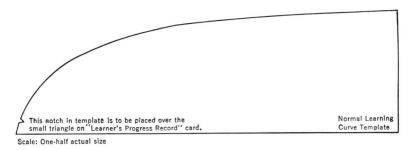

This notch in template is to be placed over the small triangle on "Learner's Progress Record" card.

Normal Learning Curve Template

Scale: One-half actual size

Fig. 342. Normal learning-curve template (celluloid). The shape of this curve has been determined from studies of hundreds of different operations made by this company over a period of years.

| OPERATOR'S NAME | Martin Harrison | CLOCK NO. | 76483 | FOREMAN | G. R. Roberts |

CODE 3041   GRADE & CLASS 11A   NEW X   CHANGED_____   INSTRUCTOR A. C. Wilson

OPERATION Punching Miscellaneous Parts (Punch Press)   BLDG. 17

Learning Time in Weeks

EFF.
110%
100%
90%
80%
70%
60%
50%
40%
30%
20%
10%
0%

Place Notch on Curve Here

LEARNER'S PROGRESS RECORD

Fig. 343. Learner's progress record—front of card. Size of card 5 × 8 inches.

the learner's progress on the job. At the time the learner begins work, his progress record card (Figs. 343 and 344) is filled out and the learning curve is constructed for the job in the following way. The normal curve (Fig. 342), a template cut from a sheet of transparent celluloid, is used for all training periods. The curve is placed with the notch exactly on the black triangle in the lower left-hand corner of the card (Fig. 343). The top of the template intersects the 100%

horizontal line at the point of intersection of the vertical line designating the number of weeks of training required for that labor grade. A pencil line is drawn which follows the contour of the template, and this line represents the learning curve for the job in question. The card is retained by the foreman or instructor until the learner becomes proficient on the job (reaches 100% efficiency) or changes jobs.

Each week the progress of the learner is recorded on the back of the card (Fig. 344) in two ways. The efficiency percentage figure

C-4 OPERATOR'S RECORD					QUALITY: E-EXCELLENT G-GOOD F-FAIR P-POOR	
					ATTITUDE: Very good. Operator	INSTRUCTOR'S CHECK
STARTING DATE: December 3, 1947					wants to learn the job.	TOOLS REQUIRED X
RECORD WEEKLY						INTRODUCTION X
QUALITY RATING ──────→						QUALITY X
EFFICIENCY PERCENTAGE ──→						SAFETY AND HOUSEKEEPING X
F	F	G	G		COMMENTS:	COOPERATION X
45	48	50	58		At end of second week:	FAIRNESS X
G	G	G	E		Operator seems slow in	RATES OF PRODUCTION X
74	79	89	83		learning job.	ALLOWED PERSONAL TIME X
E	E	E	E		At end of fourth week:	
92	98	98	103		Operator is still slow.	
					He thinks he can handle	
					the job.	FOREMAN'S CHECK
					At end of sixth week:	WAGE PAYMENT PLAN X
					Operator is progressing	POSSIBLE ADVANCEMENT X
					satisfactorily.	SAFETY X
						QUALITY X
					FOREMAN'S SIGNATURE	CONTINUALLY STRESS
					INSTRUCTOR'S SIGNATURE	SAFETY AND QUALITY X

Fig. 344. Learner's progress record—back of card.

is computed and recorded in the appropriate box, and a letter designating the quality of the work done by the learner is inserted above and to the left of the efficiency figure. At the same time this information is recorded on the back of the card, a point representing the efficiency is plotted on the front, and a straight line is drawn connecting the zero point on the curve with the efficiency at the end of the first week. In a similar manner the efficiency is recorded and plotted each week, and the quality of the work is indicated on the back of the card. The learning curve is shown to the operator when he first begins work on the job, and its meaning and purpose are discussed with him in detail. Then each week, after the learner's performance has been posted on the card, the instructor or the foreman goes over his accomplishment with him. They discuss each of the

```
                          JOB CLASSIFICATION

CODE    2706   J.R.              DATE                    DEPT.     3-7-9
OPERATION TYPE                   SUPER.                  BLDG.
INDEX NO.                        PAYMENT                 WORK STATION OR GROUP NO.
OCCUPATION: Single or Multiple Spindle Drill Press Operator
```

JOB IDENTIFICATION

Operate small size single or multiple spindle drill press to drive, tap, ream, countersink, and counterbore various parts and materials to specified location, depth, size, etc., with jigs and holding fixtures. Simple work on small and light parts of ordinary tolerances. Maintain set up by exchange of tools and simple adjustment. Check work.

TYPE OF OPERATOR DESIRED · Grammar school education. Girl of average height and size. Should be mechanically inclined.

LEARNING TIME - New operator 3 weeks.

Fig. 345. Job classification card for small single- or multiple-spindle drill.

```
                          JOB CLASSIFICATION

CODE    L2746   J.R.             DATE                    DEPT.     4
OPERATION TYPE                   SUPER.                  BLDG. 24W
INDEX NO.                        PAYMENT                 WORK STATION OR GROUP NO.
OCCUPATION: Engine Lathe Operator
```

JOB IDENTIFICATION

DESCRIPTION OF WORK - Get job assignment and secure proper materials and necessary tools for the job. Place work in machine either by chucking or placing between centers. Set up cross slide tools to make necessary cuts. Determine the depth and number of cuts to be made in machining work to drawing and specifications. Machine work as outlined above. Remove work from machine and identify if necessary.

EQUIPMENT USED · 24" Lodge & Shipley, 12' x 24" Prentice, 9' x 18" Chard, and 10' x 25" LaBlond engine lathes, machinist tools, micrometers, calipers, pin gauges, crane.

TYPE OF OPERATOR DESIRED - High school graduate or equivalent with previous lathe experience. Must be able to work to close tolerances and read blueprints.

TYPE OF WORK PERFORMED - Bearings, thrust collars, gear and pinion, etc.

LEARNING TIME - New operator 9 weeks; semi-experienced operator 6 weeks.

Fig. 346. Job classification card for engine lathe.

items listed on the back of the card, such as quality, safety, and good housekeeping. Each item is checked after it has been thoroughly explained to the learner.

The instructor records the attitude of the learner toward the job. Notes are also recorded concerning the learner's progress and any irregularities that may have occurred during the period. After the operator reaches standard performance (100% efficiency), his progress record card is sent to the supervisor of training, where it is permanently filed with other records of the employee.

The procedure described above has proved to be a very effective method of keeping the employee, as well as the foreman, informed each week as to the progress being made. In cases where an operator proves to be unsuited to the work and his output consistently falls below the expected production, it is possible to transfer him to other work without excessive loss of time.

Job Classification Card. The "job classification card" contains a rather complete description of the job. (See Figs. 345 and 346.) It shows the labor grade and the learning time for a new operator as well as for a semiexperienced operator. A new operator is defined as one who has had no experience on the particular type of work, and a semiexperienced operator as one who has had some experience on the particular job or machine.

Figure 345 is a job classification for a small-size single- or multiple-spindle drill. The learning time for a new operator is 3 weeks. Figure 346 is a job classification for an engine lathe. The learning time for a new operator is 9 weeks, and for a semiexperienced operator, 6 weeks.

CHAPTER 37

Evaluating and Controlling Factors Other Than Labor—Multi-factor Wage Incentive Plans

In some industries labor costs are small in comparison to operating costs of machines and process equipment. Likewise in some departments in a factory more can be saved by controlling quality and scrap and by increasing material utilization than by increasing labor effectiveness. For example, the loss to the company from one hour down time of a paper-making machine or a large coating machine may be greater than the wages paid to the operators of the equipment for an entire shift. The ideal incentive plan for the operators of costly machines and process equipment might be a multi-factor plan. For example, a plan might be designed to include such factors as square feet of material processed, percentage of product of acceptable quality produced, and utilization of material (Fig. 347). As industry becomes more highly mechanized, greater attention will be paid to those factors that make for low operating costs of the equipment. The goal will be lower unit cost of the end product rather than low direct labor costs alone.

Manual and Machine-Controlled Operations. An operation such as turning the outside diameter of a gear blank on an engine lathe consists of two parts: the loading and unloading of the machine, during which the operator works, and the cutting time (machine-controlled elements), during which the machine works and the operator is idle. The objective is to provide an incentive that will result in the maximum utilization of both the operator and the machine. If the lathe is a relatively inexpensive one and if the proportion of the cycle that the operator is forced to remain idle is relatively small, the usual type of measurement and financial incentive will give satisfactory results. If the machine is an expensive one to operate (high overhead rate) and

639

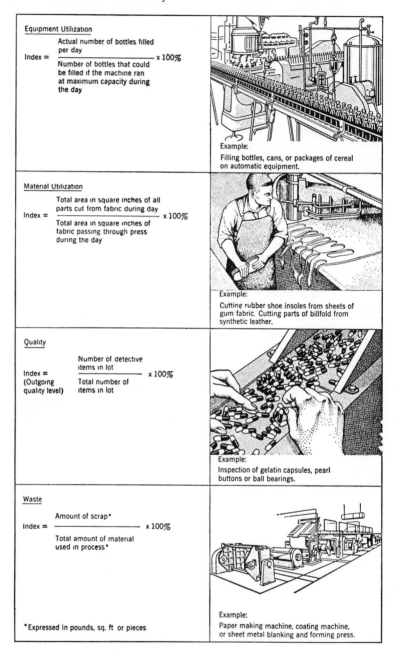

Equipment Utilization

$$\text{Index} = \frac{\text{Actual number of bottles filled per day}}{\text{Number of bottles that could be filled if the machine ran at maximum capacity during the day}} \times 100\%$$

Example:
Filling bottles, cans, or packages of cereal on automatic equipment.

Material Utilization

$$\text{Index} = \frac{\text{Total area in square inches of all parts cut from fabric during day}}{\text{Total area in square inches of fabric passing through press during the day}} \times 100\%$$

Example:
Cutting rubber shoe insoles from sheets of gum fabric. Cutting parts of billfold from synthetic leather.

Quality

$$\text{Index} = \frac{\text{Number of defective items in lot}}{\text{Total number of items in lot}} \times 100\%$$
(Outgoing quality level)

Example:
Inspection of gelatin capsules, pearl buttons or ball bearings.

Waste

$$\text{Index} = \frac{\text{Amount of scrap*}}{\text{Total amount of material used in process*}} \times 100\%$$

*Expressed in pounds, sq. ft or pieces

Example:
Paper making machine, coating machine, or sheet metal blanking and forming press.

Fig. 347.   Some factors other than labor that may affect the cost of the product.

the machine-paced portion of the cycle is large, the situation requires different treatment.

The kind and amount of incentive that must be provided to persuade the operator to keep the equipment operating is a matter that must be worked out for each individual case. We can use time study and the other measurement tools for determining the labor content of an operation, but the weight that should go to other factors, such as equipment utilization, quality, and yield, can be determined only in a more or less arbitrary manner.

For example, the operation of sorting gelatin capsules consists of removing the defective capsules (called scag) as the capsules pass in front of the inspector on a power-driven transparent belt drawn across an illuminated inspection table. It is possible to determine by time study what the standard time should be to sort 1000 capsules of a given size and color with a given percentage of defects. Time study will be of little value, however, in determining the weight that should be given to other factors in a multi-factor incentive plan that might be used on such an operation. There is no way to determine what extra alertness or attention is needed on the part of the operator to keep defective capsules from getting in with the good capsules, or to prevent good capsules from getting in with the scag. Therefore, in this particular case the weight given to these two factors was based largely on the importance of each factor to the company—costwise. In this plant a defective capsule among good capsules could cause delay in the process of filling the capsules with powder. In addition, certain kinds of defective capsules may not be detected until after the capsule has been filled, with the result that both the capsule and its contents have to be scrapped. In the case of the good capsule among the scag, the only loss is the value of the empty capsule itself. A detailed description of the multi-factor incentive plan used by one company for sorting capsules is presented on page 648.

As manufacturing processes become more highly mechanized and as process equipment becomes more complex and costly, it seems certain that there will be an increasing number of opportunities for management to reduce costs by evaluating and controlling factors other than labor. In many cases it will be impossible to measure the effort, attention, or alertness that will be required on the part of the operator to produce the desired results. Often the plan will have to be designed through trial and error and in an empirical or arbitrary manner. The conditions that permit the measurement and incentive compensation of an operator on manually controlled activities will not apply.

Perhaps the most common illustration of this is the policy of dealing with machine-paced work. It is true, of course, that working time interspersed with waiting time caused by machine-controlled elements permits the operator to work at a faster pace when he does work, inasmuch as he is able to rest while the machine works. However, the procedure for determining the "incentive opportunity" that should be allowed for the machine-paced part of the job is an arbitrary one. Such factors as the following must be considered:

1. Lowest cost of end product—including overhead and material as well as direct labor cost.
2. Sufficient incentive to encourage the worker to produce at a pace above standard and to fully utilize the machine.
3. Earning opportunities that are in line with other jobs in the plant.

The effect of machine control on the earnings of an operator is illustrated in Fig. 348. The operation is turning the outside diameter of a gear blank on a lathe. In condition A the operation is 100% manual, and consequently the operator has an opportunity to earn incentive during the entire cycle. In condition B 40% of the cycle is machine paced. The operator is idle (while the machine works), and consequently he has no incentive opportunity during this time. If the operator works at a 125% performance index during the manual part of the operation and if he is guaranteed his base hourly rate, his earnings are shown in the column at the extreme right in Fig. 348. It is obvious that in case B there is not a full incentive to encourage the operator to keep the equipment in operation, in case C he has only a small incentive opportunity, and in case D no incentive opportunity is provided.

Innumerable plans are in use to provide incentives for the operator in situations such as B, C, and D. Some plans provide a flat incentive opportunity during machine-paced portions of the cycle, from 10 or 15% to as high as 25 or 30%. Other plans provide a graduated incentive based upon the percentage of the cycle that is controlled by the machine. However, if the operator referred to in Fig. 348 were paid a 25% bonus on the machine-controlled portion of the job, and if his average performance index for the day on the manually paced portion of the job were 125%, then in cases A, B, C, and D in Fig. 348 the operator would receive the same earnings in each case. He would earn 125% of his base wage, or $24 ($8 \times \$2.40 \times 1.25 = \$24$).

The points in favor of providing an incentive opportunity during machine-paced parts of an operation are: (1) it encourages the opera-

Per Cent of Cycle During Which Operator Did Manual Work	Hours on Manual Work	Base Earnings, Manual Work	Premium Earned, Manual Work	Hours on Machine-Controlled Work	Base Earnings, Machine-Controlled Work	Total Earnings for Eight-Hour Day
A. 100% Manual (0% Machine controlled) [100%]	100% × 8 = 8	8 × $2.40 = $19.20	$19.20 × 25% = $4.80	0	0	$19.20 + $4.80 = $24.00
B. 60% Manual (40% Machine controlled) [60% 40%]	60% × 8 = 4.8	4.8 × $2.40 = $11.52	$11.52 × 25% = $2.88	40% × 8 = 3.2	3.2 × $2.40 = $7.68	$11.52 + $2.88 + $7.68 = $22.08
C. 20% Manual (80% Machine controlled) [20% 80%]	20% × 8 = 1.6	1.6 × $24.0 = $3.84	$3.84 × 25% = $0.96	80% × 8 = 6.4	6.4 × $2.40 = $15.36	$3.84 + $0.96 + $15.36 = $20.16
D. 0% Manual (100% Machine controlled) [100%]	0	0	0	100% × 8 = 8	8 × $2.40 = $19.20	$19.20

Operation: Turn outside diameter of gear blank on lathe. When power feed is used no attention is required on part of operator during cut.

Length of work day = 8 hours. Hourly base rate = $2.40.

Average effectiveness or performance index on manual work performed by the operator = 125%.

Fig. 348. The effects of a machine-paced operation on the earnings of the operator.

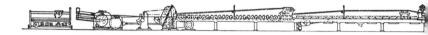

tor to increase his productivity and the productivity of the machine (where this is possible); (2) it provides incentive earnings that are more nearly in line with those of other employees working on incentives. There are two principal arguments against providing such an incentive opportunity. (1) Since the operator is doing no physical work during the machine-paced part of the cycle, some would say that he should receive his guaranteed hourly rate only—the same amount that he would receive if he were a day-work employee. The fundamental concept of a wage incentive plan for direct labor on manual tasks has been stated as follows: "The worker receives extra pay for extra effort—for extra output that is produced above standard. If there is no extra work produced, then the worker should not receive extra pay." (2) If a worker earns a premium or bonus on the machine-paced part of his job, during which he exerts no physical effort, he may feel that he must work harder to earn the same bonus when he is employed on a job that is entirely manual, and that this is unfair.

It should be emphasized again that the manner of handling the machine-paced part of the cycle is out of the realm of work measurement. Rather, a systematic and carefully thought-out company policy should be established and followed for the evaluation, control, and payment of all factors that affect cost and that are within the control of the operator.

MULTI-FACTOR INCENTIVE PLAN FOR THE MANUFACTURE OF CORRUGATED FIBERBOARD

Corrugated fiberboard is made on a corrugating machine (Fig. 349) consisting of several separate units placed in line or tandem. This integrated unit makes the corrugated fiberboard by feeding mill rolls of paper into the wet end of the machine (right side of Fig. 349), through the various units, and into the cutoff machine, which automatically cuts the board to the desired lengths and widths and stacks the cut sheets on a table. This machine can make corrugated board up to 85 inches wide and will operate satisfactorily on certain kinds

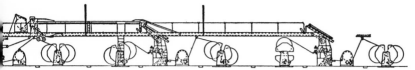

making corrugated fiberboard.

of board at a speed up to 600 feet per. minute. The machine is operated by a crew of seven men, consisting of an operator, an assistant operator, and a "roll shafter" at the wet end of the machine, and a knife man and three off-bearers at the dry end of the machine.

With a machine as expensive to operate as this one (overhead around $75 per hour) it is desirable to produce just as much corrugated board of acceptable quality as possible. Also, because of the value of the paper (approximately $1000 worth of paper is fed through the machine per hour), it is important that scrap and waste be kept as low as possible. The two factors, machine speed and waste, are both within the control of the crew. In order to encourage the seven members of the crew to operate the machine as efficiently as possible, a two-factor wage incentive plan was designed and put into effect. The following is the description of the plan.

Factor I—Lineal Feet of Corrugated Board Produced. The number of lineal feet of corrugated board produced by this machine during the day or shift is determined by multiplying the number of sheets of board produced by the length of each sheet in feet. The standard time in hours per 1000 lineal feet of board produced is shown in Table 75. These time standards were determined by time study. The total lineal feet (in thousands) of board produced by the crew during the day multiplied by the standard time in hours per 1000 lineal feet gives the total standard hours produced by the crew insofar as this factor is concerned.

Factor I = total lineal feet (in thousands) ×

standard hours per 1000 lineal feet

Factor II—Waste Produced. All paper fed into the machine either becomes corrugated board of acceptable quality or is classified as waste. Waste may result from paper scrapped in threading the machine at the start of a shift, corrugated board of unacceptable quality detected and discarded at the dry end of the machine, or defective

Table 75. Standard Time in Hours per Lineal Feet of Corrugated Board Produced

Board Combination		Standard Time, Hours per 1000 Lineal Feet
33–9–33		0.056
42–9–33	125 lb. test	0.056
42–9–42		0.056
69–9–42		0.066
69–9–69	275 lb. test	0.066
42–9–69		0.066
76–38–76 350 lb. test		0.083

Table 76. Waste Standards

Per Cent Waste	Standard Hours Added	Per Cent Waste	Standard Hours Subtracted
1.00	0.193	2.00	0.000
1.05	.183	2.05	.010
1.10	.175	2.10	.020
1.15	.165	2.15	.030
1.20	.155	2.20	.040
1.25	.145	2.25	.048
1.30	.135	2.30	.058
1.35	.127	2.35	.067
1.40	.117	2.40	.077
1.45	.107	2.45	.087
1.50	.097	2.50	.097
1.55	.087	2.55	.107
1.60	.077	2.60	.117
1.65	.067	2.65	.127
1.70	.058	2.70	.135
1.75	.048	2.75	.145
1.80	.040	2.80	.155
1.85	.030	2.85	.165
1.90	.020	2.90	.175
1.95	.010	2.95	.183
2.00	.000	3.00	.193

When the per cent waste is below 2%, the corresponding standard hours are added to Factor I (standard hours earned from lineal feet of corrugated board produced). When the per cent waste is above 2%, the corresponding standard hours afe subtracted from Factor I.

board found and discarded in subsequent operations, such as printing, cutting, and stitching. All waste is delivered to the baling room, where it is segregated, weighed, baled, and then shipped to the paper mill. Thus the total amount of waste in pounds produced during the day can be accurately determined.

It is impossible to operate a corrugating machine without producing some waste. Studies show that 2% waste can be considered "standard," and a table (Table 76) was worked out showing the relationship between waste in per cent and standard hours. As the table shows, if the crew kept their waste at 2% of the total weight of corrugated board produced during the day, it was considered satisfactory performance. If they could do better, that is, if they could keep the waste below 2%, then standard hours as indicated in the table would be added to any bonus hours they might have earned from Factor I. If they produced more than 2% waste, the standard hours shown in the table would be subtracted from the bonus hours earned from Factor I. The percentage of waste was determined in the following way:

(1) Per cent waste $= \dfrac{\text{total pounds waste produced per day}}{\text{total pounds board produced per day}}$

(2) The standard hours corresponding to per cent waste is obtained from Table 76.

(3) Factor II = standard hours (Table 76) × number of hours worked on standard

EXAMPLE. Assume that an order has been received for 500,000 cartons that will require sheets 21 × 60 inches in size, of ordinary double-faced corrugated fiberboard, 125-pound test. Since the machine will produce board up to 85 inches wide, four widths of 21 inches each can be cut from 85-inch sheet, so this width will be used.

On January 16 the seven-man crew worked an 8-hour day on the above order. Their record for the day was as follows.

Factor I—Corrugated Board Produced
  (a) Total run: Total number of lineal feet of corrugated
      board produced during the day. Footage obtained
      from counter on cutoff unit at the dry end of ma-
      chine (31,240 board × 5 feet each = 156,200 lineal
      feet)   156,200 lineal feet
  (b) Width of board = 84 inches
  (c) Total production of board in square feet =
      $\dfrac{156,200 \times 84}{12}$ 1,093,400 square feet

(d) Weight of board = 98 pounds per 1000 square feet

(e) Weight of dry run = $\dfrac{1,093,400 \times 98}{1000}$ 107,153 pounds

(f) Time standard for 125-pound combination = 0.056 hour per 1000 lineal feet (Table 75)

(g) Standard hours produced during 8-hour day =

$\dfrac{156,200 \times 0.056}{1000}$ 8.75 standard hours

Factor II—Waste Produced

(a) Pounds of waste produced during day. Weight obtained from baler operator, who bales and weighs all waste from each corrugated fiberboard machine   1,390 pounds

(b) Per cent waste = $\dfrac{1,390}{107,153}$ 1.30%

(c) Standard hours (Table 76). For per cent waste of 1.30%, the corresponding standard hours is 0.135

(d) Since the crew worked 8 hours during the day on a job for which there was a time standard, then 0.135 × 8   1.08 standard hours

This extra 1.08 standard hours was earned because the members of the crew were able to operate during the day with less than the standard amount of waste.

Summary of Bonus Computations

Factor I—Hours earned because of corrugated board produced	8.75
Factor II—Hours earned because of low waste	1 08
Time allowed for setup in the morning (standard make-ready time), from special table of allowances	0.07
Total standard hours earned	9.90

Bonus or premium hours = 9.90 − 8 = 1.90

Efficiency factor for day = $\dfrac{9.90}{8}$ = 1.24%

This means that each member of the seven-man crew will be paid a bonus or premium based upon the efficiency of the crew. In other words, each man will be paid for 9.90 standard hours produced instead of the 8 hours which he actually worked. The earnings for the individual members of the crew for the day are listed in Table 77.

MULTI-FACTOR INCENTIVE PLAN FOR CAPSULE SORTING OPERATION

Gelatin capsules are manufactured by Eli Lilly and Company on special automatic machines which operate 24 hours per day. Eight different sized capsules and more than 100 color combinations are

Table 77. Daily Earnings of Individual Crew Members

Crew Member	Hourly Base Rate	Standard Hours Earned	Total Earnings for Day
Operator	$2.40	9.90	$23.76
Assistant operator	2.00	9.90	19.80
Roll shafter	2.10	9.90	20.79
Knife man	2.30	9.90	22.77
Off-bearer	2.00	9.90	19.80

made. (See Fig. 350.) Capsules can be transparent, opaque, or a combination. The capsule machine is designed so that steel bars containing 30 highly polished pins are automatically lowered into a tank containing warm liquid gelatin. The bars are withdrawn, the gelatin

SIZE	000	00	0	1	2	3	4	5
WIDTH	.397	.341	.306	.277	.254	.233	.213	.196
LENGTH	1.03	.935	.864	.759	.703	.625	.573	.436

Fig. 350. Size number and dimensions of capsules manufactured.

cooled on the pins, and the excess gelatin trimmed off, giving the cap or the body the proper length. The cap and the body are automatically stripped from the pins, and the cap is inserted onto the body automatically. The finished empty capsule then drops into a fiber can beside the machine. These cans are 16 inches square and 20 inches deep, and the container weighs approximately 15 pounds when loaded.

The fiber cans containing the capsules are moved to another department for sorting and boxing. Among the good capsules will be found small quantities of scrap (called scag), consisting of unassembled caps and bodies, imperfect capsules (dented ends, crimps, thin spots, bubbles, etc.), and trimmings. The purpose of the sorting operation (Fig. 351) is to remove all scrap and rejects, leaving only good capsules. Ideally the operators would allow no scag to get in with the good capsules, and no good capsules in with the scag. Figure 352 shows the layout of the capsule sorting department.

Fig. 351.   Capsule sorting machine, showing capsules on transparent belt.

The sorting operator empties the capsules into a gravity-feed hopper at the rear of the inspection machine, and places the empty can under and toward the front of the machine. After being seated, the operator adjusts the scag pans on the table (usually four are used to receive the scag), opens the hopper door, adjusts the brush to the capsule size, and starts the machine. The capsules flow from the hopper, under the brush, onto the transparent belt (Fig. 351), not more than one deep. The belt speed is 10.8 lineal feet per minute. The transparent belt passes over an illuminated plate approximately 4½ inches wide and 10 inches long. Here the operator can detect and remove defective capsules. The good capsules flow past the operator and drop through a funnel into a can below the machine. During the sorting operation the brush can be further adjusted to increase or decrease the number of capsules on the belt in accordance with the scag encountered. The machine can be stopped by applying a slight pressure on the brake with the right knee.

Statistical Control of Quality. After the entire container of capsules has been sorted, the good capsules and the scag are delivered to the

Fig. 352.   One section of capsule sorting department.

Inspection Department for an analysis of quality. If capsules are acceptable, the utility operator empties them into the storage bin after verifying size, type, and color. If capsules do not meet the outgoing quality standard, they are returned to the operator and resorted until they are acceptable. The scag is sampled to determine the percentage of scag capsules among the scag removed.

An evaluation of outgoing quality—that is, the percentage of scag among good capsules (Factor II), and the percentage of scag capsules among the scag removed (Factor III)—is made at random by the Statistical Control Department to determine the efficiency of these two factors. The control operator posts the efficiency of each factor daily in the Sorting Department.

Multi-factor Incentive Plan. The pay of the employees in the capsule sorting room is based on the efficiency of the group. The multifactor incentive plan is one in which the efficiency of the group is determined by computing the individual efficiency of each of the three factors and weighting each factor by an assigned fraction of 100%. The three factors in this plan are:

Factor	Weight of Factor
I. Employee-hour effectiveness	60%
II. Outgoing quality level	30
III. Reject quality level	10
Total	100%

Factor I—Employee-Hour Effectiveness. The employee-hour effectiveness indicates the performance level of the employees in this department with regard to the amount of work performed or the number of capsules sorted in a given period of time. The number of capsules that an operator can sort per hour is affected by the following

Table 78. Table of Standard Time Values

Operation: Single-sort empty capsules on transparent belt—transparent, opaque, and combination capsules. Standard time in minutes of good capsules sorted according to per cent of scag. These are not actual standards, but are used for illustrative purposes.

Cap. Size	Per Cent of Scag					
	0.0–5.9	6.0–9.9	10.0–14.9	15.0–19.9	20.0–24.9	25.0–29.9
000	. . .	1.88	2.03	2.20	2.39	2.50
00	1.38	1.98	2.29	2.64	. . .	. . .
0	1.02	1.38	1.64	1.91	2.18	. . .
1	.901	1.29	1.49	1.78	2.07	. . .
2	.795	1.07	1.26	1.47	1.72	. . .
3	.738	1.06	1.25	. . .	. .	. . .
4	.875	1.20	1.37	1.69	. . .	. . .
5	.809	1.18	1.38	. . .	. . .	. . .

Table 79. Calculation of Standard Minutes

Operation	Units Completed	Group Standard Time in Minutes	Standard Minutes Developed
Sort size No. 000 empty caps (scag range 25–29.9%)	8,960M	2.50/M good capsules	22,400
Sort size No. 0 empty caps (scag range 0–5.9%)	22,750M	1.02/M good capsules	23,205
Sort size No. 0 empty caps (scag range 6–9.9%)	20,465M	1.38/M good capsules	28,242
Sort size No. 1 empty caps (scag range 10–14.9%)	18,250M	1.49/M good capsules	27,193
Sort size No. 3 empty caps (scag range 6–9.9%)	12,475M	1.06/M good capsules	13,224
Constant per container	2,000 cont.	3.20/cont.	6,400
Operational constant	10 days	270/day	2,700
Standard minutes developed (from capsule boxing bonus slips)			17,556
Standard minutes developed			140,720

factors: (1) size of capsule, (2) percentage of scag in capsules to be sorted, and (3) type of capsule—whether transparent, opaque, or combination.

Time standards have been established by time study for each capsule size, color, and percentage of scag. Table 78 contains one such set of data.

Calculation of Standard Minutes. The standard minutes developed by the group will be determined by multiplying each time standard by the number of units of work completed. Daily constants and allowances will be added in order to obtain the total number of standard minutes developed for the pay period.

As an example of how standard minutes developed are calculated, let us assume that the units of work shown in Table 79 were completed by the group during a 2-week period.

The standard minutes developed by the group during the pay period was 140,720.

Calculation of Actual Minutes. Actual minutes for personnel on biweekly payroll, determined by the payroll section from the employees' timecards, are reported to the Incentive Department at the end of each pay period. For the purpose of this example let us assume that the total actual minutes for the group during the pay period was 173,090.

Calculation of Employee-Hour Efficiency—Factor I. To the standard minutes developed previously, necessary allowances will be added to compensate the group for such things as clothes change and personnel constants. The sum will be the total standard minutes developed for the pay period. The per cent of efficiency for Factor I is calculated by dividing the total standard minutes developed for the pay period by the actual minutes for the pay period. The result will be the per cent of efficiency for Factor I.

The following calculation indicates how the per cent of efficiency for our example is determined for Factor I.

Standard minutes developed	140,720
Necessary allowances (clothes change, personnel constants, etc.)	31,322
Group total standard minutes	172,042

$$\text{Factor I efficiency} = \frac{172,042 \text{ group total standard minutes}}{173,090 \text{ group total actual minutes}} = 99.39\%$$

Factor II—Outgoing Quality Level. This factor has been included in the incentive plan to encourage sorters to maintain a good outgoing quality level. Outgoing quality level is defined as the percentage of scag capsules in good capsules. The efficiency of this factor will depend upon the outgoing quality level as reported by the Statistical Inspection Department. At the end of each pay period a report is received showing the average outgoing quality level for the pay period. This figure is used in determining the efficiency for Factor II as indicated in Table 80.

EXAMPLE. Let us assume an outgoing quality level of 0.50. Using this figure, Table 80 gives an efficiency of 115% for Factor II. Factor II efficiency will be weighted 30% in determining the final efficiency for incentive payment to the group.

Factor III—Reject Quality Level. This factor has been designed to encourage the sorters to maintain a good reject quality level. Reject

Table 80. Factor II—Outgoing Quality Level, Conversion Scale

(Factor weight 30%)

Index	Factor Efficiency in Per Cent
0.50 or better	115
.51 to .55	112
.56 to .60	109
.61 to .65	106
.66 to .70	103
.71 to .75	100
.76 to .80	99
.81 to .85	98
.86 to .90	97
.91 to .95	96
.96 to 1.00	95

The outgoing quality level is reported by Statistical Inspection Department.

quality level is defined as the percentage of scag capsules in the scag removed. The efficiency of this factor will depend upon the reject quality level as reported by the Statistical Inspection Department. At the end of each pay period, a report is received showing the reject quality level for the pay period. Table 81 shows how this factor is converted into an efficiency percentage.

EXAMPLE. Let us assume a reject quality level of 93.3%. Using this figure, Table 81 shows the Factor III efficiency to be 102%. This Factor III efficiency will be weighted by 10% in determining the final efficiency for the group.

Calculation of Paid Efficiency. The paid efficiency of the group will be computed from the efficiencies attained for each of the three factors described. In order to combine the three factor efficiencies, the factors are weighted to reflect the relative importance of each. Table 82 is an example of this procedure.

The efficiency at which the group operated during the sample pay period was 104.33%. However, any efficiency, falling between whole per cents of efficiency will be taken to the nearest whole per cent. When 104.33% is rounded to the nearest whole, the paid efficiency is 104%. If the efficiency had been 104.65%, the paid efficiency would have been 105%.

Table 81. Factor III—Reject Quality Level, Conversion Scale

(Factor weight 10%)

Index	Factor Efficiency in Per Cent
95.0 or better	115
94.9 to 94.0	108
93.9 to 93.0	102
92.9 to 92.0	97
91.9 to 91.0	93
90.9 or less	90

The reject quality level is reported by Statistical Inspection Department.

Table 82. Calculation of Paid Efficiency

Incentive Factor	Factor Efficiency	Factor Weight	Weighted Efficiency
I. Employee-hour effectiveness	99.39	60	59.63
II. Outgoing quality level	115.00	30	34.50
III. Reject quality level	102.00	10	10.20
Total efficiency			104.33

Calculation of Pay. The sorting operator has a base rate which is established by job evaluation. The bonus paid to the operator is determined by referring to the "bonus base rate multiplier scale" (B.B.R.M.), shown in Table 83. It may be noted that the bonus begins with a percentage of efficiency of 51%, and amounts to a bonus of 25% when the efficiency is 100%, increasing at a uniform rate.

In this example the total group efficiency was 104%. Referring to Table 83, the B.B.R.M. scale for 104% is 1.270. Let us assume that a given operator has a base rate of $3.00 per hour and works 8 hours a day for a 2-week pay period of 10 days. Her pay would be computed in the following manner.

$3.00 (base rate) $\times$ 1.270 (B.B.R.M.) = $3.81   rate per hour for the pay period

80 (hours worked) $\times$ $3.81 =

$304.80 base and incentive earnings for the pay period

To this amount will be added any additional payments for which she may be eligible, such as cost of living allowance and night bonus.

Evaluation of Capsule Sorting Operation. Evaluation of cans of capsules is made by the personnel of the Statistical Inspection Department in the Capsule Sorting Department. These evaluations determine the bonus factor for quality, Factor II, and the reject quality level, Factor III. Evaluations are made in the following manner.

Determination of Outgoing Quality Level—Factor II

1. Select at random 10 acceptable cans of capsules.

2. Remove by volumetric measure a random sample of 300 capsules from each can obtained in step 1.

3. Determine and record the number of unacceptable capsules, to the standard provided, in each sample obtained in step 2. Record date produced, machine number, can number, capsule sorter, and bulk inspector's initials.

4. When steps 1, 2, and 3 are completed, determine the total number of defectives observed, the daily average, and the range of the observations. The average is the average outgoing quality level, Factor II, which will be used in the incentive plan.

5. If more than 3 defective capsules are observed in step 2, take a second sample consisting of 600 capsules.

a. If 10 or more defective capsules are observed in the two samples, consisting of 900 capsules, return the can to the capsule supervisor for resort and exclude reading from evaluations.

b. If 9 or less defective capsules are observed in the total sample, consisting of 900 capsules, include the original capsules observed in the initial can sample in the data for daily computations.

Table 83. Lilly Wage Incentive Plan, Bonus Base Rate Multiplier Scale

Per Cent of Efficiency	B.B.R.M.	Per Cent of Efficiency	B.B.R.M.	Per Cent of Efficiency	B.B.R.M.	Per Cent of Efficiency	B.B.R.M.
50	1.000	88	1.190	126	1.380	164	1.570
51	1.005	89	1.195	127	1.385	165	1.575
52	1.010	90	1.200	128	1.390	166	1.580
53	1.015	91	1.205	129	1.395	167	1.585
54	1.020	92	1.210	130	1.400	168	1.590
55	1.025	93	1.215	131	1.405	169	1.595
56	1.030	94	1.220	132	1.410	170	1.600
57	1.035	95	1.225	133	1.415	171	1.605
58	1.040	96	1.230	134	1.420	172	1.610
59	1.045	97	1.235	135	1.425	173	1.615
60	1.050	98	1.240	136	1.430	174	1.620
61	1.055	99	1.245	137	1.435	175	1.625
62	1.060	100	1.250	138	1.440	176	1.630
63	1.065	101	1.255	139	1.445	177	1.635
64	1.070	102	1.260	140	1.450	178	1.640
65	1.075	103	1.265	141	1.455	179	1.645
66	1.080	104	1.270	142	1.460	180	1.650
67	1.085	105	1.275	143	1.465	181	1.655
68	1.090	106	1.280	144	1.470	182	1.660
69	1.095	107	1.285	145	1.475	183	1.665
70	1.100	108	1.290	146	1.480	184	1.670
71	1.105	109	1.295	147	1.485	185	1.675
72	1.110	110	1.300	148	1.490	186	1.680
73	1.115	111	1.305	149	1.495	187	1.685
74	1.120	112	1.310	150	1.500	188	1.690
75	1.125	113	1.315	151	1.505	189	1.695
76	1.130	114	1.320	152	1.510	190	1.700
77	1.135	115	1.325	153	1.515	191	1.705
78	1.140	116	1.330	154	1.520	192	1.710
79	1.145	117	1.335	155	1.525	193	1.715
80	1.150	118	1.340	156	1.530	194	1.720
81	1.155	119	1.345	157	1.535	195	1.725
82	1.160	120	1.350	158	1.540	196	1.730
83	1.165	121	1.355	159	1.545	197	1.735
84	1.170	122	1.360	160	1.550	198	1.740
85	1.175	123	1.365	161	1.555	199	1.745
86	1.180	124	1.370	162	1.560	200	1.750
87	1.185	125	1.375	163	1.565		

Note: Any efficiency falling between whole per cents shall be taken to the nearest per cent. For example: 93.5–94.4% would be 94%; 94.5–95.4% would be 95%.

Determination of Reject Quality Level—Factor III

6. Select at random 5 containers of capsule scag.

7. Remove by volumetric measure a sample consisting of 300 capsules from each container obtained in step 6.

8. Determine and record the capsule can number, the machine number, and the capsule sorter for each scag determination on permanent form. Determine the number of completely acceptable capsules, with *no* visible major or minor defects, from each container.

9. Determine the total, average, and range for each sorting day. The average is the average reject quality level for Factor III, which will be used in the incentive plan.

General Instructions

1. The average and range of Factor II and Factor III are reported for each pay period. A copy of this report is sent to the department head, Capsule Sorting Department chief, Incentives Department, and Statistical Inspection Department chief, on the first day following the end of the pay period.

2. Any observed trends or out-of-control conditions are immediately reported to the supervisory personnel of the Capsule Sorting Department and the Statistical Inspection Department.

3. Additional observations to correct training or out-of-control conditions are taken when necessary.

4. Daily plotting on the capsule-sorting control charts is made at the conclusion of each sorting work day.

CHAPTER 38

Motivation and Work

Many changes have taken place in the way goods and services have been produced over the years. Prior to the industrial revolution, skilled work was done by craftsmen who provided the skill and who, aided by relatively simple tools, produced the entire product, such as a pair of shoes, a wood carving, or suit of clothes. The craftsmen worked for wages, sold their products directly to the consumer or to a middleman for a fixed price, or perhaps were vassals of a landowner and paid their keep in labor instead of money. Some craftsmen worked in factories and were often paid on a piece-rate basis.

The Industrial Revolution. During the latter part of the eighteenth century a revolution took place in production methods and work organization. The basic change was the transfer of the skill of the worker to the machine. For a given operation, the more skill that was transferred to the machine, the less skill was required by the worker. Semiautomatic machines, which eventually came into use, needed only an unskilled person to feed material to the machine and remove the finished part and, in the case of automatic machines, no operator was needed at all. Transfer of skill, however, is not directly related to division of labor, the factory system, or the use of power in industry. In fact, factories existed from ancient times up to the time of the industrial revolution. Craftsmen worked under one roof, there was some division of labor, and waterpower was available in some factories. However, it was the concept of transfer of skill that revolutionized production methods—first in Great Britain in spinning and weaving, and then the idea spread rapidly into many other areas and into other parts of the world. Factories were equipped with steam power-driven machines, since Watt's steam engine had been perfected and put into use about this time. In many cases each machine was designed to perform a single operation, and unskilled men and women and even children were employed to operate them. Often this consisted of merely placing the material in the machine, forming or processing the part, and then removing the finished piece from the machine. The new system resulted in an enormous increase in industrial production, lower unit cost, lower selling price of manufactured products, increased consumption, and increased demand for people to work in the factories.

For over one hundred years, from around 1775 to the latter part of the nineteenth century, emphasis was placed on the invention and development of new products, the improvement of machine tools and manufacturing processes, and the expansion of the manufacturing industries. The number of industrial workers increased greatly, but more thought and effort were given to expanding and improving the physical aspects of production than to the welfare of the industrial workers.

Scientific Management. During the latter part of the nineteenth century, scientific management appeared on the scene and had a profound influence on the industrial world. Taylor, Gantt, the Gilbreths, and other pioneers used the scientific approach in work organization and work design, and their methods resulted in further increases in productivity and still lower production costs along with increased wages for the worker. Under scientific management, thorough study and experimentation was used to determine the most economical production process and the best method for doing each job. The worker was selected and trained to do the task in the prescribed manner and was paid according to his output. Piece rates had been used from early times as a method of compensation, and some form of wage incentives became an integral part of the new management system. Because of the spectacular results of scientific management, there was greater demand for people to install this system in industry than there were qualified people available. Unfortunately, incompetent and unscrupulous "engineers" and consultants entered the picture. This, along with deliberate "rate cutting" by some managers brought about many unsatisfactory applications and some failures and along with this was the opposition of some workers and labor unions to scientific management. It required a good many years to overcome the bad effects of the "efficiency expert" and to bring about an improvement in the situation. New and better methods and techniques were developed, colleges and universities provided trained people in increasing numbers and management performed its functions in a more satisfactory manner. Confidence was restored in the overall system of scientific management, including payment of wages based on results where such systems were practical.

With industrial growth and increased productivity, there was a gradual increase in benefits provided for factory workers in addition to the bonus paid for high productivity—such as better lighting, heating, and ventilation, lunch rooms and food service, and company recreation facilities and programs. Although scientific management as proposed by Taylor and his associates was broad in scope and applied to management as well as to labor, the applications during the period 1885 to the 1930's too often amounted to the measurement of labor and the use of wage incentive systems as a means of increasing productivity and reducing costs. In fact, many people believed that incentive pay was a highly satisfactory motivator for factory workers.

The Hawthorne Experiment. An investigation was started in 1927 at the Hawthorne works of the Western Electric Company that was destined to point the way to a new and different approach to motivating people at work. The Hawthorne experiment was dramatic in throwing light on some of the things that bring job satisfaction to the workers resulting in increased productivity, improvement in quality, and lower absenteeism.[1] Originally, an investigation was undertaken at Hawthorne to determine the effect of varying the intensity of illumination on the production of the workers in the factory. However, the results showed that regardless of whether the lighting was brighter, dimmer, or constant, production increased. This led to a new and more carefully designed study to investigate rest periods and the length of the working day. Special steps were taken to keep all factors constant except the one being studied. This investigation ran from 1927 to 1932, a period of five years. Five skilled young women were selected from a large group of relay assemblers and began working in a separate test room adjacent to the main relay assembly department in the factory. The relays weighed a few ounces, consisted of 40 or 50 parts, and required approximately one minute to assemble. The whole experiment was discussed with the girls, and their cooperation was sought. The results of this study were startling. Output increased at every step along the way. The length of the rest periods and the length of the working day were of minor importance to the girls compared to the motivation they received from their new en-

---

[1] T. N. Whitehead, *Leadership in a Free Society,* Harvard University Press, Cambridge, Mass., 1937; T. N. Whitehead, *The Industrial Worker,* Two Volumes, Harvard University Press, Cambridge, Mass., 1938; F. J. Roethlisberger and W. J. Dickson, *Management and the Worker,* Harvard University Press, Cambridge, Mass., 1940; Henry A. Landsberger, *Hawthorne Revisited,* Cornell University, Ithaca, N. Y., 1958.

vironment. From the moment they started work in the test room they were the center of attention. Continuous records were kept of their production, of the quality of the incoming materials and purchased parts, and the quality of the finished product. Even their conversation was monitored and recorded. They no longer had the feeling that they were just a small part of a large assembly department, subjected to the routine orders and instructions from management. They responded by increasing their output beyond anything that had been expected.

The following statements are taken from a report of the Hawthorne experiment.[2]

There has been a continual upward trend in output which has been independent of the changes in rest pauses. This upward trend has continued too long to be ascribed to an initial stimulus from the novelty of starting a special study.

The reduction of muscular fatigue has not been the primary factor in increasing output. Cumulative fatigue is not present. . . .

There has been an important increase in contentment among the girls working under test-room conditions.

There has been a decrease in absences of about 80 per cent among the girls since entering the test-room group. Test-room operators have had approximately one-third as many sick absences as the regular department during the last six months.

Observations of operators in the relay assembly test room indicate that their health is being maintained or improved and that they are working within their capacity. . . .

Important factors in the production of a better mental attitude and greater enjoyment of work have been the greater freedom, less strict supervision and the opportunity to vary from a fixed pace without reprimand from a gang boss.

The operators have no clear idea as to why they are able to produce more in the test room; but as shown in the replies to questionnaires . . . there is the feeling that better output is in some way related to the distinctly pleasanter, freer, and happier working conditions.

Motivation-Maintenance Theory. Although scientific management was well entrenched in American industry, there were a number of people who objected to certain practices such as excessive job specialization, monotonous short-cycle repetitive operations, machine-paced jobs and the use of piece rates and other forms of wage incentives

---

[2] From a privately published report by the Division of Industrial Research to officers of the Western Electric Company, May 11, 1929, pp. 34–131. Reproduced in Elton Mayo, *The Human Problems of an Industrial Civilization*, Viking Press, New York, 1960, pp. 65–67.

as motivators. Also, changes in business and industry were taking place at a more rapid rate. Methods, tools, machines, and equipment; in fact, many aspects of the employee's job were affected, and some people resented these changes and the way they were made.

During the years following the work at Hawthorne, various investigators conducted studies[3] to learn more about the whole subject of work organization, social systems, work design, group dynamics, and the motivation to work. Most of the studies contributed new knowledge and advanced our understanding of human behavior in the work environment. Because the motivation-maintenance theory has been so well validated and can be successfully applied, it will be described here.

The results of an investigation of job attitudes made by Frederick Herzberg have been very useful in identifying and better understanding specific factors that affect the motivation of people.[4] Moreover, his motivation-maintenance theory is supported by over 30 replications or variations of his original study designed to test the validity of the theory.[5] Business and industry to an increasing extent is studying and testing this theory and a number of organizations are successfully using it. The plan here is to present a statement of Herzberg's findings on motivation and then to describe some actual industrial applications.

An analysis of earlier research led Herzberg to observe that there are some things that the worker likes about his job and there are some things he dislikes—some factors are "satisfiers" and some are

[3] Chris Argyris, *Understanding Organizational Behavior,* The Dorsey Press, Homewood, Ill., 1960; L. E. Davis and Richard Werling, "Job Design Factors," *Occupational Psychology,* Vol. 34, No. 2, pp. 109–132, April, 1960; S. W. Gellerman, *Motivation and Productivity,* American Management Association, New York, 1963; Frederick Herzberg, B. Mausner, and B. B. Snyderman, *The Motivation to Work,* 2nd edition, John Wiley & Sons, New York, 1959; C. L. Hughs, *Goal Setting: Key to Individual and Organizational Effectiveness,* American Management Association, New York, 1965; Kurt Lewin, *Resolving Social Conflict,* Harper, New York, 1948, Rensis Likert, *New Patterns of Management,* McGraw-Hill Book Co., New York, 1961; Douglas McGregor, *The Human Side of Enterprise,* McGraw-Hill Book Co., New York, 1960; E. L. Trist, G. W. Higgin, H. Murray, and A. B. Pollock, *Organizational Choice,* Tavistock Publications Ltd., London, 1963; A. N. Turner and P. R. Lawrence, *Industrial Jobs and the Worker,* Harvard University, Boston, 1965; A. H. Maslow, *Motivation and Personality,* Harper and Brothers, New York, 1954.

[4] Frederick Herzberg, B. Mausner, and B. B. Snyderman, *The Motivation to Work,* 2nd ed., John Wiley & Sons, New York, 1959.

[5] Frederick Herzberg, "Work and the Nature of Man," The World Publishing Company, New York, 1966, pp. 92–167.

"dissatisfiers." His investigation was designed to test the concept that man has two sets of needs: his need as an animal to avoid pain or unpleasant situations and his need as a human to grow psychologically—his need for self-actualization in his work.

Some 200 engineers and accountants who worked for 11 different Pittsburgh firms were carefully interviewed. Each person was asked to recall an event that had occurred at work and that had either resulted in a substantial improvement in his job satisfaction or had led to marked reduction in job satisfaction. The interviewer was looking for an event or series of events or objective happenings that took place during the time in which feelings about the job were exceptionally good or exceptionally bad. The results of the interviews are shown in Fig. 353. The objective events that the engineers and accountants disclosed were classified as to satisfiers and dissatisfiers. The length of each box indicates the frequency with which this factor appeared in the events presented, and the width of the box indicates the time during which the good or bad job attitude lasted. A short duration of attitude change ordinarily lasted less than two weeks, while a long duration of attitude change may have lasted for some years.

There were five factors that strongly determined job satisfaction— these motivators were achievement, recognition, work itself, responsibility, and advancement. The maintenance or hygiene factors—the potential dissatisfiers—were company policy and administration, supervision, salary, interpersonal relations, and working conditions.

The motivators all have to do with the job—opportunity to undertake a difficult assignment; freedom to use imagination and ingenuity in doing the job; encouragement to learn new skills, to be promoted, and to earn more money as recognition for achievement. All of these things encouraged the people to be more productive, and brought more satisfaction and good feeling about the job. On the other hand, the maintenance factors had no positive effect but mainly served to provide a satisfactory environment for the motivators. The maintenance factors do not influence the worker to increase his productivity or do his job well. However, they can be dissatisfiers. Unsatisfactory supervision, inadequate pay, or poor working conditions can bring about a substantial negative attitude. This is an important point. Human relations programs, supervisory training, recreation facilities, wages, and the long list of fringe benefits do not serve as motivators. If these maintenance factors are reasonably well satisfied the setting is right for the motivators to operate.

Man has long understood and used motivators in industry; however, it seems that few managers fully recognize the sharp differentiation between motivators and maintenance factors, which recent research has revealed. This new knowledge now makes the manager's job easier. He can more intelligently organize his operations in order to maximize the satisfiers and minimize the dissatisfiers.

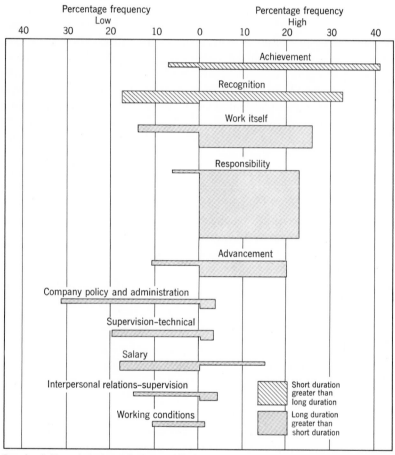

Reproduced with permission from F. Herzberg et al. The Motivation to Work.
John Wiley and Sons, New York, 1959

Fig. 353. Comparison of satisfiers and dissatisfiers.

James A. Lincoln, who founded the Lincoln Electric Company in 1895, understood how to motivate the members of his organization.[6]

[6] James F. Lincoln, *Incentive Management,* The Lincoln Electric Company, Cleveland, Ohio, 1951; James F. Lincoln, *A New Approach to Industrial Economics,* Devin-Adair, New York, 1961.

Over his lifetime he employed most of the satisfiers referred to above. His objective was to assist his people in their striving for self-realization. He states "No man wants to be just a cog in a wheel. The most insistent incentive is the development of self respect and the respect of others. Earnings that are the reward for outstanding performance, progress, and responsibility are signs that he is a man among men. The worker must feel that he is part of a worthwhile project and that the project succeeded because his ability was needed in it. Money alone will not do the job." The last sentence is of special interest inasmuch as the Lincoln Electric employees are among the highest paid people in the world. Last year the average year-end bonus was over $8000 per person. This is in addition to a base wage equal to or greater than the community rate plus incentive earnings.

Study of Motivation at Texas Instruments. In 1961 Texas Instruments Incorporated started a six-year study of motivation in their Dallas divisions.[7] Their research procedure was similar to that used in the Herzberg Pittsburgh study. Texas Instruments wanted to see whether the motivation-maintenance theory could be validly applied to their own workers. Two hundred and eighty-two subjects were selected at random from a list of representative employees distributed almost equally over five job categories of scientist, engineer, and manufacturing supervisor, and hourly paid technician and assembler. Of the 282 subjects, 52 were female hourly assemblers.

Each subject was interviewed by a competent personnel administrator beginning by explaining the general purpose of the study and the nature of the information required. Then, following Herzberg's interview pattern, the interviewer asked:

Think of a time when you felt exceptionally good or exceptionally bad about your job, either your present job or any other job you have had. This can be either "long-range" or the "short-range" kind of situation, as I have described it. Tell me what happened.

After the employee had described a sequence of events that he felt good about ("favorable"), he was asked to tell of a different time when he felt the opposite ("unfavorable") or vice versa. A total of 715 sequences were obtained from the 282 interviews. Each of the sequences was classified "favorable" or "unfavorable" and as "long-range" (strong feelings lasting more than two months) or "short-range" (strong feelings lasting less than two months). Sample

[7] M. Scott Myers, "Who Are Your Motivated Workers?" *Harvard Business Review,* Vol. 42, No. 1, pp. 73–88, January–February, 1964.

favorable and unfavorable responses to interview questions are given below.

MANUFACTURING SUPERVISOR—FAVORABLE

I was asked to take over a job which was thought to be impossible. We didn't think Texas Instruments could ship what had been promised. I was told half would be acceptable, but we shipped the entire order! They had confidence in me to think I could do the job. I am happier when under pressure.

MANUFACTURING SUPERVISOR—UNFAVORABLE

I disagreed with my supervisor. We were discussing how many of a unit to manufacture, and I told him I thought we shouldn't make too many. He said, "I didn't ask for your opinion . . . we'll do what I want." I was shocked as I didn't realize he had this kind of personality. It put me in bad with my supervisor and I resented it because he didn't consider my opinion important.

HOURLY MALE TECHNICIAN—FAVORABLE

In June I was given a bigger responsibility though no change in job grade. I have a better job, more interesting and one that fits in better with my education. I still feel good about it. I'm working harder because it was different from my routine. I am happier . . . feel better about my job.

HOURLY MALE TECHNICIAN—UNFAVORABLE

In 1962 I was working on a project and thought I had a real good solution. A professional in the group but not on my project tore down my project bit by bit in front of those I worked with. He made disparaging remarks. I was unhappy with the man and unhappy with myself. I thought I had solved it when I hadn't. My boss smoothed it over and made me feel better. I stayed away from the others for a week.

Results. Fourteen first-level factors were identified by the interviewer and the head of the project. Figure 354 shows the factors and the number of sequences grouped under each factor. *Achievement* was the largest category accounting for 33% of the sequences. Also *achievement* is comprised of almost twice as many favorable responses as unfavorable ones. On the other hand, *company policy and administration* (the employee's perception of company organization, goals, policies, procedures, practices, or rules) accounts for more than four times as many unfavorable as favorable responses.

Personality Differences. . . . the potency of any of the job factors mentioned, as a motivator or dissatisfier, is not solely a function of the nature of the factor itself. It is also related to the personality of the individual.

For most individuals, the greatest satisfaction and the strongest motivation are derived from *achievement, responsibility, growth, advancement, work itself,* and *earned recognition.* People like this, whom Herzberg terms "motiva-

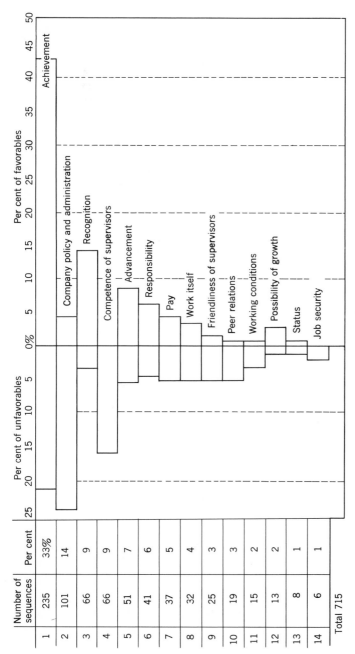

Fig. 354. First-level factors.

tion seekers," are motivated primarily by the nature of the task and have high tolerance for poor environmental factors.

"Maintenance seekers," on the other hand, are motivated primarily by the nature of their environment and tend to avoid motivation opportunities. They are chronically preoccupied and dissatisfied with maintenance factors surrounding the job, such as pay, supplemental benefits, supervision, working conditions, status, job security, company policy and administration and fellow employees. Maintenance seekers realize little satisfaction from accomplishment and express cynicism regarding the positive virtues of work and life in general. By contrast, motivation seekers realize great satisfaction from accomplishment and have positive feelings toward work and life in general.

Maintenance seekers show little interest in kind and quality of work, may succeed on the job through sheer talent, but seldom profit professionally from experience. Motivation seekers enjoy work, strive for quality, tend to overachieve, and benefit professionally from experience. . . .

Although an individual's orientation as a motivation seeker or a maintenance seeker is fairly permanent, it can be influenced by the characteristics of his various roles. For example, maintenance seekers in an environment of achievement, responsibility, growth, and earned recognition tend to behave like and acquire the values of motivation seekers. On the other hand, the absence of motivators causes many motivation seekers to behave like maintenance seekers, and to become preoccupied with the maintenance factors in their environment.

Conclusions. The study clearly shows that the fulfillment of both motivation and maintenance needs is required for effective job performance. Moreover, the study points out that the factors in the work situation which motivate employees are different from the factors that dissatisfy employees. Figure 355 shows motivational needs to include growth, achievement, responsibility, recognition, and the job itself, and are satisfied through the media grouped in the inner circle. Motivational factors focus on the individual and his achievement of personal and company goals. Maintenance needs are satisfied through media listed in the outer circle under the headings of economic, security, orientation, status, social, and physical (see Fig. 355). The study shows that maintenance factors have very little motivational value. If an environment is provided in which motivational need can be satisfied, the maintenance factors have relatively little influence either as satisfiers or dissatisfiers. An abundance of fringe benefits and actions, which overrate maintenance needs, are no substitute for a work environment rich in opportunities for satisfying motivational needs. The conclusions of the study were concisely stated as follows.

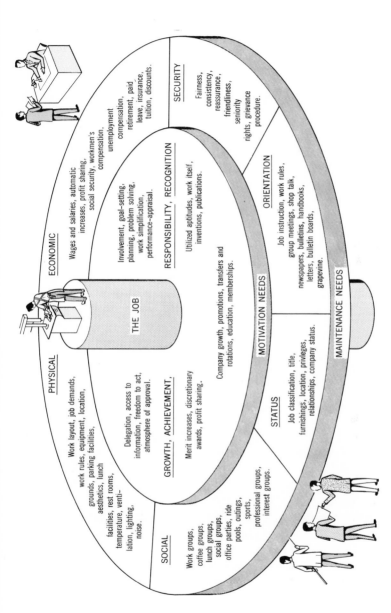

Fig. 355. Employee needs—motivational and maintenance.

What motivates employees to work effectively? A challenging job which allows a feeling of achievement, responsibility, growth, advancement, enjoyment of work itself, and earned recognition.

What dissatisfies workers? Mostly factors which are peripheral to the job—work rules, lighting, coffee breaks, titles, seniority rights, wages, fringe benefits, and the like.

When do workers become dissatisfied? When opportunities for meaningful achievement are eliminated and they become sensitized to their environment and begin to find fault.

Statement by the President of Texas Instruments. At the 1967 annual meeting of stockholders of Texas Instruments Incorporated, President Mark Shepherd, Jr. made the following statement.

. . . by far our greatest opportunity for improved performance at Texas Instruments lies in the area of increasing human effectiveness. It is our

---

Creating change—through increased human effectiveness. Technical innovation leads to useful new products and services to fill the needs of society. Innovation in how people plan and manage their own work can lead to greater realization of their individual potential. This idea to elevate jobs to more than the routine performance of set tasks is taking on real meaning among a growing percentage of men and women of Texas Instruments. These examples illustrate how the people of Texas Instruments contributed to their own and to the company's success in 1966.

Planning together, this group worked out ways to reduce time to produce radar from 130 to as low as 36 hours per system.

An operator in Germanium transistor test group wrote a training manual from the operator's viewpoint that stepped up job knowledge and effectiveness of her entire assembly and test section.

Fig. 356. Excerpts from Annual Report of Texas Instruments.

goal that the work of every person be made as meaningful as possible to the individual, and that their individual career goals and the corporate goals for growth and development are compatible.

This means that the people at Texas Instruments not only must understand what they are expected to do in a specific job assignment, but that they must have the opportunity to plan the job or influence the planning that goes into it, and that they must themselves be concerned with the measurement or evaluation of how well the job is performed.

This kind of participation in planning and managing work is being multiplied now in many operations throughout Texas Instruments. We feel that a significant step has been made in finding ways to help increase our total human effectiveness. In turn, this will enhance our ability to set and achieve our personal and our institutional goals.

The back cover of the 1966 Annual Report of Texas Instruments Incorporated contained the statements and pictures shown in Fig. 356.

CHAPTER 39

Job Enlargement— Deliberate Change

Many different approaches have been used to reduce monotony and give more meaning to industrial work. Job enlargement is one method that has become of increasing importance. However, before presenting a case for job enlargment perhaps a statement should be made explaining why we have highly repetitive operations and assembly line work.

Division of labor has been practiced for many centuries, but the present high degree of job specialization has evolved since the time of the Industrial Revolution. The constant drive to subdivide work has occurred because of the increase in labor effectiveness and the lower unit production costs which have resulted from such practice.

Economics of Specialization. The arguments in favor of division of labor are numerous:

1. A high degree of specialization enables the worker to learn the task in a short period of time.

2. A short work cycle permits rapid and almost automatic performance with little or no mental direction required.

3. Less capable people can be employed to perform highly repetitive short-cycle operations—with a lower hourly wage being paid.

4. Less supervision is required, since the operator soon learns his job, and with the standardization of materials and parts coming from preceding operations, there is little chance of interruptions during the day.

Specialization of work can be carried out with (*a*) the operation entirely within the control of the worker, or (*b*) the work paced, such as operations on an automobile assembly line. The following advantages are often given for such specialized and paced work:

1. Management can be assured of meeting production schedules, since there is a steady flow of finished products coming off the conveyor line.

2. The conveyor forces service departments and parts-supply lines to perform their functions properly; otherwise, the conveyor line would stop.

3. No one operator can work ahead of others, thus accumulating work-in-process. It is the finished part or product coming off the conveyor that is desired, not parts in various stages of completion.

4. Work performed on conveyors makes efficient use of floor space. Often overhead auxiliary supply conveyors can be employed to bring parts to the operator, thus making storage of parts at the assembly line unnecessary.

The statements listed above have been validated[1] in thousands of factories throughout the world, and there are many situations today in which labor effectiveness can be increased and unit costs and total costs reduced by division of labor. However, in some cases specialization has been carried too far.[2] Benefits resulting from job enlargment may outweight those resulting from division of labor.

Jobs may be enlarged horizontally *or* vertically, or horizontally *and* vertically.[3] If a job is expanded so that it includes a greater number or greater variety of operations, it is enlarged horizontally. Horizontal job enlargement is intended to counteract oversimplification and to give the worker an opportunity to perform a "whole natural unit of work." Vertical job enlargement involves the operator in planning, organizing, and inspection as well as the performance of his work. Vertical job enlargement brings most of the motivators into play. Jobs can also be individual or group. There is considerable

---

[1] By the use of assembly line methods, Henry Ford reduced the time to assemble a car from 12 hours and 28 minutes (September, 1913) to 1 hour and 33 minutes (April 30, 1914). This is one of the first and most spectacular uses of the assembly line. Horace L. Arnold and Fay L. Faurote, *Ford Methods and the Ford Shops,* The Engineering Magazine Co., New York, 1915.

[2] L. E. Davis, "Toward a Theory of Job Design," *The Journal of Industrial Engineering,* Vol. 8, No. 5, pp. 305–309, September–October, 1957; C. R. Walker and A. G. Walker, *Modern Technology and Civilization,* McGraw-Hill Book Co., New York, 1962, pp. 119–136; C. R. Walker and R. H. Guest, *The Man on the Assembly Line,* Harvard University Press, Cambridge, Mass., 1952; C. R. Walker, "The Problem of the Repetitive Job," *Harvard Business Review,* Vol. 28, No. 3, pp. 54–58, May, 1950; James C. Worthy, "Organizational Structure and Employee Morale," *American Sociological Review,* Vol. 15, No. 2, pp. 176–178, April, 1950.

[3] E. R. Gomersall and M. Scott Myers, "Breakthrough in On-The-Job Training," *Harvard Business Review,* Vol. 44, No. 4, p. 63, July–August, 1966.

evidence to show that, in many cases, job enlargement—horizontal or vertical, individual or group—results in improved performance and greater job satisfaction. Job enlargement may be part of a management system that more completely involves the worker in solving production problems and in setting goals.

Job Enlargement at IBM Corporation. The term "job enlargement" was first used at the International Business Machines Corporation to identify a program that was inaugurated at the suggestion of the president of the company. The president believed that the job should be enriched by making it more interesting, more varied, and more significant; that a person employed on a meaningful job would be motivated to do more and better work and would receive greater satisfaction from the job.

Walker[4] made a study of job enlargement at IBM and reported it in 1950. The plan was first introduced in a general machine shop in 1943. Several hundred people were employed as operators of the basic machine tools such as lathes, drill presses, broaching machines, milling machines, and grinders. Their work consisted mainly of placing the part in the machine and removing the piece after the cutting tool or drill had performed the machining operation. The operators needed very little skill and could be trained to do the work in a relatively short time. A specially trained setup man prepared the machine for each new job, tools were ground in a central tool room, and finished parts were checked by inspectors. Although many of the operators had been on this work for years and had acquired the skill to set up the machine and check the work, they were not permitted to do so because this was considered an inefficient method of operating the department.

Under the new plan each operator was taught how to set up his machine for each new job from the blueprints provided. He sharpened some of his tools and inspected his own work. By 1950 all setup men had been completely eliminated and inspectors were used only when an operator asked for an inspection double check. The installation of the plan resulted in the displacement of all of the setup men and most of the inspectors, inasmuch as the operators had learned the skills of these men. In line with company policy, no one lost his job or suffered a reduction in pay. Since the company's business was expanding, it was not difficult to solve this problem. For example, of the 35 displaced setup men, 21 became operators without loss

---

[4] Charles R. Walker, "The Problem of the Repetitive Job," *Harvard Business Review,* Vol. 28, No. 3, pp. 54–58, May 1950.

in pay, 12 were promoted to other jobs within the company, all but one with increases in pay, and two men voluntarily left the company.

Results. The main outlay entailed in installing the new plan consisted of the cost of training the operators in the new skills and the higher wage that they received because they were now qualified for a higher rank under the company job evaluation plan. Also there was the cost of additional inspection equipment which was provided. These outlays were more than offset by lower costs and other benefits.

1. Better product quality resulted—there were fewer defects and less scrap. Management attributed this to "greater responsibility taken by the individual operator for the quality of his work."

2. Less idle time of both operators and machines resulted because inherently it was simpler for the operator to set up and check his own work than it was to call a setup man and an inspector to do this for him. Records showed that the cost of setting up and inspecting was reduced 95 per cent.

3. Management stated that job enlargement "enriched the job for the worker." It introduced variety, interest, pride, and responsibility which had not been present before. Because of their higher skill, the men now received higher pay for their work.

Job Enlargement at the Maytag Company. The Maytag Company has had six years of satisfactory experience in changing highly mechanized operations to a method whereby each operator completes an entire cycle.[5] Twenty-five job enlargement projects, sixteen of which were formerly conveyor-paced group assemblies, have been installed. Approximately 130 employees have worked on these enlarged jobs, of whom over ninety have had experience on both this type of work and on paced conveyorized assembly lines. The people at Maytag believe that this provides a sufficient sample from which to draw reasonably valid conclusions concerning employee acceptance of job enlargement as well as the attainment of certain manufacturing objectives. They summarize the important results that have been characteristic of these installations as follows.

---

[5] Paul A. Stewart, "Job Enlargement," Monogram Series No. 3, University of Iowa, Iowa City, 1967; Irwin A. Rose, "Increasing Productivity Through Job Enlargement," *Proceedings Fifteenth Industrial Engineering Institute,* University of California, Los Angeles-Berkeley, February, 1963; E. H. Conant and M. D. Kilbridge, "An Interdisciplinary Analysis of Job Enlargement: Technology, Costs, and Behavioral Implications," *Industrial and Labor Relations Review,* Vol. 18, No. 3, pp. 377–395, April 1965. These cases reproduced with permission of The Maytag Company.

1. Quality has improved.
2. Labor costs are lower.
3. A large majority of operators came to prefer job enlargement in a relatively short time.
4. Problems inherent in paced groups have been largely eliminated. For example, realignment of each operator's job content is no longer necessary whenever production levels change, resulting in less training, higher productivity, fewer changes in production standards, reduction in grievances, etc.
5. Equipment and installation costs have been recovered by tangible savings in an average of about two years.
6. Space requirements for enlarged jobs of the type described here are comparable to that required for powered conveyor-line assembly.

Assembly of Water Pump for Automatic Washer. The assembly of the water pump for the automatic washer is a good example of the application of job enlargement concepts. The paced conveyor system with several operators performing a highly repetitive portion of the total assembly was replaced by individual work stations where each operator assembles and tests the entire component.

Original System of Group Assembly of Pump. The water pump for the automatic washing machine consists of 26 parts, is seven inches in diameter and weighs one and one-half pounds. It was previously assembled by a group of from five to seven operators, depending upon the production level. This included one man for relief, repair, and stock-up. The cycle time for the various levels ranged from 0.33 minutes to 0.44 minutes per pump, depending upon the number of operators in the group. The pump was assembled on a combination bench and power-driven slat conveyor. Figure 357 shows the arrangement of a five-man group (stock man is not shown). The operator on the left working at the bench performed the first operation by assembling five parts of the pump housing. The next operator assembled eight parts to the housing. He worked on two housings at one time, completing half of his operation at the bench; then he transferred the two pumps to the conveyor where he finished his operation. The third operator worked entirely on the moving conveyor, assemblying five parts on each pump and positioning seven screws in every other pump. The last operator positioned seven screws in the remaining pumps and drove the screws in all pumps. He removed the completed assembly and tested it in the water tank shown in the lower right corner of Fig. 357. Rejects were set aside to be repaired and acceptable pumps

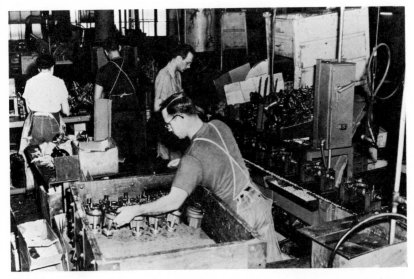

Fig. 357. Original assembly line, showing progressive assembly from bench at rear to power conveyor at right. Testing of assembly is done in tank in right foreground, and the finished pump is placed in the tote box in foreground.

were placed in the skid and taken to the main assembly department. The operators in this group were paid under the company wage-incentive system and normally produced at approximately 135 per cent performance level; that is, they earned a bonus of 35 percent.

Enlarged Job-Present One-Man Assembly Operation. A new method was to be designed that would permit one operator to have complete responsibility for assembling and testing the pump and at a lower assembly cost. Improvements in material handling and housekeeping were also desired. Studies indicated that three operators working at three individual benches or workplace areas could do the work previously done by the four operators and the one relief man. The main technical problem was to design efficient material handling to and from the three benches.

Figure 358 shows one of three enlarged work places. Notice the compact arrangement of the fixtures, tools, materials, and water tank for testing the finished assembly. The assembly operation starts at the point of delivery of the pump housing on the chute at the extreme left. Most of the work is done in the double fixture at the center, and after the unit is assembled, it is given a water-immersion test

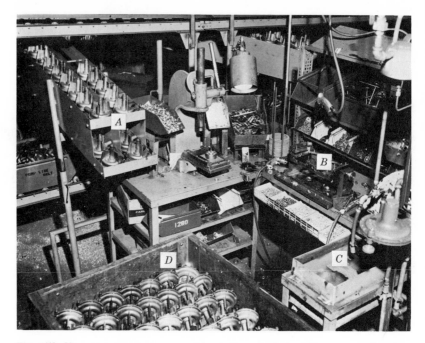

Fig. 358. New one-man pump assembly station. *A,* chute delivers pump housings from previous operation; *B,* duplicate assembly fixture; *C,* double-station testing fixture; *D,* finished pumps ready for delivery to washer assembly department.

in the fixture in the right foreground. The completed assembly is then placed in a tote box for delivery to the final washing-machine assembly department. The double-station assembly and testing fixtures were designed to permit the use of simultaneous motions for the major part of the operation. Moreover, a roller conveyor brings tote pans of components close to the assembly position, and the tote box for the finished assemblies is located directly adjacent to the water-testing tank. The operator's cycle time per unit for the enlarged job is approximately 1½ minutes, as compared to approximately 0.33 minutes on the paced line. (3.0 minutes for each cycle of two parts.) The annual reduction in cost amounted to $8400. The total cost of tooling and equipment for the new method was $4440. The reductions in "make-up" and "helper" costs, which were substantial, are not included in the savings.

A comparison of the important characteristics of the original group assembly method with the present enlarged job is given in Table 84.

Table 84. Assembly of Water Pump for Automatic Washer

Item	Original System: Group Assembly	Present System: Enlarged Job
Quality	The system results in sharing responsibility for quality with several other operators. The operator tends to lose identification of his work with the quality of the total assembly. The paced conveyor allows the operator little time to correct personal mistakes or solve problems caused by variations in the quality of material. When a new operator joins the group, there is a special problem created in maintaining quality.	The operator assembles a complete unit and tests it immediately, resulting in maximum personal identification with the quality of the total unit. This system permits the operator to correct his own mistakes; and if any parts are defective, this is known immediately and no additional assemblies will be produced until satisfactory parts are available. The present system brought about a reduction in defective pumps from 5% to less than one half of 1%.
Productivity Imbalance of work	It is impossible to establish an equal amount of work for each operator on a conveyor line. Therefore, certain idle time or waiting time is inherent in this type of system. On the pump assembly, imbalance averaged about 5% of the total labor cost.	When the individual works alone, there is no imbalance among operators.
Effects of new operators and changes in production schedules on output	When a new operator is assigned to work on a conveyor, the output of the entire group is limited to the output of the new operator, or there is the added cost of a "helper" who might be assigned to assist the new operator. When a new production schedule is put into effect, it is necessary to rebalance the conveyor line and reorganize the work for each individual on the line. Approximately $2500 was incurred for "make-up" and for "helpers" during the year before the initiation of the enlarged job.	No rebalancing of work for changes in production schedules is required. New operators are influenced to quickly attain higher output by exposure to the successful experiences of other operators performing exactly the same work.

Table 84 (Continued)

Item	Original System: Group Assembly	Present System: Enlarged Job
Operator training	Whenever the production schedule is changed, operators are either added to or removed from the conveyor and the task of each person is changed. Therefore, retraining is necessary.	It is necessary to train a new operator only when the production schedule increases, or when the method has been changed.
Time standards	Wage incentives are used in this plant; therefore, whenever the line is rebalanced, it is necessary to make a restudy and determine a new time standard for the revised method.	New time standards are required only when there is a change in method, which would be less frequent than mere changes in the production schedule.
Maintenance costs	Considerable maintenance is encountered on highly mechanized equipment. Moreover, greater down time may be expected. The original cost of the unit is high.	There is practically no maintenance cost. The fixtures and other equipment are very simple. The total cost is less than $1500 per unit.
Material handling	Inconvenient location of containers and tote boxes increases stock-up time. The paced conveyor prevents the operator from stocking parts, unless the system is shut down.	Major parts are delivered by conveyors and chutes directly to the work place. Less time is required for stocking small parts, and the need for a separate material handler is eliminated.
Cost reduction		Annual savings were $8400. This does not include the "make-up" and "helper" costs, which were substantial. Total cost to install new tooling and equipment was $4440.

Other Enlarged Jobs. In addition to the water-pump assembly, the top cover for the automatic washer was changed from a 28-operator assembly line 85 feet long to a one-station assembly operation. The operator now tests her own work and identifies it. The assembly time is reduced 10% over the previous method. Here are the comments of one girl who worked on the top-cover job.

Question: How do you like working with this unit assembly?

ANSWER: I like it very much. I'm happy here.

Question: What do you like about this job?

ANSWER: Well, I'll tell you one of the things I like is the time goes so fast; before you know it, it is noon time. Sometimes, I even forget to take a break; whereas, on the assembly line, I used to watch for the time when my relief man was due. Here, I have all these parts to put together, and I have to think all the time, or I get into trouble, and have to do my work over again. You see, after I get this all assembled, I check my assembly by pushing each of these eight buttons. Each button lights certain lights, as the chart shows, and every time I push the eight buttons and all of them have checked good, I get a thrill and great satisfaction. I put my number on the finished assembly and put it in the rack. All of these units in the rack are mine and I'm proud of them, and if I should get one back tomorrow I worry about it. You know, it is like you are creating something—maybe like painting a picture, or maybe like painting a house— you get the paint, work hard, and put it on, and when it is finished, you step back and enjoy looking at the work you have done.

Question: How does this job compare with working on the assembly line?

ANSWER: Over there, I used to beat my brains out all day and when the day was over I felt like I didn't do anything. Somebody may see all those machines, but I never did—and the time went so slow, and I was always more tired out at the end of the day, at least, I thought so. Many times, I have thought that a factory wasn't a good place for a woman to work, but on this job, I feel differently.

The automatic washer assembly line was also changed from a 1250-foot conveyor with 13 operators to a five-station activity. This resulted in a 19% savings in labor.

To summarize, several manufacturing problems identified with conveyorized groups were eliminated by the new system. Of major importance to supervision was the greater stability of production when changes were made in schedules, or when new operators were assigned. It was no longer necessary to retrain all the operators in the group when a production schedule was revised. No rebalancing of the work of the group was required. When a new operator is now assigned, the production of the other operators is not affected, and shop supervision can more accurately predict the production which can be attained. Since Maytag has a wage incentive system, the operators were pleased that their incentive pay was not dependent upon the capabilities or performance of other members of the group. The average normal earnings now range between 130% and 140% of standard, which equals or exceeds the performance under the group method. Turnover from job bidding has been substantially reduced, and grievances have been eliminated.

Job Enlargement at Texas Instruments. The Texas Instruments Corporation has made extensive use of job enlargement and they find that manufacturing processes appear to improve most through vertical enlargement involving groups united by common goals or processes.

One example of successful job enlargement at Texas Instruments began with 10 assemblers and their supervisor in a conference for solving problems and setting production goals for the manufacture of complex radar equipment. Through their initiative and creativity, assemblers improved manufacturing processes and gradually reduced production time by more than 50%, and exceeded labor standards (based on a previously approved method) by 100%. This process ultimately embraced the entire group of 700 assemblers, and it led to substantial cost reductions in the division, less absenteeism and tardiness, and fewer complaints and personnel problems. This successful group process, which granted unprecedented freedom to assemblers in managing their own work (such as rearranging their own assembly lines), also caused supervisors to begin changing their traditional authoritarian self-image to one of coordination and support.[6]

DELIBERATE CHANGE

The Procter and Gamble Company has developed a number of original and unique approaches that have proved effective in reducing costs, improving quality, and increasing profits. One of these is the principle of deliberate change.[7] In the highly competitive industry in which it operates, Procter and Gamble must exert great effort to make a profit and maintain its place in the field. It expects to obtain this profit by giving value to the customer and by making quality products the customer wants. The company not only has cost-control and cost-reduction programs, but all of management is alert to take advantage of every opportunity to save money and do things in a better way. The company not only accepts change but seeks change deliberately as a means to more effective and more profitable operation.

Philosophy of Deliberate Change. Management has fully accepted the principle of deliberate change. Deliberate change is quite different from improvement. Improvement means performing a method more effectively. Change means developing and using a new method. When an operation is perfectly done there is no more opportunity for im-

---

[6] E. R. Gomersall and M. Scott Myers, "Breakthrough in On-The-Job Training," *Harvard Business Review,* Vol. 44, No. 4, p. 63, July–August, 1966.

[7] Arthur Spinanger, "Increasing Profits Through Deliberate Methods Change," *Proceedings Seventeenth Industrial Engineering Institute,* University of California, Los Angeles-Berkeley, pp. 33–37, 1965. This material reproduced with the permission of the Procter and Gamble Company.

provement. However, there is still a potential for savings by making a deliberate change.

At Procter and Gamble, as a first step, no attempt is made to improve the operation. The reason for this is that if an examination is made of a well-run department with good methods, the conclusion is soon reached that nothing can be done or needs to be done, and so the same decision is made for all of the other departments, for the manager of each department feels that his department is well run. As a result, costs are not reduced and profits are not increased because no changes are made. Instead, the question is asked of well-run operations, "How can we change them?" The present operations may be perfect but the company cannot continue doing them that way any longer. They must be changed and continue to be changed. Of course each change must be for the better.

The following principles form the basis for the deliberate change approach to profit improvement.

1. Perfection is no barrier to change.
2. Every dollar of cost must contribute its fair share of the profits.
3. The savings potential is the full existing cost.
4. Never consider any item of cost necessary.

The following examples from Procter and Gamble illustrate these four principles.

1. *Perfection is no Barrier to Change. Cheer,* one of the company's products was shipped in containers consisting of one layer of six cartons with a support sheet in the middle. This was considered to be correctly designed and was performing perfectly, and no consideration was given to changing it. However, a packing department manager, acting on the principle that perfection is no barrier to change, suggested that another way of packing the cartons in the container would be in two layers of three cartons each. After testing this idea it was found that the support sheet could be eliminated and less cardboard would be used in the container resulting in savings of over $100,000 annually for all brands.

2. *Every Dollar of Cost Should Contribute its Fair Share of the Profits.* The second principle means that no dollars are spent unless there is a possibility of making a profit. An illustration to test this point might be, "How much profit does the company make on a storage tank?" A natural reaction to this question might be "You really can't expect a storage tank to make a profit by itself. A tank is a part of the process equipment needed to manufacture a product

or material for a product." Yet the profit question was asked, and attractive profits to the company resulted.

The company's Quincy, Massachusetts plant formerly received lye from a supplier that was located in the Great Lakes district of the United States. This particular company made lye during the summer and shipped it by barge down to New York City harbor for storage and distribution. These barge shipments terminated when the Great Lakes system and the canal system of the Hudson River froze over in the winter. During this period of the year they ship lye from New York City by tank car to customers. In fact, the Quincy plant received all of its lye by tank car from New York City. The question was raised as to why this lye should be shipped all the way down to New York City by barge and then back to Quincy by the higher-costing tank-car method. With barge unloading facilities checked out as feasible, the buying department and traffic department asked the lye supplier if he could deliver the chemical directly to Quincy by barge. The supplier was pleased with this proposal because it would simplify his method of delivering the lye. Furthermore, the supplier said that they could pay the Quincy plant rent on the tanks because they now pay rent to the owners of the New York City tank field. In addition, they said that Quincy need not pay the cost of this lye until it was actually sent to its first point of use. As a result, Quincy now rents these tanks and has, in effect, reduced working capital. This same approach was followed at the company's New York City plant. They had surplus tanks and, upon checking around with users of tanks, they found a company anxious to rent them. Thus, the New York City plant is now renting some of their tanks.

If one accepts the principle "Every dollar of cost should contribute its fair share of the profits," it tends to lead to the recognition of savings opportunities of this type.

3. *The Savings Potential is the Full Existing Cost.* As an example, *Drene* shampoo used to be packaged in a carton with a liner. Application of the elimination approach led to the elimination of the liner first, and then the carton as well. The company found that the attractive product sold better without a carton than with a carton.

4. *Never Consider Any Item of Cost Necessary.* The company used to handle its case goods on wood pallets with fork trucks.[8] This operation was quite satisfactory. However, because tens of thousands of wood pallets costing $3 to $5 each were used, the company set out to deliberately change the method and eliminate the costly wood pal-

[8] See page 320.

lets. The next step was to develop a paper pallet costing 50¢ to 70¢ along with a Pul Pac type of truck. Then the clamp truck was designed which permits the company to handle 90% of its cases *without any pallets.* An annual savings of $500,000 has resulted.

Experience shows that the deliberate change program functions best in a favorable atmosphere. Small groups take the initiative in selecting problems which they want to study and in establishing their own goals. Positive recognition for methods-change accomplishments is a definite part of the program.

CHAPTER 40

The Individual Rate—Performance Premium Payment Plan

The plan described in this chapter deals with the largest of the Eastman Kodak Company plants, the Kodak Park Works,[1] which is located in Rochester, New York. It employs about 25,000 people, is almost 6½ miles long, and includes more than 165 major manufacturing buildings with 13½ million square feet of floor area on about 1700 acres of land. Kodak Park is virtually a city within a city, with its own power plants, cafeterias, laundry, and other needed services. The men and women of Kodak Park produce some 265 kinds of film—including roll film, sheet film and motion picture film, over 375 kinds of photographic paper, about 450 kinds of photographic chemicals, more than 40 kinds of photographic plates, and some 4500 research chemicals.

History of Measurement and Incentives at Kodak Park. Around 1917, an industrial engineering department was organized and, initially, it spent much of its effort measuring work and installing and maintaining wage-incentive plans for direct labor. After a few years, a uniform plan of wage incentives for direct and indirect labor was established, which proved to be effective. Originally the system was mainly a 100 per cent premium plan with individual and group incentives and a guaranteed base wage. However, the process-type operations, of which there were many, did not lend themselves to individual incentives, and they were covered by multifactor incentive plans.

The plant maintained its standards and incentive systems during World War II, but, following the war, faced new challenges as technological improvements developed. The problem of providing satisfac-

[1] Appreciation is expressed to Eastman Kodak Company, Kodak Park Works, and, in particular, to Robert J. Rohr, Jr. for the information on which this chapter is based.

tory incentive opportunity on process-type operations, mechanized jobs, and indirect labor became more difficult. The history of measurement on these jobs has been one of continuing change and improvement.

The experience gained over the years indicated that, while money was a powerful stimulator, people gave outstanding effort to their job for other reasons as well. This led the engineers to examine the field more closely in order to determine if these other reasons could be more formally and consistently added to all measurement systems.

Around 1959, the Industrial Engineering Division began studies to determine whether or not work could be made more meaningful both from the standpoint of the individual and the industry. A senior member of the Industrial Engineering Division and one of the company psychologists began the study. This included a systematic examination of the literature in the area of motivation, group dynamics, management systems, and related fields. Published results of research were studied, visits were made to colleges and universities and other organizations where research projects were under way, and discussions were held with leaders in the field. During 1959 and 1960 the project industrial engineer and the psychologist held discussions with management at various levels to consider such matters as more precise definition of goals, different ways of stimulating further interest in meeting the company goals, ways of creating an environment for introducing change, and organizing things so that the individual's goals and the company's goals would be more nearly the same. Also, during this period a seminar was held for industrial engineering supervisors. This seminar met once per week for a period of two or three months.

Because of the growing interest in the subject, the project industrial engineer prepared a document called a "Working Theory." This amounted to -a digest of the research put in simple language. This working theory presentation was made to the supervision[2] in one of the large manufacturing departments on an experimental basis. Several other industrial engineers joined the project industrial engineer and, working together, a plan was developed. It is referred to as the Individual Rate—Performance Premium Payment Plan.

Basic Concepts of the Individual Rate-Performance Premium Payment Plan. This plan embodies a stable pay plan for nonmanagement people and replaces the present wage-incentive plans in many areas. It is not a rigid system nor a standardized program. In fact, no two applications are alike and the way in which applications are made

[2] "Supervision" refers to the division superintendent, assistant superintendent, department head, foreman and first line supervisor.

may vary. It implies a different way of looking at work and suggests a different way of planning, organizing, and operating resources of men, materials, and facilities. It provides uniformity of earnings free from the day-to-day fluctuations that often occur in conventional wage-incentive plans. It strives to establish an environment that makes it possible to bring company goals and individual goals closer together, aiming at better performance. The purpose is to provide information leading to improved efficiency and reduced costs—a plan that facilitates changes and promotes the design of more meaningful work. The following statements further amplify the plan.

1. Management and nonmanagement people, through discussion and consultation, develop a better understanding of department goals and individual goals.

2. Each person has greater opportunity for self direction. Fewer detailed instructions are necessary, and decisions are made at the lowest practicable level. Once the goals are fully understood and accepted, each person has greater freedom in using his initiative in determining the way he will perform his task as one part in achieving the goals.

3. In this plan, as in all other operations of the company, each person is recognized as an individual. His individual performance is important in his day-to-day work and in his future progress. It is assumed that each person wants to do his best and wants to be considered for advancement in accordance with his performance and as the opportunity arises.

4. Management and nonmanagement people understand that continual development and introduction of new and improved methods and processes are necessary to the successful conduct of the business; and only by utilizing these improvements can the company continue to provide employment at good wages. Under this plan, management and nonmanagement people become intimately acquainted with the analysis and design techniques and the problem-solving approach of the industrial engineer, and learn to apply these techniques as a part of their day's work. Each person has some talents for innovation and creativity, and he welcomes the opportunity to use these talents in his work.

5. Job security has long been recognized as an essential part of any plan that calls for the full cooperation of all members of the organization, and this plan continues this policy. In a dynamic organization change will occur, work methods will be modified, machines and processes will be automated, and jobs eliminated. Nevertheless,

before such improvements are made, careful attention is given to any possible effect upon the individuals concerned. Through this long-standing policy, the company adopts changes essential to its growth and at the same time endeavors to avoid any considerable hardship to individuals involved.

6. The concepts of the plan are quite different from those usually found in use in manufacturing industries today, and a longer period of time is needed to develop full understanding among all those concerned. A change in thinking on the part of the people takes place and this requires time. Experience shows that from nine to eighteen months are required before the plan can be put into effect with assurance of success.

The plan is most successful when the members of management of a division understand the plan, realize its potential, and initiate studies essential to an installation. The industrial engineer plays an important part in all of the installations. Capable and experienced industrial engineers are thoroughly grounded in the features of the new plan and therefore can contribute to its success.

Procedure. The first few installations of the plan were started because the department supervisors believed that it would help them to operate more effectively. They recognized that the plan was new and that some aspects of it probably would have to be modified with experience. By the end of the first year, all supervisors concerned in Kodak Park were familiar with the basic principles of the plan. As a result, the Industrial Engineering Division was hard pressed to provide enough men to meet the requests for new installations.

Possible Steps in Making an Installation. The following is a list of the steps that might be used in making an installation.

1. Background and work sessions. Discussions between management and the industrial engineer.
2. Briefing sessions with nonmanagement people by the department head and industrial engineer.
3. Development of a time table and more detailed plans for the installation.
4. Individual discussions with each person in the department conducted by the industrial engineer.
5. Group discussions and individual discussions between management and nonmanagement people.
6. Presentation of the pay stability features.
7. Putting the plan into effect.

1. *Background and Work Sessions*. When the decision is made to being the installation of the plan in a department, one or more industrial engineers are assigned to the department on a full-time basis. The industrial engineer usually will spend at least one year in the department and may remain for a longer period of time. He meets with members of management individually and in groups to discuss background theory and to present detailed information about the application of the plan in other departments. A start is made on designing a version of the plan that will best fit the needs of the department. At the same time, the industrial engineer studies all phases of operation of the department, the machines and processes, raw materials, quality requirements, scheduling procedures, and the current wage-incentive plans. He also becomes acquainted with the line and staff people in the department, learning their duties, responsibilities, and methods of operation.

2. *Briefing Sessions with Nonmanagement People*. Short group meetings are held with all nonmanagement people in the department. These sessions are conducted by the department head assisted by the industrial engineer. The plan is described in general terms. The objectives of the plan are stated. Then an explanation is made as to the method of installation, the results that are expected, and how they will be involved in the plan. At this initial meeting the people are told that their pay will not be adversely affected, although the details of the plan are not discussed at this time.

3. *Development of a Time Table and More Detailed Planning for the Installation*. As the managers and supervisors become better acquainted with the plan, consideration is given to detailed procedures for implementing the plan, which phases will be handled by the people in the department and which by the industrial engineer, and when each phase will be started and completed. Also, a target date will be set for the changeover from the wage-incentive plan to the new plan.

4. *Individual Discussions*. Before individual discussions are held, the department head holds group meetings and explains to the people the reasons for individual discussions, how these discussions fit into the overall plan, and how it is hoped they will be helpful to the people themselves. The industrial engineer then arranges to spend from one to two hours with each person. He is getting acquainted with the person, is asking for ideas, and is giving him an opportunity to discuss any and all aspects of the department's operations. The following might be a list of typical questions for discussion with the people in a manufacturing department.

(a) With regard to the job you perform, how do you feel in general about doing this type of work?

(b) If you had an opportunity to change your job to improve it, what are some of the things you would consider changing?

(c) How do you feel about the amount of control you have over what you do? Do you feel you have the necessary freedom to do the job the way it needs to be done?

(d) How do you feel about the amount of information you get regarding your job?

(e) Overall, what do you like most about your job?

(f) What would you say you like least about your job?

(g) Taking all things into consideration, what is the most important thing in your job?

(h) What are the most difficult parts of your job?

(i) Just what are the major objectives of your job? How do you know how well you are doing with respect to these objectives?

(j) Is there anything we haven't discussed that you would like to add?

5. *Group Discussions and Individual Discussions between Management and Nonmanagement People.* During this phase, supervisors and the industrial engineer hold group discussions with the people in the department and work individually with them to define goals, develop better methods, redesign jobs, modify equipment, improve planning and scheduling, reduce downtime of machines, and increase overall effectiveness. The object is to reduce costs, maintain or improve quality, increase output and develop more meaningful jobs, and provide ways of recognizing individuals and groups. The suggestions received from discussions are given serious consideration in the development of the plan.

In one case, a new building was to be constructed to house a rapidly growing department. During the planning phase, both management and nonmanagement people were consulted about the operating arangement of the building. For example, each person had an opportunity to discuss with the engineers the particular location of his machine and the arrangement of the materials and auxiliary equipment. Many new and valuable ideas were received and incorporated in the early planning stage and the people looked forward enthusiastically to the move and to using the new facilities.

Under the new plan the industrial engineer is serving not only as a professional who is there to study the problem and recommend solutions but also as a consultant and a helper for the operating supervision. In some departments, several months may be spent on this phase. All members of the organization are now learning the full meaning of the plan. In many cases the people had not been as familiar with departmental goals as they now are. Rather, each person was expected to do his specific job according to the prescribed method, for which he would be paid an amount based upon units he produced. If on a particular forenoon, for example, things did not go well with him and it appeared that he could not earn an incentive premium for the day, he might be tempted to work at a moderate speed for the rest of the day and simply earn his guaranteed base wage for the day. Under the new plan the daily goals and the weekly goals for the department are known and in some departments hourly output figures are posted on a bulletin board in the department. Now if a person has difficulty or falls behind expectations, he often is helped by one who has met his goals. The goal now is much broader. Moreover, the individual now can depend upon uniform earnings on a day-to-day basis. Department goals, therefore, are more meaningful.

6. *Presentation of the New Pay Plan.* When the new plan is to be put into effect, the department head meets with each person and discusses the stable pay plan with him and explains just how it will affect him. This involves information about his individual rate and how he can move to the next step in his present wage grade and opportunities for advancement to higher wage grades are pointed out to him. And so each aspect of the pay plan is explained. The individual sees a sample of his monthly earning sheet and knows what his weekly earnings will be. He has an opportunity to ask questions and is invited to discuss any aspect of the whole plan which now has been in the process of development for some months.

7. *Putting the Plan into Effect.* Payment under the former wage-incentive plan is discontinued. Under the new plan, if production is maintained according to established goals, the pay remains constant week after week. The individual, therefore, is in a much better position to manage his personal finances.

The Kodak Park Works of Eastman Kodak Company has been engaged in work measurement and wage incentives for many years and has an outstanding reputation in this field. The company uses individual incentives, group incentives, and multifactor plans. Approximately 10,000 people were paid under some form of incentives at the time the new plan was introduced in 1961. The average performance level was approximately 100 per cent, which means that the average operator earned a premium of 15 per cent in return for performance of 15 per cent above standard.[3]

Base Wage. Wage grades within the company are determined by means of a job evaluation plan. This is a point plan with 15 *wage grades* for manufacturing personnel. The factors such as skill, responsibility, physical application, and job conditions are used in determining wage grades. Wage grades cover a range from entry level jobs to skilled craftsmen. Each wage grade has a five-step range of rates. These cover a commensurate performance range of low, fair, good, excellent, and outstanding. Low is a hiring step for new people; progress through the range is determined by the individual's performance. Advancement from one wage grade to a higher grade can occur when an individual, through training, meets the specifications and an opening is available in the next highest grade. Under the new plan, the person's base rate remains with him regardless of the job he might be doing. If he works on a lower task temporarily he still receives his own base wage.

Work Measurement. The conventional methods of work measurement are used in establishing time standards for work that lends itself to this type of measurement. However, the unit of time is an eight-hour day instead of a minute. Instead of a standard being stated as so many "minutes per piece" it is expressed as "number of pieces per eight-hour day." Another unique feature is that in some cases the people, the supervisors, and the industrial engineer assist in arriving at the time standard to be used. The objective is not to arrive at a standard that would be correct if all conditions were ideal, but rather a realistic standard that represents the output in units that the people in the department or production area can turn out in an eight-hour day. For example, on work involving machines or process equipment each time standard is made up of three parts.

1. The major task, that is, the main part of the job or the repetitive part of the cycle.
2. The auxiliary part of the job—those elements that are definitely a part of the work but not directly related to the main task.
3. Utilization factor or delay factor. Using work sampling studies or from an analysis of past records, information is obtained indicating the percentage of the day that the operator cannot work because of unexpected delays.

[3] At Kodak Park, 100 per cent performance level is a so-called high task or incentive level, which pays 15 per cent premium.

In most departments, the 480-minute day is reduced to 430 minutes available for work. Because items 2 and 3 represent "nonproductive time," these are fertile areas for cost reduction. The new plan minimizes resistance to change, presents no difficulties to undertaking experimentation. For example, in one department there was a battery of 24 semiautomatic machines operating on three eight-hour shifts. Some machines were more productive than others, some seemed to have more downtime than others. It had been the policy to rotate the operators so that no one would be assigned permanently to the low producers. Under the new plan, the supervisor asked three of the best operators to take over the three poorest machines to help find out what was causing the trouble. Machine adjustors, mechanical and industrial engineers, and supervision working together with the operators determined the cause of the trouble and corrected the difficulty.

Pay Plan. One of the basic concepts of the new plan is that the right kind of people, properly trained and working in the right kind of environment, will give improved performance. To be more specific, it is expected that their performance will be equal to the average pace or 100 per cent performance level. Therefore, each person under the new plan receives his base wage plus about 15 per cent or a total wage of 115 per cent of his base pay. This is paid on a weekly basis for the hours worked.

Standards are established for each operation or task and a performance index is determined for each operator on a weekly basis. In a few cases, at the start of the plan, the index was determined on a daily basis. The information is assembled each month and a summary sheet is prepared for each operator and is given to him by his supervisor. The information sheet gives the worker his performance index in per cent, states the equivalent standard day's work produced, the equivalent actual days worked, and also the equivalent days worked on tasks for which there is no time standard. The supervision knows through the performance report how each person is doing and, if the performance index should drop, the supervision would look into the matter immediately to determine the cause and take corrective action. The assumption is made that the person wants to do a good job and that he will look beyond the day's output. If trouble develops one day and production drops, then when the difficulty is corrected and things go well, output will be high to make up for the low production. Because the people know what the schedule is and have day-to-day information about progress in meeting the schedules, they are in a position to take individual action as well as group action to achieve the goals.

The design of the plan is made up of positive factors. The individual base rate plus the 15 per cent premium is paid to all people on a weekly basis for the time worked. It does not fluctuate with variations in output. If the performance index for an individual drops, the supervision expects the cooperation of the individual in locating the difficulty and in correcting it. However, if an unusual situation should develop, the plan has flexibility. If a person deliberately restricts output or fails to perform his work properly, the supervisor could reduce or eliminate the 15 per cent premium, reduce the basic individual rate, or use other disciplinary methods. It should be added that before any corrective steps are taken, the superintendent, supervision, industrial engineer, and industrial relations representative would consider all aspects of the problem. During the several years the plan has been in effect, the number of cases in which such corrective steps has been used has been minimal.

SCRAP COLLECTION—AN APPLICATION OF THE NEW PLAN

Discussions were started with the superintendent and supervision of the Scrap Collection Department with the idea of bringing the truck drivers and their helpers into the plan. Several manufacturing departments were successfully using the plan, and it seemed to be a good time to study a different type of activity. The department head and supervision were very receptive to the idea, and full discussions took place between the industrial engineer and division supervision and then the truck driver—helper group.

Some years earlier a study had been made of all phases of scrap collection by the Industrial Engineering Department in cooperation with the department head and the supervisors. Truck routes were revised, schedules were changed, and methods of handling scrap were improved resulting in a reduction in the number of trucks, drivers, and helpers needed in the department. The work was measured and the performance index and incentive pay for the truckers was based on the number of loads of scrap delivered to the incinerator and the paper bailers, which were located in a remote part of Kodak Park property. The Scrap Collection Department seemed to be operating in a satisfactory manner, the truckers earnings were good, the scrap was being taken care of properly, and supervision was satisfied with the quality of the work. Although costs had been reduced substantially when the original study was made, supervision still felt that the scrap collection costs were too high. The people disliked the method of keeping records of the number of truck loads of scrap moved per day and often complained about it. A checker was located near the incinerator and paper bailers, and he inspected and recorded each truckload of scrap delivered. The truckers felt that this inspection and record keeping was a form of policing,· and they objected to it.

In line with the plan, the truckers in small groups came into the conference room, and supervision and the industrial engineer discussed the objectives of the new plan. They were told that when the plan was put into effect, each trucker would receive his base pay plus a 15 per cent premium or his "average earnings during the previous three-month period," whichever was greater. There would be no day-to-day fluctuations in earnings, as sometimes occurred under the incentive plan. In line with this approach, the checker would not be needed, and no longer would a record be kept of the truckloads of scrap moved per day. The point was made that supervision and truckers, with the assistance of the industrial engineer, would work toward determining the goals of the Scrap Collection Department and determine how best to achieve these goals. Also it was indicated that supervision was seeking ways of decreasing the cost of scrap collection, and that an attempt would be made to find some acceptable method of measuring the effectiveness of scrap collection. The industrial engineer then proceeded to talk with each trucker. These discussions lasted from one to two hours, and a summary report was produced covering all information obtained. Each trucker was asked what he thought the goals or purposes of the Scrap Collection Department were and how best to achieve these goals. The engineer also was looking for suggestions to improve the operation of the department, ideas for better equipment, better methods and ways to reduce costs and to make the job better for the truckers. After the discussions had been completed, more meetings were held between supervision and the truckers and, gradually, valuable ideas emerged. For example, there was uniform agreement that the purpose of the Scrap Collection Department was to "keep Kodak Park clean" and that the number of truckloads of scrap handled per day was not necessarily a true index or measure of the objective. In fact the truckers revealed that some days when scrap output was low, they would load their trucks with empty boxes and throw some loose scrap over them and haul them to the paper bailer in order to earn their customary premium.

Once the new plan went into effect, and the truckers saw that they really were on the plan and that their assistance was needed by supervision to reduce costs, they responded in a positive way. On the average, the number of trips per day was reduced by 25 per cent. One driver suggested that he did not need a helper in the forenoons and that this man could be used on other work. Another driver stated that only occasionally would he need a helper and, if he could get a man when he needed him, this would be satisfactory. A driver who worked on the night shift and who had a helper stated that he did not need the helper. Moreover he said he had time and could move the empty four-wheel scrap containers from his truck onto the loading platform and into the factory building, thus making it unnecessary for the material handlers to interrupt their own work to do this job.

After further study and consultation, supervision, the truckers, and the industrial engineer worked out a rather simple system of determining a performance index based on tonnage of scrap moved per month. This was calculated on a monthly basis and seemed to be directly related to the goal of "keeping Kodak Park clean." If plantwide production increased or decreased, or if new manufacturing processes were put into operation, the index number would reflect the load on the trucks and on the truckers.

The department head and the supervisor spent some time each day driving over the Kodak Park area—around the buildings and warehouses in order to appraise special problems that might occur in their efforts to "keep Kodak Park clean." An old factory building might be demolished to make room for a new one, a roadway might be relocated, a new sewer line might be installed, or outside contractors might create special problems. The department head and the supervisor, as a part of their job, were well aware of the activities of each trucker. In fact the department head had at one time been a truck driver in the Scrap Collection Department. There was excellent communication between the truckers and the supervisor and, when a special problem developed involving other departments or activities, the truckers knew that their supervisor would be on the job to effect a solution that would be in the best interest of the company, and they felt assured that they would not be adversely affected whatever the outcome of the situation might be. They had experienced the satisfaction of working out countless problems together with supervision, and their jobs had become more meaningful because they had a part in designing them. They were recognized for their ability to perform an important and necessary function in the Kodak Park facility.

As the summary in Table 85 shows, the plan assisted supervision in making it possible for 4 helpers to do the work of 17. One new truck was added to the fleet to handle scrap. In line with company policy, the helpers who were no longer needed in this department were transferred to other jobs.

Table 85. Summary of the Number of Truckers and Helpers in the Scrap Collection Department Before and After the Installation of the New Plan

Kind of Scrap—Type of Truck	Before		After	
	Trucks and Drivers	Helpers	Trucks and Drivers	Helpers
1. General scrap dump trucks	8	8	8	2
2. General trash and scrap—bucket handling type trucks	3	0	3	0
3. Incinerator scrap "packer" type trucks	3	6	4	2
4. Bailer scrap "stake" and covered trucks	3	3	3	0
	17	17	18	4
Totals	34		22	

Results. The first installation of the plan was made in 1961 and by 1964 five installations had been completed and four more were in progress. Since that time, changeover has continued at an increasing rate. The system provides operating people with information to assist them in:

1. Making decisions that will maintain and in some instances reduce costs.
2. Maintaining or improving productivity per man hour.

The following statements can be made regarding the results of the plan.

1. Paper work has been reduced—in some departments as much as 75 per cent.

2. People are earning at least as much as they did under the incentive plan and there is no day-to-day fluctuation in pay. Each person is encouraged to learn new skills and become more versatile.

3. The supervisor does less clerical work and now has more time available to communicate with his people, knows them better, and takes a more active part in training them for better jobs. He also can concentrate on methods change, work arrangement, equipment utilization, waste elimination, and other areas of operation that result in improved quality, greater productivity, and lower costs. Likewise, the industrial engineer is no longer simply the professional practitioner in the usual sense. Instead he is a consultant assisting supervision effectively in this area of responsibility. He is dedicated to the principles of the plan and is most successful when his efforts result in the growth, increased effectiveness, and greater personal satisfaction of the people working under the new plan.

4. The plan encourages communication both ways in the organization so that both supervision and the people are better informed.

5. It has been observed that quality of workmanship has improved through supervision stressing this aspect of production. The people responded to this stimulation instead of trying for more premium earnings.

6. Industrial engineering service required for the maintenance of a measurement-incentive system is reduced in many instances, thus making these skills available to production people in the solution of other problems.

7. The work is more meaningful. There is greater opportunity for self-direction. There is less resistance to job changes and assignment of work. The people now work on jobs they help arrange, on production standards they assist in establishing, and in an environment they help create.

CHAPTER 41

The Lakeview Plan

The ABC Company produces a wide variety of consumer products in 15 plants located in various parts of the country. Although the industry is highly competitive, ABC has been a leader for many years. New factories are built as needed to provide for the increasing demands for the company's products. Over the years there has been much discussion concerning the size and organization of an ideal plant and, in 1961, a committee was formed to study this matter and to assist with the design of a new factory scheduled to be built the following year. The committee consisted of the Director of Industrial Engineering, the Chief Engineer, the Manager of Manufacturing, and the Director of Industrial Relations. The city of Lakeview, with a population of 50,000 was selected for the location of the plant. The plant was designed to manufacture 10 different kinds of consumer products.

In 1927 the ABC Company organized an industrial engineering department and installed work measurement and wage incentives in all of its plants. Over the years the functions of the industrial engineering division were enlarged and expanded. Industrial engineering has had a continuous record of outstanding leadership and has contributed in a very important way to the profits of the company. The company still maintains a highly successful system of wage incentives and cost control in its manufacturing plants.

During the period 1958 to 1961 the Director of the Industrial Engineering Division, some members of his staff, and other members of management began studying and discussing among themselves the whole matter of more precise definition of plant goals, different ways of motivating operators and managers to meet the company goals, ways of creating an environment for introducing change, and organizing things so that the worker's goals and the company goals would be more nearly the same. Over a period of several years these people became well acquainted with the research in the whole area of motivation. Therefore, in 1961 when the committee proposed that the Lakeview plant follow a radically different plan of organization from that used in the other plants, the members of management were kowledgeable in this area.

In February 1961, a company division manager with long experience was given charge of the Lakeview plant design. He, working with the planning committee and new plant management, created a design for the Lakeview plant that incorporated the best equipment and processes in the industry. Construction of the plant was completed in January 1963, and approximately 165 people were employed by the end of the first year of operation.

The warehouse and shippping department had been constructed first, and were put into operation as a distribution center for company products about six months before the manufacturing plant was completed. A carefully designed program of selecting and training workers was started in October 1962. No nonmanagement people were transferred from other company plants to Lakeview.

Development of a Philosophy of Management

One of the objectives at Lakeview was to introduce unique methods of work organization and management systems in order to establish relationships and to create an environment in which fewer orders and instructions would be given and people would have greater opportunity for self-direction. Another objective was to bring the company's goals and the individual's goals as close together as possible. This was referred to as a "a common objective approach." Many of the ideas now in effect evolved during the plant design and construction period, and some were added or modified after the plant went into operation in 1963.

At Lakeview the people were given greater freedom to make decisions. There was greater opportunity for self-direction. Provision was made for effective communications. The importance of teamwork was emphasized. The goals of the company were carefully explained to all management and nonmanagement people, and discussions were frequently held to consider this and other aspects of the Lakeview operations. Each person was considered a unique individual. It was taken for granted that he was mature, intelligent, and honest, that he wanted a meaningful job, that he would improve his skills on the job, that he wanted to work to his full capacity, and that he would assume responsibility. Furthermore, it was assumed that when he did his job well, he would be recognized by his fellow workers and by his supervisor, and that he wanted to be in line for a better job because of his success in this present work. At Lakeview, management and nonmanagement people worked together to define problems, to develop solutions, and to appraise results. When a failure occurred there was the opportunity to determine just what went wrong and to try again. In the early days of operation at the Lakeview plant, the phrase "every man his own manager" evolved.

Department Organization

Each department is self-contained and is operated with a minimum of direction from above. The decisions are made at the lowest possible level. The department operates as a team with definite daily and weekly goals, but with great freedom as to how they will organize their facilities and personnel to meet the goals. In the conference room, wall charts show goals that the people have set for themselves and their progress toward achieving them.

Packing Department

For example, the packing-line mechanic might relieve the line operator for short periods; the people in the department might help decide which lines to run, determine packing-lines speeds, and decide how to handle the situation in case of a major breakdown. Two years ago the operators and mechanics on the packing lines worked out a way to stagger their lunch and break periods in order to keep the packing line operating the entire eight-hour shift and thus to produce more product. The men figured out how they could do this and came to the supervisor requesting permission to try it out.

With ten different products and twelve different package sizes being produced it is necessary to make changeovers at frequent intervals. There is a difference in the capacity, reliability, and flexibility of the various packing lines and also a difference in the time required to clean the equipment when a different product is to be packaged or when the package size is changed. With all the members of management and nonmanagement working together as a team, they have an opportunity to use their combined ingenuity, imagination, and effort to maximize the utilization of the equipment. They are challenged to make the changeovers as quickly as possible, to anticipate equipment failures, to reduce the possibility of shutting down the lines because of lack of product or packaging materials, and to maintain a high level of quality. Preventive maintenance is a part of every machinist's and operator's job. Operators may carry tools and make minor adjustments and repairs. They also work on a rotating basis to do line cleanup at night, since this is important to the efficient operation of the department. If the cartons are in especially good condition and if the packing line is well adjusted, the operators increase the line speed, thus increasing the output of the packing line for the shift. Recently, when all six packing lines broke all previous production records, the department manager provided free coffee and doughnuts for the department. It is not unusual for an individual or the people in a production unit to be recognized in some special way.

Communications

The Lakeview plan requires a high level of communications. Each department has a well-appointed conference room where the manager or supervisor can talk with his people in the right kind of environment. Department managers hold regular meetings with their people. Subjects discussed range from the consideration of department-operating problems to economics, business forecasts, civic responsibilities, comparative wage rates, fringe benefits, and companywide business-operating problems. The department manager has a private discussion with each of his people at least every six months about his progress. This also provides an opportunity for the employee to make suggestions or discuss things that are on his mind.

Indirect Labor

It has been possible to reduce the amount of indirect labor at Lakeview. Janitor service for locker rooms and offices is contracted out. The plant opened with eleven office and clerical people, which is less than one third the number in the best of other similar plants. For example, the cashier handles payroll, weekly and monthly effectiveness pay calculations, first aid, sale of safety shoes, and the payment of bills. The administration of the salary and the effectiveness pay plan is so simple that it requires only one half the time of the cashier and one half the time of the one-plant industrial engineer. Line management is responsible for setting correct measures for the effectiveness pay plan in their department. The plant industrial engineer is responsible for the overall pay plan, measure maintenance, and pay calculations. He advises and consults, trains, guides, and helps the department managers to make decisions for setting correct measures in their departments.

Training

Great care has been used in selecting and training management and nonmanagement people. As a result the plant is staffed with outstanding, young, cooperative, enthusiastic, hard-working people, many of whom have those qualitites that bring rapid promotion. Some operators and machinists perform certain management functions. They gain authority through their knowledge and make decisions because of their know-how. At the present time, approximately 50 per cent of the plant supervisors have been promoted to this position from the operator position. The ages of these men range between 22 and 45, with half of the group under 25 years old. College graduates go almost immediately into supervisory positions. At the present time, six are in this position, and eleven more are department managers. One of these men moved into this position after but one year as a supervisor and another after two years' experience. Promotion is not based on seniority or on effort expended, but rather on results.

During the past three years some 80 operators, machinists, and clerks have received formal training in methods and work simplification. The course, with ten or twelve in a group, has been taught by the department managers and the group managers with staff managers serving as coordinators. A project engineer, safety and training specialist, and the accounting manager have served as coordinators. The course is held two hours per day for a ten-day period. In addition, each participant completes at least one methods-improvement project. The payout for this training came quickly in actual methods savings made by the people on their jobs. But also they learn to think like managers in selecting, evaluating, and installing changes, and better understand that improvement is part of everybody's job.

Managers and supervisors make certain that each person joining the organization fully understands that the Lakeview method of operation is possible only because the company can supply the customer with products of uniformly high quality at low cost, delivered to the customer in good condition when he wants them. At the present time, a new program is under way in which members of the organization in groups of three or four will spend two weeks learning about the various functions of the business such as purchasing, quality standards determination and control, distribution, and costs. Each member of management, including the plant manager, discusses his duties responsibilities, and typical day-to-day decisions with these people It is expected that this will broaden the outlook of the people at Lakeview, resulting in their greater interest and enthusiasm, and will provide wider opportunities for their contribution to the success of the business—that is, higher output, better quality, lower costs, and more profits. This in turn will bring greater satisfaction and greater financial returns to the people.

Superior Performance from Superior People

The absence of time clocks, the introduction of unlocked tool rooms and storerooms, and then in 1966 the advent of the weekly salary provide evidence that the people at Lakeview are trusted. However, this is not a philanthropic organization. Goals are high, quality standards must be maintained, and delivery schedules are rigid. There is a big job to be done everday. Each person works with his mind as well as with his hands, and only the best qualified can survive at Lakeview. This is a place for superior people who do a superior job and who are highly rewarded. Although the financial rewards are excellent, the great motivator seems to be the opportunity to be a member of the unique and dynamic Lakeview organization, knowing that each person's very best creative efforts are wanted and needed, joining with others in using imagination and ingenuity in achieving worthwhile goals, and being employed by a company where every person is important and is recognized and rewarded for his accomplishments.

Results. It is not easy to evaluate the results of Lakeview's unique management system and pay plan. However, the following statements can be made.

1. The Lakeview plant is producing more product than the factory and equipment were designed to produce.
2. The start-up was more rapid and less costly than is customary for such a plant.
3. The productivity per man-hour is equal to or greater than any other plant in the company making similar products, and it is increasing steadily as people gain experience.
4. The quality is equal to or better than that of other plants.

The company is highly pleased with the plant's operation and the people at Lakeview are enthusiastic about the results thus far. They look forward to new and exciting opportunities ahead, since the plant is now being expanded and new facilities are being added.

THE LAKEVIEW EFFECTIVENESS PAY PLAN

The effectiveness pay plan consists of a base salary plus extra compensation in the form of a monthly cash payment when superior results are achieved.

Salary

It is the policy of the company to maintain a base pay that is equal to or higher than the average pay of similar companies in the community. A wage survey is made at frequent intervals and base wage adjustments are made as necessary. Although there have never been time clocks at Lakeview, wages were paid for time worked until June 1966 when a weekly salary plan was introduced. Now, qualified management and nonmanagement people alike are paid a salary. The new employee is paid only for the hours he works for the first three months or until he is qualified for the weekly salary. Each qualified person receives his full weekly salary even when absent, provided that there is a good reason for his absence and provided that he is not absent more than four days in any week. There are only five job classifications in the plant, which makes for flexibility in the assignment of people and simplifies pay scales and promotion procedures.

Effectiveness Pay

The effectiveness pay plan provides up to 30 per cent compensation in addition to the base salary for group accomplishments in the areas of cost reduction, cost control, and quality achievements depending upon the plant performance. These areas or categories are divided into several factors, each of which is weighted according to its contribution to the profitableness of the Lakeview operations.

Effectiveness pay takes the form of a separate monthly check, which is given to each person by his supervisor. This pay is not considered to be a wage incentive in the usual sense. However, it is a constant reminder that superior performance results in higher pay. It is not easy to reach the maximum bonus of 30 per cent. The plan is designed to eliminate month to month fluctuation in pay inasmuch as some of the factors in the plan are based on running averages of from 4 to 12 months. The effectiveness pay together with the base salary provide a total compensation for each person that is proper for a superior job. It is somewhat higher than could be earned elsewhere in the community. Everyone at Lakeview except the plant manager participates in the pay plan on the same basis.

The following criteria were followed in designing the effectiveness pay plan.

1. The application of the pay plan must be in harmony with other important factors affecting motivation, such as respect, fairness, challenge, opportunity, goal setting, and self-realization.

2. Measurement is made of results, not of effort expended.

3. Emphasis is placed on change, which is considered a basic responsibility of all.

4. Management and nonmanagement people are on the same percentage basis of measurement.

5. People of different skills or pay classification, who have common objectives, should be on the same basis of performance measurement (mechanics and operators in the same department, for example).

6. Each combination of measures used to determine effectiveness pay attempts to balance opposing needs. This includes balancing between end results (as they might be seen in the end product) and partial results strongly influenced by the participant. There is an attempt to attain a balance between long-range importance and early feedback of results.

EFFECTIVENESS PAY CALCULATIONS

Lakeview Plant—Month of May. The package-filling department (Department 3) will be used to illustrate how the effectiveness pay plan functions. The department is equipped with high-speed packing lines similar to the one shown in Fig. 359. The product is fed into the filling head of the packing lines from bins located on the floor above. The finished product is either stacked into unit loads from the end of the case-sealing machines by hand or by an automatic stacking machine, and is removed by a clamp truck.

The procedure required to determine the effectiveness factor and the effectiveness pay will be described briefly. The three areas or categories included in the plan are (1) cost reduction, (2) cost control, and (3) quality. The Plant Summary Statement for the month of May is shown in Fig. 360.

1. Cost reduction

    a. Plant Methods
    b. Plant Target Reduction
    c. Department Target Reduction

At the Lakeview plant, the management and nonmanagement people consider cost reduction a regular part of their job. The effectiveness pay program (EPP) is designed to recognize and reward these people for success in this area. Cost reduction is divided into three categories: (*a*) Plant Methods, (*b*) Plant Target Reduction, and (*c*) Department Target Reduction.

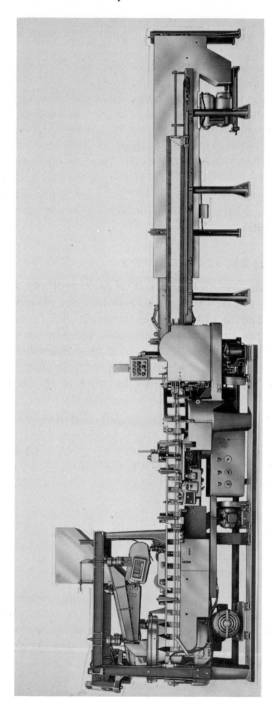

Fig. 359. Typical high-speed packing line.

Plant Methods. Each year a target is established for cost reduction through better plant methods. For the current calendar year the target is $20,000 per budgeted member of management. This is the same for all plants in the company. Since the Lakeview plant has 30 budgeted members of management, the total target or expected reduction in costs for the current year is $600,000 ($20,000 × 30 = $600,000). An "Effectiveness Factor" (E.F.) or results factor is computed each month, and it is based upon the relationship of the target to the actual reduction in costs for the month. The effectiveness factor for the month of May is combined with the E.F. for the preceding eleven months to obtain a running twelve-month average. As Fig. 360 shows, this E.F. was .96. This refers to *plant* methods and applies to management and nonmanagement people in all departments.

Example. If a certain raw material could be purchased in larger containers resulting in lower freight costs and lower handling costs, this saving would fall under the "Cost Reduction-Plant Methods" category.

Plant Target Reduction. This factor is intended to encourage management and nonmanagement people to make improvements and to revise the targets and bring them in line with improvements in costs that have been made. The targets cover the controllable items of manufacturing expense such as management and nonmanagement salaries, mechanical labor, employee benefits, repair materials and expenses, contract services such as guard and janitorial people, telephones, utilities, etc. Not included are taxes, insurance, depreciation, and expenses in connection with new equipment installations.

The improvement or cost reduction is the actual reduction in costs of the controllable items made during the month of May. The effectiveness factors for the Plant Target Reduction and the Department Target Reduction are determined by taking a twelve-month running average of target reduction for the plant (or department) in dollars and comparing it with the budgeted dollars for the controllable items for the same twelve-month period. There is a separate effectiveness factor for each department. The plant E.F. is the weighted average of the effectiveness factor of all departments plus the general plant overhead. As Fig. 360 shows, the Department 3 Plant Target Reduction Effectiveness Factor was .86 and the Department Target Reduction E.F. was 1.09.

Examples. The purchase and use of an automatic floor-scrubbing machine, which reduces the target from two men to one man, illustrates both a plant methods change and a plant and department target reduction.

Alertness and teamwork on the part of the managers, operators, and mechanics resulting in increased output of the automatic filling and casing machines is another illustration of a change in plant and department target reduction.

2. Cost control

a. Plant Product System
b. Department Operations

Cost control is concerned with the cost of all direct and indirect labor, repairs and expenses, utilities, losses, and rework. Taxes, depreciation, and insurance are not considered.

Targets or standards for direct labor are determined by conventional work measurement methods. Budgets are used for indirect labor such as clerks, machinists, and managers—also for repairs and expenses, utilities, losses, and rework. The dominant factor in cost control is manufacturing efficiency. The effectiveness factor for the department is determined by comparing the actual costs with the budgeted or target costs. The calculations are made weekly and are accumulated for one month. The effectiveness factor for the month of May for Department 3 was .98 and for the plant it was .94 (see Fig. 360).

3. Quality

Quality standards are stated in per cent of the product not meeting specifications and they are established for each department.

Examples

Product-Manufacturing Department—Department 1. Finished product quality index is based upon variation from specifications.

Package-Filling Department—Department 3. Quality index is affected by variation from specified weight of product in the package and volume or density of product in the package.

Warehouse and Shipping—Departments 4 and 5. Quality index is based on the per cent of the product that must be scrapped or reworked due to improper handling.

The effectiveness factor for quality is calculated as a six-month running average. For the month of May the Quality-Plant Product System E.F. was 1.00, and for Department 3 it was .89 (see Fig. 360).

The total weighted average effectiveness factor for Department 3 for May was .959 which is equivalent to 25.9 per cent effectiveness

EFFECTIVENESS PAY PROGRAM

Lakeview Plant Summary Statement for Month of May

	Department 1 Product Manufacturing		Department 2 Package Filling		Department 3 Package Filling		Department 4 Warehouse		Department 5 Shipping		Plant Staff	
	Per cent	E.F.	Per cent	E.F.	Per cent	E.F.	Per cent	E.F.	Per cent	E.F.	Per cent	E.F.
1. COST REDUCTION												
Plant methods	10	.96	10	.96	10	.96	10	.96	10	.96	10	.96
Plant target reduction	10	.86	10	.86	10	.86	10	.86	10	.86	20	.86
Department target reduction	10	.78	10	.91	10	1.09	10	.96	10	.82		
2. COST CONTROL												
Plant product system	20	.94	25	.94	25	.94	30	.94	25	.94	55	.94
Department operations	20	.87	30	.88	30	.98	30	1.09	25	1.04		
3. QUALITY												
Plant product system	10	1.00	05	1.00	05	1.00	05	1.00	05	1.00	15	1.00
Department production	20	1.00	10	.86	10	.89	05	.84	05	.84		
Department control									10	.75		
WEIGHTED AVERAGE E.F.	100	.922	100	.908	100	.959	100	.979	100	.926	100	.935
EFFECTIVENESS PAY PER CENT	22.2		20.8		25.9		27.9		22.6		23.5	

Fig. 360. Lakeview plant summary statement for month of May.

pay. Thus all management and nonmanagement people in Department 3 earned additional pay based on this effectiveness factor of 25.9%. A check for 25.9% of each person's base salary for May was given to him in person by his supervisor. Ordinarily this payment is made around the 20th of the following month—in this case, by the 20th of June.

Example. John Smith, an operator in Department 3, worked during the entire month of May. His base salary is $128 per week or $554.67 for the month of May $\left(\dfrac{128.00 \times 52}{12} = 554.67\right)$.

$$\text{Effectiveness Pay} = \$544.67 \times 25.9\% = \$143.66$$

Base salary for month of May $554.67
Effectiveness Pay   143.66
Total earnings for May $698.33

The effectiveness pay percentage for Department 1 was 22.2%, for Department 2, 20.8%, for Department 4, 27.9%, for Department 5, 22.6%, and for the staff personnel (all group managers and laboratory, engineering, office, industrial relations, and cost accounting people) it was 23.5%.

APPENDIX A

Time Study Manual

I. Time Study Department Responsibility
A. The determination and administration of all factory wages.
1. Hourly day-work rates established by job evaluation.
2. Standard times and unit piecework rates established by time study, elemental data, motion-time data, or work sampling.
3. Determination and maintenance of the wage-payment system.

B. The coordination of the development of all factory production methods.

C. The preparation and maintenance of plant layout.

D. The determination of the proper kinds and quantity of new and replacement equipment.

II. Definition of Time Study
Time study is the analysis of a job for the purpose of determining the time that it should take a qualified person, working at a normal pace, to do the job, using a definite and prescribed method. This time is called the *standard time* for the operation.

III. Purposes of Time Study
A. A basis for determining time standards and establishing piecework rates.

B. A basis for establishing a "standard day's work" for jobs paid on a day-work basis.

C. An aid in improving methods.

D. Production planning and control purposes.

E. Cost-control purposes.

IV. Requests for Time Study
A. *Request for time study by foreman when a new job is put into production.* The foreman will request the time study department in writing, on Time Study Form TS 101 (Fig. 361) provided by the time study department, to make a study of a new operation when he considers the job ready. Before requesting a time study, the foreman should make certain that the following conditions have been met.

715

```
┌─────────────────────────────────────────────────────────────────────┐
│                    REQUEST FOR PIECE RATE                           │
│                                                                     │
│                                                                     │
│  To: Time Study Department              Department_____│
│                                                                     │
│                                                                     │
│                                    _____has been set up in my department. │
│                    (operation name)                                 │
│  The following points have been checked:                            │
│                                                                     │
│  Number of operators involved___ ___(    )   Operator is qualified and experienced on │
│  Machine is working properly____ __(     )   this job _____ (     )         │
│  Materials are to specification __ ___ (  )   Operator has been notified that operation │
│  Best work area is utilized__ __ __ _(   )   is to be timed _____ (     )     │
│                                              Necessary tools and equipment are available (   ) │
│  I believe this job is ready for time study.                        │
│                                                                     │
│  Date_____Signed_____ │
│                                                    Foreman          │
│  NOTE: The information on the bottom of sheet to be filled in by Time Study Deparment. │
│                                                                     │
│  Date received_____Date checked_____ │
│  Date studies taken_____Observer_____ │
│  Date rate effective_____│
│  Other disposition_____. _____│
│                                                                     │
│                              Signed by_____│
│  TS 101                                                             │
└─────────────────────────────────────────────────────────────────────┘
```

Fig. 361.   Request for piece rate—Form TS 101.

Preparation to Be Made by Foreman Before Study Is Requested

1. A satisfactory method of performing the operation must be developed. This method may not be the very best obtainable, but it should take into consideration such factors as sequence and economy of motions, distance materials have to be moved, including arrangements for delivery and removal of supplies by service personnel, and workplace layout.

2. The machinery and equipment must be running at the correct speed and must be in good working order. Tools, dies, fixtures, or any auxiliary equipment must be functioning properly and must be adapted to the job.

3. Materials must meet the specifications set by the pattern department and laboratory. A study will not be made while abnormal stock conditions exist. Restudies will not be made if the quality of materials during the rate trial period fluctuates beyond normal specification limits.

4. The operator must be trained to perform the operation, using the method, machine, tools, and equipment that have been specified, and

should have gained sufficient skill through experience on the job to be studied to display efficient performance. It is not advisable to make a time study of an inexperienced operator. In most cases it will be found that an inexperienced operator has so many fumbles, delays, and hesitations that it is practically impossible to segregate the true element time from the fumbles and delays.

5. The foreman must discuss the job with the operator and point out the reason why a time study is being requested.

B. *Request for time study or production study by foreman when a time standard and piecework rate is already in effect.* The foreman will request in writing, on the standard form, that the time study department make a re-study of the operation.

1. A time study will be made if there has been a change in method, work-place layout, materials, or tools and equipment used on the operation.
2. A production study will be made if the operator who was performing the job at the time the time study was made, or a different operator, is unable to reach the day-work standard of performance after a reasonable period of time while using the prescribed method and materials and exhibiting normal effort.

C. *Request for time study or production study by other persons.* In some cases persons other than the foremen, such as the factory manager, chief chemist, pattern and design, cost, purchasing, or sales departments, may, for the purpose of securing information pertaining to their particular function in management, request that a time study be made. These requests will be made on a Form TS 102, which is obtainable in the time study department.[1] In such cases the time study analyst will contact the foreman, explaining the reason for making the special time study.

V. Time Study Procedure

A. *Contact the foreman.* The time study analyst will contact the foreman upon entering the department. The foreman will show the time study analyst the location of the job, and will check the operation to see that the proper method is being used.

B. *Make contact with the operator to be timed.* In no case should a time study analyst start making a time study without the operator's knowledge. If there are several operators doing the same job, the person who is giving the nearest to a normal performance should be studied. In fact, two or more operators might be studied if this seems advisable. Under no circumstances should standards be established on the basis of time studies made of in-experienced operators or operators who are unwilling to cooperate.

C. *Check operation for method.* When a new item is put into production or when a new piece of equipment is installed, a number of people may be involved in the development of a method. The time study department should

---

[1] This form not shown here.

be consulted in such matters. It is the time study department's responsibility to check the method for possible improvement before setting a time standard for the job. The time study analyst may only suggest possible changes; he will not inaugurate them unless requested to do so. Before making a time study of a job, the time study analyst should have the foreman approve the method in use. This will include an examination of the elements of the job to be timed, and approval of their completeness.

D. *Obtain all necessary information.* The time study analyst should obtain and record on the time study observation sheet, Form TS 103 [see Fig. 241 in Chapter 25], all the information about the job, machine, and materials that he needs to fully complete his study. A drawing or layout of the work place should be made, showing the location of the operator, materials, tools, etc. Whenever necessary, a process chart showing the location of the particular operation in the process should be made. A sketch of the part should be included with the time study whenever it seems advisable.

E. *Divide the operation into elements.* The operation should be divided into elements as short in duration as can be accurately timed. The beginning and ending points of these elements usually are easily determined because they come at natural break points in the operation. It is important that each element be carefully defined so that the starting and stopping point will be exactly the same in each cycle timed. Handling time should be separated from machine time, and constant elements should be separated from variable elements wherever possible.

F. *Record the time.* The purpose of timing the operation is to obtain the representative time taken for each element of work in the operation. It is the policy, therefore, to carefully time each and every part of the operation. If, for example, a "book leaf" must be turned once for every ten pairs of parts cemented, such information should be recorded on the time study sheet, and a sufficient number of cycles including this element should be timed so that the representative time for this element can be obtained.

When foreign elements occur they should be timed and recorded on the time study sheet. These elements may or may not be included in the time standard, depending on their nature. It is necessary on the time study observation sheet to account for all the time consumed by the operator while the time study is being made. The foreign elements must be very carefully reviewed to determine if they should be incorporated in the time standard or if they are unnecessary delays caused by the operator. Personal time and time for rest and some unavoidable delays are incorporated in time study allowances and should not be included as elements in the time study, as this would be a duplication.

The time study analyst should record the time of day the study was started and the time it was finished, thus obtaining the elapsed time. The total number of units finished during the study should also be recorded.

G. *Rate operator's performance.* We all know that there is a difference in the effort or speed at which different people naturally work. For example, a few people usually walk at a slow pace and a few people at a very fast pace, while most walk at a pace somewhere between these two extremes. So in the factory some people work at a slow pace while others work at an excellent pace. The normal day-work pace is given an index number of 100 points in making a time study. A qualified operator who is trained to work correctly with the specified materials, tools, and equipment, and who is working at the pace expected from an individual being paid by the hour and therefore without incentive, is said to be working at a 100-point pace. For the purposes of comparison, it is expected that a few especially fast operators might reach a pace of 130 to 150 points when on incentive. The operator's performance or pace is rated when the time study is made, and the rating index is applied to the time study data in order to determine the standard time for the job.

VI. Computation of the Time Standard and Piecework Rate

A. *Compute the normal time.* The representative time for each element should be determined and recorded in its proper place on the observation sheet. This representative time should be multiplied by the rating factor to obtain the *normal* time for the element.

B. *Prepare the computation sheet, Form TS 104* [Fig. 243 in Chapter 25].
1. Transfer the element name and its normal time from the time study observation sheet to the computation sheet. The elements should be listed in the sequence of performance.
2. At this point, other studies of similar operations contained in the files and any standard data that are available should be reviewed to supplement the information contained in the new time study.
3. In the fourth column, headed "Units per Element," is placed the number of units that are completed in the element. The unit referred to is the unit of physical count—one yard, one pair, one batch, etc. Example: 8 pairs (16 pieces) of heel pieces are placed on a "leaf of a book" and the next leaf is turned. In this instance "8 pairs" is recorded in column 4.
4. In the fifth column, headed "Occurrence of Element per (. . . . . .)," is to be recorded the number of times this element occurs per 100 pairs or per other unit that may be used as a base.
5. Multiply the normal time per element by the occurrence of the element per 100 pairs (or per other unit that is used as a base) and record the result in column 6 on the computation sheet.
6. Obtain the total normal time for all elements by adding together the normal times of each element.
7. Add allowances for fatigue, personal needs, and delay to the total normal time of all elements to obtain the total standard time for the operation.
8. Divide the total standard time for the operation into 60 minutes and multiply by 100 to obtain the day-work hourly production. This then is the number of pieces or amount of work that has been

established by time study to be the hourly task that an operator should complete when working at a normal pace, that is, at a day-work pace and without incentive.

C. *Compute the piecework rate.* To compute the piecework rate the basic hourly wage or day-work rate which has been assigned to the job is divided by the day-work hourly production. Piecework rates are usually expressed in dollars and cents per unit or 100 units.

VII. Preparation for Rate Installation

A. *Discuss the time standard with the foreman.* At this point the foreman is contacted and all phases of the time study are discussed with him. Sufficient discussion time will be taken so that the foreman will be completely familiar with all phases of the time study and will therefore be in a position to describe it to the operator in a constructive manner, and he will be able to answer any questions that the operator may have.

B. *Determine the method of application of the piecework rate.*

1. Determine the exact manner in which production and time are to be measured and recorded. Design any forms necessary to report the amount of work finished per day.
2. Whenever necessary, prepare a statement of the payroll procedure to be followed in computing workers' earnings.

VIII. Putting Piecework Rate into Effect

A. *Several copies of the piecework rate sheets will be made.* Two copies will contain signatures of approval by the superintendent and the head of the time study department. All copies will contain the effective date, basic hourly wage, day-work hourly production, and the piecework unit rate, as well as the operation name, number, and the department.

B. A copy of the standard element sheet, TS 105 (Fig. 362), and a copy of the piecework rate sheet, TS 106 [Fig. 244 in Chapter 25], will be given to the foreman. Information about each new piecework rate installation will be supplied by the time study department to the payroll department and to interested persons in the other departments.

IX. Follow-up of the Piecework Rate Application

Soon after an operator begins working on incentive, a check will be made of his production by the foreman or by the time study analyst, separately or together. The foreman will in every case make a production check once each hour during the trial period. These production records are to be turned in on Form TS 107 [2] to the time study department at the end of each day for analysis. At least once during the first ten-day period following the installation of the piecework rate, the time study analyst and the foreman will jointly compare the operation with the standard element sheet. If further

---

[2] This form not shown here.

ELEMENTS OF JOB

DEPARTMENT Shoe Room   **FOREMAN**   W.M. Wilson

OPERATION   Assemble and Cement Heel Plugs on Swing Boot Insoles

PIECE WORK JOB NO. 16-15

DAY WORK JOB NO. 16-16   **DAY WORK HOURLY PROD.** 237 Pr.

NO.	ELEMENTS OF JOB
1	Get Supply of Heel Plugs
2	Get Supply of Insoles
3	Get, Loosen, and Lay Out Insoles In 15 Piles
4	Get, Pick, and Spot Heel Plugs on Insoles
5	Get Brush of Cement, Cement, and Aside Brush
6	Stack Completed Work
7	Mark Size on Stack
8	Aside Completed Work
9	Get Cement Supply
10	Empty and Clean Cement Pan
11	Clean Up Work Place and Cover Work
12	Record Production
13	
14	
15	
16	
17	
18	
19	
20	

Note: Whenever any one of the original conditions or the above elements are changed or eliminated, or when the worked place lay out is changed In any manner, the job must be checked by the Time Study Department.

TS 105

Fig. 362.   Elements of job—Form TS 105.

checks are required, production studies will be made by the time study department. Whenever any methods, materials, tools, or related equipment are changed in any way, the time study department must be notified so that they may have an opportunity to determine if a change in rate is necessary.

This manual was prepared by Earl L. Frantz, with the assistance of James A. Kenyon and Robert J. Parden.

APPENDIX B

Wage Incentive Manual

The Maytag Company has prepared a *Wage Incentive Manual*, which they use in connection with a time study training program given to all foremen and supervisors. This manual also serves as a handbook of methods and procedures pertaining to time study and wage incentives. The first six pages and the last three pages from this manual are reproduced here.

THE MAYTAG COMPANY
EXECUTIVE OFFICES
NEWTON, IOWA

FRED MAYTAG II
PRESIDENT

TO: MAYTAG MANAGEMENT

Over the years since the Maytag Wage Incentive Plan
was introduced in 1946, the benefits to our employees, our customers, and our Company, have become increasingly evident. Our
employees have attained higher wages than would have otherwise
been possible. Our customers have been able to buy Maytag
products at the lowest cost.

This has been due in great part to the high level
of productivity resulting from the installation of the best
manufacturing methods, and exertion of the best skill and
productive effort by our employees. With such productivity
and competitive costs, our Company has grown to provide new
jobs and new opportunities for each of us.

The future success of our organization, and therefore your success, is to a large degree dependent upon the continuing improvement of methods, and of effort and skills.

You, a Maytag Supervisor, have a great responsibility
for the success of one of the best tools of productivity, -- the
Wage Incentive Plan. Your thorough understanding and enthusiastic
support of the principles, application and administration of the
Incentive Plan, can help assure the many benefits to our employees,
our customers, and our Company.

This Manual has been prepared to assist you in the
application and administration of the Incentive System in your
Department. I urge you to become thoroughly familiar with the
material in the Manual. I am sure that such knowledge will
enable you to take an active part in the Wage Incentive program
with self assurance and confidence. In turn this should lead to
matter-of-fact acceptance, and provide a firm basis for our future
establishment of even better methods and lower costs.

My congratulations to you for your past good work,
and a wish for continuing success.

Sincerely,

Fred Maytag II

President

Fred Maytag II:hb

The **MAYTAG**
STANDARD HOUR INCENTIVE PLAN
is designed to help you in . . . . .

YOUR JOB
of PROPERLY
UTILIZING

raw materials

equipment

supplies

and most importantly, of managing

people

to **BUILD**
BETTER PRODUCTS
at **LOWER COSTS**

the following pages explain the
MAYTAG WAGE INCENTIVE PLAN

wage
incentives

The Maytag Wage Incentive Plan, which is based on thorough analyzation of each job, establishing the best method of performing the work, training the operator to use the best method, and proper application of Labor Standards, provides many benefits to our Employees, our Customers, and our Company

The **PURPOSES OF WAGE INCENTIVES** include:

- **Increasing employee's earnings.**
- **Establishing the most economical manufacturing costs.**
- **Providing greater utilization of machines and equipment.**
- **Scheduling production.**
- **Planning changes in manufacturing methods, and estimating costs.**
- **Budgeting and controlling costs.**

Although wage incentives require determining the necessary time to perform a job and result in extra pay for extra effort and skill of the operator, the job study which must be made before a Labor Standard can be established requires analyzing the job to PROVIDE:

- The most effective **EQUIPMENT.**
- The proper **MATERIALS.**
- The most effective **TOOLING.**
- The best manual **METHOD.**
- The best **FLOW OF MATERIALS.**
- Proper **WORKING CONDITIONS.**
- Adequate **SAFETY CONTROLS.**
- Adequate **QUALITY CONTROLS.**
- Proper **SELECTION AND TRAINING OF THE OPERATOR.**

wage incentives

wage incentives.....
make your job easier

1. **Careful study of jobs results in better and simpler methods of doing the work.** Industrial Engineering and other staff departments will assist you in developing good work methods. These methods must be developed before a labor standard is determined.

2. **Your employees will work more efficiently.** You are assured of producing according to schedule because your workers will want to exceed the standard in order to increase their pay. Your job of supervising becomes easier when the workers are so motivated.

3. **The detailed job instructions on the labor standard sheet helps you train the worker how to do the job.** The job instructions are written to provide a detailed description of the method, which will result in quality production and will reflect safe operating practices.

4. **Labor Standards help you plan your production.** When you know how many pieces per hour can be produced on a job, it is easy to determine how many men and machines you will need on the job to produce a specified number of units in a certain period of time. Also, materials needed for the job can be scheduled into your department systematically.

5. **Your labor turnover is reduced.** You do not have so many new workers to train because high earnings influence your experienced workers to stay at Maytag.

The help which wage incentives will provide you depends upon your knowledge of the Incentive Plan, your participation in establishing standards, and your proper administration of the incentive program in your department.

wage incentives

Labor Standards are the basis of the Maytag Wage Incentive Plan

successful application of Labor Standards depends on . .

1. Development of the **Best Practical Method.**

2. **Training the Workers** to do the job using the best practical method.

3. **Accurate Measurement** of the manual work and machine time by means of job study, considering method, effort, most economical equipment operation, and application of proper allowances.

4. **Participation by the Employee** in changes affecting his job, and his thorough understanding of the Labor Standard.

5. **Follow Up** after Labor Standards are issued to assure proper application, acceptance and adequate performance by the employee, and to keep the standards up-to-date.

In order to establish Labor Standards promptly and accurately, let's follow through, step by step, the procedure for wage incentive application.

to be taken in determining and maintaining a Labor Standard

prepare the job

The foreman, with cooperation of staff departments, prepares the job for study by the Industrial Engineering Department.

request the Labor Standard

The foreman requests the Industrial Engineering Department to study the operation.

study the operation

The industrial engineer studies the operation after checking the details of the job with the foreman and the operator.

compute the Labor Standard

The industrial engineer computes the labor standard.

apply the Labor Standard

The industrial engineer writes up the "Labor Standard Sheet," including detailed job instructions, and after necessary approvals, issues it to the shop. The labor standard is then thoroughly explained to all operators, and any questions which they may have are answered.

follow-up the Labor Standard

Both foreman and industrial engineer observe the job closely on initial application of labor standard to be sure of operator acceptance and adequate performance, and periodically thereafter to assure that the job is being performed in accordance with the requirements of the "Labor Standard Sheet."

Let's consider these *steps in detail*

administration
of the
WAGE INCENTIVE
PLAN

During the time since the Maytag Wage Incentive Plan was introduced, its benefits to our employees, our customers and our Company have become increasingly
evident. New and better manufacturing methods and increasing skill and productive effort of our employees has made it possible for our customers to buy
quality products at the lowest prices. With the increased productivity, for which
the wage incentive program has played an important part, and wider distribution of our products, our company has grown to provide new job opportunities
and greater security.

As a Maytag Supervisor you have a great responsibility for the success of the
Wage Incentive Plan. Only with your thorough understanding and active participation in the application and administration of wage incentives can the many
benefits to our employees, customers and company continue.

Your responsibilities for proper administration of the Wage Incentive Plan
require:

1. **A thorough knowledge of Maytag Wage Incentive Plan.**

2. **Active participation in the incentive program.**

3. **Proper application of wage incentives.**

4. **Adequate communication with employees.**

5. **Proper administration of the incentive sections of the Labor
   Agreement.**

KNOWLEDGE of the MAYTAG WAGE INCENTIVE PLAN

This manual has been prepared to help you in understanding the purposes of wage incentives
and the procedures for establishing the best methods and applying labor standards. A thorough understanding of this material and the incentive sections of the Labor Agreement should
provide you the knowledge to take an active part in the wage incentive program with self
assurance and confidence.

administration of the
Wage Incentive Plan
(CONTINUED)

2 PARTICIPATION in the INCENTIVE PROGRAM

Although the industrial engineers have had the specialized training for determining the best job methods, for measuring work and applying and administering wage incentives, the success of the wage incentive program is largely dependent on your active participation and skill in handling your many responsibilities in establishing labor standards and for the administration of incentives in your area.

This manual explains most of your responsibilities from preparing the job for study through the follow-up of labor standards and administration of the incentive plan. With adequate knowledge of these responsibilities, active participation in the incentive program and proper utilization of the assistance of industrial engineering and other staff departments, you will find that the many advantages of wage incentives can be obtained with a minimum of problems.

3 PROPER APPLICATION of WAGE INCENTIVES

It is the Company's policy to provide fair and equitable labor standards and apply them to jobs which, in the opinion of the Company, can properly be placed on incentive. With such a policy, labor standards should be applied to as many jobs as practical. However, for various reasons it is not always practical to establish standards for all operations. IT MUST ALWAYS BE REMEMBERED THAT THE FUNDAMENTAL CONCEPT OF WAGE INCENTIVES IS BASED ON EXTRA PAY FOR EXTRA EFFORT. Some operations, because of machine or process control, include substantial observation time and therefore do not provide an opportunity for extra effort. For such operations labor standards should not be used merely as a device to increase the operator's pay, but you should provide the additional work necessary, by combination of operations or the addition of other required work, in order to make the application of a labor standard practical. When new operations are first started in production, tooling or equipment problems often make it impossible to immediately establish a standard. Adequate pre-planning can often eliminate such problems, but if adverse conditions do exist they should be immediately corrected so that a standard can be established and the operator provided an opportunity for incentive earnings. If you question the advisability of applying a standard, it is always a good idea to confer with the industrial engineer. He may have suggestions which will be helpful in preparing the job for standard more quickly.

Preliminary Estimates play an import part in the wage incentive program. They are designed to be used until a labor standard can be established for new operations, or where major changes are made on existing operations and time allowances or disallowances are not applied. A P.E. provides the operator an opportunity to attain higher earnings and usually results in higher production than if the job were run on daywork. In order to attain these advantages you should arrange to have the P.E. established before the operator is assigned to the job. For most operations the P.E. can be established in advance of production if the industrial engineer is provided information on the job conditions, and the machine feeds and speeds or cycle times are known. A P.E. is intended to be a temporary method of incentive payment and should be replaced by a labor standard within 40 hours of operation of the job, or sooner if possible.

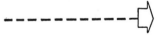

The most important responsibility which you, as the shop supervisor, have for proper application of wage incentives is to see that your operators are not compensated by standards inapplicable to the job because of changes in the manual method or job conditions. Remember, this responsibility requires that you frequently check each operation to determine if it is running exactly the same as when the standard was established and if it is not, to immediately report the changes to the Industrial Engineering Department so that the standard can be properly revised.

4 ° COMMUNICATION with EMPLOYEES

Often many of the problems encountered in the application of wage incentives result from misunderstandings among employees, shop supervisors and industrial engineers, which could have been eliminated through better communications. Incentive employees should know how labor standards are established and how the wage incentive plan operates. Naturally, they look to their supervisors for such information. Sometimes, because of the technical implication of their questions, it may be necessary for you to ask the industrial engineer to assist in the answer. Whatever may be required, it is important that you provide your incentive employees the answers to any legitimate questions that they may have concerning wage incentives. With such an approach much of the so-called "mystery" of wage incentives, and many of your problems, can be eliminated.

The Employee Attitude Survey conducted by the University of Michigan showed conclusively that the employees were most satisfied with the incentive system when they saw their foremen doing a good job of explaining wage incentives and the reasons for changes that affected their jobs. It is important to the employees that their foremen do a good job of communications.

It is equally important that you do not, through careless remarks, give your employees an improper impression of wage incentives. For example, if you ever say to an employee, or otherwise give him the impression that you believe a standard is "too tight," you are making a generalized conclusion that has little meaning to anyone, and can be interpreted as criticism of the Company's incentive program. Any such generalized criticism can soon lead employees to believe that they are being unfairly treated. On the other hand, specific criticism of a labor standard can be helpful. If, for example, you say, "This standard does not provide for the increase in the machine-cycle time and will be checked for any necessary change in the standard," you are properly carrying out your responsibility for keeping standards current. Remember, you will be criticized for negative, generalized remarks, but you will be doing your job if you make constructive criticisms intended to correct specific errors.

5 ADMINISTRATION of the LABOR AGREEMENT

The incentive provisions of the Labor Agreement provide the employees assurance of fair treatment in the application and administration of the wage incentive program, and also provides the requirements for a sound wage incentive plan that will continue to provide the many benefits to the employees, our customers and our Company. Your knowledge and proper application of these contractual requirements can have a great affect on the continuing success of the Maytag Wage Incentive Plan.

Problems

CHAPTER 1

1. Define motion and time study according to (*a*) Taylor, (*b*) Gilbreth, and (*c*) Farmer.[1]

2. Explain fully the meaning of the phrases "most economical way of doing work"; "finding the preferred method."

3. The motion and time study function is often a part of the industrial engineering department. Draw an organization chart of a typical medium-size manufacturing company and show the location of the industrial engineering department in the organization.[2]

4. Name two "tools" in the broad field of motion and time study that in your opinion come closest to filling the specifications of Lord Kelvin as outlined in the following statement of his. Give reasons for your choice.

"I often say that when you can measure what you are speaking about, and express it in numbers, you know something about it; but when you cannot measure it, when you cannot express it in numbers, your knowledge is of a meagre and unsatisfactory kind; it may be the beginning of knowledge, but you have scarcely, in your thoughts, advanced to the stage of science, whatever the matter may be."

5. The field of motion and time study has changed during the past 25 years. Describe some of the changes that have taken place during this period.

CHAPTER 2

6. After reading *Scientific Management* by Frederick W. Taylor, give a summary of the life of Taylor.[3]

[1] Eric Farmer, "Time and Motion Study," Industrial Fatigue Research Board, *Report* 14, H. M. Stationery Office, London, 1921. Also see C. S. Myers, *Industrial Psychology in Great Britain*, Jonathan Cape, London, 1926; M. S. Viteles, *Industrial Psychology*, W. W. Norton & Co., New York, 1932.

[2] National Industrial Conference Board, "Industrial Engineering Organization and Practices," *Studies in Business Policy, No.* 78, New York, 1956. Also see R. A. Forberg, "Administration of the Industrial Engineering Activity," *Proceedings Twelfth Industrial Engineering Institute,* University of California, Los Angeles-Berkeley, pp. 22–30, February, 1960.

[3] Frederick W. Taylor, *Scientific Management*, Harper & Bros., New York, 1947.

7. Compare motion and time study as it is understood and used today with Taylor's concept of it.

8. Summarize Taylor's investigations of (*a*) handling pig iron, (*b*) cutting metals, (*c*) shoveling.

9. Give a sketch of the life of Frank B. Gilbreth.

10. Summarize Gilbreth's investigations of (*a*) bricklaying, (*b*) work at the New England Butt Company,[4] (*c*) work for the handicapped.[5]

11. State the chief criticisms [6] of motion and time study and evaluate each criticism in light of the best present-day practice.

12. Discuss the broadening scope of industrial engineering. Describe some changes in the field of industrial engineering that are likely to occur with the increasing use of mathematics, statistics, and electronic data-processing equipment.

13. List some of the activities that the industrial engineering department is directly responsible for administering today.[7]

14. To whom in the organization does the chief industrial engineer report?[7]

CHAPTER 3

15. Make a list of all possible ways of holding together two 8½″ x 11″ sheets of 20-pound-weight bond paper.

16. What are the essential requirements for a successful brainstorming session?

17. Formulate any one of the following problems: (*a*) raise and cure tobacco, (*b*) protect your house against fire, (*c*) design the form and contour of a single unit telephone, that is, handset, dial, bell, and auxiliary equipment all in one unit. In stating the problem include what is known, what is unknown, and what is desired.

18. Develop a plan for the care of your lawn. This should include preparation of the soil, planting, watering, fertilization, mowing, and disposing of the grass clippings.

19. Discuss the following statement by Niccolo Machiavelli: "There is nothing more difficult to take in hand, more perilous to conduct, or more

[4] John G. Aldrich, "The Present State of the Art of Industrial Management," discussion, *Transactions of the ASME*, Vol. 34, pp. 1182–87, 1912.

[5] F. B. and L. M. Gilbreth, *Motion Study for the Handicapped*, George Routledge & Sons, London, 1920.

[6] William Gomberg, *A Trade Union Analysis of Time Study*, 2nd ed., Prentice-Hall, Englewood Cliffs, N. J., 1955; R. F. Hoxie, *Scientific Management and Labor*, D. Appleton & Co., New York, 1915; Ralph Presgrave, *The Dynamics of Time Study*, McGraw-Hill Book Co., New York, 1945; Richard S. Uhrbrock, *A Psychologist Looks at Wage Incentive Methods*, American Management Association, New York, 1935.

[7] Ralph M. Barnes, *Industrial Engineering Survey*, University of California, Los Angeles, 1967.

uncertain in its success, than to take the lead in the introduction of a new order of things."

CHAPTER 4

20. What factors affect the extent to which motion and time study may be profitably used?

21. Indicate the extent of a motion and time stuly investigation for a department in a plant with which you are familiar.

22. Explain the "law of diminishing returns," in relation to the desirable elaborateness of a motion and time study program in a plant.

23. Obtain information concerning the actual installation of a piece of production equipment. Determine the number of years required for the savings in operating costs to pay for the initial cost of the equipment. Also compute the rate of return in per cent per year on the investment.

CHAPTERS 5, 6, AND 7

24. Determine the specifications, design a reel to hold 100 feet of 16-mm. motion picture film, and determine the manufacturing processes for this product. Assume an annual volume for the next 5 years of 500,000 of each reel, *A* and *B* below. Six months' time is available for the design and testing of equipment and production methods.

A. Reel for unexposed film (see Fig. 363)

The distance between the two flanges must be held to specified tolerances. If the distance between the flanges is too small, the film will be damaged when it is wound onto the spool. If the distance is too great, then light may leak in between the edges of the film and the flanges and fog the film. The specifications should also include thickness and diameter of the flanges and barrel, dimensions of the square hole and round hole, and the kind and quality

Fig. 363. Parts of reel for un-exposed film.

Fig. 364. Parts of reel for processed film.

of the lacquer coating on the flanges and barrel. Of course, the outside dimensions of the reel must permit the reel to fit into the motion picture camera.

B. Reel for processed film (see Fig. 364)

This reel serves to contain the film for storage and for showing on the projector.

25. Suggest changes that could be made which would reduce the time and effort required to get ready to water the garden. (See Figs. 23 and 24.) Make a process chart and a flow diagram of your proposed method.

26. Construct a process chart and flow diagram for a dentist performing some activity that requires him to use his workroom or laboratory as well as working with the patient in the chair.

27. You have volunteered to help a boys' club make 5000 wood boxes (Fig. 365) and fill them with candy. These boxes will be sold at a local fair

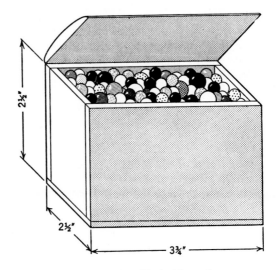

Fig. 365.   Box filled with candy.

as a means of raising money. The sides, ends, and bottom of each box will be made of ¼-inch plywood, and the top from clear white pine. The top will be hinged to the body of the box by means of a strip of scotch tape. The box will not be painted or finished in any way. Develop the most economical method for each operation, and design the necessary jigs and fixtures for making each part and for assembling the box and filling it with small candy beans. Construct a process chart and flow diagram.

28. Construct a process chart and flow diagram for each of the following:

(*a*) Writing a letter and mailing it.

(*b*) Making a cheese sandwich.

(c) Making a small gear from a gray iron casting.

(d) Dressing, having breakfast, and leaving the house in the morning.

(e) Washing a bundle of clothes in a commercial or home laundry.

29. Work out improvements in problem 28 (a), (b), (c), (d), and (e), and construct a process chart and flow diagram of each new method.

CHAPTER 8

30. Draw a man and machine chart showing one man (the operator) operating two semiautomatic lathes. The cycle consists of *load machine,* ½ minute; *machine part,* 1 minute (the machine stops at the end of cut); and *unload machine,* ¼ minute. The two machines are alike, and each completes the machining operation and stops automatically. The man and machine chart is to show the operator starting the machines in the morning with both machines empty, and continues until each machine has completed two cycles; that is, until it has machined two pieces. The man and machine chart should have one column for the man, and a column each for machine No. 1 and machine No. 2.

Fig. 366. Electric toaster.

31. The electric toaster shown in Fig. 366 is hand-operated, each side being operated independently of the other. A spring holds each side of the toaster shut, and each side must be held open in order to insert bread. In toasting three slices of bread in the above toaster, what method would you recommend to obtain the best equipment utilization—that is, the very shortest over-all time? Assume that the toaster is hot and ready to toast bread.

The following are the elemental times necessary to perform the operations. Assume that both hands can perform their tasks with the same degree of efficiency.

Place slice of bread in either side of toaster	3 seconds
Toast either side of bread	30 seconds
Turn slice of bread on either side of toaster	1 second
Remove toast from either side of toaster	3 seconds

Make a man and machine chart of this operation.

32. Make a man and machine chart of washing a bundle of clothes in a home laundry or a commercial laundry.

33. Make a man and machine chart of (a) an operator cutting ½-inch pieces from a 1-inch steel bar on a power-driven hack saw; (b) one person operating two hack saws like that referred to in (a).

34. One hundred thousand studs ½″ x 2½″ are manufactured each year for use in airplane motors. These studs must be accurately machined, and after they are threaded, all burrs must be removed. This is now done by brushing each end of the stud by rotating it by hand against a buffing wheel, as shown in Fig. 367. Develop a better method for performing this operation.

Fig. 367.   Brushing studs for airplane engines—old method.

35. Strips of wood molding ¾" x 3½" x 16' are given a priming coat of white paint in the following manner. The strips of molding are brought to the paint department on a special truck. The operator places three strips on a table, and then, using a spray gun, walks back and forth three times, spraying one strip each time. He then turns the strips over and sprays the other side, making three more trips. The molding is returned to the shelves on the truck, where it is allowed to dry. Develop a better method for painting the strips of molding. There is sufficient volume to keep one man occupied full time on this job, using the present method.

CHAPTER 9

36. Make a left- and right-hand operation chart of each of the following:
 (a) Lighting a cigarette.
 (b) Punching a sheet of paper in a three-hole paper punch.

(c) Loading a piece in the chuck of an engine lathe.

(d) Drilling a hole in the end of a square bar of steel.

(e) Assembling a mounting spring for a refrigerator (Fig. 368).

37. Make a list of the motions of the left hand and of the right hand used in opening a bottle with the conventional type of bottle opener. The left hand reaches to back of table, gets bottle, carries it to front edge of table in convenient position for opening. The right hand already has the opener, moves it up to cap of bottle, and removes the cap, opening the bottle.

38. Work out a better method for assembling the rope clips (page 112). Assume that there is sufficient production to keep two operators employed on this job 40 hours per week during the next year. Make a left- and right-hand operation chart of your proposed method.

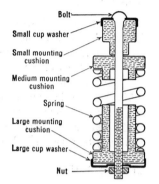

Bolt

Small cup washer

Small mounting cushion

Medium mounting cushion

Spring

Large mounting cushion

Large cup washer

Nut

Fig. 368. Mounting spring for refrigerator.

CHAPTERS 10 AND 11

39. Why has micromotion study been used at such an accelerated rate in recent years?

40. The Gilbreths made motion pictures at speeds faster than 2000 frames per minute and at speeds as slow as one frame every 10 minutes. Where might it be desirable to make motion pictures at each of these extreme speeds?

41. Illustrate each of the 17 therbligs by means of an operation with which you are familiar.

CHAPTER 12

42. Examine three different makes of motion picture cameras and projectors and evaluate the important features of each for use in motion study, memomotion study, and micromotion study.

43. Explain the relationship between "speed" and "f. setting" of a camera.

44. Describe a wide-angle, a zoom, and a telephoto lens, and state the circumstances under which each might be used.

45. Some motion picture cameras are equipped with a variable shutter. Describe how this shutter functions, and indicate the conditions under which it might be desirable to have a motion picture camera equipped with a variable shutter.

46. Describe the use of an exposure meter.

CHAPTER 13

47. Make a motion picture at normal speed of:

(a) Drilling a ¼-inch hole in the end of a small steel shaft.

(b) Picking up a pen and writing.

(c) Inserting a letter in an envelope and sealing it.

(d) Stapling together two 3″ x 5″ cards.

48. Make a motion picture at 50 frames per minute of:

(a) A crew pouring concrete.

(b) Three or four men laying a brick wall.

(c) A labor gang repairing a hole in the street.

(d) Operators at the check-out counter of a self-service grocery.

CHAPTER 14

49. Make an analysis sheet of the following operations. List the therbligs for both hands, omitting the time values.

(a) Assembling the parts of a ball-point pen.

(b) Sharpening pencil in pencil sharpener.

(c) Drilling hole in block of wood.

50. Analyze the film of the operations in problem 47 and record data on an analysis sheet similar to that shown in Fig. 100.

51. Make a simo chart of the operations listed in problem 47. Use a form similar to that shown in Fig. 101.

CHAPTER 15

52. Make a study of three different methods of collating sets of eight pages of 8½″ x 11″ sheets of paper.

53. Determine the time required to fill the pinboard shown in Fig. 87 under each of the following conditions: (a) pins with bullet nose down are inserted into bevel holes in the board, using simultaneous motions of the two hands; (b) pins with square end down are inserted into holes without bevel, using simultaneous motions of the two hands.

54. Study a spray-gun operator and determine the percentage of time that he sprays the object being painted and the percentage of time that the gun is "spraying air."

55. Describe equipment that might be used to measure the time for fundamental hand motions for research purposes.

CHAPTER 16

56. State in detail the procedure that should be used in designing the front seat of an automobile to provide the greatest comfort (especially for the driver) for cross-country driving.

57. Design the ideal (a) alarm clock, (b) can opener, (c) cork screw, or (d) control panel for an automatic passenger elevator.

58. Study the design of the operating controls and visual displays of any one of the following: (a) clock radio, (b) power lawnmower, (c) motor boat, (d) cement mixer, or (e) television receiver. Prepare a written report on the good and poor design features and recommend improvements.

59. Design an experiment for the evaluation of the two different methods of filling the pinboard as described on page 143. Assume that this is a regular operation performed by people in a factory.

CHAPTERS 17, 18, AND 19

Determine the most economical method of performing the operations described below. Prepare an instruction sheet of the proposed method, showing the motions of the two hands. Include a layout of the work place.

60. An electrical appliance manufacturer has received an order for 50,000 connection plugs for attaching the cord to an electric iron. The plugs are to be shipped in equal installments of 2000 per day. Determine the most economical method of making the final assembly of this unit.

61. The Miller Refrigerator Company has received an order for 100,000 mounting springs similar to the one shown in Fig. 368. Determine the most economical method of assembling the parts.

62. A cabinet manufacturer uses several round-headed wood screw and washer assemblies (Fig. 369) in the final assembly of one of his products. Orders on hand show that 100,000 of these assemblies will be needed each month for the next 6 months. On this basis (a total of 600,000 screw and washer assemblies) determine the most economical method for making the assemblies.

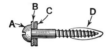

Fig. 369. Wood screw and washer assembly: A, wood screw $\frac{3}{16}'' \times 2\frac{1}{4}''$; B, steel washer $\frac{1}{2}''$ outside diameter, $\frac{7}{32}''$ diameter of hole, $\frac{1}{16}''$ thick; C, fiber washer; D, point of the screw coated with beeswax for $1''$.

63. A total of 500 toy tops similar to the one shown in Fig. 370 is to be manufactured on ordinary power-driven woodworking tools. The parts are assembled by forcing the center pin through the hole in the two body pieces. Friction holds the parts together. Determine the method, design the necessary jigs and fixtures, and make an operation sheet for the manufacture of the three parts, and for the assembly of the parts, as follows:

(*a*) Center pin—Part No. 100: Material, ¼-inch dowel in 3-foot lengths.

(*b*) Small body piece—Part No. 105: Material, 1-inch dowel in 3-foot lengths; ⁷⁄₃₂-inch center hole to be drilled in each piece.

(*c*) Large body piece—Part No. 110: Material, 2⅜-inch awning pole in 8-foot lengths; ⁷⁄₃₂-inch center hole to be drilled in each piece.

64. An order has been received for 10,000 toy ships similar to the one shown

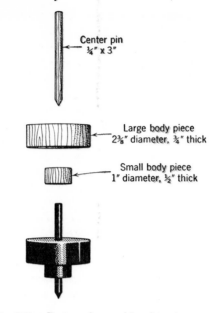

Fig. 370. Parts and assembly of toy top.

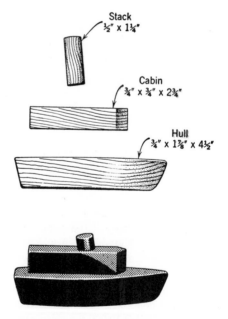

Fig. 371. Parts and assembly of toy ship.

in Fig. 371. These ships are to be made of wood and are to be delivered without paint or other finish.

(a) Design the jigs and fixtures needed to manufacture the three parts and to make the assembly of these parts. Assume that ordinary power-driven woodworking tools are available. The assembly is performed by forcing the dowel (stack) through the hole in the cabin and into the hull. Friction holds the parts together. Material for the stack will be supplied in the form of dowels ½ inch in diameter, in 3-foot lengths; pine strips for the cabin, ¾″ x ¾″ x 10′ long; and pine strips for the ship hull, ¾″ x 2¾″ x 10′ long. All exposed surfaces of all parts are to be sanded, and all sharp edges are to be lightly sanded.

(b) Construct an operation sheet for manufacturing each of the three parts of the top ship, and for the assembly of the parts.

(c) Construct the jigs and fixtures which you have designed in (a).

(d) Using time study, determine the standard time for each operation and compute the direct labor cost for manufacturing the toy ship, using the prevailing hourly wage rate in your community for this type of work.

(e) Determine the total factory cost for manufacturing the toy ship, using the labor cost obtained in (d). Material costs are as follows: ½-inch dowel, 2 cents per foot; ¾″ x ¾″ pine strip, 4 cents per foot; and ¾″ x 2¾″ pine strip, 11 cents per foot. The overhead or burden rate is 100% of the direct labor cost.

65. At the present time, empty glass jars (pint size) to be filled with pickles are placed on a narrow belt conveyor and passed through a machine which automatically adds a measured amount of vinegar to each jar. The jars then pass in front of the first group of two operators, who add chopped onions with a measuring spoon from a container located directly in front of them. The partly filled jars then pass in front of a second group of two operators, who add spices. These operators work with both hands and add spices to two jars at a time. The jars then pass on to an automatic machine where they are filled with pickles and are sealed. Assume that there is enough work to keep two lines of two operators supplied continuously for 6 months during the year, 8 hours per day, 5 days per week. Develop a better method of adding chopped onions and spices to the jars to be filled with pickles.

66. Zinc alloy is cast into bars weighing approximately 11 pounds each. Eighty-eight of these bars are stacked on a skid platform and strapped into place with steel strapping for delivery to the customer. The zinc alloy bars are used mainly for making die castings. The customer returns the empty skids to the supplier. Devise a method that will eliminate the need for the skid platforms.

67. Make an eye-hand simo chart of an operation filmed in problem 47.

CHAPTER 20

68. Visit a factory or office where automatic or semiautomatic equipment has been installed during the past year or two. Obtain facts concerning the following: (*a*) the extent to which the manual part of the operation had been improved before it was mechanized; (*b*) the number of different kinds or makes of equipment considered before the purchase was made; (*c*) the method of evaluation of (1) production or output per hour, (2) unit cost of product, (3) cost of maintenance of equipment, and (4) obsolescence cost or depreciation of equipment; (*d*) the kind of information given to employees before, during, and after installation of equipment and the manner in which this information was given; (*e*) if equipment has been installed long enough, the immediate and long-range effect of the introduction of the new equipment on employment in the department and in the plant.

69. Obtain information for a specific factory operation from some company or trade association, showing the relationship between (*a*) output per man-hour, (*b*) hourly base rate for the worker on this operation, and (*c*) the unit labor cost for the operation at intervals during the past 25 or 50 years. Determine insofar as possible the extent to which the changes noted have resulted from mechanization.

CHAPTER 21

70. Prepare a written standard practice for problems 60, 61, and 62.

71. The London plant of an American company is about to begin manufacturing metal boxes similar to the one shown in Fig. 73. The manager of the London plant has asked for a description of the method used by the parent plant. (*a*) Outline the essential information that should be included in a motion picture of a simple "blank and draw" operation required for making such a can. (*b*) Prepare the supplementary written data that should accompany the film.

CHAPTER 22

72. Give the arguments that are often presented in favor of time wage rather than an incentive wage for paying factory workers.

73. State the several ways in which a company might benefit from a good wage incentive system for direct labor. The company is well managed and employs approximately 1500 people on direct factory work. For the most part the operations are short cycle, repetitive, and operator controlled.

74. Interview five of your acquaintances who are now working in a factory or an office. Analyze their comments on the subject of motion and time study.

75. You are the foreman of the assembly department of a plant manufacturing electrical supplies. A program of motion and time study has been in successful operation in your department for 2 years. One of your best employees asks you if the results of this work will not mean fewer jobs and less work for the employees in the plant. How will you answer?

76. Fifty people are employed in one department on short-cycle manual operations of several different kinds. An individual wage incentive plan is used and there are no restrictions on output. Time standards are established by stop-watch time study. Thirty per cent of the operators in this department regularly have index numbers (performance level) of from 175 to 200% (100% = normal performance). The superintendent suggests that these high index numbers are due to incorrect time standards.

(a) State three other fatcors that might account for the high performance index of this group. Discuss each fully.

(b) What procedure might be used to determine whether the time standards are incorrect?

77. The manager of a local nonunionized farm implement plant employing 150 factory workers recently stated, "We want our employees to feel that they 'belong' to the company team. We are now encouraging the individual worker to determine his own work methods and we are encouraging groups (the employees in a department) to rearrange the equipment in the department and to change other work details to suit themselves, the objective being to give the employees real participation in the operation of the plant."

(a) Discuss fully the advantages and disadvantages to management and to the workers of a plan whereby factory workers "manage their own jobs," in comparison to a plan whereby trained specialists (staff men) play a major part in "engineering the job." Present your own views in answering this question.

(b) From your knowledge of F. W. Taylor, how do you think he would have answered the question?

CHAPTERS 23, 24, 25, AND 26

Make a stop-watch study of the following operations Use the "average" method of selecting the time, and include proper allowances. Make an instruction sheet for the operation.

78. Assembling parts of some small article such as a plug for attaching an electric iron.

79. Drilling a hole in a small piece held in a jig.

80. Turning a piece in a lathe.

81. Milling a piece strapped to the table.

82. Time 10 cycles of any one of the operations referred to above. Using Table 14, determine the number of readings required for each element of the time study. Use ±5% precision and 95% confidence level.

83. Time 32 cycles of the operation referred to above. Using the curves in Fig. 224, determine the number of readings required for each element.

84. From the alignment chart in Fig. 227 determine the number of readings required for each element of the study in problem 83.

85. Draw a control chart for each element of the study in problem 83.

86. The time study shown below was made of the operation "assemble and rivet flanges to hub of metal spool." This operation consisted of two elements as follows:

Element 1—Assemble Flanges to Hub. Pick up hub in right hand, flange in left hand. Position hub to flange and place assembly over pin on fixture. Pick up second flange and position to hub with both hands. Remove assembly and dispose to turntable.

.14	.15	.14	.20*	.15	.20	.18	.17	.19	.18	.14	.17
.19	.13	.15	.17	.17	.19*	.14	.17	.18	.16	.14	.16
.13	.19	.14	.13	.14	.17	.12	.13	.14	.18*	.14	

* Flange sticking.

Element 2—Rivet. Reach to spool on turntable with right hand as left removes and disposes spool to chute. Pick up spool with right hand and move to left hand. Grasp with left hand. Positions on hammer with both hands and rivet with foot-controlled riveting machine.

.09	.06	.06	.06	.05	.07	.06	.09	.07	.07	.06	.07
.07	.07	.07	.08	.07	.07	.07	.07	.07	.07	.07	.06
.06	.07	.06	.07	.06	.07	.07	.07				

Use the alignment chart in Fig. 227 to determine the number of observations required for each element of this study—desired precision ±5%, confidence level 95%.

87. Make a pinboard and pins according to the drawing shown in Fig. 372 and try the experiment described below.

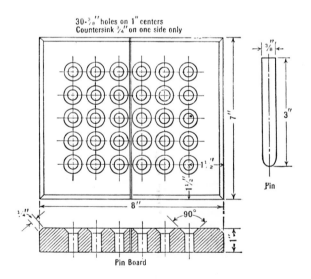

Fig. 372. Details for making pinboard and pins.

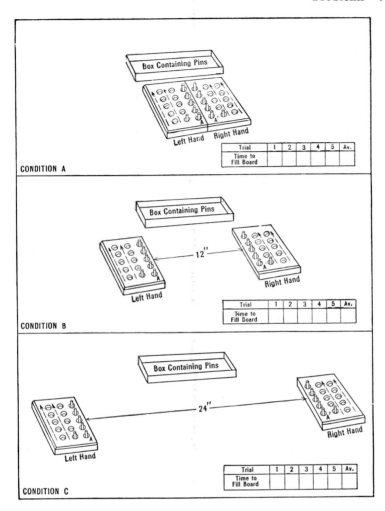

Fig. 373. Arrangement of work place for pinboard study.

(a) Determine the time required to fill the 30 holes in the board with 30 pins under each of the three conditions indicated in Fig. 373. Time five consecutive cycles and take the average.

(b) Determine the number of pinboards that could be filled in an 8-hour day under each of the three conditions. Assume that an operator could maintain the pace used in the experiment and that no fatigue or delay allowances were made.

(c) Calculate in percentage how much more time was required to fill the pinboard under condition B than A; under condition C than A.

(*d*) Compute the total distance in feet through which the two hands would move in filling 1000 pinboards under each of the three conditions.

(*e*) Calculate in percentage how much farther the hands would move under condition *B* than *A*; under condition *C* than *A*.

88. Make a study of three or more different types of self-service retail grocery store check-out counters found in your community. Decide on a "standard" or "average" order size. For each of the different types of check-out counter, determine the time for the following: (1) sort, (2) ring up, (3) take money, (4) make change, (5) position bag (6) bag merchandise, and (7) other work.

(*a*) Determine the production in orders per hour for one person working on the check-out counter, for two persons, and for three persons.

(*b*) Using the prevailing hourly wage rate for these people in the community, determine the cost in cents per order for each of the conditions in (*a*).

89. Draw a frequency distribution curve of stature of a typical group of men. Use "number of men" in each height range as ordinate, and "height in inches" (in groups of 1 inch) as abscissa.

Height in Inches	Number of Men	Height in Inches	Number of Men
59	1	69	141
60	3	70	118
61	5	71	90
62	11	72	58
63	22	73	30
64	41	74	15
65	69	75	6
66	103	76	3
67	133	77	1
68	150		

90. Determine the actual walking speed for men and women. Measure off 50 feet on a smooth, level sidewalk. Then from a point where you can see clearly this 50-foot section of sidewalk, with a decimal-minute stop watch determine the time required by individuals to walk this 50-foot distance. Obtain data on people walking singly rather than in groups. Also make the following classifications of your data: first, men and women; second, three age groups—15 to 18, 18 to 50, 50 to 70; and then under each of these age groups further subdivide the people as to height—short, medium, and tall. The data might be recorded on the form shown in Fig. 374.

91. Make an 8-hour time study of a janitor, setup man, material handler, or other person on indirect labor.

92. The Merit Toy Company installed a power-driven turntable on its bench sander on July 20. This turntable was designed to enable the operator to sand 90 toy ship hulls per hour instead of 60 using the former hand method

Men ☐   Women ☐   Place   Date   Temperature   Humidity

Age 15 to 18			Age 18 to 50			Age 50 to 70		
Short	Medium	Tall	Short	Medium	Tall	Short	Medium	Tall

Fig. 374. Data form for Problem 90.

(see problem 64). This attachment consists of an electric motor-driven turntable rotating at 1 r.p.m. The turntable or dial is 19 inches in diameter and has three pockets or cavities which hold the hull of the toy ship. On July 25 a new time standard was put into effect for the new operation.

Old Method. Previously the operator sanded the two sides and stern of the hull on the disk sander, and sanded the sharp edges all around on both top and bottom of the hull on the belt sander.

Improved Method. The foreman thought of the idea of using a dial attachment, and had the master mechanic build this attachment and install it on the sander. The operator now sits beside the sander and removes and inserts a ship hull in the dial cavity as the dial rotates before him. Since the hull is sanded on one side at a time, the hull must be turned end for end after the first side is sanded. Two small auxiliary motor-driven sanders (hand power drill with rotating sander attachment) are mounted above and below the dial. These two sanders remove the sharp edges on the top and bottom of the hull as it moves past the disk sander. The operator still must sand the stern and break the edges of the top and bottom of the stern end of the hull by hand on the belt sander. However, this operation can be performed during the interval between the loading and unloading of the dial.

The Union has just protested the new time standard and has presented a grievance to the effect that: (1) there has not been a change in method; (2) the new time standard is incorrect; (3) management is not justified in changing the time standard on this operation.

The contract which the Company has with the Union contains the following clauses: (*a*) The Union accepts the principle that incentive pay is compensation voluntarily offered by the Company for extra production of acceptable quality beyond that considered a standard day's work at the basic hourly rate of pay. (*b*) The Union accepts the Company's incentive plan, including all established work standards and elemental standard times, and the Company's principles, techniques, and procedures of industrial engineering. (*c*) The Company shall make no change in work standards, except when there has been a definite change in methods, tools, equipment, specifications, or materials, which affects the time standard in excess of 5%, or by mutual agreement with the Union. (*d*) The question to be determined by an arbitrator shall be limited to whether the new or revised work standard established by the Company for the operation in question was properly established under the Company's industrial engineering principles, techniques, and procedures, or if not, in what respect errors were made thereunder in job elements, basic timing, or calculations. The arbitrator shall only be concerned with variations in excess of 5% from the proposed standard, and shall accept all of the Company's established elemental standard times. The Union shall have the burden of proof.

You, as head of the Methods and Standards Department of this company have been asked by the General Manager to obtain the facts and prepare a report in reply to the Union's grievance. Although your contract with the Union calls for a three-step procedure before a grievance goes to arbitration, this particular grievance has come to the attention of the General Manager in step one and he is especially interested in it. This operation is now running in the shop, so that the present time standard, which was established by men in your own department, can be checked. Since the Union did not protest the time standard on the old method of sanding and since it had been in effect for several years, it can be assumed that the time standard on the old method was correct.

State the procedure you would follow to obtain the facts for the report requested by the General Manager.

93. (*a*) Obtain information about the IBM Automatic Production Recording System."[8] Compare the use of such equipment with other methods of performing a given function in a factory with which you are familiar. (*b*) List the advantages and disadvantages of the IBM Automatic Production Recording System.

94. Visit a factory using the IBM Autorate System. Describe and evaluate the system.

CHAPTER 27

95. Determine the standard time for drilling the part shown on the sketch at the bottom of the observation sheet (Fig. 217) if the piece is 1.750 inches in diameter and the actual drilling time is 0.94 minute. Use time-setting tables for the sensitive drill.

96. Determine the time required to mill the hexagon, using a gang mill, on part 612W-377A (Fig. 256) if the dimension *A* (length of the hexagon) is 1.125 inches and all other dimensions are as shown.

CHAPTER 28

97. Calculate time for cutting teeth on index change gear similar to part 1670 AG (Fig. 259) if length of face is 1.150 inches; diametral pitch (D.P.),

---

[8] "Now—Record Keeping Goes *Really* Automatic," *Factory Management and Maintenance*, Vol. 114, No. 10, pp. 94–97, October, 1956.

16; number of teeth (N), 60; diameter of bore, 1.250 inches; material, 4620; hob HBG 573. Ground-tooth spur gear. Size of order 50 gears.

98. Determine the standard time required for the operation "solder the side seam" of rectangular can with the following dimensions: length, 8.125 inches; width, 1 inch; depth, 8.5 inches.

CHAPTER 29

99. Determine the standard time required to perform operation 4, "work out shape through die block," for the blank for the part shown in Fig. 262. The blank is 2.00 inches square with round corners of $\frac{1}{4}$-inch radius. Quality required is Class B.

CHAPTERS 30 AND 31

100. Make a comparison of motion-time data and stop-watch time study as methods for establishing time standards for use as the basis of a wage incentive plan for direct factory labor on short-cycle repetitive operations.

101. Refer to Fig. 293. If an opening were made in the work bench directly between the fixture and the material container, and if the finished parts were disposed into this opening instead of into the disposal chute at the front edge of the bench, how would this affect the time for this operation? Show the calculations.

102. The apparatus shown in Fig. 375 is used by the Maytag Company for demonstration purposes in connection with training programs in the field of motion and time study. Using any one of the systems of motion-time data, determine the time required per cycle to obtain one dowel and place it in the hopper. Present your results in tabular form.

There are four positions of the hopper along the chute, and there are also four conditions of *get* and *place*, as follows: (1) dowels in box—small opening; (2) dowels in box—large opening; (3) dowel against peg—small opening; (4) dowel against peg—large opening.

Size of wood dowel is $\frac{3}{8}'' \times 3''$, bullet nose on one end. Distance of center of hopper from grasp point: position 1, 30 inches; position 2, 24 inches; position 3, 18 inches; and position 4, 12 inches.

103. Using any one of the systems of motion-time data, determine the standard time to load and unload the fixture shown in Fig. 376. The cast-iron blocks are center-punched for drilling, weigh 3 pounds, and fit loosely in fixture.

104. (*a*) Using any one system of motion-time data, determine the standard time required to "assemble two plates, two washers, bolt, and nut," as shown in Fig. 377. Each plate weighs 1 pound. The arrangement of the work place is shown in Fig. 378, and the sequence of motions for the right and the left hand is given in Fig. 379.

(*b*) If two plates were grasped and transported in one hand, what difference would it make in the assembly time?

(*c*) Would it be advantageous to mount the driver under the bench and drive the nut from below? How might such a setup be arranged?

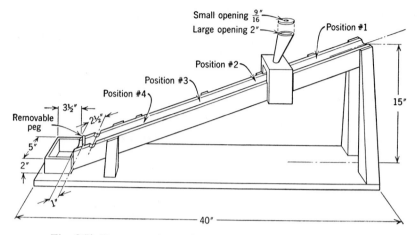

Fig. 375. Demonstration unit—time required for hand motions.

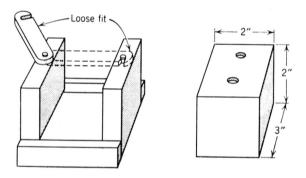

Fig. 376. Fixture for drill-press work.

(d) In the present layout the washer is placed over the hole on the plate. Would it be quicker and easier to assemble the washer to the bolt first? Design a fixture which would make such an arrangement possible. Determine the standard time.

105. Determine the standard time for assembling the two cast-iron plates using the method shown in Figs. 245 and 246 (Chapter 25).

106. (a) Determine the standard time required to "assemble heating element to grid," as shown in Fig. 380. Use any system of motion-time data. The arrangement of the work place is shown in Fig. 381; and the sequence of motions for the right hand and the left hand is given in Fig. 382.

(b) If tray E containing the screws were placed on the left-hand side of the work place, alongside covers at C, what difference would it make

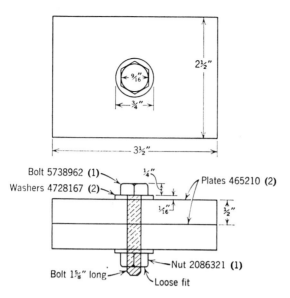

Fig. 377. Plate assembly.

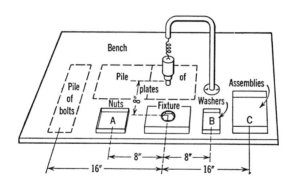

Fig. 378. Arrangement of the work place for the plate assembly.

in the total assembly time? Make chart showing motion sequence with this rearrangement of parts.

CHAPTER 32

107. Work sampling is to be used to measure the down time of a group of presses. A preliminary study shows that the down time is likely to be around 30%. Determine the number of observations required for this study, for a 95% confidence level and a desired accuracy of ±5%. Use the formula in solving this problem.

Operation—Assemble 2 plates, 2 washers, bolt, and nut

Left Hand	Right Hand
Get nut from Bin A T. E. + G. Place nut in fixture (nest). Loose fit T. L. + P. + R. L.	Get washer from Bin B T. E. + G. U. D. Place washer in fixture (nest) over nut T. L. + P. + R. L.
Get plate from pile T. E. + G. Place plate in fixture T. L. + P. + R. L.	Get plate from pile T. E. + G. U. D. Place plate in fixture T. L. + P. + R. L.
Get bolt from pile T. E. + G. U. D. Place bolt in hole. Loose fit T. L. + P. + R. L. U. D.	Get washer from Bin B T. E. + G. Place washer on plate over hole T. L. + P. + R. L. Get power driver T. E. + G. Place driver on bolt head T. L. + P. + R. L.
Get completed assembly T. E. + G. Place aside into Bin C T. L. + P. + R. L.	Use time—drive bolt Dispose of driver T. L. + P. + R. L.

Fig. 379. Sequence of motions for the plate assembly.

108. By means of work sampling determine the working time, and the average performance index of a group of people performing some nonrepetitive activity.

109. *Time* magazine reported[9] that a city councilman made the record of activities of a seven-man crew of electrical maintenance workers shown in Fig. 383. The councilman, believing the operation of the City Electrical Department was inefficient, followed the crew for an entire day to get this information. Using the random sampling technique, determine the percentage of nonworking time for the crew shown in Fig. 383.

---

[9] "Let There Be Light," *Time*, Vol. 60, No. 26, p. 15, December 29, 1952.

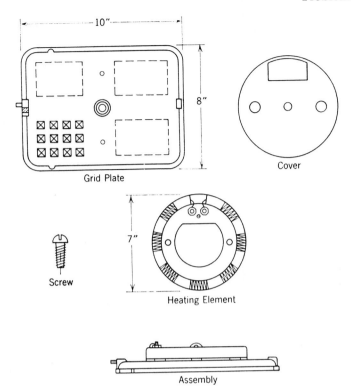

Grid Plate

Cover

Screw

7"

Heating Element

Assembly

Fig. 380. Parts for heating-element assembly.

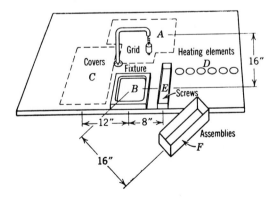

Grid

Heating elements

Covers
C

Fixture

Screws

Assemblies
F

16"

12"   8"

16"

Fig. 381. Arrangement of the work place for the assembly of heating elements.

Operation—Assemble heating element to grid

Left Hand	Right Hand
Get aluminum grid from pile A on bench	Get aluminum grid from pile A on bench
T. E. + G.	T. E. + G.
Place in fixture B. Loose fit	Get heating element from pile D on bench
	T. E. + G.
Get cover from pile C on bench	Place heating element in position on aluminum grid
T. E. + G.	T. L. + P. + R. L.
Place cover over heating element. Loose fit	Get screw from tray E
T. L. + P. + R. L.	T. E. + G.
Get power driver	Place screw in hole. Loose fit
T. E. + G.	T. L. + P. + R. L.
Place driver on screw	(Guide screw)
T. L. + P. + R. L.	
	Use time—drive screw
Place driver aside	Get assembled part
T. L. + P. + R. L.	T. E. + G.
U. D.	Place assembled part aside in tote box F
	T. L. + P. + R. L.

Fig. 382. Sequence of motions for heating-element assembly.

110. One company occasionally asks the foremen of maintenance crews to make what is called a "period study." In making a period study, the foreman observes a group of two, three, or four men working together on a job. The foreman observes the members of the crew for ten or fifteen consecutive 5-minute periods, and records on a simple data sheet whether each member of the crew is working or idle more than half of each 5-minute period. For example, if man No. 1 worked more than $2\frac{1}{2}$ minutes during a 5-minute period, the foreman would place a tally mark for this man under "working"; if the man was idle more than $2\frac{1}{2}$ minutes during the 5-minute period, he would receive a tally marked under "idle." The foreman tries to determine from observation (he does not use a watch) whether each member of the crew worked more than half of each 5-minute period. The percentage of the total study period that each member of the crew is working and is idle is determined in the same way as for a regular work sampling study. The main purposes of the period study are to cause the foreman to really look at the activities of each member of a crew, and to better determine the proper crew size.

Make an investigation to compare this type of period study with the conventional work sampling study and with a continuous time study of some activity similar to maintenance work, involving two or three crew members.

111. Make a performance sampling study of the workers in one department of a factory for a period of a week.

112. What can be done to convince employees that work sampling can be used satisfactorily for measuring working time and nonworking time of men and machines?

CHAPTERS 33 AND 34

113. Determine the physological cost of walking, using heart rate in beats per minute before and after walking on a smooth level surface at four different speeds.

(a) Have subject sit in chair for 5 minutes. Record his heart rate in beats per minute for a period of $\frac{1}{2}$ minute.

(b) Have subject walk $2\frac{1}{2}$ miles per hour for a period of 5 minutes, then have subject sit in chair and record his heart rate in beats per minute for the second half of the first minute, the second half of the second minute, and the second half of the third minute after walking ends.

(c) Repeat (b) at walking speeds of 3, $3\frac{1}{2}$, and 4 miles per hour. Allow time between trials for heart rate to return to normal.

Plot curves showing "Heart rate in beats per minute" as the ordinate and "Minutes after work" as the abscissa.

114. Determine the physiological cost of handling brick at four different speeds, using heart rate in beats per minute.

(a) Have subject sit in chair for 5 minutes. Record his heart rate in beats per minute for a period of $\frac{1}{2}$ minute.

(b) Have subject pick up bricks one at a time from the floor and stack them on a bench 33 inches high at a speed of 16 bricks per minute. Have subject work for a period of 5 minutes, then have subject sit in chair and record his heart rate in beats per minute for the second half of the first minute, the second half of the second minute, and the second half of the third minute after work ends.

(c) Repeat (b) at working speeds of 22, 28, and 34 bricks per minute. Allow time between trials for heart rate to return to normal.

Plot curves showing "Heart rate in beats per minute" as the ordinate, and "Minutes after work" as the abscissa.

115. What are the main criticisms of the common definition of fatigue?[10]

[10] E. Mayo, *The Human Problems of an Industrial Civilization,* Macmillan Co., New York, 1933.

	ELAPSED MINUTES	MEN WORKING	MAN-MINUTES OF WORK	PER CENT NON-PRODUCTIVE TIME
8:30 a.m.: Starting time. 8:34 a.m.: First maintenance truck leaves city garage with two men aboard. 9:40 a.m.: Truck stops at 2020 West Cullerton.	70	2	140	71
One man apparently siphons gas from truck into gas can and puts it in another car.	17	0	0	100
9:57 a.m.: Truck proceeds to Maplewood & Flournoy, meets five men, who drive up in own cars.	5	2	10	71
10:02 a.m.: Two men put ladder against pole. Others do nothing.	3	2	6	71
10:05 a.m.: One man ascends pole to attach rope at top. Others do nothing.	18	1	18	86
10:23 a.m.: One man starts painting base of pole. Man on pole erects pulley arrangement to enable him to get can of paint to top of pole without carrying it.	6	2	12	71
10:29 a.m.: Three men go to other car and drive off. Only one man working.	18	1	18	86
10:47 a.m.: Three men come back in their car.	7	1	7	86
10:54 a.m.: Cars and truck leave.	31	2	62	71
11:25 a.m.: New location, School & Ravenswood.	1	2	2	71
11:26 a.m.: Equipment unloaded. One man digging, others watching.	19	1	19	86
11:45 a.m.: Six off for lunch. Go to tavern nearby.	15	1	15	86
12:00 a.m.: Last man to lunch.	24	0	0	100
30 MINUTE LUNCH PERIOD				
12:54 p.m.: Six men return from tavern and one resumes digging.	9	1	9	86
1:03 p.m.: One man ascends pole and detaches electric wire. One man digs.	5	2	10	71
1:08 p.m.: One descends pole.	29	1	29	86
1:37 p.m.: Two men working to remove pole.	12	2	24	71
1:49 p.m.: Two men remove pole, using pulley; put in new pole.	13	2	26	71
2:02 p.m.: Two men tamp dirt.	2	2	4	71
2:04 p.m.: Two men go to tavern.	11	0	0	100
2:15 p.m.: One man on top of pole, attaching wire.	6	1	6	86
2:21 p.m.: One man on top of pole, one painting base of pole.	4	2	8	71
2:25 p.m.: Man on pole working, others in truck.	15	1	15	86
2:40 p.m.: One man painting pole—three in tavern—three in truck.	8	1	0	86
2:48 p.m.: Man descends pole, puts ladder back in truck. Nobody working.	4	1	4	86
2:52 p.m.: Another man leaves for tavern; no one working.	2	0	0	100
2:54 p.m.: Last three men leave for tavern. All seven men in tavern now. Truck unattended though motor is running, as it has been all day.	18	0	0	100
3:12 p.m.: Mass exodus from tavern.	7	0	0	100
3:19 p.m.: Truck drives off. Other men get in cars and leave.	71	2	142	71
4:30 p.m.: Quitting time.	480		594	81%

Fig. 383. All-day study made by a city councilman showed 81% nonproductive time for the members of a seven-man crew of electrical maintenance workers. (Reproduced, with permission, from *Time*, December 29, 1952.)

116. Obtain information showing the relationship between heart rate in beats per minute, oxygen consumption, and pulmonary ventilation, before, during, and immediately following heavy physical exercise.[11]

117. Study the reports of (a) the Industrial Health Research Board[12] and (b) the National Institute of Industrial Psychology[13] in Great Britain, and present a summary of the nature of the work of these two organizations.

CHAPTER 36

118. Work out a motion and time study training program for foremen and supervisors for a specific plant in your locality.

119. Make a rating study of walking.

Object

To obtain group practice in rating operator performance.

Equipment and materials needed.

1. Decimal-minute stop watch, steel tape, chalk, string.

2. Rating forms: B204 (Fig. 384).

Place

Select a room with smooth level floor, or use a smooth level sidewalk.

Procedure for conducting rating study.

(1) Measure off 50 feet of unobstructed floor space, marking a starting and a stopping line on the floor and allowing 10 or 15 feet of additional space at either end for the operator to start and stop. Tie a string to the back of one chair and throw it loosely across the back of another chair so that the string is stretched directly above the starting line on the floor. Also place a string across two chairs at the other end of the 50 feet. These strings are to help the operator to start his watch and read it at the proper instant.

(2) Have someone (this person will be called the operator) practice walking the 50 feet at exactly 3 miles per hour. This practice should take place before the group assembles. The operator should take 0.189 minute to walk the 50 feet. After a little practice it will not be difficult for the operator to walk the 50 feet in this time or at a speed of 3 miles per hour. Table 86 shows the times needed for the operator to walk 50 feet at other speeds.

(3) Provide each person with a data sheet similar to Form B204 (Fig. 384). Have him fill in his name and the date at the bottom of this form. No stop watch or clock is to be used by member of the group in this study.

(4) The operator then walks at 3 miles per hour, and the group is told that this speed represents a rating of 100%. Two or three trials are made

[11] Peter V. Karpovich, *Physiology of Muscular Activity*, 4th ed., W. B. Saunders Co., Philadelphia, 1953.

[12] Reports published by H. M. Stationery Office, London.

[13] Official monthly publication, *The Human Factor*, Aldwych House, London, W.C. 2.

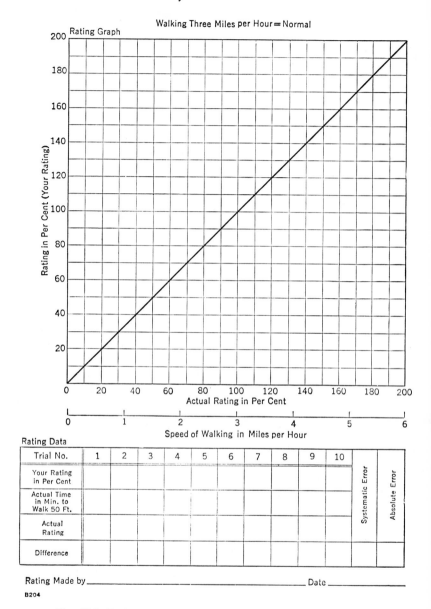

Fig. 384. Performance rating data sheet and graph for walking.

Table 86. Coversion of Watch Readings to Walking Speeds and
Performance Ratings

Time in minutes to walk 50 ft.	.120	.125	.130	.135	.140	.145	.150	.155	.160	.165	.170	.175	.180	.185
Actual speed in miles per hour	4.72	4.54	4.35	4.20	4.05	3.91	3.78	3.66	3.54	3.44	3.34	3.24	3.15	3.06
Rating in % (3 m.p.h. = 100%)	158	151	145	140	135	130	126	122	118	115	111	108	105	102

Time in minutes to walk 50 ft.	.189	.190	.195	.200	.205	.210	.215	.220	.225	.230	.235	.240	.245	.250
Actual speed in miles per hour	3.00	2.98	2.91	2.84	2.77	2.70	2.64	2.58	2.52	2.47	2.41	2.36	2.31	2.27
Rating in % (3 m.p.h. = 100%)	100	99	97	95	92	90	88	86	84	82	80	79	77	76

at this speed. The operator times himself, and if he takes more or less than 0.189 minute to walk the 50 feet, he informs the group of that fact and immediately determines the actual speed in per cent, which he gives to the group. The group, of course, makes no record of these preliminary trials.

(5) The operator then walks the 50 feet ten different times, called trials, varying his speed at random. At the end of each trial he records on his data sheet the actual time it took him to walk the 50 feet, and the corresponding rating in per cent. This information ordinarily is not given to the group until all ten trials are finished, although the correct ratings may be announced immediately after each trial. The walking speeds should fall between approximately 2.5 miles per hour (85% of normal) and 4.5 miles per hour (150% of normal), inasmuch as working speeds in practice are usually within these limits. It is considered more difficult to rate accurately when extremes are encountered.

(6) Each person watches the operator walk the 50 feet and rates him, using 100% = 3 miles per hour as normal. Each trial is recorded in per cent on the first horizontal line at the bottom of Form B204.

(7) Then read the correct ratings in per cent, and ask each person to copy these ratings on the third horizontal line at the bottom of Form B204.

(8) Ask each person to plot his ratings on the rating graph. Each person should then draw a straight line through the average position of these points.

(9) Compute the systematic error, mean deviation, and absolute error for each person and for the group.[14]

[14] Ralph M. Barnes, *Work Measurement Manual,* 4th ed., College Book Company, Los Angeles, Calif. 90024, 1951, pp. 91–97.

(10) Repeat this walking experiment each week until there is no further improvement. Use the same person as the subject or "operator" throughout the experiment.

120. Make rating study of dealing cards.

Object

To obtain practice in rating operator performance.

Equipment and materials needed

1. Decimal minute stop-watch, deck of cards, card table.
2. Rating forms: B205 (Fig. 385).

Place

Select a room large enough to accommodate the group of people who will participate in the study.

Procedure for conducting rating study

(1) Have someone (this person will be called the operator) practice dealing the deck of cards in four equal piles in ½ minute. Another person, called the timer, will by means of a stop watch time the operator and record the total time for dealing the deck. If the operator takes more or less than ½ minute to deal the deck, the timer informs the group of this fact and immediately determines the actual speed in per cent, which he gives to the group. The group, of course, makes no record of the preliminary trials. The operator is seated and deals a standard deck of 52 cards in the following way. The deck is held in the left hand and the top card is positioned with the thumb and index finger of the left hand. The right hand grasps the positioned card, carries it, and tosses it onto the table. The four piles of cards are arranged on the four corners of a 1-foot square. The only requirements are that the cards shall all be face down and that each of the four piles shall be separate from the others. Care should be used to make certain that the method does not deviate from this as the speeds are varied. After a little practice, the operator can deal the cards in exactly ½ minute or at 100% rating. Table 87 shows the time required to deal the cards at other speeds.

(2) Provide each person with a data sheet similar to Form B205 (Fig. 385). Have him fill in his name and the date at the bottom of this form. No stop watch or clock is to be used by members of the group in this study.

(3) The operator then deals the deck of cards in ½ minute, and the group is told that this speed represents a rating of 100%. Two or three trials are made at this speed.

(4) The operator then deals the deck ten different times, called trials, varying the speed at random. At the end of each trial the timer records the time and shows it to the operator, but does not give this information to the group until after all ten ratings have been made. The speeds should fall between approximately 85% of normal (dealing deck in 0.588 minute)

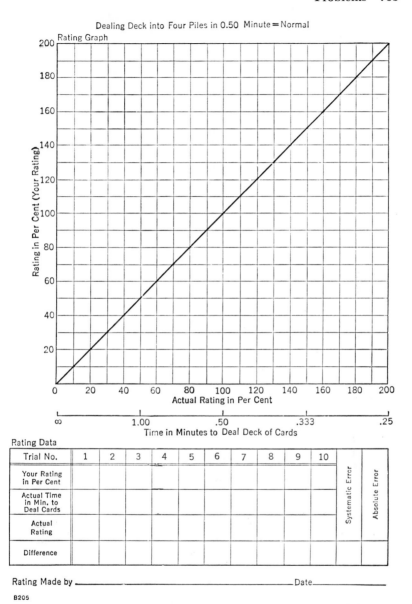

Fig. 385. Performance rating data sheet and graph for dealing cards.

Table 87. Conversion of Watch Readings to Performance Ratings for Dealing Cards

Time in minutes to deal deck of cards	.313	.314	.316	.318	.321	.323	.325	.327	.329	.331	.333	.336	.338
Rating in per cent .500 min. = 100 %	160	159	158	157	156	155	154	153	152	151	150	149	148

Time in minutes to deal deck of cards	.340	.342	.345	.347	.350	.352	.355	.357	.360	.362	.365	.368	.370
Rating in per cent .500 min. = 100 %	147	146	145	144	143	142	141	140	139	138	137	136	135

Time in minutes to deal deck of cards	.373	.376	.379	.382	.385	.388	.390	.394	.397	.400	.403	.407	.410
Rating in per cent .500 min. = 100 %	134	133	132	131	130	129	128	127	126	125	124	123	122

Time in minutes to deal deck of cards	.413	.417	.420	.424	.427	.431	.435	.439	.442	.446	.450	.455	.459
Rating in per cent .500 min. = 100 %	121	120	119	118	117	116	115	114	113	112	111	110	109

Time in minutes to deal deck of cards	.463	.467	.472	.476	.481	.485	.490	.495	.500	.505	.510	.515	.521
Rating in per cent .500 min = 100 %	108	107	106	105	104	103	102	101	100	99	98	97	96

Time in minutes to deal deck of cards	.526	.532	.538	.543	.549	.556	.562	.568	.575	.581	.588	.595	.602
Rating in per cent .500 min. = 100 %	95	94	93	92	91	90	89	88	87	86	85	84	83

Time in minutes to deal deck of cards	.610	.617	.625	.633	.641	.649	.658	.667	.676	.685	.694	.704	.714
Rating in per cent .500 min. = 100 %	82	81	80	79	78	77	76	75	74	73	72	71	70

and 150% of normal (dealing deck in 0.33 minute), inasmuch as working speeds in practice are usually within these limits. It is considered more difficult to rate accurately when extremes are encountered.

(5) Each member of the group watches the operator deal the deck of cards and rates him, using 100% = ½ minute as normal. Each trial is recorded in per cent on the first horizontal line at the bottom of Form B205.

(6) Then read the correct ratings in per cent and ask each person to copy these ratings on the third horizontal line at the bottom of Form B205.

(7) Ask each person to plot his ratings on the rating graph. Each person should then draw a straight line through the average position of these points.

(8) Compute the systematic error, mean deviation, and absolute error for each person and for the group.

(9) Repeat this card-dealing experiment each week until there is no further improvement. Use the same person as the subject or "operator" throughout the experiment, and make certain that the same method of dealing the cards is used.

CHAPTER 36

121. Discuss the advantages and the disadvantages of training the operator at the machine versus training in a separate training school.

122. Indicate the training that would be given to a new employee beginning work on operations described in problems 60, 61, and 62.

123. What would be the ultimate effect on the personnel of an organization of a training program in motion study and methods design for every new employee?

CHAPTER 37

124. Obtain the following information from two or three factories which have workers on incentive on operations containing machine-paced elements.

(a) How are the workers paid for the machine-paced part of the operation?

(b) What plan would each company prefer to use if it was to install an entirely new work measurement and wage incentive system?

125. (a) Determine the total factory cost per hour to operate one unit or one battery of automatic machines.

(b) Determine the cost per piece or per unit of product produced for each of the following: (1) direct labor, (2) indirect labor, (3) direct material, and (4) all other factory costs including overhead.

126. The crew referred to on page 647 produced 162,400 lineal feet of fiberboard, 125-pound test, with 1.25% waste in one 8-hour day. Time allowed for setup in the morning was 0.07 hour. Determine the earnings for each man for the day, using the base hourly rate and the standard data given in the text.

127. (a) Determine the total efficiency for the Capsule Sorting Department for one pay period during which the following conditions existed:

(1) Group total standard minutes for the pay period, 171,160.
(2) Group total actual minutes for the pay period, 172,230.
(3) Factor II, outgoing quality level index, 0.53.
(4) Factor III, reject quality level index, 93.9.

(b) Determine the earnings for Helen Smith, a sorting operator, for the pay period referred to in (a). Use that table and other data given in the text.

(1) Hourly base rate of Helen Smith, $2.50.
(2) Hours worked by Helen Smith during pay period, 80.

CHAPTER 38

128. Give a résumé of the results of the "Hawthorne Experiment" at the Western Electric Company.[15]

129. Review and evaluate the contributions made by Douglas McGregor[16] and Rensis Likert.[17]

130. Make a critical evaluation of the current program of a company which has a formal plan for motivating its factory workers. If you are not able to make a first hand study of such an organization make your evaluation from available literature.[18]

CHAPTER 39

131. Make a study of five different paced assembly lines and determine the imbalance, that is, the "waiting time" for the operator having the lightest load or shortest task. Present several specific ways of reducing the imbalance in each case.

132. Study and evaluate the several arguments which have been made against division of labor and job specialization in the factory and the office.

133. Investigate an actual situation where a paced assembly line was installed and make a comparison of each cost item before and after the installation.

134. Make a study of the method and time required to wash a car using three different systems or degrees of mechanization. Determine the preferred system and prepare an appropriate report.

135. Design a Job Enlargement Program for (a) a supermarket, (b) a commercial bank, or (c) telephone installers and maintenance personnel.

[15] T. N. Whitehead, *Leadership in a Free Society,* Harvard University Press, Cambridge, Mass., 1937; T. N. Whitehead, *The Industrial Worker,* Two Volumes, Harvard University Press, Cambridge, Mass., 1938; F. J. Roethlisberger and W. J. Dickson, *Management and the Worker,* Harvard University Press, Cambridge, Mass., 1940; Henry A. Landsberger, *Hawthorne Revisited,* Cornell University, Ithaca, N. Y., 1958.

[16] Douglas McGregor, *The Human Side of Enterprise,* McGraw-Hill Book Co., New York, 1960.

[17] Rensis Likert, *New Patterns of Management,* McGraw-Hill Book Co., New York, 1961; Rensis Likert, *The Human Organization,* McGraw-Hill Book Co., New York, 1967.

[18] M. Scott Myers, "Who Are Your Motivated Workers?" *Harvard Business Review,* Vol. 42, No. 1, pp. 73–88, January–February, 1964; E. R. Gomersall and M. Scott Myers, "Breakthrough in On-The-Job Training," *Harvard Business Review,* Vol. 44, No. 4, pp. 62–72, July–August, 1966.

CHAPTERS 40 AND 41

136. Which phases of "The Individual Rate-Performance Premium Payment Plan" and "The Lakeview Plan" make use of the motivators and which make use of the maintenance factors.

137. If you were to design an ideal management system and employee payment plan for a company such as the Kodak Park Works of Eastman Kodak, how would it differ from The Individual Rate-Performance Premium Payment Plan?

138. If you were an employee what do you think you would like most and least about (a) The Individual Rate-Performance Premium Payment Plan and (b) The Lakeview Plan?

139. State some of the main differences between the worker-supervision relationship in the Test Room-Hawthorne Experiment of the Western Electric Company and at the ABC Company's operation under The Lakeview Plan.

140. State some of the main reasons why industry is moving at such a slow rate in applying the research findings of social and behavioral scientists in the field of motivation.

Illustrations from Other Books by the Author

The following illustrations were taken from *Work Methods Manual*, by Ralph M. Barnes, published by John Wiley & Sons, Inc., New York: Figures 17, 18, 25, 26, 27, 30, 56, 57, 58, 59, 60, 61, 62, 64, 65, 68, 71, 72, 73, 84, 85, 86, 87, 88, 130, 131, 132, 133, 134, 135, 141, 142, 152, 153, 154, 156, 157, 158, 166, 167, 178, 179, 245.

From *Work Methods Training Manual*, 2nd ed., by Ralph M. Barnes, published by Wm. C. Brown Co., Dubuque, Iowa: Figures 235 and 327.

From *Work Measurement Manual*, 3rd ed., Ralph M. Barnes, published by Wm. C. Brown Co., Dubuque, Iowa: Figures 319, 320, 322, 323.

Bibliography

Abruzzi, Adam, *Work, Workers and Work Measurement*, Columbia University Press, New York, 1956, 318 pages.

Abruzzi, Adam, *Work Measurement*, Columbia University Press, New York, 1952, 290 pages.

Aitken, H. G. J., *Taylorism at Watertown Arsenal*, Harvard University Press, Cambridge, Mass., 1960, 269 pages.

Alford, L. P., *Henry Laurence Gantt*, American Society of Mechanical Engineers, New York, 1934, 315 pages.

Allen, C. R., *The Instructor, the Man and the Job*, J. B. Lippincott Co., Philadelphia, 1919, 373 pages.

Amar, Jules, *The Human Motor*, George Routledge & Sons, London, 1920, 470 pages.

Amar, Jules, *The Physiology of Industrial Organization and the Reemployment of the Disabled*, Macmillan Co., New York, 1919, 371 pages.

Amber, G. H., and P. S. Amber, *Anatomy of Automation*, Prentice-Hall, Englewood Cliffs, N. J., 1962, 245 pages.

American Institute of Industrial Engineers, *Industrial Engineering Terminology Manual*, *Journal of Industrial Engineering*, Vol. 16, No. 6, November–December, 1965.

American Institute of Industrial Engineers, *Proceedings of the Annual Conference*, annually since 1955, 345 East 47th Street, New York, N. Y. 10017.

American Society of Mechanical Engineers, *Fifty Years Progress in Management*, New York, 1960, 329 pages.

American Society of Mechanical Engineers, *ASME Standard Industrial Engineering Terminology*, New York, 1955, 48 pages.

American Society of Mechanical Engineers, *Manual on Cutting of Metals*, 2nd ed., New York, 1952.

American Society of Mechanical Engineers, *ASME Standard Plant Layout Templates and Models*, New York, 1949.

American Society of Mechanical Engineers, *ASME Standard Operation and Flow Process Charts*, New York, 1947.

Anyon, G. J., *Collective Agreements on Time and Motion Study*, Society for Advancement of Management, New York, 1954, 46 pages.

Apple, J. M., *Plant Layout and Materials Handling*, Ronald Press Co., New York, 1950, 367 pages.

Argyris, Chris, *Organization and Innovation*, Richard D. Irwin, Homewood, Ill., 1965, 274 pages.

Argyris, Chris, *Integrating the Individual and the Organization*, John Wiley & Sons, New York, 1964, 330 pages.

Argyris, Chris, *Interpersonal Competence and Organizational Effectiveness*, The Dorsey Press, Homewood, Ill., 1962, 292 pages.

Babbage, Charles, *On the Economy of Machinery and Manufactures*, 4th ed., Charles Knight, Pall Mall, East, London, 1835, 408 pages.

Backman, Jules, *Wage Determination*, D. Van Nostrand Co., Princeton, N. J., 1959, 316 pages.

Bailey, G. B., and Ralph Presgrave, *Basic Motion Timestudy*, McGraw-Hill Book Co., New York, 1958, 195 pages.

Bailey, N. R., *Motion Study for the Supervisor*, McGraw-Hill Book Co., New York, 1942, 111 pages.

Bainbridge, F. A., *The Physiology of Muscular Exercise*, 3rd ed., rewritten by A. V. Bock and D. B. Dill, Longmans, Green & Co., New York, 1931, 272 pages.

Barish, Norman N., *Systems Analysis for Effective Administration*, Funk & Wagnalls Co., New York, 1951, 316 pages.

Barnes, Ralph M., *Industrial Engineering Survey*, University of California, Los Angeles, 1967, 16 pages.

Barnes, Ralph M., *Motion and Time Study Problems and Projects*, 2nd ed., John Wiley & Sons, New York, 1961, 232 pages.

Barnes, Ralph M., *Motion and Time Study Applications*, 4th ed., John Wiley & Sons, New York, 1961, 188 pages.

Barnes, Ralph M., "Industrial Engineering," in *McGraw-Hill Encyclopedia of Science and Technology*, McGraw-Hill Book Co., New York, 1960, pp. 82–83.

Barnes, Ralph M., *Work Sampling*, 2nd ed., John Wiley & Sons, New York, 1957, 283 pages.

Barnes, Ralph M., *Work Measurement Manual*, 4th ed., College Book Company, Los Angeles 90024, 1951, 297 pages.

Barnes, Ralph M., *Work Methods Training Manual*, 3rd ed., College Book Company, Los Angeles 90024, 1950, 337 pages.

Barnes, Ralph M., "Industrial Engineering Survey," *Industrial Engineering Bulletin* 101, University of Iowa, Iowa City, 1949, 48 pages.

Barnes, Ralph M., "Work Measurement Project," *Industrial Engineering Bulletin* 201, University of Iowa, Iowa City, 1948, 48 pages.

Barnes, Ralph M., *Work Methods Manual*, John Wiley & Sons, New York, 1944, 136 pages.

Barnes, Ralph M., "An Investigation of Some Hand Motions Used in Factory Work," *University of Iowa Studies in Engineering, Bulletin* 6, 1936, 63 pages.

Barnes, Ralph M., *Industrial Engineering and Management*, McGraw-Hill Book Co., New York, 1931, 366 pages.

Barnes, Ralph M., and Robert B. Andrews, *Performance Sampling*, University of California, Los Angeles, 1955, 58 pages.

Barnes, Ralph M., and N. A. Englert, *Bibliography of Industrial Engineering and Management Literature*, College Book Company, Los Angeles 90024, 1946, 136 pages.

Barnes, Ralph M., and J. L. McKenney, *Industrial Engineering Survey*, University of California, 1957.

Barnes, Ralph M., and M. E. Mundel, *University of Iowa Studies in Engineering:* "Studies of Hand Motions and Rhythm Appearing in Factory Work," *Bulletin* 12, 1938, 60 pages; "A Study of Hand Motions Used in Small Assembly Work," *Bulletin* 16, 1939, 66 pages; "A Study of Simultaneous Symmetrical Hand Motions," *Bulletin* 17, 1939, 36 pages.

Barnes, Ralph M., M. E. Mundel, and John M. MacKenzie, "Studies of One- and Two-Handed Work," *University of Iowa Studies in Engineering, Bulletin* 21, 1940, 67 pages.

Barnes, Ralph M., J. S. Perkins, and J. M. Juran, "A Study of the Effect of Practice on the Elements of a Factory Operation," *University of Iowa Studies in Engineering, Bulletin* 22, 1940, 95 pages.

Barnes, Ralph M., and J. B. Sullivan, *Production Management Survey*, University of California, Los Angeles, 1950, 21 pages.

Bartlett, F. C., *The Problem of Noise*, Cambridge University Press, London, 1934, 87 pages.

Bartley, S. H., and E. Chute, *Fatigue and Impairment in Man*, McGraw-Hill Book Co., New York, 1947, 429 pages.

Belcher, David W., *Wage and Salary Administration*, 2nd ed., Prentice-Hall, New York, 1962, 598 pages.

Benedict, F. G., and E. P. Cathcart, "Muscular Work, a Metabolic Study with Special Reference to the Efficiency of the Human Body as a Machine," Carnegie Institution of Washington, *Publication* 187, 1913, 176 pages.

Bennis, W. G., E. H. Schein, D. E. Berew, and F. I. Steel, *Interpersonal Dynamics*, The Dorsey Press, Homewood, Ill., 1964, 763 pages.

Bills, Arthur, *Psychology of Efficiency*, Harper & Brothers, New York, 1943.

Birn, S. A., R. M. Crossan, and R. W. Eastwood, *Measurement and Control of Office Work*, McGraw-Hill Book Co., New York, 1961, 318 pages.

Blackburn, J. M., "The Acquisition of Skill: an Analysis of Learning Curves," Industrial Health Research Board, *Report* 73, His Majesty's Stationery Office, London, 1936, 92 pages.

Bolz, H. A. (editor), *Materials Handling Handbook*, Ronald Press Co., New York, 1960, 1740 pages.

Bowman, E. H., and R. B. Fetter, *Analysis for Production Management*, 3rd ed., Richard D. Irwin, Homewood, Ill., 1967, 870 pages.

Brisco, Norris A., *Economics of Efficiency*, Macmillan Co., New York, 1921, 385 pages.

Brouha, Lucien, *Physiology in Industry,* Pergamon Press, New York, 1960, 145 pages.

Brown, A. Barrett, *The Machine and the Worker,* Nicholson & Watson, London, 1934, 215 pages.

Buffa, E. S. *Operations Management: Problems and Models,* 2nd ed., John Wiley & Sons, New York, 1968.

Buffa, E. S., *Production-Inventory Systems: Planning and Control,* Richard D. Irwin, Homewood, Ill., 1968, 457 pages.

Buffa, E. S., *Modern Production Management,* 2nd ed., John Wiley & Sons, New York, 1965, 758 pages.

Buhl, Harold R., *Creative Engineering Design,* Iowa State University Press, Ames, Iowa, 1960, 195 pages.

Bullen, A. K., *New Answers to the Fatigue Problem.* University of Florida Press, Gainsville, Fla., 1956, 176 pages.

Bureau of the Budget, *A Work Measurement System: Development and Use,* U. S. Government Printing Office, Washington, D. C., 1950, 44 pages.

Burns, Tom, and G. M. Stalker, *The Management of Innovation,* Quadrangle Books, Chicago, 1962, 269 pages.

Burtt, H. E., *Psychology and Industrial Efficiency,* D. Appleton & Co., New York, 1929, 395 pages.

Carlson, A. J., and V. Johnson, *Machinery of the Body,* 4th ed., University of Chicago Press, Chicago, 1953, 663 pages.

Carroll, Phil, *Overhead Cost Control,* McGraw-Hill Book Co., New York, 1964, 314 pages.

Carroll, Phil, *How to Chart Data,* McGraw-Hill Book Co., New York, 1960, 260 pages.

Carroll, Phil, *Better Wage Incentives,* McGraw-Hill Book Co., New York, 1957, 230 pages.

Carroll, Phil, *How Foremen Can Control Costs,* McGraw-Hill Book Co., New York, 1955, 301 pages.

Carroll, Phil, *Timestudy for Cost Control,* 3rd ed., McGraw-Hill Book Co., New York, 1954, 299 pages.

Carroll, Phil, *How to Control Production Costs,* McGraw-Hill Book Co., New York, 1953, 272 pages.

Carroll, Phil, *Timestudy Fundamentals for Foremen,* 2nd ed., McGraw-Hill Book Co., New York, 1951, 209 pages.

Carson, Gordon B. (editor), *Production Handbook,* 2nd ed., Ronald Press Co., New York, 1958.

Carter, R. M., "Labor Saving Through Farm Job Analysis—I. Dairy Barn Chores," Vermont Agricultural Experiment Station, *Bulletin* 503, Burlington, Vermont, June, 1943, 66 pages.

Cathcart, E. P., *The Human Factor in Industry,* Oxford University Press, London, 1928, 105 pages.

Chane, G. W., *Motion and Time Study,* Harper & Brothers, New York, 1942, 88 pages.

Chapanis, A., *Research Techniques in Human Engineering*, Johns Hopkins Press, Baltimore, Md., 1959, 316 pages.

Chapanis, A., W. R. Garner, and C. T. Morgan, *Applied Experimental Psychology: Human Factors in Engineering Design*, John Wiley & Sons, New York, 1949, 434 pages.

Chestnut, Harold, *Systems Engineering Methods*, John Wiley & Sons, New York, 1967, 392 pages.

Clark, Wallace, *The Gantt Chart*, Ronald Press Co., New York, 1923, 157 pages.

Close, G. C., *Work Improvement*, John Wiley & Sons, New York, 1960, 388 pages.

Cohen, A., *Time Study and Common Sense*, MacDonald & Evans, London, 1947, 112 pages.

Copley, F. B., *Frederick W. Taylor, Father of Scientific Management*, Vols. 1 and 2, Harper & Brothers, New York, 1923.

Cox, J. W., *Manual Skill*, Cambridge University Press, London, 1934, 247 pages.

Cox, J. W., *The Economic Basis of Fair Wages*, Ronald Press Co., New York, 1926.

Crossan, R. M., and H. Nance, *Master Standard Data*, McGraw-Hill Book Co., New York, 1962, 257 pages.

Crowden, G. P., *Muscular Work, Fatigue and Recovery*, Sir Isaac Pitman & Sons, London, 1932, 74 pages.

Currie, R. M., *Work Study*, Sir Isaac Pitman & Sons, London, 1959, 232 pages.

Damon, A., H. W. Stoudt and R. McFarland, *The Human Body in Equipment Design*, Harvard University Press, Cambridge, Mass., 1966, 360 pages.

Dana, R. T., and A. P. Ackerman, *The Human Machine in Industry*, Codex Book Co., New York, 1927, 307 pages.

Davidson, H. O., *Functions and Bases of Time Standards*, American Institute of Industrial Engineers, New York, 1952, 403 pages.

Derse, Joseph C., *Machine Operation Times for Estimators*, Ronald Press Co., New York, 1946, 156 pages.

Dickinson, Z. Clark, *Compensating Industrial Effort*, Ronald Press Co., New York, 1937, 479 pages.

Dickson, W. J., and F. J. Roetlisberger, "Management and the Worker—Technical vs. Social Organization in an Industrial Plant," Graduate School of Business Administration, Division of Research, *Business Research Studies* 9, Harvard University, Boston, 1934, 17 pages.

Diebold, John, *Beyond Automation*, McGraw-Hill Book Co., New York, 1964, 220 pages.

Diebold, John, *Automation: The Advent of the Automatic Factory*, D. Van Nostrand Co., Princeton, N. J., 1952, 175 pages.

Dixon, W. J., and F. J. Massey, Jr., *Introduction to Statistical Analysis*, McGraw-Hill Book Co., New York, 1951.

Dreyfuss, Henry, *The Measure of Man*, Whitney Library of Design, New York, 1960.

Dreyfuss, Henry, *Designing for People*, Simon & Schuster, New York, 1955, 240 pages.

Drury, H. B., *Scientific Management; A History and Criticism*, Columbia University Press, New York, 1922, 271 pages.

Eastman Kodak Co., *How to Make Good Movies*, Eastman Kodak Co., Rochester, New York.

Elliott, Jaques, *The Changing Culture of a Factory*, Tavistock Publications Ltd., London, 1951, 341 pages.

Emerson, H., *The Twelve Principles of Efficiency*, 5th ed., Engineering Management Co., New York, 1917, 423 pages.

Emerson, H., *Efficiency as a Basis for Operation and Wages*, Engineering Magazine Co., New York, 1912, 254 pages.

Enrick, N. L. (editor), *Time Study Manual for the Textile Industry*, Interscience Publishers, Inc., New York, 1960, 216 pages.

Farmer, E., "Motion Study in Metal Polishing" (Metal Series 5), Industrial Fatigue Research Board, *Report* 15, His Majesty's Stationery Office, London, 1924, 65 pages.

Farmer, E., "Time and Motion Study," Industrial Fatigue Research Board, *Report* 14, His Majesty's Stationery Office, London, 1921, 63 pages.

Feigenbaum, A. V., *Total Quality Control*, McGraw-Hill Book Co., New York, 1961, 627 pages.

Florence, P. S., *Economics of Fatigue and Unrest and the Efficiency of Labor in English and American Industry*, Henry Holt & Co., New York, 1924, 426 pages.

Florence, P. S., *Use of Factory Statistics in the Investigation of Industrial Fatigue*, Columbia University Press, New York, 1918, 153 pages.

Floyd, W. F., and A. T. Welford (editors), *Symposium on Human Factors in Equipment Design*, H. K. Lewis & Co., London, 1954, 132 pages.

Floyd, W. F., and A. T. Welford (editors), *Symposium on Fatigue*, H. K. Lewis & Co., Ltd., London, 1953, 196 pages.

Forrester, Jay W., *Industrial Dynamics*, John Wiley & Sons, New York, 1961, 464 pages.

Friedmann, Georges, *The Anatomy of Work*, The Free Press of Glenco, New York, 1961, 203 pages.

Gantt, H. L., *Work, Wages and Profits*, Engineering Management Co., New York, 1913.

Gellerman, S. W., *Motivation and Productivity*, American Management Association, New York, 1963, 304 pages.

Geppinger, H. C., *Dimensional Motion Times*, John Wiley & Sons, New York, 1955, 100 pages.

Ghiselli, Edwin E., and C. W. Brown, *Personnel and Industrial Psychology*, 2nd ed., McGraw-Hill Book Co., New York, 1955, 492 pages.

Gilbreth, F. B., *Primer of Scientific Management*, D. Van Nostrand Co., Princeton, N. J., 1914, 108 pages.

Gilbreth, F. B., *Motion Study*, D. Van Nostrand Co., Princeton, N. J., 1911, 116 pages.

Gilbreth, F. B., *Bricklaying System*, Myron C. Clark Publishing Co., Chicago, 1909, 321 pages.

Gilbreth, F. B., and L. M., *Motion Study for the Handicapped*, George Routledge & Sons, London, 1920, 165 pages.

Gilbreth, F. B., and L. M., *Fatigue Study*, 2nd ed., Macmillan Co., New York, 1919, 175 pages.

Gilbreth, F. B., and L. M., *Applied Motion Study*, Sturgis & Walton Co., New York, 1917, 220 pages.

Gilbreth, Lillian M., *The Psychology of Management*, Sturgis & Walton Co., New York, 1914, 344 pages.

Gilbreth, Lillian M., and A. R. Cook, *The Foreman in Manpower Management*, McGraw-Hill Book Co., New York, 1947.

Gilbreth, Lillian M., Orpha Mae Thomas, and Eleanor Olymer, *Management in the Home*, Dodd, Mead, and Co., 1954, 241 pages.

Gillespie, James J., *Dynamic Motion and Time Study*, Chemical Publishing Co., Brooklyn, N. Y., 1951, 140 pages.

Gilmour, R. W., *Industrial Wage and Salary Control*, John Wiley & Sons, New York, 1956, 261 pages.

Goetz, B. E., *Quantitative Methods*, McGraw-Hill Book Co., New York, 1965, 541 pages.

Goldmark, Josephine C., *Fatigue and Efficiency*, Charities Publication Committee, Russell Sage Foundation, New York, 1912, 591 pages.

Gomberg, W., *A Trade Union Analysis of Time Study*, 2nd ed., Prentice-Hall, Englewood Cliffs, N. J., 1954.

Grabbe, E. M. (editor), *Automation in Business and Industry*, John Wiley & Sons, New York, 1957, 611 pages.

Grant, Eugene L., *Statistical Quality Control*, 3rd ed., McGraw-Hill Book Co., New York 1964, 610 pages.

Grillo, E. V., and C. J. Berg, *Work Measurement in the Office*, McGraw-Hill Book Co., New York, 1959, 200 pages.

Hadden, A. A., and V. K. Geneer, *Handbook of Standard Time Data—For Machine Shops*, Ronald Press Co., New York, 1954, 473 pages.

Haggard, Howard W., *A Physiologist and a Statistician Look at Wage Incentive Methods*, American Management Association, New York, 1937, 26 pages.

Haggard, Howard W., and L. A. Greenberg, *Diet and Physical Efficiency*, Yale University Press, New Haven, 1935, 180 pages.

Haire, Mason (editor), *Organization Theory in Industrial Practice*, John Wiley & Sons, New York, 1962, 173 pages.

Haire, Mason, *Psychology in Management*, McGraw-Hill Book Co., New York, 1956, 212 pages.

Hansen, B. L., *Work Sampling for Modern Management*, Prentice-Hall, Englewood Cliffs, N. J., 1960, 263 pages.

Heiland, R. E., and W. J. Richardson, *Work Sampling*, McGraw-Hill Book Co., New York, 1957, 243 pages.

Heiner, M. K., and H. E. McCullough, "Kitchen Cupboards That Simplify Storage," *Cornell Extension Bulletin* 703, New York State College of Home Economics at Cornell University, Ithaca, 1947, 32 pages.

Hendry, J. W., *A Manual of Time and Motion Study*, 3rd ed., Sir Isaac Pitman & Sons, London, 1950, 217 pages.

Herzberg, F., *Work and the Nature of Man*, The World Publishing Company, New York, 1966, 203 pages.

Herzberg, F., B. Mausner, and B. B. Snyderman, *The Motivation of Work*, 2nd ed., John Wiley & Sons, New York, 1959, 157 pages.

Herzberg, F., B. Mausner, R. O. Peterson, and D. F. Capwell, "Job Attitudes: Review of Research and Opinion," *Psychological Service of Pittsburgh*, Pittsburgh, Pa., 1957, 279 pages.

Hill, A. V., *Living Machinery*, Harcourt, Brace & Co., New York, 1927, 306 pages.

Hill, A. V., *Muscular Movement in Man; the Factors Governing Speed and Recovery from Fatigue*, McGraw-Hill Book Co., New York, 1927, 104 pages.

Hill, A. V., *Muscular Activity*, Williams & Wilkins, Baltimore, 1926, 115 pages.

Hill, W. A., and D. M. Egan, *Readings in Organization Theory: A Behavioral Approach*, Allyn and Bacon, Inc., Boston, 1966, 746 pages.

Hodnett, Edward, *The Art of Problem Solving*, Harper & Brothers, New York, 1955, 202 pages.

Holmes, W. G., *Applied Time and Motion Study*, revised, Ronald Press Co., New York, 1945, 383 pages.

"Hours of Work, Lost Time and Labour Wastage," Industrial Health Research Board, *Emergency Report 2*, His Mapesty's Stationery Office, London, 1943.

Hoxie, R. F., *Scientific Management and Labor*, D. Appleton & Co., New York, 1915, 302 pages.

Handbook of Human Engineering Data, 2nd ed., Institute for Applied Experimental Psychology, Tufts College, Medford, Mass., 1951.

Hughs, Charles L., *Goal Setting: Key to Individual and Organizational Effectiveness*, American Management Association, New York, 1965, 157 pages.

Hunt, E. E. (editor), *Scientific Management Since Taylor; a Collection of Authoritative Papers*, McGraw-Hill Book Co., New York, 1924, 263 pages.

Immer, John R., *Materials Handling*, McGraw-Hill Book Co., New York, 1953, 570 pages.

Immer, John R., *Layout Planning Techniques*, McGraw-Hill Book Co., New York, 1950, 430 pages.

Industrial Engineering Institute Proceedings, University of California, Los Angeles-Berkeley, annually, 1950–1965.

Industrial Management Society, *Proceedings of the Time and Motion Study Clinic*, Chicago, annually since 1938.

Ireson, W. G., *Factory Planning and Plant Layout*, Prentice-Hall, Englewood Cliffs, N. J., 1952, 359 pages.

Ireson, W. G., and Eugene L. Grant (editors), *Handbook of Industrial Engineering and Management*, Prentice-Hall, Englewood Cliffs, N. J., 1955, 1203 pages.

Juran, J. M., *Quality Control Handbook*, 2nd ed., McGraw-Hill Book Co., New York, 1962, 1220 pages.

Juran, J. M., *Bureaucracy, a Challenge to Better Management*, Harper & Brothers, New York, 1944, 138 pages.

Karger, D. W., and F. H. Bayna, *Engineered Work Measurement*, The Industrial Press, New York, 1957, 635 pages.

Karpovich, Peter V., *Physiology of Muscular Activity*, 4th ed., W. B. Saunders Co., Philadelphia, 1953, 340 pages.

Kennedy, Van Dusen, *Union Policy and Incentive Wage Methods*, Columbia University Press, New York, 1945, 260 pages.

Knox, F. M., *Design and Control of Business Forms*, McGraw-Hill Book Co., New York, 1952, 226 pages.

Koepke, C. A., and L. S. Whitson, "Power and Velocity Developed in Manual Work," Institute of Technology, *Technical Paper* 18, University of Minnesota, Minneapolis, 1940.

Kosma, A. R., *The A.B.C.'s of Motion Economy*, Institute of Motion Analysis and Human Relations, Newark, N. J., 1943, 133 pages.

Kuriloff, Arthur H., *Reality in Management*, McGraw-Hill Book Co., New York, 1966, 247 pages.

Krick, E. V., *Methods Engineering*, John Wiley & Sons, New York, 1962, 530 pages.

Laban, R. von, *Effort*, MacDonald & Evans, London, 1947, 88 pages.

Landsberger, Henry A., *Hawthorne Revisited*, Cornell University, Ithaca, N. Y., 1958, 119 pages.

Langsner, Adolph, and H. G. Zollitsch *Wage and Salary Administration*, South-West Publishing Co., Cincinnati, 1961, 726 pages.

Lee, F. S., *The Human Machine and Industrial Efficiency*, Longmans, Green & Co., New York, 1919, 119 pages.

Leffingwell, W. H., and E. M. Robinson, *Textbook of Office Management*, 3rd ed., McGraw-Hill Book Co., New York, 1950, 649 pages.

Lehrer, R. N., *The Management of Improvement*, Reinhold Publishing Co., New York, 1965, 415 pages.

Lehrer, R. N., *Work Simplification*, Prentice-Hall, Englewood Cliffs, N. J., 1957, 394 pages.

Lesieur, Frederick G., *The Scanlon Plan*, John Wiley & Sons, New York, 1958, 173 pages.

Lesperance, J. P., *Economics and Techniques of Motion and Time Study*, William C. Brown Co., Dubuque, Iowa, 1953, 258 pages.

Levenstein, Aaron, *Why People Work,* The Crowell-Collier Press, 1962, 320 pages.

Levin, H. S., *Office Work and Automation,* John Wiley & Sons, New York, 1956, 203 pages.

Lewin, Kurt, *Resolving Social Conflict,* Harper, New York, 1948, 230 pages.

Lichtner, W. O., *Time Study and Job Analysis,* Ronald Press Co., New York, 1921, 397 pages.

Likert, Rensis, *The Human Organization: Its Management and Value,* McGraw-Hill Book Co., New York, 1967, 258 pages.

Likert, Rensis, *New Patterns of Management,* McGraw-Hill Book Co., New York, 1961, 279 pages.

Lincoln, James F., *A New Approach to Economics,* The Lincoln Electric Co., Cleveland, Ohio, 166 pages.

Lincoln, J. F., *Incentive Management,* Lincoln Electric Co., Cleveland, Ohio, 1952, 280 pages.

Lincoln, J. F., *Lincoln's Incentive System,* McGraw-Hill Book Co., New York, 1946, 192 pages.

Lindquist, E. F., *A First Course in Statistics,* Houghton Mifflin Co., Boston, 1942.

Littlefield, C. L., and F. M. Rachel, *Office and Administrative Management: Systems Analysis, Data Processing, and Office Services,* 2nd ed., Prentice-Hall, Inc., Englewood Cliffs, N. J., 1964, 577 pages.

Louden, J. K., and J. W. Deegan, *Wage Incentives,* 2nd ed., John Wiley & Sons, New York, 1959, 227 pages.

Lowry, S. M., H. B. Maynard, and G. J. Stegemerten, *Time and Motion Study,* 3rd ed., McGraw-Hill Book Co., New York, 1940, 432 pages.

Luckiesh, M., *Seeing and Human Welfare,* Williams & Wilkins Co., Baltimore, 1934, 193 pages.

Lytle, C. W., *Job Evaluation Methods,* 2nd ed., Ronald Press Co., New York, 1954, 507 pages.

Lytle, C. W., *Wage Incentive Methods,* Ronald Press Co., New York, 1942, 462 pages.

Malcolm, D. G., A. J. Rowe, and L. F. McConnell (editors), *Management Control Systems,* John Wiley & Sons, New York, 1960, 375 pages.

Malcolm, J. A., Jr., W. J. Frost, R. E. Hannan, and W. R. Smith, *Ready Work-Factor Time Standards,* WOFAC Corporation, Haddenfield, N. J., 1966.

Mallick, R. W., and A. T. Gaudreau, *Plant Layout: Planning and Practice,* John Wiley & Sons, New York, 1951, 391 pages.

Marrow, A. J., *Making Management Human,* McGraw-Hill Book Co., New York, 1957, 241 pages.

Marrow, A. J., D. G. Bowers, and S. E. Seashore, *Management by Participation,* Harper & Row, New York, 1967, 264 pages.

Maslow, A. H., *Motivation and Personality,* Harper and Brothers, New York, 1954, 411 pages.

Mathewson, Stanley B., *Restriction of Output Among Unorganized Workers*, Viking Press, New York, 1931, 212 pages.

Maynard, H. B. (editor), *Top Management Handbook*, McGraw-Hill Book Co., New York, 1960, 1236 pages.

Maynard, H. B. (editor), *Industrial Engineering Handbook*, 2nd ed., McGraw-Hill Book Co., New York, 1963.

Maynard, H. B., G. J. Stegemerten, and J. L. Schwab, *Methods-Time Measurement*, McGraw-Hill Book Co., New York, 1948, 229 pages.

Maynard, H. B., and G. J. Stegemerten, *Guide to Methods Improvement*, McGraw-Hill Book Co., New York, 1944, 82 pages.

Maynard, H. B., and G. J. Stegemerten, *Operation Analysis*, McGraw-Hill Book Co., New York, 1939, 298 pages.

Mayo, Elton, *The Human Problems of an Industrial Civilization*, 2nd ed., Harvard University Press, Cambridge, 1946.

Mayo, Elton, *The Social Problems of an Industrial Civilization*, Division of Research, Graduate School of Business Administration, Harvard University, Boston, 1945, 150 pages.

McGregor, Douglas, *The Human Side of Enterprise*, McGraw-Hill Book Co., New York, 1960, 246 pages.

McGregor, Douglas, *The Professional Manager* (edited by Caroline McGregor and W. F. Bennis), McGraw-Hill Book Co., New York, 1967, 202 pages.

Meister, David, and G. F. Rabideau, *Human Factors Evaluation in System Development*, John Wiley and Sons, New York, 1965, 307 pages.

McLachlan, N. W., *Noise*, Oxford University Press, London, 1935 148 pages.

Melman, Seymour, *Dynamic Factors in Industrial Productivity*, John Wiley & Sons, New York, 1956, 238 pages.

Merrick, Dwight V., *Time Studies as a Basis for Rate Setting*, Engineering Magazine Co., New York, 1920, 366 pages.

Metzger, R. W., *Elementary Mathematical Programming*, John Wiley & Sons, New York, 1958, 246 pages.

Michael, L. B., *Wage and Salary Fundamentals and Procedures*, McGraw-Hill Book Co., New York, 1950.

Miles, G. H., *The Problem of Incentives in Industry*, Sir Isaac Pitman & Sons, London, 1932, 58 pages.

Miles, L. D., *The Will to Work*, George Routledge & Sons, London, 1929, 80 pages.

Miles L. D., *Techniques of Value Analysis and Engineering*, McGraw-Hill Book Co., New York, 1961, 267 pages.

Moede, W., *Arbeitstechnik* (*Work Technique*), Ferdinand Enke Velag, Stuttgart, 1935, 267 pages.

Mogensen, A. H., *Common Sense Applied to Motion and Time Study*, McGraw-Hill Book Co., New York, 1932, 228 pages.

Moore, J. M., *Plant Layout and Design*, Macmillan Co., New York, 1961, 644 pages.

Morrow, R. L., *Motion Economy and Work Measurement*, Ronald Press, New York, 1957, 468 pages.

Morse, P. M., and G. E. Kimball, *Methods of Operations Research*, John Wiley & Sons, New York, 1951, 158 pages.

Mosso, A., *Fatigue*, G. P. Putnam's Sons New York, 1904, 334 pages.

Mundel, M. E., *Motion and Time Study*, 3rd ed., Prentice-Hall, Englewood Cliffs, N. J., 1960, 690 pages.

Mundel, M. E. *Systematic Motion and Time Study*, Prentice-Hall, Englewood Cliffs, N. J., 1947, 232 pages.

Münsterberg, H., *Psychology and Industrial Efficiency*, Houghton Mifflin Co., New York, 1913, 321 pages.

Murrell, K. F. H., *Human Performance in Industry*, Reinhold Publishing Co., New York, 1965, 496 pages.

Musico, B., *Lectures on Industrial Psychology*, George Routledge & Sons, London, 1920, 300 pages.

Muther, Richard, *Systematic Layout Planning*, Industrial Educational Institute, Boston, Mass., 314 pages.

Muther, Richard, *Practical Plant Layout*, McGraw-Hill Book Co., New York, 1955, 384 pages.

Muther, Richard, *Production-Line Technique*, McGraw-Hill Book Co., New York, 1944, 320 pages.

Myers, C. S. (editor), *Industrial Psychology*, Thornton Butterworth, London, H. Holt & Co., New York, 1930, 252 pages.

Myers, C. S., *Industrial Psychology in Great Britain*, Jonathan Cape, London, 1926, 164 pages.

Myers, C. S., *Mind and Work*, G. P. Putnam's Sons, New York and London, 1921, 175 pages.

Myers, H., *Human Engineering*, Harper & Brothers, New York, 1932, 318 pages.

Myers, H. J., *Simplified Time Study*, Ronald Press Co., New York, 1944, 140 pages.

Nadler, Gerald, *Work Systems Design: The Ideals Concept*, Richard D. Irwin, Homewood, Ill., 1967, 183 pages.

Nadler, Gerald, *Work Design*, Richard D. Irwin, Homewood, Ill., 1963, 837 pages.

Nadler, Gerald, *Work Simplification*, McGraw-Hill Book Co., New York, 1957, 292 pages.

Nadler, Gerald, *Motion and Time Study*, McGraw-Hill Book Co., New York, 1955, 612 pages

Nadworny, Milton, *Scientific Management and the Unions*, Harvard University Press, Cambridge, Mass., 1955, 187 pages.

National Industrial Conference Board, "Industrial Engineering Organization and Practices," *Studies in Business Policy*, No. 78, New York, 1956, 56 pages.

National Office Management Association, *Manual of Practical Office Time Savers*, McGraw-Hill Book Co., New York, 1957, 256 pages.

National Research Council, *Fatigue of Workers: Its Relation to Industrial Production*, Reinhold Publishing Corp., New York, 1941, 165 pages.

National Time and Motion Study Clinic, *Proceedings of the Time and Motion Study Clinic*, annually since 1938, Industrial Management Society, Chicago.

Nickerson, J. W., and J. H. Eddy (compilers), *A Handbook on Wage Incentive Plans*, Management Consultant Division, War Production Board, U. S. Government Printing Office, Washington, D. C., 1945.

Niebel, B. W., *Motion and Time Study*, 4th ed., Richard D. Irwin, Homewood, Ill., 1967, 628 pages.

Nordhoff, W. A. *Machine Shop Estimating*, 2nd ed., McGraw-Hill Book Co., New York, 1960, 528 pages.

Nyman, R. C., and E. D. Smith, *Union-Management Cooperation in the Stretch Out*, Yale University Press, New Haven, 1934, 210 pages.

Pappas, F. G., and R. A. Dimberg, *Practical Work Standards*, McGraw-Hill Book Co., New York, 1962, 223 pages.

Parton, J. A., *Motion and Time Study Manual*, Conover-Mast Publications, New York, 1952, 400 pages.

Patton, John A. (editor), *Manual of Industrial Engineering Procedures*, William C. Brown Co., Dubuque, Iowa, 1955, 144 pages.

Payne, Matthew A., *The Fatigue Allowance in Industrial Time Study*, Matthew a. Payne, 1949, 66 pages.

Pear, T. H., *Fitness for Work*, University of London Press, London, 1928, 187 pages.

Pear, T. H., *Skill in Work and Play*, E. P. Dutton & Co., New York, 1924, 107 pages.

Poffenberger, A. T., *Principles of Applied Psychology*, Appleton-Century-Crofts, New York, 1942.

Presgrave, Ralph, *The Dynamics of Time Study*, 2nd ed., McGraw-Hill Book Co., New York, 1945, 238 pages.

Presgrave, Ralph, and G. B. Bailey, *Basic Motion Timestudy*, McGraw-Hill Book Co., New York, 1958, 195 pages.

Quick, J. H., J. H. Duncan, and J. A. Malcolm, Jr., *Work-Factor Time Standards*, McGraw-Hill Book Co., New York, 1962, 458 pages.

Reed, R., *Plant Layout: Factors, Principles, and Techniques*, Richard D. Irwin, Homewood, Ill., 1961, 459 pages.

Rice, A. K., *Productivity and Social Organization: The Ahmedabad Experiment*, Tavistock Publications Ltd., London, 1958, 298 pages.

Rice, W. B., *Control Charts in Factory Management*, John Wiley & Sons, New York, 1947, 149 pages.

Riegel, J. W., *Management, Labor, and Technological Change*, University of Michigan Press, Ann Arbor, 1942, 187 pages.

Riegel, J. W., *Wage Determination*, University of Michigan Press, Ann Arbor, revised 1941, 138 pages.

Riegel, J. W., *Salary Determination*, University of Michigan Press, Ann Arbor, 1940, 278 pages.

Roethlisberger, F. J., *Management and Morale*, Harvard University Press, Cambridge, 1941, 194 pages.

Roethlisberger, F. J., and W. J. Dickson, *Management and the Workers*, Harvard University Press, Cambridge, 1940, 615 pages.

Ryan, T. A., *Work and Effort*, Ronald Press Co., New York, 1947, 323 pages.

Society for Advancement of Management, *A Fair Day's Work*, New York, 1954, 84 pages.

Sayles, L. R., and George Strauss, *Human Behavior in Organizations*, Prentice-Hall, Englewood Cliffs, N. J., 1961, 500 pages.

Sayles, L., *Behavior of Industrial Work Groups: Prediction and Control*, John Wiley & Sons, New York, 1958.

Sampter, H. C., *Motion Study*, Pitman Publishing Co., New York, 1941, 152 pages.

Schein, E. H., and W. G. Bennis, *Personal and Organizational Change Through Group Methods*, John Wiley & Sons, Inc., New York, 1965, 376 pages.

Schell, E. H., and F. F. Gilmore, *Manual for Executives and Foremen*, McGraw-Hill Book Co., New York, 1939, 185 pages.

Schutt, W. H., *Time Study Engineering*, McGraw-Hill Book Co., New York, 1943, 426 pages.

Scott, M. G., *Analysis of Human Motion*, A. S. Barnes & Co., New York, 1942.

Sisson, R. L., and R. G. Canning, *A Manager's Guide to Computor Processing*, John Wiley & Sons, New York, 1967, 124 pages.

Shaw, Anne G., *An Introduction to the Theory and Application of Motion Study*, Harlequin Press, London, 1953, 37 pages.

Shaw, Anne G., *Purpose and Practice of Motion Study*, 2nd ed., Columbine Press, London, 1960, 324 pages.

Shevlin, J. D., *Time Study and Motion Economy for Supervisors*, National Foremen's Institute, Deep River, Conn., 1945, 73 pages.

Shubin, J. A., and H. Madenheim, *Plant Layout*, Prentice-Hall, Englewood Cliffs, N. J., 1961, 433 pages.

Shumard, F. W., *A Primer of Time Study*, McGraw-Hill Book Co., New York, 1940, 519 pages.

Slichter, Sumner, *Union Policies and Industrial Management*, Brookings Institution, Washington, D. C., 1941.

Smalley, H. E., *Hospital Industrial Engineering*, Reinhold, New York, 1966, 460 pages.

Smalley, H. E., *Motion and Time Study Laboratory Manual*, William C. Brown Co., Dubuque, Iowa, 1948.

Smith, E. D., *Technology and Labor*, Yale University Press, New Haven, 1939, 222 pages.

Smith, May, *The Handbook of Industrial Psychology*, Philosophical Library, New York, 1944.

Society for Advancement of Management, *Collective Agreements in Time and Motion Study*, New York, 1954, 46 pages.

Society for Advancement of Management, *Glossary of Terms Used in Methods, Time Study and Wage Incentives,* New York, 1952.

Spriegel, William R., and C. E. Myers (editors), *The Writings of the Gilbreths,* Richard D. Irwin, Homewood, Ill., 1953, 513 pages.

Starr, M. K., and D. W. Miller, *Inventory Control: Theory and Practice,* Prentice-Hall, Englewood Cliffs, N. J., 1962, 354 pages.

Stewart, Paul A., "Job Enlargement," Monogram Series No. 3, University of Iowa, Iowa City, 1967, 64 pages.

Stivers, C. L., "Experience in Retraining on the Dvorak Keyboard," American Management Association, *Supplementary Office Management Series, No. 1,* New York, 1941, 12 pages.

Stocker, H. E., *Materials Handling,* 2nd ed., Prentice-Hall, Englewood Cliffs, N. J., 1951, 330 pages.

Strong, E. P., *Increasing Office Productivity,* McGraw-Hill Book Co., New York, 1962, 287 pages.

Sylvester, L. A., *The Handbook of Advanced Time-Motion Study,* Funk & Wagnalls Co., New York, 1950, 273 pages.

Tannenbaum, A. S., *Social Psychology of the Work Organization,* Wadsworth Publishing Co., Belmont, Calif., 1966, 136 pages.

Taylor, F. W., *Scientific Management; Comprising Shop Management, Principles of Scientific Management, and Testimony before Special House Committee,* Harper & Brothers, New York, 1947.

Taylor, F. W., *The Principles of Scientific Management,* Harper & Brothers, New York, 1911, 144 pages.

Taylor, F. W., "On the Art of Cutting Metals," *Transactions of the ASME,* Vol. 28, pp. 31–350, 1907.

Taylor, F. W., *Shop Management,* Harper & Brothers, New York, 1919, 207 pages, reprinted from *Transactions of the ASME,* Vol. 24, pp. 1337–1480, 1903.

Taylor Society, *Scientific Management in American Industry,* Harper & Brothers, New York, 1929, 472 pages.

Thompson, C. B. (editor), *Scientific Management,* Harvard University Press, Cambridge, 1914, 878 pages.

Thuesen, H. G.. *Engineering Economy,* 2nd ed., Prentice-Hall, Englewood Cliffs, N. J., 1957, 581 pages.

Tiffin, Joseph, and E. J. McCormick, *Industrial Psychology,* 5th ed., Prentice-Hall, Englewood Cliffs, N. J., 1965, 682 pages.

"Time and Motion Study, Investigation of German Radio and Associated Industries," *B.I.O.S. Final Report* 943, Items 1, 7, and 9, British Intelligence Objectives Sub-Committee, 37 Bryanston Square, London, W. 1, 1946, 110 pages.

Tippett, L. H. C., *Technological Applications of Statistics,* John Wiley & Sons, New York, 1950, 184 pages.

Trist, E. L., G. W. Higgin, H. Murray, and A. B. Pollock, *Organizational Choice,* Tavistock Pulications Ltd., London; 1963, 332 pages.

Turner, A. N., and P. R. Lawrence, *Industrial Jobs and the Worker*, Harvard University, Boston, 1965, 177 pages.

Uhrbrock, R. S., "A Psychologist Looks at Wage-Incentive Methods," *Institute of Management Series, No.* 15, American Management Association, New York, 1935, 32 pages.

University of California, *Proceedings of Industrial Engineering Institute*, University of California, Los Angeles-Berkeley, annually, 1950–1965.

Urwick, L. (editor), *The Golden Book of Management*, Newman Neame, London, 1956, 208 pages.

Urwick, L., and E. F. L. Brech, *The Making of Scientific Management*, Vol. 1, *Thirteen Pioneers*, Management Publications Trust, London, 1945, 196 pages.

Van Doren, H. L., *Industrial Design*, 2nd ed., McGraw-Hill Book Co., New York, 1954, 379 pages.

Vaughan, L. M., and L. S. Hardin, *Farm Work Simplification*, John Wiley & Sons, New York, 1949, 145 pages.

Vernon, H. M., *The Health and Efficiency of Munitions Workers*, Oxford University Press, London, 1940, 138 pages.

Vernon, H. M., *The Shorter Working Week*, G. Routledge & Sons, London, 1934, 201 pages.

Vernon, H. M., *Industrial Fatigue and Efficiency*, G. Routledge & Sons, London, 1921, 264 pages.

Viteles, M. S., *Motivation and Morale in Industry*, Norton & Co., New York, 1953, 510 pages.

Viteles, M. S., *The Science of Work*, Norton & Co., New York, 1934, 442 pages.

Viteles, M. S., *Industrial Psychology*, Norton & Co., New York, 1932, 652 pages.

Von Fange, E. K., *Professional Creativity*, Prentice-Hall, Englewood Cliffs, N. J., 1959, 260 pages.

Von Neumann, J., and O. Morgenstern, *Theory of Games and Economic Behavior*, 3rd ed., Princeton University Press, Princeton, N. J., 1953, 641 pages.

Vroom, V., *Work and Motivation*, John Wiley & Sons, New York, 1964.

Walker, Charles R., "The Problem of the Repetitive Job," *Harvard Business Review*, Vol. 28, No. 3, pp. 54–58, May 1950, 331 pages.

Walker, C. R., and R. H. Guest, *The Man on the Assembly Line*, Harvard University Press, Cambridge, 1952, 180 pages.

Walker, C. R., R. H. Guest, and A. N. Turner, *The Foreman on the Assembly Line*, Harvard University Press, Cambridge, 1956, 197 pages.

Walker, C. R., and A. G. Walker, *Modern Technology and Civilization*, McGraw-Hill Book Co., New York, 1962, 469 pages.

Watson, W. F., *Machines and Men*, Allen & Unwin, London, 1935, 226 pages.

Watson, W. F., *The Worker and Wage Incentives*, Hogarth Press, London, 1934, 46 pages.

Watts, F., *An Introduction to the Psychological Problems of Industry*, Allen & Unwin, London, 1921, 240 pages.

Wechsler, David, *The Range of Human Capacities*, 2nd ed., Williams & Wilkins Co., Baltimore, 1952, 190 pages.

Welch, H. J., and G. H. Miles, *Industrial Psychology in Practice*, Sir Isaac Pitman & Sons, London, 1932, 249 pages.

Welch, H. J., and C. S. Myers, *Ten Years of Industrial Psychology—an Account of the First Decade of the National Institute of Industrial Psychology*, Sir Isaac Pitman & Sons, London, 1932, 146 pages.

Whitehead, T., *Leadership in a Free Society*, Harvard University Press, Cambridge, 1947, 266 pages.

Whitehead, T., *The Industrial Worker*, Harvard University Press, Cambridge, 1938, two volumes.

Whiting, C. S., *Creative Thinking*, Reinhold Publishing Co., New York, 1958, 168 pages.

Whyte, W. F., *Men at Work*, Richard D. Irwin, Homewood, Ill., 1961, 393 pages.

Whyte, William F., *Money and Motivation: An Analysis of Incentives in Industry*, Harper & Brothers, New York, 1955, 268 pages.

Wiener, Norbert, *The Human Use of Human Beings, Cybernetics and Society*, revised ed., Houghton Mifflin Co., Boston, 1954, 199 pages.

Wiener, Norbert, *Cybernetics: Control and Communication in the Animal and the Machine*, 2nd ed., John Wiley & Sons, New York, 1961, 212 pages.

Woodson, W. E., and D. W. Conover, *Human Engineering Guide for Equipment Designers*, 2nd ed., University of California Press, Berkeley, 1964.

Yost, Edna, *Frank and Lillian Gilbreth: Partners for Life*, Rutgers University Press, New Brunswick, N. J., 1949, 372 pages.

PERIODICALS

Advanced Management Journal, quarterly publication, Society for Advancement of Management, 16 West 40th Street, New York, New York, 10018.

American Behavioral Scientist, monthly publication, Sage Publications, Inc., 275 South Beverly Drive, Beverly Hills, California, 90212.

Automation, monthly publication, the Penton Publishing Co., Cleveland, Ohio 44113.

Data Processing Magazine, monthly publication, North American Publishing Co., 134 North 13th Street, Philadelphia, Pennsylvania 19107.

Ergonomics, bimonthly publication, Ergonomics Research Society and the International Ergonomics Association, Taylor & Francis Ltd., Red Lion Court, Fleet Street, London, E. C. 4, England.

Factory, monthly publication, McGraw-Hill Publishing Co., Inc., 330 West 42nd Street, New York, New York 10036.

Harvard Business Review, bimonthly publication, Graduate School of Business Administration, Harvard University, Boston, Massachusetts 02163.

Human Organization, Society for Applied Anthropology, Lafferty Hall, University of Kentucky, Lexington, Kentucky 40506.

Industrial Quality Control, monthly publication, American Society for Quality Control, Inc., 161 West Wisconsin Avenue, Milwaukee, Wisconsin 53202.

International Journal of Production Research, quarterly publication, The Institution of Production Engineers, 10 Chesterfield Street, London, W. 1, England.

Journal of Applied Psychology, bimonthly publication, American Psychological Association, 1200 Seventeenth Street, N. W., Washington, D. C. 20036.

Journal of Industrial Engineering, monthly publication, American Institute of Industrial Engineers, Inc., 345 East 47th Street, New York, New York 10017.

Journal of Industrial Psychology, quarterly publication, Elias Publications, Inc., P. O. Box 3194, Margate, New Jersey 08402.

Management Record, monthly publication, National Industrial Conference Board, 460 Park Avenue, New York, New York 10022.

Material Handling Engineering, monthly publication, Industrial Publishing Corp., 812 Huron Road, Cleveland, Ohio 44115.

Mechanical Engineering, monthly publication, American Society of Mechanical Engineers, 345 East 47th Street, New York, New York 10017.

Mill and Factory, monthly publication, Conover-Mast Corp., 205 East 42nd Street, New York, New York 10017.

Modern Materials Handling, monthly publication, Cahners Publishing Co., 221 Columbus Avenue, Boston, Massachusetts 02116.

N. A. A. Bulletin, monthly publication, National Association of Cost Accountants, 505 Park Avenue, New York, New York, 10017.

Operations Research, bimonthly publication, Operations Research Society of America, Mt. Royal and Guilford Avenues, Baltimore, Maryland 21202.

Operational Research Quarterly, Pergamon Press, Ltd., Headington Hill Hall, Oxford, England.

Personnel Administration, bimonthly publication, Society for Personnel Administration, 1221 Connecticut Avenue, N. W., Washington, D. C. 20036.

Personnel Journal, monthly publication, Personnel Journal, Inc., 100 Park Avenue, Swarthmore, Pennsylvania 19081.

Personnel Psychology, quarterly publication, Mt. Royal & Guilford Avenues, Baltimore, Maryland 21202.

Psychological Abstracts, monthly publication, American Psychological Association, 1200 17th Street, N. W., Washington, D. C. 20036.

SIAM Journal on Applied Mathematics, bimonthly publication, 33 South 17th Street Philadelphia, Pennsylvania 19103.

Sociometry, quarterly publication, The American Sociological Association, The Boyd Printing Company, 49 Sheridan Avenue, Albany, New York 12210.

Work Study, monthly publication, Sawell Publications, Ltd., 4 Ludgate Circus, London, E. C. 4, England.

FOREIGN TRANSLATIONS OF BOOKS BY RALPH M. BARNES

Manual de Metodos de Trabajo. Translated by Saturnino Alvarez. Introduction by Fermin de la Sierra. Published by Aguilar, S. A. de Ediciones, Juan Bravo 38, Madrid, Spain, 1950, 161 pages.

Étude des Mouvements et des Temps, troisieme edition. Translated by le Bureau des Temps Élémentaires. Published by Les Editions d'Organization, 8 Rue Alfred de Vigny, Paris 8e, France, 1953, 560 pages.

Studio dei Movimenti e dei Tempi. Translated by Giorgio Deangeli. Published by Edizioni di Communità, Via Manzoni 12, Milan, Italy, 1955, 380 pages.

Estudio de Movimientos y Tiempos. Translated by Carlos Paz Shaw. Published by Aguilar, S. A. de Ediciones, Juan Bravo 38, Madrid, Spain, 1956, 575 pages.

Practique des Observations Instantanées. Translated by le Bureau des Temps Élémentaires. Published by Les Editions d'Organisation, 8 Rue Alfred de Vigny, Paris 8e, France, 321 pages, 1958.

Motion and Time Study, 4th ed. Translated into Japanese by Mayumi Otsubo. Published by Nikkan Kogyo Shimbun-Sha, No. 1, 1-chome Iidamachi, Chiyoda-Ku, Tokyo, Japan, 1960, 658 pages.

Étude des Mouvements et des Temps, quatrieme edition. Translated by M. Maze-Sencier and le Bureau des Temps Elementaires. Published by Les Editions d'Organization, 8 Rue Alfred de Vigny, Paris 8e, France, 1960, 749 pages.

Work Sampling, 2nd ed. Translated into Japanese by Masakazu Tamai. Published by Nikkan Kogyo Shimbun-Sha, No. 1, 1-chome Iidamachi, Chiyoda-Ku, Tokyo, Japan, 348 pages, 1961.

Industrial Engineering. Translated into Japanese and published by Nippon Noritsu Kyokai, New Ohtemachi Building, 2–4, Ohtemachi, Chiyoda-Ku, Tokyo, Japan, 1961, 115 pages.

La Tecnica del Muestreo Aplicada a la Medida del Trabajo, translated into Spanish by Anselmo Calleja Siero, Actuario Matemático Estadistico Facultativo del Instituto Nacional de Estadística. Published by Aguilar, S. A. de Ediciones, Madrid, Spain, 1962, 301 pages.

Motion and Time Study, 4th ed., translated into Serbo-Croatian language by Dmitar Culíc-Jugoslovenska Autorska Agencija, Zegreb, Yugoslavia. Published 1964, 726 pages.

Estudo de Movimentos E de Tempos: Projeto E Medida do Trabalho, translated into Portuguese by Sérgio Luiz Oliveira Assis, José S. Guedes Azevedo and Arnaldo Pallotta. Published by Editõra Edgard Blücher Ltda., Editõra da Universidade de São Paulo, São Paulo, Brazil, 1966, 744 pages.

Estudio de Movimientos y Tiempos, 5th ed., translated into Spanish by Ricardo Garcia-Pelayo Alonso. Published by Aguilar, S. A. de Ediciones, Juan Bravo, 38 Madrid, Spain, 1966, 746 pages.

L'Analisi del Lavoro con il Metodo del Campionamento, translated into Italian by Olinto Praturlon, published by Etas-Kompass, S.p.A., Via Mantegna 6, Milan, Italy, 1967, 361 pages.

Motion and Time Study: Design and Measurement of Work, 5th ed., in press. Translated into Italian, published by Etas-Kompass, S.p.A., via Mantegna 6, Milan, Italy.

FOREIGN PUBLICATIONS IN THE ENGLISH LANGUAGE

Motion and Time Study, 4th ed., Modern Asia Edition. Published by the Charles E. Tuttle Company, 15 Edogawa-Cho, Bunkyo-ku, Tokyo, Japan, 1960, 665 pages.

Motion and Time Study: Design and Measurement of Work, International Edition, John Wiley & Sons, Inc., London, 1965, 739 pages.

MOTION PICTURE FILMS

Color, Sound 16 mm.

THE FOREMAN DISCOVERS MOTION STUDY, Running Time 16 minutes.

INTRODUCTION TO WORK SAMPLING, Running Time 19 minutes.

MAKING A WORK SAMPLING STUDY, Running Time 23 minutes.

ESTABLISHING WORK STANDARDS BY SAMPLING, Running Time 25 minutes.

Above films sold and rented by University of California Extension, Media Center, 2223 Fulton Street, Berkeley, California 94720.

Black and White 16 mm.

UNIT I WORK MEASUREMENT FILMS, Five Reels.

UNIT II WORK MEASUREMENT FILMS, Six Reels.

Above films sold by College Book Company, 1002 Westwood Blvd., Los Angeles, Calif. 90024.

Index

Abnormal time values, 374
Absolute error, 523
Accumulative timing, 356
Accuracy, degree of, 517
Activity chart, 97
Alignment chart, for time study, 370
  for work sampling, 522, 532
Allowances, 395
  delay, 398
  fatigue, 396
  personal, 395
  table for, 397
Alternatives, evaluation of, 29
American Hard Rubber Company, 194
American Institute of Industrial Engineers, 20
American Management Association, 20
American Society of Mechanical Engineers, 18, 19, 64
Amos Tuck School, 19
Analysis of hand motions, 170
Anthropometric data, 214
Anxiety, reduction of, 622
Arm rest, 286
Armstrong Cork Company, 157, 582, 585, 609, 612
Arrangement of equipment, 258, 264
Assemble, disassemble, and use (therblig), check list for, 206
  definition of, 139
  example of, 205
  symbol for, 136
Assembling and cementing operation, 405
Assembly operations, time standards for, 474
Audio-visual instruction, 621
Auditing methods, time standards and incentive plans, 414
Automatic electric timer, 208
Automation, 307
Automeasurement, 422
AUTORATE, 424
Average incentive pace, 387
Average method of selecting time, 374

Bailey, G. B., 473, 502
Ballistic movements, 245
Barber-Colman gear hobber, 442
Basic control data, 219
Basic display data, 218
Basic Motion Timestudy, 502
Basis for rating, 385
Bates, Guy J., 416
Bedaux, Charles E., 376
Bell System, 55
Bench, packing, 266
Bench grinder case, 333
Benson, Bernard S., 28
Bethlehem Steel Works, 14
Bibliography, 769
Biggane, J. F., 634
Bins, study of, 271
  types of, 271
Blakelock, R. M., 130
Blank and perforate, time standard for, 468
Blanking dies, plain, 459
Block-tossing operation, 383
Board, observation, 348
Body bending, physiological cost of, 237
Boeing Airplane Company, 540
Bolt and washer assembly, analysis sheet of, 174
  description of, 223
  film analysis of, 170
  operation charts of, 113, 114
  pictures of, 172
  simo charts of, 177, 178
Booth, projection, 161
Breaking operation into its elements, 353
Brick handling methods, 237
Bricklaying, improved method of, 16
Bridgeport Plant of General Electric, 473, 474
Brouha, Lucien, 549, 566, 575
Brush, push, 126
Buffing wheels, recoating, 66
Burring small parts, 272
Bushings, inserting pins in, 200

California citrus and lettuce packing operations, 25, 55
Cameras, motion picture, 149, 150, 346, 539
Candling eggs, 27, 308
Candy dipping, 244
Can making, method of, 75, 450
  process chart of, Fig. 41
Capital cost, 36
Capsule sorting, 641, 648
Card dealing, rating of, 604, 762
Carnegie Foundation for the Advancement of Teaching, 300
Carton, code dating, 56
  fiberboard, 44
  folding, 228
  opening, 290
Cast iron plate assembly, 411
Caterpillar Tractor Company, 52
Cathcart, E. P., 578
Cementing operation, 405
  computation sheet, 408
  observation sheet, 406
  piecework rate sheet, 409
Chair of proper height, 283
Change direction, time for, 241
Charts, activity, 97
  alignment, 368
  gang process, 83, 87
  left- and right-hand, 111, 113, 114, 116
  man and machine, 102, 105, 106
  operation, 111, 113, 114, 116
  possibility, 188
  process, 67, 70, 72, 77, 79, 80, 81, 84, 87, 88
  simo, 176, 177, 178, 183, 184
Checking overall time, 409
Check sheet for operation analysis, 112
Cheddar cheese, packing, 185
Chronocyclegraph, 18, 131
Cincinnati vertical milling machine, 438
Citrus fruit packing, 25
Classification, of hand motions, 235
  of types of dies, 459
  of work done on plain blanking dies, 460
Clean-up work, 121
Collecting data for formulas, 428

Colonial Radio Company, 619
Color, use of, 190
Compensation, 330, 469, 639, 688, 702
Compound blank and perforate dies, 470
Computation sheet, 326, 408
Computations of time standards, 398
Conditions of get, 477
Confidence level, 514
Containers, types of, 268
Continuous motions, 241
Continuous timing, 355
Control charts, 371, 531, 533
Cooperation resulting from motion study, 594
Core-making operation, 400
Corrugated fiberboard, incentive plan for manufacture of, 644
Cost-reduction report, 39
Cox, C. H., 265
Crossbars, design of, 301
Curved motions, 241
Curves for setting time standards on die and tool work, 465
Cutting lumber to length, 235
Cyclegraph, 18, 131

Dartmouth College, 19
Data, for formulas, 426, 441
  standard, 428, 441
Delay, allowances for, 398
Deliberate change, 674, 684
Demonstration panel for work sampling, 513
Design, of product, 50
  of tools and equipment, 289
Developing a better method, 50
Die and tool work, time standards for, 456
Die block, operations on, 463
Dill, D. B., 578
Dipping candy, 244
Disassemble, assemble, and use (therblig), check list for, 206
  definition of, 139
  example of, 139
  symbol for, 136
Dishwashing department, 92
Dispose, 477
Distribution center, equipment for, 36

Distribution curves, 340, 384, 388, 390, 415, 514
Division of operation into elements, 354
Donnelley, R. R., 418, 421
Dragline, 37
Dreyfuss, Henry, 213
Drill press, for burring, 272
standard data for, 429
Drop deliveries, use of, 271
Duncan, James H., 473
Du Pont Company, 134, 569
Durnin, J. V. G. A., 575
Dvorak, A., 300
Dvorak-Dealey simplified keyboard, 300

Earnings curve, 390
Eastman Kodak Company, 36, 134, 558, 688
EDP system, 422
Effect of practice, 607, 628
Effectiveness pay plan, 707
Eggs, grading and packing, 308
processing plant, 27
Electronic data processing, 422
organization for, 423
Electronic timer, 207
Elemental time data, 428
Elements, breaking job into, 353
foreign, 358
of job, 721
Elevator scales, 91
Eli Lilly and Company, 203, 604, 605, 648
Elimination approach, 53
Employee's Idea Plan, 592
Energy expenditure table, 576
Engstrom, Harold, 473, 474
Equipment, microchronometer, 155
motion-picture film, 157
motion study cameras, 149, 150, 346, 539
projectors, 158, 603
stop watches, 345
stop-watch time study board, 348
time recording machines, 347, 418
Errors in time values, 409
Extractor, operation of, 107, 108
Eye-hand simo chart, 255

Eye-hand simo chart, coordination, 251
Eye movements, 249
Eyestrain, relief of, 275

Farmer, Eric, 13
Farm work simplification, 22, 73
Fatigue, 563
allowances, 395
effect on rhythm, 249
Feeding silage, 73
Female, measurements of, 216
File for standard data, 413
Filling mailing envelopes, 227
Film analysis, 169
Film, indexing and storing, 157
kinds of, 157
loops, 387
multi-image, 387
rating, 387
Fine assembly work, 275
Fingers, capacities of, 298
Fixation movements, 245
Fixations, two and three, 251
Fixed stations for tools and materials, 256
Fixtures for painting metal containers, 119
Flagler, L. A., 121
Flow diagram, 68, 69, 74, 75, 76, 78, 82, 83
steps used in making, 96
Folder-gluer, 48
Folding paper, method of, 242
Folding paper cartons, 228
Foot-operated tools, 289
Foot pedal, design of, 292
study of, 295
Foot rest, 286
Forberg, Richard A., 414, 587, 589
Force platform, 568
Ford Motor Company, 95, 121, 297
Foreign elements, 358
Fork-lift truck, 320
Forms, for computation sheet, 408, 480
for cost reduction, 39
for instruction sheets, 326, 610, 611, 612
for motion-analysis data, 171, 174
for motion-picture data, 166
for process chart, 77

Forms, for rating, 760, 763
for simo charts, 177, 178, 183, 184
for standard practice, 322, 323, 324, 326
for stop-watch studies, 350, 401, 406
Formulas, for determining number of observations, 359, 514
for gear hobbing, 439
for milling, 437
for soldering cans, 453
Frantz, Earl L., 721
Frequency distribution, 383, 415
curve, bell, 388
curve before and after incentives, 340
curve for block tossing, 385
curve for lathe operators, 384
Full-hook grasp, 191
Fundamental hand motions, 135
measurement of, 206
use of, 190

Gang process chart, 83
Gauging hard rubber washers, 194
Gelatin capsules, sorting, 641, 648
General Electric Company, 60, 130, 157, 473, 474, 584, 587, 593, 618
General Motors Corporation, 41, 52, 61, 157, 291, 332, 386, 416
Geppinger, H. C., 473
Get, elements of, 137
time for, 476
Get ready, discussion of, 100
Gilbreth, F. B., and L. M., 3, 15, 18, 64, 130, 135, 137, 156, 164, 221
Gisholt lathes, 89, 302
Grading eggs, 308
Grasp (therblig), check list for, 192
definition of, 138
example of, 192
four types of, 476
full hook, 191
pressure, 191
symbol for, 136
Greenwald, D. U., 196
Guaranteed time standard, 413

Hable, O. W., 291
Hand motions, abrupt changes in, 241
classification of, 235

Hand motions, continuous curved, 241
fundamental, 135
Hand wheels, design of, 304
Harvard Fatigue Laboratory, 552
Harvesting time for peas, 22
Hawthorne experiment, 662
Health of Munition Workers Committee, 573
Heart rate measurement, 550
Heating factory buildings, 577
Height, of chair, 286
of table top, 284
of work place, 284
Herzberg, Frederick, 664, 666
History of motion and time study, 10
Hobbing gears, standard time for, 441
Hold (therblig), check list for, 199
definition of, 138
example of, 223
symbol for, 136
Hollen, E. H., 291
Holmes, W. G., 473
Hotel, rearrangement of departments in, 90
Hours of work, 573
Hughes Aircraft Company, 621
Human engineering, 209

Illumination, adequate, 273
Incentives, administration of, 729
for die and tool work, 469
for manufacture of fiberboard, 644
for sorting gelatin capsules, 648
relation of motion and time study to, 330
Individual differences, 380, 554
Individual rate-performance premium payment plan, 688
Industrial engineering, definition of, 20
Industrial Engineering Center, 582
Industrial Engineering Survey, 5, 338, 386, 391, 605
Industrial revolution, 660
Industrial Welfare Commission of California, 284
Information, securing, 351, 352
Inspect (therblig), check list for, 203
definition of, 138
example of, 202
reaction time for, 203

Inspect (therblig), symbol for, 136
Inspection, of bottles, 612
  of cloth, 274
  of electric meter mechanisms, 275
  of metal spools, 277, 318
  of small parts, 57
Instruction sheet, for inspecting bottles, 612
  for lacing tennis shoes, 611
  for packing chocolates, 610
  for turret lathe work, 326, 608
International Business Machines Corporation, 78, 419, 420, 424, 676
Iowa, University of, 593

Janitor work, 121
Jigs, principles of motion economy related to design of, 289
Jigs and fixtures, 289
Job, specifications for, 321
Job analysis, 3, 63
Job classification card, 637
Job Conditions Form, General, 323
  Standard, 322
Job enlargement, 674
Jones and Lamson Machine Company, 325, 328
Judgment in time study, 375

Kenyon, James A., 721
Kitchen layout, 90
Koch, B. C., 78
Kodak Park, 688, 697
Koehler, R. E., 473
Kymograph, 206

Labels, inspection of, 202
Labor, value of, 331
Labor accomplishment, 330
Laboratory, motion study, 60, 62, 156
Lakeview Plan, 702
Lathe, controls, 302
  semiautomatic, 381
  turret, 325
Laundry work, 106
Layout of building, for assembly, and inspection of small parts, 58, 59
  for feeding silage, 73
  for making magnet armature, 78
Lazarus, I. P., 473

Learner's progress record, 635
Learning curve, for mechanism assembly, 617
  for pinboard operation, 630
  for punch-press operation, 629
  for showing effect of anxiety reduction, 626
  for showing effect of fumbles and delays, 631
Left- and right-hand chart, for bolt and washer assembly, 112
  for cast iron plate assembly, 411
  for rope clip assembly, 112
  for signing letter, 111
Lemons, packing, 25
Length, of rest periods, 574
  of working day, 573
Lettuce, packing in cartons, 55
Level of performance, 375
Levers, study of, 302
Lighting, 165, 577
Lincoln, James A., 666
Link-forming operation, 179
Load clamp truck attachment, 319
Louden, J. A., 473
Lowry, Hotel, 90

Machine paced operations, 639
Machines, arrangement of, 89, 264
MacKenzie, John M., 251
McLandress, R. D., 41, 61
Macy's Department Store, 170
Magnet armature, drawing of, 82
  manufacture of, 78
Making motion pictures, 162
Making the time study, 349
Malcolm, J. A., 473
Male, measurements of, 214
Management, principles of, 12
Man and machine chart, 100, 102, 105, 106
Man-machine systems, 209
Manual, time study, 715
  wage incentives, 722
Manual and machine controlled operations, 639
Master tables of time data, 422, 424, 428, 441
Max Planck Institute, 549, 552
Maximum working area, 261

Maynard, H. B., 473
Mayo, E., 578
Maytag Company, 364, 592, 593, 677
Measurement, of heart rate, 550
  of labor, 10, 330
  of oxygen consumption, 552
  of therblig time, 206
Mechanization, 307
Memomotion study, 130
Merck & Company, 265
Metal boxes, methods of painting, 118
Metal spools, inspection of, 277, 318
Methods and devices for measuring
  work, 343
Methods change program, 587
Methods design, 3, 4
  work sheet for, 22, 24
Methods development program, 580,
  584
Methods-Time Measurement, 496
Microchronometer, 155
Micromotion study, aid in improving
  methods, 128
  aid in teaching, 128
  definition of, 17
  equipment for, 149
  purposes of, 128
  training in, 128
Midvale Steel Works, 10
Milling, data for, 432
  machine, 438
Modal method of selecting time, 374
Mogensen, A. H., 594
Momentum, effective use of, 237
  in dipping candy, 244
Mop, specifications for, 122
Mopping, methods of, 123
Motion and time study, definition of,
  4
  effect on worker, 332
  extent of use, 32
  history of, 10
  part of industrial engineering, 20
  techniques, 32
  training programs, 580
  types of, 32, 33
Motion economy, principles of, 220
Motion-minded, 130
Motion pictures, "before" and "after,"
  162
  cameras for making, 149, 346, 539

Motion pictures, data sheet for, 166
  for improving rating ability, 387
  making, 162
  procedure for making, 167
  projectors for, 158, 159, 160, 603
  records, 329
  uses of, 162
Motions, classes of hand, 235
  continuous, 241
  controlled, 245
Motion study, applied to every member of organization, 593
  beginning of, 16
  cooperation resulting from, 594
  definition of, 3
  Gilbreth's use of, 16
  laboratory, 60, 583
  left- and right-hand charts for, 111
  use by General Electric, 584
Motion-time data, systems of, 471, 487
  for assembly work, 474
Motivation and work, 660
Motivation-maintenance theory, 663
Mullee, W. R., 194, 286
Multi-factor incentive plans, 639

New York State Labor Laws, 283
New York University, 593
Noise, reduction of, 577
Nonsymmetrical motions, 232
Norem, Bert H., 251
Normal distribution curves, 340, 384, 385
  388, 390, 513
Normal pace, 387
Normal time, 395, 404
Normal working area, 259
Number of cycles to be timed, 358
Nut runner, 297

Objective rating, 378
Observation board, 347, 348
Observations, number of, 359, 523
Observation sheet, 348, 350, 401, 406
Office procedure, chart of, 76, 78
Office work, 9, 74
One- and two-handed work, 233
Operation analysis, 110
Operator, selection of, 162
  training of, 607
Operator method, control of, 41

Organizations, 19
Overall time, 331
Oxygen consumption, rate of, 549

Pace, average incentive, 387
  normal, 387
Packaging small parts, 251
Packard Electric Division of General
  Motors, 61
Packing, citrus fruit, 25
  lettuce, 55
Packing bench, 265
Packing papers, description of opera-
  tion, 242
  simo chart of, 184
Painting with spray gun, container
  covers and bottoms, 118
  rectangular cans, 75
  refrigerator units, 205
Paper folding, 242
Parden, Robert J., 721
Passmore, R., 575
Peas, harvesting, 22
Pedal design, 292
Perception, 273
Performance, rating, 380
  sampling, 511
Persing, L. P., 585, 587, 618, 619
Personal allowance, 395
Photoelectric cells, 207
Physiological changes resulting from
  work, 563
Physiological cost of doing work, 549
Physiological evaluation of performance
  level, 379, 560
Physiology, work, 549
Piece rate, request for, 715
Pinboard, 143, 746
Pin sets, 200
Pipe bridge, installation of, 93
Place, times for, 477
Placing the camera, 163
Planning, 41
Plant layout, 78
Points, 376, 391
Polishing, silverware, 92, 249
  typewriter parts, 248
Position and pre-position (therbligs),
  check list for, 201
  definition of, 138
  examples of, 200

Position and pre-position (therbligs),
  symbols for, 136
Possibility charts, 185
Posture, 286
Power shear operation, 558
Practice, effect of, 607, 628
Pratt and Whitney Aircraft Division,
  422
Preferred method, 4
Pre-position, see Position
Pre-production, 44
Presgrave, Ralph, 473, 502
Pressure grasp, 191
Principles of management, Taylor's, 12
Principles of motion economy:
  1. The two hands should begin as
     well as complete their motions
     at the same time, 222
  2. The two hands should not be
     idle at the same time except dur-
     ing rest periods, 222
  3. Motions of the arms should be
     made in opposite and symmetri-
     cal directions and should be
     made simultaneously, 222
  4. Hand and body motions should
     be confined to the lowest classi-
     fication with which it is possible
     to perform the work satisfac-
     torily, 235
  5. Momentum should be employed
     to assist the worker wherever
     possible, and it should be re-
     duced to a minimum if it must
     be overcome by muscular effort,
     237
  6. Smooth continuous motions of
     the hands are preferable to
     straight-line motions involving
     sudden and sharp changes in di-
     rection, 241
  7. Ballistic movements are faster,
     easier, and more accurate than
     restricted (fixation) or "con-
     trolled" movements, 245
  8. Work should be arranged to per-
     mit easy and natural rhythm
     wherever possible, 247
  9. Eye fixations should be as few
     and as close together as possible,
     249

10. There should be a definite and fixed place for all tools and materials, 256

11. Tools, materials, and controls should be located close to the point of use, 258

12. Gravity feed bins and containers should be used to deliver the material close to the point of use, 268

13. "Drop deliveries" should be used wherever possible, 271

14. Materials and tools should be located to permit the best sequence of motions, 273

15. Provisions should be made for adequate conditions for seeing. Good illumination is the first requirement for satisfactory visual perception, 273

16. The height of the work place and the chair should preferably be arranged so that alternate sitting and standing at work are easily possible, 283

17. A chair of the type and height to permit good posture should be provided for every worker, 286

18. The hands should be relieved of all work that can be done more advantageously by a jig, fixture, or foot-operated device, 289

19. Two or more tools should be combined wherever possible, 295

20. Tools and materials should be pre-positioned wherever possible, 297

21. Where each finger performs some specific movement, as in typewriting, the load should be distributed in accordance with the inherent capacities of the fingers, 298

22. Levers, crossbars, and hand wheels should be located in such positions that the operator can manipulate them with the least change in body position and with the greatest mechanical advantage, 301

Printed labels, inspection of, 202
Printer-slotter, 47
Problems, 733
Process analysis, 63
Process charts, 63
  gang, 83
  steps for making, 96
  symbols for, 64, 65
Procter and Gamble Company, 53, 93, 121, 290, 320, 415, 587, 589, 590, 591, 684
Production curves, 337, 339, 572, 573
Production studies, 409, 412
Projector, motion picture, 158, 159, 160, 603
Pul-Pack method, 320, 687
Pulse rate and physical work, 550
Purchasing coffee, chart of, 101

Quality classification on die and tool work, 466
Quick, J. H., 473

Radio Corporation of America, 270
Random number tables, 533, 535
Range of human capacities, 380
Rating, accuracy of, 600
  Bedaux system, 376
  objective, 378
  performance, 380
  physiological evaluation of performance level, 379
  scales for, 392
  skill and effort, 376
  synthetic, 377
Rating, systems of, 376
  training in, 603
  Westinghouse system, 376
Rating factor, application of, 393
  definition of, 375
Reaction time, 203
Reading, eye movements in, 276
Recoating buffing wheels, chart of, 66
Recording stop-watch readings, 357
Relation of time standards to wage incentives, 330
Release load (therblig), check list for, 199

Release load (therblig), definition of, 138, 199
    examples of, 199
    symbol for, 136
Repetitive timing, 356
Request for time study, 349
Rest periods, 574
Rethreading machine, 291
Rhythm, 247
Rohr, Jr., Robert J., 688
Rope clips, assembly of, 112
Rubber washers, gauging, 194

Saginaw Steering Gear, Division of General Motors, 291
Sandblasting castings, chart of, 97
Satisfiers and dissatisfiers, 666
Scales in elevator, 91
Schaefer, M. G., 473
Schwab, J. L., 473
Scientific management, 12, 661
Scope, 6
Scrap collection at Kodak Park, 697
Screwdriver, attachments, 295
    electric motor-driven, 290, 298
Screws, packaging, 252
Seabrook Farms, 22
Search (therblig), definition of, 136
    symbol for, 136
Securing time study information, 352
Seeing, adequate conditions for, 273
    time for, 276
Segur, A. B., 473
Select (therblig), check list for, 191
    definition of, 137
    example of, 190
    symbol for, 136
Selecting the operator, for micromotion study, 162
    for stop-watch study, 371
Selecting time values, 372
Semiautomatic lathe operation, 381, 384
Sensitive drills, standard data for, 429
Sequence of motions, best, 273
Service Bureau Corporation, 424
Servis recorder, 347
Shea, W. J., 473
Shepherd, Jr., Mark, 672
Shipping cartons, opening, 290

Shipping-department operations, 265
Shoveling, Taylor's investigation of, 14
Signing letter, chart of, 140
Silage, feeding, 73
Simplification on farms, 22, 73
Simultaneous motion-cycle chart, 177, 178, 183, 184
    eye-hand, 255
    modified, 187, 188
Simultaneous motions, 222
Skill and Effort Rating, 376
Slitting, chart of, 105, 106
    machine, 103
Smith, E. J., 231
Society for Advancement of Management, 20, 386
Soda crackers, baking, chart of, 80
Soldering cans, time standards for, 450
Soldering iron, foot-operated, 289
Specialization, 674
Spectacles, use of, 276
Speed and method, 393
Spencer, F. R., 456
Spenser, P. R., 246
Spinanger, Arthur, 53, 590, 684
Splicing insulated wire, 55
Spools, inspection of, 277, 318
Spray gun painting, 75, 118, 205
Spray painting, of container covers and bottoms, 118
    of rectangular cans, 75
    of refrigerator unit, 205
Standardization, 321
Standard job conditions, 322
Standard practice, 321
Standard times for get and place, 474
Stats, H. E., 90
Stegemerten, G. J., 473
Steps in making stop-watch observations, 372
Stop watches, 344
Stop-watch study, determination of allowances, 395
    determination of time standard, 400
    division of operation into elements, 353
    equipment required, 344
    forms for, 350, 401, 406
    rating, 369, 376
    recording and filing data, 413

Stop-watch study, securing information, 352
  selection of operator for, 371
  timing and recording data, 355
  time-recording machines, 346
Swing saw, operation of, 235
Symbols, therblig, 136
Synthetic rating, 377
Systematic approach to problem solving, 4, 21
Systems of predetermined motion-time data, for assembly work, 475
  Basic Motion Timestudy, 502
  Methods-Time Measurement, 496
  summary of facts, 472
  Work Factor System, 487

Table, inspection, 275, 278, 282
  packing, 266
  shipping room, 257
Taylor, F. W., 3, 10, 12, 14, 19, 549
Team-goal approach to methods improvement, 589
Tempo, 540
Texas Instruments Inc., 622, 667, 672, 684
Therbligs, best sequence of, 273
  colors, 136
  definition of, 136
  symbols, 136
  time values for, 471, 487
  use of, 190
Thornthwaite, C. W., 23
Timer, automatic, 208
Time recorder, U.C.L.A., 208
  Donnelley, 418, 421
Time standard, determination of, 5, 395
  guaranteed, 413
Time standards, for constant elements, 428
  for die and tool work, 456
  for drill press work, 429
  for gear hobbing, 441
  for light assembly work, 474
  for milling, 432
  for soldering cans, 450
  for variable elements, 431

Time study, *see also* Motion and time study
  conferences, 594
  definition of, 342
  equipment, 344
  extent of profitable use, 32
  mechanized, 418
  motion-picture camera for, 346
  narrow interpretation of, 19
  rating films, 387
  request for, 349
  staff activity, 415
  surveys, 5, 338, 386, 391, 605
  uses of, 344
Time values, abnormal, 374
  computing, 398
  establishing by formula, 428
Tippett, L. H. C., 511
Tiredness, feeling of, 563
Tool and die work, standards for, 456
Tool chest, 258
Tractor train, 37
Training, in colleges and universities, 593
  inspectors of metal spools, 277
  operator, 6, 607
  training to reduce anxiety, 622
Transport distances, corrections for, 475
Transport empty and transport loaded (therblig), check list for, 198
  definition of, 137
  effect of eye movements on, 196
  example of, 193
  length of, 193
Trays for bolts and nuts, 268
Trucks, tool-chest, 258
Tubeless tires, 54
Turnbull, T. R., 383
Typewriter keyboard, 298

Unavoidable delay, 139, 398
United Aircraft Corporation, 422
United California Bank, 543
University of California, 208, 318
University of Iowa, 593
Unnecessary work, elimination of, 53
Use, assemble, and disassemble (therblig), check list for, 206

Use, definition of, 139
  example of, 205
  symbol for, 136
  warehouse, utilization of space in, 81

Vacuum, cleaning with, 126
Variable elements, time standards for, 431
Ventilated suits, 566
Ventilation, 577
Vibration, reduction of, 577
Visual perception, 249, 273

Waffle-iron assembly, 477
Wage-incentive applications, on die and tool work, 456
Walker, C. R., 676
Walking, on level, 386
  rating of, 604, 759
Warehouse handling methods, 319
Warner and Swasey standard practice sheet, 324
Washers of varying thicknesses, grasping, 192
Washington University studies, 300
Watering garden, chart of, 66
Water pump assembly, 678

Wechsler, D., 380
Western Electric Company, 473, 662
Westinghouse system of rating, 376
WETARFAC, 418
Window washing, 127
Wink, definition of, 170
Work design, 3
Work-Factor System, 488
Working area, maximum, 261
  normal, 259
Working conditions, 573
Work measurement films, 597, 788
Work methods design, 40, 49, 50, 212
  physiological measurements in, 558
Work sampling, advantages and disadvantages of, 546
  alignment charts, 522, 532
  control charts, 531, 533
  demonstration panel, 513
  for determining time standards, 543
  formula, 514
  motion picture films, 788
  tables, 520, 524, 528, 530
Work simplification, 9, 587, 592, 594, 706
Work study, 3
Written standard practice, 5, 321

Youde, L. F., 614